THE GEOMETRY, TOPOLOGY AND PHYSICS OF MODULI SPACES OF HIGGS BUNDLES

LECTURE NOTES SERIES
Institute for Mathematical Sciences, National University of Singapore

Series Editors: Chitat Chong and Adrian Röllin
Institute for Mathematical Sciences
National University of Singapore

ISSN: 1793-0758

Published

Vol. 36 *The Geometry, Topology and Physics of Moduli Spaces of Higgs Bundles*
edited by Richard Wentworth & Graeme Wilkin

Vol. 35 *Combinatorial and Toric Homotopy: Introductory Lectures*
edited by Alastair Darby, Jelena Grbić, Zhi Lü & Jie Wu

Vol. 34 *White Noise Analysis and Quantum Information*
edited by Luigi Accardi, Louis H Y Chen, Takeyuki Hida, Masanori Ohya, Si Si & Noboru Watanabe

Vol. 33 *Sets and Computations*
edited by Sy-David Friedman, Dilip Raghavan & Yue Yang

Vol. 32 *Mathemusical Conversations:*
Mathematics and Computation in Music Performance and Composition
edited by Jordan B L Smith, Elaine Chew & Gérard Assayag

Vol. 31 *Geometric Analysis Around Scalar Curvatures*
edited by Fei Han, Xingwang Xu & Weiping Zhang

Vol. 30 *Modular Representation Theory of Finite and p-Adic Groups*
edited by Wee Teck Gan & Kai Meng Tan

Vol. 29 *Forcing, Iterated Ultrapowers, and Turing Degrees*
edited by Chitat Chong, Qi Feng, Theodore A Slaman, W Hugh Woodin & Yue Yang

Vol. 28 *Slicing the Turth:*
On the Computable and Reverse Mathematics of Combinatorial Principles
by Denis R Hirschfeldt, edited by Chitat Chong, Qi Feng, Theodore A Slaman, W Hugh Woodin & Yue Yang

Vol. 27 *E-Recursion, Forcing and C*-Algebras*
edited by Chitat Chong, Qi Feng, Theodore A Slaman, W Hugh Woodin & Yue Yang

Vol. 26 *Notes on Forcing Axioms*
by Stevo Todorcevic, edited by Chitat Chong, Qi Feng, Theodore A Slaman, W Hugh Woodin & Yue Yang

*For the complete list of titles in this series, please go to
http://www.worldscientific.com/series/LNIMSNUS

Lecture Notes Series, Institute for Mathematical Sciences,
National University of Singapore Vol. 36

THE GEOMETRY, TOPOLOGY AND PHYSICS OF MODULI SPACES OF HIGGS BUNDLES

Editors

Richard Wentworth
University of Maryland, USA

Graeme Wilkin
National University of Singapore, Singapore

World Scientific

NEW JERSEY • LONDON • SINGAPORE • BEIJING • SHANGHAI • HONG KONG • TAIPEI • CHENNAI • TOKYO

Published by

World Scientific Publishing Co. Pte. Ltd.
5 Toh Tuck Link, Singapore 596224
USA office: 27 Warren Street, Suite 401-402, Hackensack, NJ 07601
UK office: 57 Shelton Street, Covent Garden, London WC2H 9HE

Library of Congress Cataloging-in-Publication Data
Names: Wentworth, Richard A., editor. | Wilkin, Graeme, editor.
Title: The geometry, topology, and physics of moduli spaces of Higgs bundles /
editors, Richard Wentworth (University of Maryland, USA), Graeme Wilkin (NUS, Singapore).
Description: New Jersey : World Scientific, 2018. | Series: Lecture notes series /
Institute for Mathematical Sciences, National University of Singapore ; volume 36 |
Includes bibliographical references.
Identifiers: LCCN 2017059460 | ISBN 9789813229082 (hardcover : alk. paper)
Subjects: LCSH: Vector bundles. | Geometry, Algebraic. | Moduli theory.
Classification: LCC QA612.63 .G46 2018 | DDC 514/.224--dc23
LC record available at https://lccn.loc.gov/2017059460

British Library Cataloguing-in-Publication Data
A catalogue record for this book is available from the British Library.

For any available supplementary material, please visit
http://www.worldscientific.com/worldscibooks/10.1142/10683#t=suppl

Contents

Foreword

The Institute for Mathematical Sciences (IMS) organizes thematic programs of longer duration as well as shorter workshops and public lectures. The themes are selected from among areas at the forefront of current research in the mathematical sciences and their applications. Each volume of the *IMS Lecture Notes Series* is a compendium of papers based on lectures and tutorials delivered at the IMS. The aim is to make original papers and expository articles on a subject of current interest available to the international research community. These volumes also serve as a record of activities that took place at the IMS. We hope that through regular publication of these Lecture Notes the Institute will achieve, in part, its objective of reaching out to the community of scholars in the promotion of research in the mathematical sciences.

Chitat Chong
Adrian Röllin
Series Editors

Preface

This volume is based on the minicourses from the program "The geometry, topology and physics of moduli spaces of Higgs bundles" held at the Institute for Mathematical Sciences in Singapore from 7 July to 29 August, 2014. The goal of the program was to bring together leading researchers in the field to share their results and expertise, and to introduce graduate students and young researchers to the latest developments and open problems. The successful event attracted over 100 participants from 26 countries.

In his seminal paper of 1987, Nigel Hitchin introduced the two-dimensional reduction of the self-duality equations of four-dimensional Yang-Mills theory. The resulting equations involved a connection and an auxiliary field, which he called the *Higgs field*. The terminology of *Higgs bundle* is due to Carlos Simpson, who generalized the theory to higher dimensional Kähler manifolds. Superficially, one might guess that this has something to do with the famous Higgs boson which plays a crucial role in the Standard Model of particle physics and is currently the subject of intense investigation at CERN. Properly interpreted, the mathematical notion indeed corresponds to the particle searched for in the accelerators outside of Geneva, although somewhat amazingly the relationship of Higgs bundles to physics goes far deeper into, for example, notions of quantum gravity. In fact, Higgs bundles capture aspects of integrable systems, conformal field theories, and super-symmetric gauge theories all at the same time. From the purely mathematical point of view, Higgs bundles may be regarded as a nonabelian generalization of Hodge theory, which in part gives a way to characterize algebro-geometric data in terms of linear or algebraic structures. For these reasons, Higgs bundles continue to play an important role for mathematicians and physicists working in very diverse subjects. In Hitchin's original paper on the subject, he observed that "... the moduli space of all solutions turns out to be a manifold with an extremely rich geometric structure". The aim of the lecture notes collected in this volume is to introduce some of this intricate geometric structure in a way that is accessible to graduate students with a background in geometry.

The first set of lectures by Olivier Guichard gives a detailed description of the construction of the moduli space of Higgs bundles and the nonabelian Hodge correspondence from the general viewpoint developed by Simpson, after the theorem of Simon Donaldson, Karen Uhlenbeck, and Shing-Tung Yau. The author has compiled a number of useful exercises for the reader who wants a deeper understanding of the theory.

An important feature of the moduli space is that it admits the structure of a completely integrable system, known as the "Hitchin system". This fact has surprising connections to mirror symmetry and the Langlands program. The lectures by Laura P. Schaposnik introduce this integrable system and the associated spectral data for both real and complex Lie groups. Once again, the reader will find a number of useful exercises throughout the notes.

The lectures by Olivia Dumitrescu and Motohico Mulase explain an exciting new relationship between two seemingly disparate subjects. The first is the theory of topological recursion as introduced by Bertrand Eynard and Nicolas Orantin. This has its origin in random matrix theory and has been effectively applied to many enumerative geometry problems. The second is the quantization of the spectral curves associated to Hitchin systems, as mentioned above.

The language of algebraic stacks is a central tool in the study of moduli problems, but it can be abstract and somewhat forbidding to those outside the subject. The lectures by Sebastian Casalaina-Martin and Jonathan Wise give a pedagogical introduction to stacks with the goal of constructing the moduli stack of Higgs bundles over an algebraic curve. In addition, they describe the relationship between the moduli stack and the moduli space constructed by Nitsure using Geometric Invariant Theory. These notes will serve as a useful survey to those interested in how stacks are used in the theory of Higgs bundles.

There are a number of people and institutions who helped make the program successful. Firstly, we would like to thank the Institute for Mathematical Sciences for the excellent facilities and significant financial support for the program, and we would also like to thank the director Chong Chi Tat for his guidance throughout the program. We also gratefully acknowledge support from the National University of Singapore Science Faculty Research grant R-146-000-152-133 "Geometry and topology of moduli spaces of Higgs bundles and quiver varieties", the U.S. National Science Foundation grants DMS 1107452, 1107263, 1107367 "RNMS: GEometric structures And Representation varieties" (the GEAR Network) and the European Research

Council under the European Community's seventh Framework Programme (FP7/2007-2013)/ERC grant agreement no. FP7-246918. We would also like to thank all of the staff at the IMS, especially Eileen Tan and Emily Chan Ee Cheng, for their help in making the program run smoothly and providing an environment conducive to collaboration.

September 2016

Richard Wentworth
University of Maryland

Graeme Wilkin
National University of Singapore

Volume Editors

An Introduction to the Differential Geometry of Flat Bundles and of Higgs Bundles

Olivier Guichard

Institut de Recherche Mathématique Avancée, UMR 7501
Université de Strasbourg et CNRS
7 rue René-Descartes, 67000 Strasbourg, France
olivier.guichard@math.unistra.fr

This chapter collects the notes of the lectures given on that subject during the introductory school of the program "The Geometry, Topology and Physics of Moduli Spaces of Higgs Bundles" (7 July - 29 August 2014). The main purpose is to state the correspondence between flat bundles and Higgs bundles and some of its features.

Contents

Partially supported by the European Research Council under the European Community's seventh Framework Programme (FP7/2007-2013)/ERC grant agreement n° FP7-246918.

1. Introduction

As its title indicates, this chapter aims to give an introduction on vector bundles over Riemannian and Kählerian manifolds. More specifically these notes cover the following topics:

- the necessary backgrounds in differential geometry: Riemannian, symplectic and Kählerian manifolds.
- the theory of flat bundles over (compact) manifolds: representations of the fundamental group; local systems (below hidden under the concept of "flat structures" on a given bundle, this is an ad-hoc definition whose use should be restricted to these notes); connections and curvature.
- the theory of holomorphic bundles over complex (or analytic) manifolds: interpretation in terms of pseudo-connections with vanishing pseudo-curvature.
- the theory of Hermitian bundles over manifolds and, in the particular case when the base manifold is a complex manifold, the correspondence between pseudo-connections and Hermitian connections (via the introduction of Chern connections).

Those necessary tools need to be complemented with the notion of stability of holomorphic vector bundles over Kählerian manifolds. For this, the Chern-Weyl theory for constructing differential forms representing the Chern classes is recalled and afterward come the degree and the slope of vector bundles over Kählerian manifolds.

The slopes at hand, the notion of stability can be explained as well as some of its variants (semi-stability and poly-stability) and the theorem of Narasimhan and Seshadri is stated, i.e. the one-to-one correspondence between stable holomorphic bundles and irreducible projectively flat unitary bundles over a Riemann surface. In fact the more general result due to Uhlenbeck and Yau where the base is a compact Kählerian manifold is given.

The proof of this correspondence, as well as the Hitchin-Kobayashi correspondence, involves some remarkable identities over Kählerian manifolds that lead to automatic cancellations of the curvature of a connection as soon as only "one part" of the curvature is zero and under some topological conditions. For the purpose of this text those cancellations are called "opportune cancellations" and are consequences, among other things, of the Kähler identities. An explanation of the Kähler identities is here accomplished with the help of symplectic geometry and the corresponding "symplectic Kähler identities".

At the heart of the Narasimhan-Seshadri correspondence and of the Hitchin-Kobayashi correspondence, are notions of the "best" Hermitian metrics. In one direction of the correspondence they are called Yang-Mills metrics.

In the other direction, it is the notion of harmonic metrics that appears: it means the best metric in a L^2-sense. A result of Corlette says that harmonic metrics exist on semisimple flat bundles over compact Riemannian manifolds (Donaldson gave also a proof of a particular case of that result, Labourie proved it in a more general setting).

Over a Kählerian manifold, the connection of an harmonic bundle can be decomposed furthermore with respect to the bidegree of forms: the $(0,1)$-component of the unitary connection together with the $(1,0)$-component of the symmetric part give rise to a Higgs bundle. This notion of Higgs bundles is due to Hitchin, a companion definition in a somewhat different setting was introduced by Griffiths under the name "variation of Hodge structure".

The statement of the Hitchin-Kobayashi correspondence is disconcertingly similar to the above correspondence: on a bundle with vanishing Chern classes over a compact Kählerian manifold, there is a one-to-one bijection between the stable Higgs bundle structures and the simple flat bundle structures. Section 10 illustrates the necessity of the (poly)stability and the (semi)simplicity assumptions and explains the strategies for proving the existence of the sought-for metrics (harmonic metrics and Yang-Mills metrics).

Section 9 gives examples of Higgs bundles, and sometimes of the corresponding representations, over Riemann surfaces. The construction of the Hitchin component for $\mathrm{SL}_n(\mathbf{R})$ is given there and also a proof of the Milnor-Wood inequality. This section is moreover the opportunity to illustrates how one can detect representations with values in some subgroups of the general linear groups: $\mathrm{SL}_2(\mathbf{C})$, $\mathrm{SL}_2(\mathbf{R})$, $\mathrm{Sp}_{2n}(\mathbf{R})$ or $\mathrm{U}(p,q)$.

No attempts to build a proper bibliographical section or to systematically account for proper contributions were made in those notes. The unforgivable omission concerns the moduli spaces aspects of the correspondence (analytical or algebraical structures, hyper-Kählerian structure, $\mathbf{C}^*$-action, etc.). The notes [23] by Richard Wentworth beautifully remedies those lacunas.

The present notes are based on exercises, many proofs are left to the reader in the form of exercises. I am glad to thank the participants of the summer school, their active participation helped tremendously to improve the following text. I am also grateful to the referees for their numerous suggestions.

2. Riemannian, symplectic, complex and Kählerian manifolds

This section is mainly here to fix the notation that will be used in the rest of this text. It also introduces different tools subsequently used through the chapter: L^2-metrics, L^2-adjoints (only on spaces of differential forms) and the Kähler identities (with their proofs using symplectic geometry).

2.1. *Riemannian manifold*

A Riemannian structure on an n-dimensional manifold M is a smooth section of $\mathrm{Sym}^2_{>0}T^*M \subset \mathrm{Sym}^2T^*M \subset T^*M \otimes T^*M$. More explicitly, the data of a Euclidean scalar product $\langle \cdot, \cdot \rangle$ or $\langle \cdot, \cdot \rangle_m$ (or sometimes $\langle \cdot, \cdot \rangle_{TM}$) is given on T_mM, for all m in M, and it varies smoothly with m. This smoothness property can be expressed saying that the function

$$m \longmapsto \langle X_m, Y_m \rangle_m$$

is $\mathcal{C}^\infty$ whenever X and Y are $\mathcal{C}^\infty$ vector fields on M.

In local coordinates $(x^1, \ldots, x^n)$ on M, the Riemannian metric is given by:

$$\sum_{i,j} g_{ij} dx^i \otimes dx^j$$

i.e. $g_{ij} = \langle \frac{\partial}{\partial x^i}, \frac{\partial}{\partial x^j} \rangle = g_{ji}$. It is sometimes convenient in calculations to work with:

Definition 2.1: A *normal coordinates system* at m in M is a system of coordinates around m such that

$$\begin{array}{ll} \forall i,j & g_{ij}(m)=\delta_{i,j} \\ \forall i,j,k & \frac{\partial}{\partial x^k} g_{ij}(m)=0. \end{array}$$

Other times it is even more convenient to work with:

Definition 2.2: A *normal frame* (or sometimes an *orthonormal frame*) is a family $(\epsilon_1,\ldots,\epsilon_n)$ of vector fields (defined on an open subset $U \subset M$) such that

$$\langle \epsilon_i, \epsilon_j \rangle = \delta_{i,j} \quad \forall i,j.$$

Remark 2.3: Usually normal coordinates systems are the coordinates obtained by the exponential map $\exp : T_m M \to M$ (hence they are exponential coordinates systems). Here is allowed a little more generality.

An Euclidean structure on a vector space V (here V is the tangent space $T_m M$) induces an Euclidean structure on every vector space constructed from V: V^*, $\bigwedge^p V$, $\bigwedge^p V^* \simeq (\bigwedge^p V)^*$, $\bigotimes^p V$, etc. Hence a Riemannian structure on M induces a scalar product on every bundle constructed from TM: T^*M, $\bigwedge^p T^*M$, etc.

In particular, given α and β two forms of degree p and q respectively, $\langle \alpha, \beta \rangle$ denotes the function $m \mapsto \langle \alpha_m, \beta_m \rangle_m$ with $\langle \cdot, \cdot \rangle_m$ still denoting the scalar product on $\bigwedge^\bullet T^*M = \bigoplus_p \bigwedge^p T^*M$, $\alpha_m \in \bigwedge^p T^*_m M$ and $\beta_m \in \bigwedge^q T^*_m M$. Of course by construction (or by convention), $\langle \alpha, \beta \rangle = 0$ if $p \neq q$.

2.2. *Orientation, volume form*

When M is furthermore oriented, there is a preferred volume form vol on M; it is

(1) a top dimensional form: $\mathrm{vol}_m \in \bigwedge^n T^*_m M$ ($\forall m \in M$).
(2) defining the orientation: $(\epsilon_1,\ldots,\epsilon_n)$ is an oriented basis of $T_m M$ if and only if $\mathrm{vol}_m(\epsilon_1,\ldots,\epsilon_n) > 0$.
(3) of norm 1: $\langle \mathrm{vol}, \mathrm{vol} \rangle = 1$.

2.3. *The Hodge star*

Using the scalar product $\langle \cdot, \cdot \rangle$ on forms together with the volume form (hence the orientation), one defines the Hodge star $\star\alpha$ of a form α as the unique form satisfying[a]:

$$\star\alpha \wedge \beta = \langle \alpha, \beta \rangle \mathrm{vol}, \ \forall \beta.$$

If the degree of α is p, then the degree of $\star\alpha$ is $n-p$.

For example, $\star\mathbf{1} = \mathrm{vol}$, $\star\mathrm{vol} = \mathbf{1}$ ($\mathbf{1}$ is the constant function equal to 1 in those equalities).

Also

$$\langle \star\alpha, \star\beta \rangle = \langle \alpha, \beta \rangle \text{ and } \star\star\alpha = (-1)^{p(n-p)}\alpha.$$

Remark 2.4: The later equality will be used only when M is even dimensional, it then becomes

$$\star\star\alpha = (-1)^{-p^2}\alpha = (-1)^p\alpha.$$

2.4. *Symplectic manifold*

A symplectic form (or a symplectic structure) on a manifold M is a closed 2-form ω (i.e. $d\omega = 0$) such that $\omega_m \in \bigwedge^2 T_m^* M$ is a non-degenerate symplectic form on the vector space $T_m M$ for every m in M. This forces the dimension of M to be even: $\dim_{\mathbf{R}} M = 2n$. A standard reference on symplectic manifolds is the book [10].

An important aspect of symplectic manifolds is the uniqueness of the local model:

Theorem 2.5: *[Darboux] At every point m of a symplectic manifold (M, ω) there are local coordinates $(x^1, \dots, x^n, y^1, \dots, y^n)$ such that*

$$\omega = \sum_i dx^i \wedge dy^i.$$

A symplectic manifold has a canonical volume form:

$$\mathrm{vol} = \frac{\omega^n}{n!}$$

(hence a canonical orientation) and the symplectic structure induces a bilinear form on every vector bundle constructed from the tangent space; this bilinear form is symplectic on odd forms and symmetric on even forms.

Exercise 2.1: If ω is a symplectic form on a vector space V ($\dim_{\mathbf{R}} V = 2n$), then[b] ω induces a non-degenerate symmetric bilinear form on $\bigwedge^{2k} V$ ($k \leq n$). Calculate the signatures.

---

[a]This is not the most standard choice for the Hodge star (different choices differ by signs), it is the one adopted in the book [1].

[b]Often the indications "Show that" or "Prove that" will be missing in the exercises.

2.5. *The symplectic star*

With those two ingredients, one can define the symplectic star $\star_s\alpha$ of a form α by the equality:

$$\star_s\alpha \wedge \beta = \omega(\alpha, \beta)\mathrm{vol}, \quad \forall\, \beta.$$

For the symplectic star:

$$\omega(\star_s\alpha, \star_s\beta) = \omega(\alpha, \beta) \quad \forall\alpha, \beta; \quad \star_s\star_s = \mathrm{id}.$$

2.6. *The operator L*

We denote by L the operator defined on forms by wedging with the symplectic form:

$$\begin{aligned} L : \Omega^\bullet(M) &\longrightarrow \Omega^\bullet(M) \\ \alpha &\longmapsto \omega \wedge \alpha \end{aligned}$$

where $\Omega^\bullet(M) = \bigoplus_p \Omega^p(M)$ is the graded algebra of differential forms.

2.7. *Symplectic adjoints*

The adjoint (for the symplectic structure) of L is the unique operator:

$$L^{\star_s} : \Omega^\bullet(M) \longrightarrow \Omega^\bullet(M)$$

such that

$$\int_M \omega(L^{\star_s}\alpha, \beta)\mathrm{vol} = \int_M \omega(\alpha, L\beta)\mathrm{vol}$$

for all α and β.

Similarly the symplectic adjoint of the exterior differential d is denoted by $d^{\star_s}$.

Remark 2.6: Those adjoints (and the ones to be defined below in a Riemannian setting) are called *formal adjoints*. This means that they are defined to be the unique differential operators satisfying the above adjunction formula when evaluated on $\mathcal{C}^\infty$ forms but they are a priori not adjoint operators in the sense of functional analysis.

The symplectic star can be used to give formulas for $L^{\star_s}$ and $d^{\star_s}$:

$$\begin{aligned} L^{\star_s} &= \star_s L \star_s \\ d^{\star_s}\alpha &= -(-1)^{\deg(\alpha)} \star_s d \star_s \alpha. \end{aligned}$$

Exercise 2.2: Derive those identities.

2.8. *Symplectic Kähler identities*

Theorem 2.7: *Let (M, ω) be a symplectic manifold of dimension $2n$. The following equalities hold:*

$$[L^{\star_s}, L]\alpha = (L^{\star_s}L - LL^{\star_s})\alpha = (n - \deg(\alpha))\alpha$$
$$[L, d^{\star_s}]\alpha = d\alpha$$
$$[L^{\star_s}, d]\alpha = -d^{\star_s}\alpha.$$

Proof: [sketch] Note first that it is enough to check those equalities locally, hence, by Darboux's theorem, it is enough to prove the result for M a symplectic vector space.

The second observation is that if the equalities hold for (M_1, ω_1) and (M_2, ω_2), they also hold for $(M_1 \times M_2, \omega_1 + \omega_2)$.

The last step is therefore to perform the calculation for $M = \mathbf{R}^2$ and $\omega = dx \wedge dy$, in view of the formulas for the symplectic adjoints it suffices to understand the symplectic star in that case: $\star_s\alpha = \alpha$ if $\deg(\alpha) = 1$, $\star_s f = f\omega$, $\star_s f\omega = f$ for any function f. □

Exercise 2.3: Check the steps in the proof of the symplectic Kähler identities. (You may need to prove the equality $\bigwedge^\bullet(V \oplus W) = \bigwedge^\bullet V \otimes \bigwedge^\bullet W$ where $\bigwedge^\bullet = \bigoplus_k \bigwedge^k$ is the exterior algebra.)
By the same procedure, $\star_s\star_s = \mathrm{id}$ and $\star_s e^\omega = e^\omega$ (i.e. $\forall k$, $\star_s \frac{\omega^k}{k!} = \frac{\omega^{n-k}}{(n-k)!}$).

Exercise 2.4: For κ a differential form and $K : \Omega^\bullet(M) \to \Omega^\bullet(M) : \alpha \mapsto \kappa \wedge \alpha$, $[K, d] = 0$ if and only if $d\kappa = 0$. In particular $[L, d] = 0$, also $[L^{\star_s}, d^{\star_s}] = 0$.

Exercise 2.5: For X a vector field, denote by $\iota_X : \Omega^\bullet(M) \to \Omega^\bullet(M)$ the interior product by X: for α a p-form, $\iota_X\alpha$ is the $(p-1)$-form defined by $\iota_X\alpha(X_2, \ldots, X_p) = \alpha(X, X_2, \ldots, X_p)$. The symplectic gradient of X is then[c] $\xi = -\iota_X\omega$. In this exercise no new notation is introduced for the operator $\xi : \Omega^\bullet(M) \to \Omega^\bullet(M) : \alpha \mapsto \xi \wedge \alpha$.

Then $(\iota_X)^{\star_s} = \xi$, $[L, \iota_X] = \xi$ and $[L^{\star_s}, \xi] = -\iota_X$. For a 1-form ϕ, $[L, \phi^{\star_s}] = \phi$.

2.9. *Complex (analytic) manifolds*

A complex manifold is locally modeled on $\mathbf{C}^n$ and the transition functions are biholomorphic. Such a complex manifold is of course a manifold (in the "real" sense) and the (real!) tangent space T_mX, for every $m \in X$, inherits a structure of complex vector space.

[c]The minus sign is chosen so that subsequent formulas have a nicer form. (I.e. fewer minus signs.)

The multiplication by $\sqrt{-1}$ on this space is usually denoted by J: $J \in \mathrm{End}_{\mathbf{R}}(TX)$, $J^2 = -\mathrm{id}$.

Remark 2.8: The data of such an endomorphism J on the tangent bundle of a (real) manifold with $J^2 = -\mathrm{id}$ is called a pseudo-complex structure. Not every pseudo-complex structure comes from a complex structure.

As before, this endomorphism J of TX induces naturally an endomorphism on every bundle constructed from TX, in particular on the exterior algebra $\bigwedge^\bullet T^*X$. However the standard notation for this extension is C, C will be mainly used on forms:

$$C : \Omega^p(X;\mathbf{R}) \longrightarrow \Omega^p(X;\mathbf{R}) \quad C^2 = (-1)^p,$$

and its "complexification":

$$C : \Omega^\bullet(X;\mathbf{C}) \longrightarrow \Omega^\bullet(X;\mathbf{C}).$$

Remark 2.9: From now on, a distinction will be (hopefully systematically) made between real valued forms (i.e. sections of $\bigwedge^p T^*X$) and complex valued forms (i.e. sections of $(\bigwedge^p T^*X) \otimes \mathbf{C} \simeq \bigwedge^p(T^*X \otimes \mathbf{C})$).

The complexification of J is diagonalizable and the bundle $TX \otimes_{\mathbf{R}} \mathbf{C}$ decomposes according to the eigenspaces:

$$TX \otimes_{\mathbf{R}} \mathbf{C} = T^{1,0}X \oplus T^{0,1}X = \ker(J - \sqrt{-1}) \oplus \ker(J + \sqrt{-1}),$$

also $\overline{T^{1,0}X} = T^{0,1}X$. This decomposition of $TX \otimes_{\mathbf{R}} \mathbf{C}$ induces a decomposition of $T^*X \otimes_{\mathbf{R}} \mathbf{C}$ and its exterior algebra:

$$\begin{aligned} T^*X \otimes_{\mathbf{R}} \mathbf{C} &= T^{*1,0}X \oplus T^{*0,1}X \\ \textstyle\bigwedge^\bullet(T^*X) \otimes_{\mathbf{R}} \mathbf{C} &= \textstyle\bigwedge^\bullet(T^*X \otimes_{\mathbf{R}} \mathbf{C}) \\ &= \textstyle\bigwedge^\bullet(T^{*1,0}X \oplus T^{*0,1}X) \\ &= \textstyle\bigwedge^\bullet(T^{*1,0}X) \otimes \bigwedge^\bullet(T^{*0,1}X) \\ &= \textstyle\bigoplus_{p,q} \bigwedge^p T^{*1,0}X \otimes \bigwedge^q T^{*0,1}X. \end{aligned}$$

Subsequently, the space of complex differential forms acquires a bigrading refining the grading by the degree:

$$\Omega^\ell(X;\mathbf{C}) = \bigoplus_{p+q=\ell} \Omega^{p,q}(X).$$

The operator C is equal to the multiplication by $(\sqrt{-1})^{p-q}$ in restriction to $\Omega^{p,q}(X)$.

Exercise 2.6: In local holomorphic coordinates $(z^1, \ldots, z^n)$ on X, give the expressions of forms in $\Omega^{p,q}(X)$. Use this to show that the "type" (p,q) is independant of the coordinates chart.

With respect to this bigrading, the exterior differential d decomposes (uniquely) as the sum of ∂, of bidegree $(1,0)$ and of $\bar{\partial}$, of bidegree $(0,1)$:

$$\begin{aligned} d &= \partial + \bar{\partial} \\ \partial(\Omega^{p,q}(X)) &\subset \Omega^{p+1,q}(X) \\ \bar{\partial}(\Omega^{p,q}(X)) &\subset \Omega^{p,q+1}(X). \end{aligned}$$

A function f on X is holomorphic if and only if $\bar{\partial} f = 0$.

Exercise 2.7: Give conditions on the function f so that the form $\alpha = f dz^{i_1} \wedge \cdots \wedge dz^{i_p} \wedge \overline{dz^{j_1}} \wedge \cdots \wedge \overline{dz^{j_q}}$ satisfy $\bar{\partial}\alpha = 0$. What happens for $q = 0$?

2.10. *Kähler manifold*

A manifold X is *Kähler* if it is Riemannian, symplectic and complex and if the 3 structures $\langle \cdot, \cdot \rangle$, $\omega(\cdot, \cdot)$ and J cohabit nicely:

$$\begin{aligned} \langle Jv, Jw \rangle &= \langle v, w \rangle, \ \forall v, w \in T_m X \\ \omega(Jv, Jw) &= \omega(v, w) \ (\text{equivalently } C\omega = \omega) \\ \omega(v, Jw) &= \langle v, w \rangle. \end{aligned}$$

Definition 2.10: In this context, an *orthogonal frame* $(\epsilon_i)_{i=1,\ldots,n}$ is an orthogonal basis of the Hermitian vector space $T^{1,0}X$. (Compare with Definition 2.2.)

Exercise 2.8: Then $(\bar{\epsilon}_i)_{i=1,\ldots,n}$ is an orthogonal basis of $T^{0,1}X$.

What is the expression of the symplectic form ω using those 2 bases (and their duals)?

Remark 2.11: The last condition is chosen so that the Hermitian form:

$$(\cdot, \cdot) = \langle \cdot, \cdot \rangle + \sqrt{-1}\omega(\cdot, \cdot)$$

on (TX, J) is $\mathbf{C}$-linear in the second variable.

Note that any two of the above structures determine the third. Also it can be shown that the "integrability" condition on J (i.e. the requirement that it comes from a complex structure) is equivalent to the "integrability" condition on ω (i.e. $d\omega = 0$).

A more relevant way to assert the compatibility of the 3 structures is to say that the $(0,1)$ part of the (complexified) Levi-Civita connection is equal to the $\bar{\partial}$ operator.

2.11. *Adjoints and stars*

Just as in the symplectic setting (see Section 2.7) the Riemannian structure induces a Euclidean structure on $\bigwedge^\bullet T^*X$ as well as a volume form and hence a L^2-scalar product on $\Omega^\bullet(X;\mathbf{R})$ (and also a L^2-Hermitian product on $\Omega^\bullet(X;\mathbf{C})$). The (formal) adjoints of differential operators for that structure will be denoted by a $\star$ in superscript: $d^\star$, $L^\star$, etc. (Those exist already in the more general case of M an oriented Riemannian manifold.)

As suspected, the symplectic star and the Hodge star are compatible on a Kähler manifold, in fact:

$$\star_s = \star C = C\star.$$

Note also that the Riemannian volume form and the symplectic volume form are equal. In turn the symplectic adjoint and the Riemannian adjoint are related:

$$d^\star = -C^{-1}d^{\star_s}C = Cd^{\star_s}C^{-1}.$$

(The last equality follows from $C^2\alpha = (-1)^p\alpha$ and $C^2d^{\star_s}\alpha = (-1)^{p-1}d^{\star_s}\alpha$ if α is a p-form.) Similarly

$$L^\star = C^{-1}L^{\star_s}C = CL^{\star_s}C^{-1} = L^{\star_s}.$$

($L^{\star_s}$ —and hence $L^\star$— commutes with C since $C\omega = \omega$.)

Exercise 2.9: Prove the above formulas, and $\star\star = (-1)^{\deg}$. The adjoint $C^\star$ is equal to C^{-1}.

Exercise 2.10: For A a 1-form, $A^\star = CA^{\star_s}C^{-1} = -C^{-1}A^{\star_s}C$.

For α a $(1,0)$-form, $\sqrt{-1}[L,\alpha^\star] = \bar{\alpha}$. (Hint: Exercise 2.5 with $\phi = \alpha + \bar{\alpha}$ and decomposition into "types".)

2.12. *Kähler identities*

From the symplectic Kähler identities and the previous section, one gets

$$[L^\star, d] = C^{-1}d^\star C \quad \text{and} \quad [L, d^\star] = -C^{-1}dC$$

or, decomposing following the bidegree:

$$\begin{aligned} &[L^\star,\bar\partial] = C^{-1}\partial^\star C = \sqrt{-1}\partial^\star && [L,\bar\partial^\star] = -\sqrt{-1}\partial \\ &[L^\star,\partial] = -\sqrt{-1}\bar\partial^\star && [L,\partial^\star] = \sqrt{-1}\bar\partial. \end{aligned}$$

Remark 2.12: It is frequent to see the notation Λ for the operator $L^\star$. Direct proofs of the Kähler identities can be found in the literature, e.g. [9] or [13].

3. Vector bundles

In this section, the basics of vector bundles are introduced. This includes their trivializations and changes of trivializations as well as their spaces of sections and differential forms, the gauge transformations and the gauge group. Real and complex vector bundles will be treated in an uniform way.

3.1. *Trivializations*

A vector bundle is a smooth family of vector spaces. More precisely, if M is a manifold, a vector bundle over M is a manifold E together with a submersion $p : E \to M$ such that every fiber $p^{-1}(m)$, $m \in M$, has a structure of a $\mathbf{K}$-vector space ($\mathbf{K} = \mathbf{R}$ or $\mathbf{C}$) "varying smoothly" with m. The shortest way to make precise this last property is to assume *local triviality*: given $m \in M$, there is an open neighborhood U of m in M and a diffeomorphism (a *trivialization*)

$$\psi_U : p^{-1}(U) \to U \times \mathbf{K}^d,$$

such that

(1) $\mathrm{pr}_1 \circ \psi_U = p$ (where $\mathrm{pr}_1 : U \times \mathbf{K}^d \to U \mid (m', v) \mapsto m'$)
(2) for each m' in U, ψ_U restricts to a linear isomorphism between $p^{-1}(m')$ and $\mathrm{pr}_1^{-1}(m') = \{m'\} \times \mathbf{K}^d$.

3.2. *Changes of trivializations*

If ψ_U and ψ_V are two trivializations defined on $p^{-1}(U)$ and $p^{-1}(V)$ respectively (and U and V are open subsets of M) then there exists a smooth map (the *change of trivializations*)

$$g_{U,V} : U \cap V \longrightarrow \mathrm{GL}_d(\mathbf{K})$$

such that

$$\begin{aligned} \psi_U \circ \psi_V^{-1} : U \cap V \times \mathbf{K}^d &\longrightarrow U \cap V \times \mathbf{K}^d \\ (m, v) &\longmapsto (m, g_{U,V}(m) \cdot v). \end{aligned}$$

Exercise 3.1: Prove that $\psi_U \circ \psi_V^{-1}$ has the claimed form. Why is it more correct to write $(\psi_U|_{p^{-1}(U\cap V)}) \circ (\psi_V|_{p^{-1}(U\cap V)})^{-1}$?

The change of trivializations $g_{V,U}$ is the inverse of $g_{U,V}$: $\forall m \in M$, $g_{V,U}(m) = g_{U,V}(m)^{-1}$. Also $g_{U,U}(m) = \mathrm{id}$ for all m in U.

If ψ_W is a third trivialization, then the 2 new changes of trivializations $g_{U,W}$ and $g_{V,W}$ are part of an obvious compatibility condition (a "cocycle" condition):

$$g_{U,V}(m)g_{V,W}(m) = g_{U,W}(m) \quad \forall m \in U \cap V \cap W. \tag{COMP}$$

This follows from the equality $\psi_U \circ \psi_V^{-1} \circ \psi_V \circ \psi_W^{-1} = \psi_U \circ \psi_W^{-1}$.

3.3. *From cocyles to bundles*

Conversely, suppose given an open covering $\mathcal{U}$ of M ($\mathcal{U} \subset \mathcal{P}(M)$ —maybe it would be better to say there is a map $\mathcal{U} \to \mathcal{P}(M)$— every U in $\mathcal{U}$ is open and $\cup_{U \in \mathcal{U}} U = M$) and for every U and V in $\mathcal{U}$ a smooth map $g_{U,V} : U \cap V \to \mathrm{GL}_d(\mathbf{K})$ such that the relation (COMP) is satisfied for all U, V and W in $\mathcal{U}$, then there exists a (essentially unique) vector bundle E over M with trivializations $\{\psi_U\}_{U \in \mathcal{U}}$ such that the changes of trivializations are precisely the maps $g_{U,V}$.

Exercise 3.2: The relation (COMP) implies that

$$\begin{aligned} g_{U,U}(m) &= \mathrm{id} & &\forall m \in U \\ g_{V,U}(m) &= g_{U,V}(m)^{-1} & &\forall m \in U \cap V. \end{aligned}$$

Exercise 3.3: [Construction of E] Given $\mathcal{U}$ and the $g_{U,V}$ (U, V in $\mathcal{U}$) satisfying (COMP) construct an equivalence relation $\sim$ on the disjoint union $\coprod_{\mathcal{U}} U \times \mathbf{K}^d$ so that $E = (\coprod_{\mathcal{U}} U \times \mathbf{K}^d)/\!\sim$ is the sought for bundle.

3.4. *Linear algebra*

Every construction (or maybe most constructions) in linear algebra has its counterpart for vector bundles. We will mainly use:

- E^* the dual of E,
- $\mathrm{End}(E)$, endomorphisms
- $\overline{E}$, the complex conjugate (when $\mathbf{K} = \mathbf{C}$)
- $E \otimes_{\mathbf{R}} \mathbf{C}$ the complexification ($\mathbf{K} = \mathbf{R}$)
- $\overline{E}^* \otimes E^*$ and inside it the subset of Hermitian forms,
- $E \otimes F$ tensor product, etc.

Exercise 3.4: For a complex vector space V, what is $\overline{V}$? [More generally, for V a k-vector space and $\sigma \in \mathrm{Aut}(k)$, what is V^σ?]

Give the changes of trivializations for the bundle $\overline{E}$. What happens when $\overline{E}$ is isomorphic to E?

3.5. *The gauge group*

In the category of vector bundles (over a base manifold M say) there are natural notion of morphisms, isomorphisms, etc.

Exercise 3.5: A bundle morphism $E \to F$ is the same thing as a section of the vector bundle $\mathrm{Hom}(E, F)$.

A *gauge isomorphism* is an isomorphism of the bundle E.

Exercise 3.6: Why can a gauge isomorphism be seen either as a map $E \to E$ or as a section of $\mathrm{Aut}(E)$? What is the nature of the bundle $\mathrm{Aut}(E)$? (i.e. is there an additional structure on its fibers? etc.)

The group of gauge automorphisms is denoted by $\mathcal{A}ut(E)$. A gauge isomorphism g induces a gauge isomorphism on any bundle constructed from E but also on the bundles $\bigwedge^p T^*M \otimes E$, it will therefore induce an action on the space of sections of those bundles (see Section 3.6).

Exercise 3.7: In particular the isomorphism induced on $\mathrm{End}(E)$ is

$$\begin{aligned} \mathrm{End}(E) &\longrightarrow \mathrm{End}(E) \\ A &\longmapsto gAg^{-1}. \end{aligned}$$

(What is the meaning of the empty sign for the composition law involved in the notation gAg^{-1}?) Work out other examples: $\mathrm{End}(E^*)$, sesquilinear forms, etc.

Exercise 3.8: Express the trivializations of $\mathrm{Aut}(E)$ and the changes of trivializations in terms of those of E. Give a gauge isomorphism in local coordinates (i.e. in the "charts" given by the trivializations).

Remark 3.1: There is of course a strong link between trivializations and gauge isomorphism. In physics literature and sometimes in mathematics, a trivialization is called a "choice of gauge" or "fixing a gauge" and a change of trivialization is a "change of gauge". This link is quite apparent when comparing the formulas for the action of a gauge transformations with the formulas for changes of trivializations.

3.6. *Spaces of sections*

A *section* of E is a smooth map $\sigma : M \to E$ such that $p \circ \sigma = \mathrm{id}_M$. A *local* section is a smooth map $\sigma : U \to E$ with $p \circ \sigma = \mathrm{id}_U$ defined on open subset U of M. The space of sections is a topological $\mathbf{K}$-vector space denoted by $\Gamma(M; E)$ or sometimes $\Omega^0(M; E)$. We already met some of theses spaces when we mentioned differential forms:

$$\Omega^p(M; \mathbf{K}) = \Gamma\Big(M; \bigwedge^p T^*M \otimes_{\mathbf{R}} \mathbf{K}\Big),$$

or when X is a complex manifold:

$$\Omega^{p,q}(X) = \Gamma\Big(X; \bigwedge^{p} T^{*1,0}X \otimes \bigwedge^{q} T^{*0,1}X\Big).$$

Generally the sections of $\bigwedge^p T^*M \otimes E$ are called the p-forms with coefficients in E (or sometimes with "values" in E) and their space is denoted by

$$\Omega^{p}(M;E) = \Gamma\Big(M; \bigwedge^{p} T^*M \otimes E\Big)$$

and for a complex vector bundle E on an analytic manifold X:

$$\Omega^{p,q}(X;E) = \Gamma\Big(X; \bigwedge^{p} T^{*1,0}X \otimes \bigwedge^{q} T^{*0,1}X \otimes E\Big).$$

Exercise 3.9: [$\Omega^\bullet(M;E)$ is a graded $\Omega^\bullet(M;\mathbf{K})$-module] Construct (or define)

$$\begin{aligned} \wedge : \Omega^{p}(M;\mathbf{K}) \times \Omega^{p'}(M;E) &\longrightarrow \Omega^{p+p'}(M;E) \\ (\alpha,\sigma) &\longmapsto \alpha \wedge \sigma. \end{aligned}$$

This endows $\Omega^\bullet(M;E)$ with the structure of a graded $\Omega^\bullet(M;\mathbf{K})$-module. As a module it is generated by $\Omega^0(M;E)$.

Remark 3.2: It is sometimes a good thing to omit the wedge $\wedge$ symbol to avoid overloaded formulas (or even ambiguous formulas). Therefore this will be yet another composition law denoted by the empty sign.

4. Flat bundles

In this section flat bundles are introduced using also the point of view of local triviality (hence the objects given here are really "local systems"). The correspondence with linear representations of the fundamental group is detailed. Connections will be presented later.

4.1. *Flat structures*

A *flat structure* on a vector bundle E (or for short a *flat bundle*) is a family $\{\psi_U\}_{U\in\mathcal{U}}$ of trivializations, where $\mathcal{U}$ is an open cover of M, such that the changes of trivializations $g_{U,V}$ are locally constant.

Note that not every bundle has a flat structure and that the same bundle can have non-equivalent flat structures. In a sense the study of equivalence classes of flat structures is the very subject of this chapter.

There is a natural notion of isomorphism between flat bundles.

Exercise 4.1: If $(E, \{\psi_U\}_{U\in\mathcal{U}})$ and $(E', \{\psi_{U'}\}_{U'\in\mathcal{U}'})$ are two flat bundles, what is a morphism between E and E'? An isomorphism?

4.2. *Flat bundles and representations of the fundamental group*

—We suppose from now on that the base manifold M is connected.—

Every flat bundle $(E, \psi_U, g_{U,V})$ gives rise to a representation $\rho : \pi_1(M, m_0) \to \mathrm{GL}_d(\mathbf{K})$ as follow:

- Once for all, fix $U \in \mathcal{U}$ containing m_0.
- For any loop $\gamma : [0,1] \to M$, $\gamma(0) = \gamma(1) = m_0$ there are:
  - a subdivision $0 = t_0 \leq t_1 \leq \cdots \leq t_N \leq t_{N+1} = 1$ of $[0,1]$,
  - U_0, U_1, ..., U_N belonging to $\mathcal{U}$ such that
    1. $U_0 = U_N = U$
    2. and, for all $i = 0, \ldots, N$, $\gamma([t_i, t_{i+1}]) \subset U_i$.
- We then define:

$$\rho(\gamma) = g_{U_N,U_{N-1}}(\gamma(t_N)) \cdots g_{U_2,U_1}(\gamma(t_2)) g_{U_1,U_0}(\gamma(t_1)).$$

Exercise 4.2: Show that $\rho(\gamma)$ does not depend on the choices of the sequence $(U_0, \ldots, U_N)$ or of the subdivision. I.e. for another subdivision $(t'_i)_{i \leq N'}$ and another sequence $(U'_i)_{i \leq N'}$ as above then the product $g_{U'_{N'},U'_{N'-1}}(\gamma(t'_{N'})) \cdots g_{U'_1,U'_0}(\gamma(t'_1))$ is equal to $\rho(\gamma)$. (Hint: first reduce to the case where the 2 subdivisions are the same, then use the cocycle property (COMP) to change "step by step" the sequence (U'_i) into the sequence (U_i).)

Exercise 4.3: Every flat bundle on $M = [0,1]$ is (isomorphic to) the trivial flat bundle.

It is easy to see that $\rho(\gamma)$ does not change under deformations of γ (this is where the fact that the bundle is flat is involved) and hence depends only on the homotopy class of γ. Therefore, associated with the flat bundle E, is defined a homomorphism:

$$\rho : \pi_1(M, m_0) \longrightarrow \mathrm{GL}_d(\mathbf{K}), \quad \rho \in \mathrm{Hom}\big(\pi_1(M, m_0), \mathrm{GL}_d(\mathbf{K})\big).$$

Exercise 4.4: Define the law of composition on $\pi_1(M, m_0)$ so that ρ is a homomorphism.

In fact this representation ρ, called the *holonomy representation*, depends also of the choice of U: changing U to $U' \in \mathcal{U}$ (again with $U' \ni m_0$) amounts to changing the representation ρ to:

$$\begin{aligned} \pi_1(M, m_0) &\longrightarrow \mathrm{GL}_d(\mathbf{K}) \\ \gamma &\longmapsto g\rho(\gamma)g^{-1} \end{aligned}$$

with $g = g_{U,U'}(m_0)$.

More generally if E and E' are two isomorphic flat bundles, the two corresponding holonomy representations are only conjugated.

One can summarize this discussion with the following map:

$$\frac{\{\text{flat bundle of rank } d \text{ over } M\}}{\text{isomorphism}} \xrightarrow{\text{Hol}} \frac{\text{Hom}(\pi_1(M, m_0), \text{GL}_d(\mathbf{K}))}{\text{conjugacy}}.$$

4.3. *From representations to flat bundles*

It turns out that Hol is a bijection. This can be seen via the following construction which provides an inverse to the map Hol. Let $\rho : \pi_1(M, m_0) \to \text{GL}_d(\mathbf{K})$ be a morphism, then the trivial bundle

$$\widetilde{M} \times \mathbf{K}^d$$

over the universal cover $\widetilde{M}$ of M has an obvious flat structure together with an action of $\pi_1(M, m_0)$ preserving the flat structure:

$$\gamma \cdot (\tilde{m}, v) = (\gamma\tilde{m}, \rho(\gamma)v) \quad \forall \gamma \in \pi_1(M, m_0), \tilde{m} \in \widetilde{M}, v \in \mathbf{K}^d.$$

Here $\tilde{m} \mapsto \gamma\tilde{m}$ is the natural action of $\pi_1(M, m_0)$ on the universal cover $\widetilde{M}$. The quotient of the trivial bundle $\widetilde{M} \times \mathbf{K}^d$ by this action is denoted

$$E_\rho = \Gamma \backslash (\widetilde{M} \times \mathbf{K}^d),$$

it is naturally a flat bundle over the base $M = \Gamma \backslash \widetilde{M}$.

Exercise 4.5: E_ρ depends only (up to isomorphism) on the conjugacy class of ρ. The holonomy representation of E_ρ is ρ.

Exercise 4.6: Let E be a flat bundle over M. Then the pull back $\widetilde{E}$ over $\widetilde{M}$ is a flat bundle together with an action of $\pi_1(M, m_0)$. The flat bundle $\widetilde{E}$ is the trivial flat bundle: $\widetilde{E} \simeq \widetilde{M} \times \mathbf{K}^d$ (Hint: use a method similar to the construction of the holonomy representation to extend a fixed trivialization around $\tilde{m}_0$ first along paths, then everywhere by showing the desired homotopy invariance) and the action of $\pi_1(M, m_0)$ has the form $\gamma \cdot (m, v) = (\gamma m, \rho(\gamma)v)$ for some representation ρ. Furthermore $\rho = \text{Hol}(E)$ and at last E is isomorphic to E_ρ.

Conclude that the map Hol is a bijection.

Exercise 4.7: Prove that $\text{Hom}(\pi_1(M, m_0), \text{GL}_d(\mathbf{K}))$ is in bijection with

$$\frac{\{(E, (\epsilon_i)) \mid E \xrightarrow{p} M \text{ flat bundle}, (\epsilon_i) \text{ is a basis of } p^{-1}(m_0)\}}{\text{isomorphism}}.$$

We will later characterize flat bundles (or flat structures on a given bundle) with the help of connections. We only take note for the moment that there is, for a (non-flat!) bundle E, no natural way to differentiate sections of E or more generally forms with coefficients in E. Connections are here to fill this gap.

5. Holomorphic vector bundles

The investigation of the different families of vector bundles continues in this section with the study of vector bundles on a complex manifold. Here the correspondence between holomorphic structures on a given manifold and integrable pseudo-connections (those are a special class of differential operators) is explained. The gauge group acts on the space of holomorphic structures as well as on the space of pseudo-connections and this action is compatible with the mentioned correspondence.

5.1. *Holomorphic bundles and their trivializations*

When the base space X is a complex manifold, it makes sense to consider when $p : E \to X$ is a holomorphic map between complex manifolds (where E is a $\mathbf{C}$-vector bundle). One can show that in this situation there are holomorphic trivializations:

$$\psi_U : p^{-1}(U) \longrightarrow U \times \mathbf{C}^d.$$

Consequently, the change of trivializations:

$$g_{U,V} : U \cap V \longrightarrow \mathrm{GL}_d(\mathbf{C})$$

is holomorphic (meaning that every coordinates entry of the matrix $g_{U,V}$ is a holomorphic function on $U \cap V$).

Conversely given an open cover $\mathcal{U}$ of X and holomorphic maps $g_{U,V} : U \cap V \to GL_d(\mathbf{C})$ satisfying the compatibility condition (COMP), then a holomorphic vector bundle can be constructed with those changes of trivializations.

One could also speak of a *holomorphic structure* on a given smooth bundle $p : E \to X$. It is likely that not every smooth bundle has a holomorphic structure (but it may be as well an open question); a given vector bundle can have different holomorphic structure (this is related to the theorem of Narasimhan and Seshadri presented below).

5.2. *Pseudo-connection*

A holomorphic structure $\mathcal{E}$ on a smooth vector bundle E can be characterize with the help of an operator:

$$\bar{\partial}^{\mathcal{E}} : \Omega^{\bullet,\bullet}(X; E) \longrightarrow \Omega^{\bullet,\bullet}(X; E)$$

(1) of degree $(0,1)$: $\bar{\partial}^{\mathcal{E}}\sigma$ is of degree $(p, q+1)$ if σ is of degree (p, q).

(2) satisfying the Leibniz rule: if $\alpha \in \Omega^{p,q}(X)$ and $\sigma \in \Omega^{k,l}(X;E)$

$$\bar{\partial}^{\mathcal{E}}(\alpha \wedge \sigma) = (\bar{\partial}\alpha) \wedge \sigma + (-1)^{p+q}\alpha \wedge \bar{\partial}^{\mathcal{E}}\sigma.$$

(3) A section σ of E (i.e. $\sigma \in \Omega^0(X;E) = \Omega^{0,0}(X;E)$) is holomorphic if and only if $\bar{\partial}^{\mathcal{E}}\sigma = 0$.

Definition 5.1: An operator $\bar{\partial}^E : \Omega^\bullet(X;E) \to \Omega^{\bullet,\bullet}(X;E)$ satisfying 1 and 2 above is called a *pseudo-connection.*

Exercise 5.1: Express a pseudo-connection in a trivialization.

It is easy to see that a pseudo-connection is entirely determined by the map:

$$\bar{\partial}^E : \Omega^0(X;E) \longrightarrow \Omega^{0,1}(X;E).$$

When $\mathcal{E}$ is a holomorphic structure on E, the pseudo-connection in degree 0 is defined by the formula

$$\bar{\partial}^{\mathcal{E}}(f\sigma) = \bar{\partial}f \wedge \sigma$$

when σ is a holomorphic section of $\mathcal{E}$ and f is any function.

Exercise 5.2: Check the details of the construction of $\bar{\partial}^{\mathcal{E}}$. Alternatively show the existence of $\bar{\partial}^{\mathcal{E}}$ by checking compatibility of the constructions in trivializations.

5.3. *Pseudo-curvature*

It is not true that every pseudo-connection on E comes from a holomorphic structure. Given a pseudo-connection $\bar{\partial}^E$ on E it means that there are not enough local sections σ of E for which $\bar{\partial}^E\sigma = 0$, i.e. sections of $p^{-1}(U)$ over an open subset $U \subset X$ (those should be the holomorphic local sections). This lack of enough local solutions to $\bar{\partial}^E\sigma = 0$ (that could be called "default of integrability") is precisely quantified by the pseudo-curvature:

Lemma 5.2: *The operator*

$$\begin{aligned}(\bar{\partial}^E)^2 : \Omega^{\bullet,\bullet}(X;E) &\longrightarrow \Omega^{\bullet,\bullet+2}(X;E)\\ \sigma &\longmapsto \bar{\partial}^E(\bar{\partial}^E\sigma)\end{aligned}$$

is $\mathcal{C}^\infty$-linear ($(\bar{\partial}^E)^2(\alpha \wedge \sigma) = \alpha \wedge (\bar{\partial}^E)^2\sigma$ for all $\alpha \in \Omega^\bullet(X;\mathbf{C})$ and all $\sigma \in \Omega^\bullet(X;E)$).

This implies that there is a (unique) $(0,2)$-form with coefficients in $\mathrm{End}(E)$ *—called the* pseudo-curvature *of $\bar{\partial}^E$, denoted by $G = G(\bar{\partial}^E)$— such that*

$$(\bar{\partial}^E)^2\sigma = G\sigma \quad \forall \sigma \in \Omega^\bullet(X;E).$$

Remark 5.3: The empty sign for the composition law in the notation "$G\sigma$" involves 2 operations: the wedge product on forms and the evaluation of endomorphisms (elements in $\mathrm{End}(E)$) on sections of E. Explicitly if $G = \sum_i \alpha_i A_i$, $\alpha_i \in \Omega^{0,2}(X)$, $A_i \in \Omega^0(X; \mathrm{End}(E))$ and if $\sigma = \sum_j \beta_j \sigma_j$, $\beta_j \in \Omega^{p,q}(X)$, $\sigma_j \in \Omega^0(X; E)$ then $G\sigma = \sum_{i,j} \alpha_i \wedge \beta_j A_i(\sigma_j)$.

Exercise 5.3: Using (3 times) the Leibniz rule —and the fact that $\bar{\partial}^2 = 0$ on differential forms— prove the $\mathcal{C}^\infty$-linearity of $(\bar{\partial}^E)^2$.

Alternatively express $(\bar{\partial}^E)^2$ in a trivialization and conclude the existence of the pseudo-curvature (checking the compatibility under change of trivializations).

Given a pseudo-connection $\bar{\partial}^E$ on a smooth vector bundle E, all the vector bundles constructed from E have also a natural pseudo-connection: E^*, $\mathrm{End}(E)$, etc. All those pseudo-connections are such that the natural maps between those bundles must be "$\bar{\partial}^E$-holomorphic". For example for E^* the evaluation map $E^* \times E \to \mathbf{C} \mid (f, e) \mapsto f(e)$ has to be holomorphic and for $\mathrm{End}(E)$ this is the map $\mathrm{End}(E) \times E \to E \mid (A, e) \to A(e)$ that is required to be holomorphic.

Exercise 5.4: Let $\bar{\partial}^E$ be a pseudo-connection on E. There is a unique operator $\bar{\partial}^{E^*}$ such that for every section σ of E and ϕ of E^* (so that $\phi(\sigma)$ is a function) one has

$$\bar{\partial}(\phi(\sigma)) = (\bar{\partial}^{E^*} \phi)(\sigma) + \phi(\bar{\partial}^E(\sigma)).$$

Furthermore $\bar{\partial}^{E^*}$ is the sought for pseudo-connection on E^*.

Similarly, for A a section of $\mathrm{End}(E)$ and σ a section of E, the formula

$$\bar{\partial}^E(A(\sigma)) = (\bar{\partial}^{\mathrm{End}(E)} A)(\sigma) + A(\bar{\partial}^E \sigma)$$

defines the pseudo-connection on $\mathrm{End}(E)$.

Exercise 5.5: [Bianchi identity for pseudo-connection] $\bar{\partial}^{\mathrm{End}(E)} G(\bar{\partial}^E) = 0$.

The following theorem makes the link between holomorphic structure and pseudo-connections with vanishing pseudo-curvature. (See [8] for a proof.)

Theorem 5.4: *Let $\bar{\partial}^E$ be a pseudo-connection on E over X.*

There exists a holomorphic structure $\mathcal{E}$ on E such that $\bar{\partial}^E = \bar{\partial}^{\mathcal{E}}$ if and only if the pseudo-curvature $G(\bar{\partial}^E)$ vanishes identically.

Exercise 5.6: If $\mathcal{E}$ is a holomorphic structure on E, then $G(\bar{\partial}^{\mathcal{E}}) = 0$.

Exercise 5.7: [Uniqueness in the above theorem] If $\mathcal{E}$ and $\mathcal{E}'$ are 2 holomorphic structures on E such that $\bar{\partial}^{\mathcal{E}} = \bar{\partial}^{\mathcal{E}'}$ then $\mathcal{E} = \mathcal{E}'$.

Hence there is a bijection:

$\{\text{holomorphic structure on } E\} \leftrightarrow \{\text{pseudo-connection } \bar{\partial}^E \text{ with } (\bar{\partial}^E)^2 = 0\}.$

It can be checked that this bijection is equivariant with respect to the gauge group action on these 2 spaces.

5.4. *The space of pseudo-connections*

Lemma 5.5: *Let $\bar{\partial}_1$, $\bar{\partial}_2$ be 2 pseudo-connections on E.*

Then the operator $A = \bar{\partial}_2 - \bar{\partial}_1$ is $\mathcal{C}^\infty$-linear ($A(\alpha \wedge \sigma) = \alpha \wedge A(\sigma)$) and hence can be identified with a $(0,1)$-form with coefficients in $\mathrm{End}(E)$.

Exercise 5.8: Prove the lemma.

Remark 5.6: We will frequently make no distinctions (albeit notational) between a form with coefficients in $\mathrm{End}(E)$ and the corresponding operator acting on $\Omega^\bullet(X; E)$. This introduces further notational ambiguities that can be only overcome with practice. The absence of different signs (in fact the absence of any signs) for the different laws of compositions or the action of operators does not help clearing those ambiguities.

Conversely

Lemma 5.7: *If $\bar{\partial}_1$ is a pseudo-connection and $A \in \Omega^{0,1}(X; \mathrm{End}(E))$ then $\bar{\partial}_2 = \bar{\partial}_1 + A$ is a pseudo-connection.*

Exercise 5.9: Prove the lemma. Give an expression for $G(\bar{\partial}_1 + A)$. How does this relates to the expression in local trivializations? (See Exercise 5.1.)

Corollary 5.8: *The space of pseudo-connections on E is an affine space over the vector space $\Omega^{0,1}(X; \mathrm{End}(E))$.*

Therefore the space of all pseudo-connections has the most simple topology. The space of *integrable* pseudo-connections can nevertheless have a quite complicated topology.

5.5. *The action of the gauge group*

Any gauge transformation $g \in \mathcal{A}ut(E)$ acts on the space of sections of E (by composition) as well as on the forms with coefficients in E and also with coefficients in bundles constructed from E. Explicitly, if σ is a section of E then $g\sigma$ is the (fiberwise) image of σ by g, and if $\sum \alpha_i \sigma_i$ is a differential form with coefficients in E then $g(\sum \alpha_i \sigma_i) = \sum \alpha_i g \sigma_i$.

For a pseudo connection $\bar{\partial}^E$ another pseudo-connection $g\cdot\bar{\partial}^E$ is obtained by

$$g \cdot \bar{\partial}^E = \text{``}g\bar{\partial}^E g^{-1}\text{''} : \Omega^\bullet(X;E) \longrightarrow \Omega^\bullet(X;E)$$
$$\sigma \longmapsto g(\bar{\partial}^E(g^{-1}\sigma)).$$

Explanation of the notation:

- $g^{-1}\sigma$ is the action of g^{-1} on $\Omega^\bullet(X;E)$.
- $\bar{\partial}^E(g^{-1}\sigma)$ is the image of $g^{-1}\sigma$ by the operator $\bar{\partial}^E$.
- $g(\bar{\partial}^E(g^{-1}\sigma))$ is again the action of g.

Exercise 5.10: $g \cdot \bar{\partial}^E$ is a pseudo-connection.
This defines an action of $\mathcal{A}ut(E)$ on the space of pseudo-connections.

Lemma 5.9: *One has the relation:*

$$g \cdot \bar{\partial}^E - \bar{\partial}^E = g\big(\bar{\partial}^{\mathrm{End}(E)}(g^{-1})\big) = -\big(\bar{\partial}^{\mathrm{End}(E)}(g)\big)g^{-1}.$$

Explanation: We wrote here the equality between an operator (the difference between 2 pseudo-connections) and a $(0,1)$-form with coefficients in $\mathrm{End}(E)$ —c.f. the above Remark 5.6. Here g^{-1} is considered as a section of $\mathrm{End}(E)$ and $\bar{\partial}^{\mathrm{End}(E)}$ (the extension of $\bar{\partial}^E$ to $\mathrm{End}(E)$) acts on this section. The "product" $g(\bar{\partial}^{\mathrm{End}(E)}g^{-1})$ is the composition of an element of $\Omega^0(X;\mathrm{End}(E))$ with an element of $\Omega^{0,1}(X;\mathrm{End}(E))$ leading to an element in $\Omega^{0,1}(X;\mathrm{End}(E))$, this of course involves the composition of endomorphisms $\mathrm{End}(E)\times\mathrm{End}(E)\to\mathrm{End}(E)$.

Exercise 5.11: Prove the lemma.

Remark 5.10: This leads, forgetting the superscripts and the parenthesis, to the abysmal identity:

$$g\bar{\partial}g^{-1} = \bar{\partial} + g\bar{\partial}g^{-1}!!!$$

Exercise 5.12: Give explicitly the action $(g,\mathcal{E}) \to g\cdot\mathcal{E}$ of the gauge group onto the space of holomorphic structures on E. Prove that $g\cdot\bar{\partial}^{\mathcal{E}} = \bar{\partial}^{g\cdot\mathcal{E}}$.

Lemma 5.11: $G(g\cdot\bar{\partial}^E) = gG(\bar{\partial}^E)g^{-1}$.

Having in mind that the space of pseudo-connections is an affine space, the following result is easily proved.

Lemma 5.12: *The tangent space at $\bar{\partial}^E$ to the space of pseudo-connections is*

$$\Omega^{0,1}(X;\mathrm{End}(E)).$$

The tangent space to the orbit $\mathcal{A}ut(E) \cdot \bar{\partial}^E$ *is the image of*

$$\bar{\partial}^{\mathrm{End}(E)} : \Omega^0(X; \mathrm{End}(E)) \longrightarrow \Omega^{0,1}(X; \mathrm{End}(E)).$$

(As above, the induced pseudo-connection on $\mathrm{End}(E)$ *is denoted* $\bar{\partial}^{\mathrm{End}(E)}$*.)*

Exercise 5.13: Prove the lemmas.

6. Flat bundles and connections

We have characterized (or parametrized) the holomorphic structures with the help of pseudo-connections. Our objective is now to perform the similar program for flat bundles and (flat!) connections. The vanishing of the curvature will be the key notion. Moreover connections (and sometimes non-flat connections) will have their importance as they appear in the Hitchin-Kobayashi correspondence. An instance of this will be the relation between pseudo-connections and unitary connections performed via the Chern connections.

6.1. *Connections*

Let us start with flat bundles or with flat structures on a given bundle. Given a smooth $\mathbf{K}$-vector bundle E over a base M (a manifold) ($\mathbf{K} = \mathbf{R}$ or $\mathbf{C}$), the notation $\mathbb{E}$ will be used for a flat structure on E.

Lemma 6.1: *Let* $\mathbb{E}$ *be a flat structure on* E*. Then there exists a (unique) operator:*

$$D^{\mathbb{E}} : \Omega^\bullet(M; E) \longrightarrow \Omega^\bullet(M; E)$$

(1) *of degree 1,* $D^{\mathbb{E}}\sigma$ *is of degree* $p+1$ *if* $\sigma \in \Omega^p(M; E)$*.*
(2) *satisfying the Leibniz rule:*

$$D^{\mathbb{E}}(\alpha \wedge \sigma) = (d\alpha) \wedge \sigma + (-1)^{\deg(\alpha)} \alpha \wedge D^{\mathbb{E}}\sigma.$$

(3) *A section* $\sigma \in \Omega^0(M; E)$ *is flat (i.e. in the trivializations of the flat structure* $\mathbb{E}$*, the section* σ *becomes a (locally) constant map* $U \to \mathbf{K}^d$*) if and only if* $D^{\mathbb{E}}\sigma = 0$*.*

Definition 6.2: An operator $D : \Omega^\bullet(M; E) \to \Omega^\bullet(M; E)$ satisfying 1 and 2 of the above lemma is called a *connection*.

Exercise 6.1: Every bundle admits (at least) one connection.

Remark 6.3: In 3 the statement should be: the restriction of σ to some open set $U \subset M$ is flat if and only if the restriction of $D^{\mathbb{E}}\sigma$ to U is zero.

One can check that a connection D is completely determined by its action in degree 0:

$$D : \Omega^0(M; E) \longrightarrow \Omega^1(M; E)$$

(that satisfies $D(f\sigma) = df \wedge \sigma + fD\sigma$).

Any connection on E induces a connection on any bundle constructed from E —the relevant bundle will usually be indicated with a superscript, however sometimes this convention could not be respected. Also a connection $D^{E_1} = D_1$ on E_1 and a connection $D^{E_2} = D_2$ on E_2 induces a connection $D^{E_1 \otimes E_2} =$ "$D_1 \otimes D_2$" on $E_1 \otimes E_2$ (see the Exercise below).

Exercise 6.2: "Define" the induced connection on E^*, show that

$$d\phi(\sigma) = (D^{E^*}\phi)(\sigma) + \phi(D^E\sigma) \quad \forall \sigma \in \Omega^0(M; E),\ \phi \in \Omega^0(M; E^*).$$

Give formula for the connection $D^{E_1 \otimes E_2}$. Why is the notation $D_1 \otimes D_2$ inappropriate? (Some care is in order here, the operators that will appear here act naturally on $\Omega^\bullet(M; E_1) \otimes \Omega^\bullet(M; E_2)$ but *not* on $\Omega^\bullet(M; E_1 \otimes E_2)$ —of course there is a map $\Omega^\bullet(M; E_1) \otimes \Omega^\bullet(M; E_2) \to \Omega^\bullet(M; E_1 \otimes E_2)$.)

6.2. *Curvature*

The default of integrability ("lack of flatness") of a connection is measured by its curvature.

Lemma 6.4: *Let D^E be a connection on E. Then the operator*

$$(D^E)^2 : \Omega^\bullet(M; E) \longrightarrow \Omega^{\bullet+2}(M; E)$$

is $\mathcal{C}^\infty$-linear ($(D^E)^2(\alpha \wedge \sigma) = \alpha \wedge (D^E)^2\sigma$ for $\alpha \in \Omega^\bullet(M; \mathbf{R})$ and $\sigma \in \Omega^\bullet(M; E)$).

Hence there exists a (unique) 2-form $F = F(D^E)$ with coefficients in $\mathrm{End}(E)$ *so that $(D^E)^2\sigma = F\sigma$ for all σ.*

Exercise 6.3: Give local expressions for D^E and $F(D^E)$.

Exercise 6.4: Express $F(D_1 \otimes D_2)$ with $F(D_1)$ and $F(D_2)$. (See Exercise 6.2.)

Exercise 6.5: If $D^{\mathbb{E}}$ is the connection of a flat structure on the bundle E, then $F(D^{\mathbb{E}}) = 0$.

Exercise 6.6: [Bianchi identity] $D^{\mathrm{End}(E)}F(D^E) = 0$.

Theorem 6.5: *A connection D^E is flat (i.e. there exists a flat structure $\mathbb{E}$ on E such that $D^E = D^{\mathbb{E}}$) if and only if $F(D^E) = 0$.*

Exercise 6.7: If $\mathbb{E}_1$ and $\mathbb{E}_2$ are 2 flat structures on E such that $D^{\mathbb{E}_1} = D^{\mathbb{E}_2}$ then $\mathbb{E}_1 = \mathbb{E}_2$.

Therefore there is a one-to-one correspondence:

$$\{\text{flat structures on } E\} \leftrightarrow \{\text{connections } D \text{ such that } F(D) = 0\}$$

and this bijection is equivariant with respect to the gauge group action on these 2 spaces.

Exercise 6.8: Describe the action of $\mathcal{A}ut(E)$ on the space of flat structures as well as on the space of connections. Show the equivariance of the above correspondence.

We have previously seen that the space of isomorphism classes of flat bundles is in bijection with conjugacy classes of representations. To isolate the representations corresponding to the $\mathcal{C}^\infty$-vector bundle E, let us denote by

$$\mathrm{Hom}_E\big(\pi_1(M, m_0), \mathrm{GL}_d(\mathbf{K})\big)$$

the subset of representations $\rho : \pi_1(M, m_0) \to \mathrm{GL}_d(\mathbf{K})$ such that E_ρ is smoothly isomorphic to E, i.e. there is an isomorphism of *smooth* bundles between E_ρ and E.

Exercise 6.9: $\mathrm{Hom}_E(\pi_1(M), \mathrm{GL}_d(\mathbf{K}))$ is open in $\mathrm{Hom}(\pi_1(M), \mathrm{GL}_d(\mathbf{K}))$. Is it closed? (Hint: you need more tools to do this exercise:

(1) $\mathrm{Hom}(\pi_1(M, m_0), \mathrm{GL}_d(\mathbf{K}))$ is locally path-connected and
(2) homotopic bundles are isomorphic: if E is a bundle on $M \times [0,1]$ then the restrictions of E to $M \times \{0\}$ and $M \times \{1\}$ are isomorphic. See [21].)

Then:

$$\frac{\mathrm{Hom}_E(\pi_1(M, m_0), \mathrm{GL}_d(\mathbf{K}))}{\text{conjugacy}} \leftrightarrow \frac{\{D \text{ connection on } E \text{ with } F(D) = 0\}}{\mathcal{A}ut(E)}.$$

Exercise 6.10: Let $E_0 = p^{-1}(m_0)$, describe the map $\varepsilon : \mathcal{A}ut(E) \to \mathrm{GL}(E_0)$. Let $\mathcal{A}ut(E)_\varepsilon$ be the kernel of this morphism. Prove that

$$\mathrm{Hom}_E(\pi_1(M, m_0), \mathrm{GL}_d(\mathbf{K})) \leftrightarrow \frac{\{D \text{ connection on } E \text{ with } F(D) = 0\}}{\mathcal{A}ut(E)_\varepsilon}.$$

Lemma 6.6:

(1) The difference between 2 connections is a 1-form with coefficients in $\mathrm{End}(E)$.
(2) The sum of a connection and (the operator associated with) a 1-form with coefficient in $\mathrm{End}(E)$ *is a connection.*

This means that the space of all connections on E has a natural structure of an affine space.

Exercise 6.11: Give the expression for $F(D^E + A)$.

The gauge group action is the following:

$$\begin{aligned} g \cdot D : \Omega^\bullet(M; E) &\longrightarrow \Omega^\bullet(M; E) \\ \sigma &\longmapsto g(D(g^{-1}\sigma)). \end{aligned}$$

One has

$$\begin{aligned} g \cdot D^E - D^E &= gD^{\mathrm{End}(E)}(g^{-1}) = -(D^{\mathrm{End}(E)}g)g^{-1} \\ F(g \cdot D^E) &= gF(D^E)g^{-1}. \end{aligned}$$

7. Chern connections, stability, degree

The expression of the Hitchin-Kobayashi correspondence will be given in the next section "explicitly" in terms of operators (connections, pseudo-connections and operators induced by differentials forms with values in the endomorphisms bundle) and different constructions involving the Hermitian metric (decomposition unitary+Hermitian) and the complex structure (decomposition according to the bidegree). This section explains at length the construction of the Chern connection which (once a Hermitian metric is fixed) can be seen as a bijective correspondence between pseudo-connections and unitary connections on a complex vector bundle E.

Another ingredient is the notion of stability of holomorphic vector bundles. In order to define that notion, the degree and the Chern characters of vector bundles are presented.

7.1. *Hermitian structure*

An Hermitian structure H on a complex vector bundle E is a smooth family of Hermitian scalar products on E (linear in the second variable and conjugate-linear in the first variable). The Hermitian structure H can be seen as a section of $\overline{E}^* \otimes E^*$, consequently H can be seen as a section of $\mathrm{Hom}(E, \overline{E}^*)$, etc.

Exercise 7.1: The gauge group $\mathcal{A}ut(E)$ acts transitively on the set of Hermitian structures on E.

The subgroup stabilizing H is the unitary gauge group $\mathcal{U}(E, H)$ or $\mathcal{U}(E)$.

Exercise 7.2: There exists (at least) one Hermitian structure on E.

A connection D on (E,H) is called *unitary* if $DH = 0$. In this last equation D is the induced connection on $\overline{E}^* \otimes E^*$. It would be wiser to say that D^E is a connection on E, $D^{\overline{E}^*\otimes E^*}$ its extension to $\overline{E}^* \otimes E^*$and write $D^{\overline{E}^*\otimes E^*}H = 0$.

Exercise 7.3: Let D^E be a connection on E ("extended" to E^*, $\overline{E}$, etc.). For K in $\Omega^0(M; \overline{E}^* \otimes E^*)$ and σ, σ' in $\Omega^0(M;E)$, one has

$$dK(\sigma,\sigma') = D^{\overline{E}^*\otimes E^*}K(\sigma,\sigma') + K(D^E\sigma,\sigma') + K(\sigma, D^E\sigma').$$

The connection D^E is H-unitary if and only if for all σ, σ' in $\Omega^0(M;E)$

$$dH(\sigma,\sigma') = H(D^E\sigma,\sigma') + H(\sigma, D^E\sigma').$$

Remark 7.1: For sections σ and σ' the function $H(\sigma,\sigma')$ will sometimes be written $(\sigma,\sigma')_H$ or even (σ,σ') (it is the evaluation of the Hermitian scalar product H on those 2 sections of E). This procedure is also extended to forms with coefficients in E via $H(\alpha\sigma,\alpha'\sigma') = \bar{\alpha}\wedge\alpha' H(\sigma,\sigma')$.

Lemma 7.2: *Any connection D on (E,H) decomposes (uniquely) as the sum of a unitary connection D_H and an H-Hermitian 1-form ψ_H with coefficients in* $\mathrm{End}(E)$ *(i.e. $(\psi_H\sigma,\sigma')_H = (\sigma,\psi_H\sigma')_H$ for all sections σ, σ').*

Exercise 7.4: Prove the lemma. What is the relation between ψ_H and $D^{\overline{E}^*\otimes E^*}H$?

There is always (at least) one unitary connection.

7.2. *Unitary connections over a complex manifold*

Any connection D on E a complex vector bundle over a complex manifold X decomposes ("into types") as the sum of 2 operators

$$\begin{aligned} D : \Omega^0(X;E) &\longrightarrow \Omega^{1,0}(X;E) \oplus \Omega^{0,1}(X;E) \\ \sigma &\longmapsto \partial^E\sigma + \bar{\partial}^E\sigma. \end{aligned}$$

Exercise 7.5: The operator $\bar{\partial}^E$ is automatically a pseudo-connection.

In the presence of an Hermitian structure a converse construction exists.

Lemma 7.3: *Let $E \to X$ be a complex vector bundle equipped with an Hermitian structure H and a pseudo-connection $\bar{\partial}^E$.*

Then there exists a unique unitary connection D on E whose $(0,1)$ *part is $\bar{\partial}^E$.*

Exercise 7.6: Prove the lemma.

The unitary connection produced by this lemma is called the *Chern connection.* As a consequence of this lemma the gauge group $\mathcal{A}ut(E)$ acts on the space of unitary connections on E (the "natural" group acting on the space of unitary connections is $\mathcal{U}(E)$).

7.3. L^2-*metrics*

When M is a Riemannian manifold and E is an Hermitian vector bundle over M, any of the bundle $\bigwedge^p T^*M \otimes E$ gets equipped with an Hermitian structure. Hence, by integrating on M with respect to the volume form, there are corresponding L^2-scalar products on the spaces $\Omega^p(M;E)$. The (formal) adjoints with respect to those L^2-metrics will be again denoted with a $\star$ in superscript.

7.4. *Another kind of adjunction*

Still the presence of an Hermitian scalar product H on a vector space V leads to the construction of an adjoint for endomorphisms:

$$\begin{aligned} \mathrm{End}(V) &\longrightarrow \mathrm{End}(V) \\ A &\longmapsto A^{*_H}. \end{aligned}$$

In terms of matrices in an orthonormal frame, this operation is simply the transpose-conjugate. The map $A \to A^{*_H}$ is conjugate-linear ($(\lambda A)^{*_H} = \bar{\lambda} A^{*_H}$) and involutive ($(A^{*_H})^{*_H} = A$). Performing this operation "in family" we get a similar construction for an Hermitian bundle (E,H), there is a map

$$*_H : \Omega^0(M;\mathrm{End}(E)) \longrightarrow \Omega^0(M;\mathrm{End}(E))$$

(1) characterized by

$$(A\sigma,\sigma')_H = (\sigma, A^{*_H}\sigma')_H$$

for all $A \in \Omega^0(M;\mathrm{End}(E))$, σ, σ' in $\Omega^0(M;E)$.

(2) conjugate-linear: for all f in $\Omega^0(M;\mathbf{C})$ —i.e. f is a function— and for all A in $\Omega^0(M;\mathrm{End}(E))$, $(fA)^{*_H} = \bar{f}A^{*_H}$.

(3) and involutive $(A^{*_H})^{*_H} = A$.

There is only one way to extend $*_H$ to a map:

$$*_H : \Omega^\bullet(M;\mathrm{End}(E)) \longrightarrow \Omega^\bullet(M;\mathrm{End}(E))$$

satisfying the following property: for all α in $\Omega^\bullet(M;\mathbf{C})$ and all A in $\Omega^\bullet(M;\mathrm{End}(E))$, $(\alpha \wedge A)^{*_H} = \bar{\alpha} \wedge A^{*_H}$. This operator is involutive and will appear later in formulas.

Exercise 7.7: Characterize $*_H$ on forms with coefficients in $\mathrm{End}(E)$ with the help of an "adjunction" property.

When the base manifold is X a complex manifold, observe that $\bar{\alpha} \in \Omega^{q,p}(X)$ if $\alpha \in \Omega^{p,q}(X)$. This implies that $A^{*_H} \in \Omega^{q,p}(X;\mathrm{End}(E))$ if $A \in \Omega^{p,q}(X;\mathrm{End}(E))$.

A form A with coefficients in $\mathrm{End}(E)$ is called *Hermitian* if $A^{*_H} = A$ and *antihermitian* if $A^{*_H} = -A$.

Exercise 7.8: Every Hermitian (resp. antihermitian) 1-form with coefficients in $\mathrm{End}(E)$ can be decomposed uniquely as $B + B^{*_H}$ (resp. $A - A^{*_H}$) where B (resp. A) is a $(1,0)$-form.

7.5. *Kähler manifold*

The analogue of the Kähler identities are now established for (non-necessarily unitary) connection on (E,H). Note that the 2-form ω induces again an operator of degree 2

$$L : \Omega^\bullet(X;E) \longrightarrow \Omega^{\bullet+2}(X;E)$$

and its (formal) adjoint is denoted by $L^\star$ —nothing new hence.

Proposition 7.4: [18] *Let X be a Kähler manifold and (E,H) be an Hermitian vector bundle over X. Let also D be a connection on E, decompose now D:*

$$D = D_H + \psi_H$$

as the sum of a unitary connection and an Hermitian 1-form (see Lemma 7.2). Decompose furthermore into type:

$$D_H = \partial_H + \bar{\partial}_H \quad \textit{and} \quad \psi_H = \psi^{1,0} + \psi^{0,1}$$

(see Section 7.2).

Then $(\psi^{1,0})^{*_H} = \psi^{0,1}$ and setting

$$D' = \partial_H + \psi^{0,1} \quad D'' = \bar{\partial}_H + \psi^{1,0},$$

one has:

$$[L, D'^\star] = \sqrt{-1}D'' \quad [L^\star, D'] = -\sqrt{-1}D''^\star$$
$$[L, D''^\star] = -\sqrt{-1}D' \; [L^\star, D''] = \sqrt{-1}D'^\star.$$

To prove this proposition it is enough to work locally, hence choosing an orthonormal frame for E one can work with the trivial bundle equipped with the "trivial" Hermitian scalar product (but not the trivial connection). In this trivialization the unitary connection has the form

$$d + A^{1,0} + A^{0,1}$$

and $A^{1,0*_H} = -A^{0,1}$ (see Exercise 7.8).

Since the result is known for $d = \partial + \bar{\partial}$, to prove it for $D = D' + D''$ with $D' = \partial + A^{1,0} + \psi^{0,1}$ and $D'' = \bar{\partial} + A^{0,1} + \psi^{1,0} = \bar{\partial} - A^{1,0*_H} + \psi^{0,1*_H}$ it is enough to establish the following

Lemma 7.5:

(1) If $A \in \Omega^{1,0}(X; \mathrm{End}(E))$ then

$$[L, A^\star] = -\sqrt{-1}A^{*_H}.$$

(This is an equality between operators.)

(2) If $A \in \Omega^{0,1}(X; \mathrm{End}(E))$ then

$$[L, A^\star] = \sqrt{-1}A^{*_H}.$$

In fact one can prove this lemma quite easily using the symplectic (formal) adjoints and the operator C (as above, Section 2.9).

Lemma 7.6: $A^{\star_s} = -CA^\star C^{-1}$ *(compare with Exercise 2.10).*

If $A \in \Omega^1(X; \mathrm{End}(E))$, then $[L, A^{\star_s}] = A^{*_H}$ (compare with Exercise 2.5).

7.6. *Chern characters*

We will define only the first and second Chern characters. They will be given as differential forms and the proof of their properties will be left as exercises.

Let E be a complex vector bundle over a manifold M. The first and second Chern characters of E are the forms:

$$\mathrm{ch}_1(E) = \mathrm{ch}_1(E, D^E) = \frac{\sqrt{-1}}{2\pi}\mathrm{tr}F(D^E) \in \Omega^2(M; \mathbf{C})$$
$$\mathrm{ch}_2(E) = \mathrm{ch}_2(E, D^E) = \left(\frac{\sqrt{-1}}{2\pi}\right)^2 \mathrm{tr}\big(F(D^E) \wedge F(D^E)\big) \in \Omega^4(M; \mathbf{C}).$$

The notation suggests that they do not depend on the connection D^E whose curvature $F(D^E)$ is involved in the formulas. The 4-form $F(D^E) \wedge F(D^E)$ has coefficients in $\mathrm{End}(E)$; this wedge product involves the wedge product on differential forms and the composition of endomorphisms. Finally for a form with coefficients in $\mathrm{End}(E)$ its trace is the differential form obtained by applying the traces of endomorphisms: $\mathrm{tr}(\sum_i \alpha_i \sigma_i) = \sum_i \alpha_i \mathrm{tr}(\sigma_i)$ if $\alpha_i \in \Omega^\bullet(M; \mathbf{C})$ and σ_i are sections of $\mathrm{End}(E)$.

The needed properties concerning those Chern characters are summarized as follows

Lemma 7.7: *The forms* $\mathrm{ch}_1(E)$ *and* $\mathrm{ch}_2(E)$ *are closed.*

For any 2 connections D and D' on E, the difference $\mathrm{ch}_i(E, D') - \mathrm{ch}_i(E, D)$ *($i = 1, 2$) is exact.*

The first property follows from the Bianchi identity ($DF = 0$) and its consequences (i.e $D(F \wedge F) = 0$) as well as the formula for differentiating traces: $d\mathrm{tr}F = \mathrm{tr}DF$.

The second follows from the relation $F(D') = F(D) + DA + A \wedge A$ (Exercise 6.11) and the fact that $\mathrm{tr}A \wedge A = 0$.

Exercise 7.9: Provide the details of the construction of the Chern characters.

Exercise 7.10: Express $\mathrm{ch}_1(E \otimes F)$ and $\mathrm{ch}_1(E \oplus F)$.

The content of the lemma is that the forms $\mathrm{ch}_i(E)$ define uniquely cohomology classes in $H^{2i}(M; \mathbf{C})$. Using (for examples) unitary connections one can show that these cohomology classes belong to $H^{2i}(M; \mathbf{R})$. The relations with the (more common) Chern classes are

$$\begin{aligned}\mathrm{ch}_1(E) &= c_1(E)\\ \mathrm{ch}_2(E) &= c_2(E) - \tfrac{1}{2}c_1(E)^2.\end{aligned}$$

Exercise 7.11: For a vector bundle E of rank d and D a connection on E, express $\bigwedge^d D$ on $\bigwedge^d E$ in terms of D and give the relation between $F(\bigwedge^d D)$ and $F(D)$ —you may notice first that for the line bundle $L = \bigwedge^d E$, the bundle $\mathrm{End}(L)$ is the trivial bundle $\mathrm{End}(L) \simeq \mathbf{C}$ so that $F(L)$ as a form with coefficient in $\mathrm{End}(L)$ is in fact a differential form. The equality $\mathrm{ch}_1(\bigwedge^d E) = \mathrm{ch}_1(E)$ holds.

Exercise 7.12: For this exercise you need to know about

- group cohomology $H^\bullet(\Gamma; A)$ (see [2] or [17])
- the relation between central extensions $A \to E \to \Gamma$ and $H^2(\Gamma; A)$, [2]
- the natural map $\Theta : H^2(\pi_1(M); A) \to H^2(M; A)$.

Let $E \to M$ be a $\mathbf{C}$-vector bundle and $L = \bigwedge^d E$ the corresponding line bundle. Denote by $L - \{0\}$ the complement of the zero section.

(1) $\pi_1(L - \{0\}) \to \pi_1(M)$ is a central extension by a cyclic group A (i.e. there is a surjection $\mathbf{Z} \to A$). Let κ be the class in $H^2(\pi_1(M); A)$ corresponding to this central extension.

(2) The image $\Theta(\kappa)$ in $H^2(M;A)$ is the image of the (reduction of) the Chern class $\mathrm{ch}_1(E)$ in $H^2(M;A)$ (there is a natural map $H^2(M;\mathbf{Z}) \to H^2(M;A)$). [You may need a "functorial" approach of the first Chern class to prove this.]

7.7. *Degree*

For a vector bundle E over a Kähler manifold X and if ω designates the symplectic form on X,

Definition 7.8: The *degree* of E (with respect to ω) is

$$\deg_\omega E = \frac{1}{n!}\int_X \mathrm{ch}_1(E) \wedge \omega^{n-1}.$$

7.8. *Stability*

This is a key notion and its relevance will be fully justified later in Section 10.3. We refer the reader to [13] for more background on stability.

Definition 7.9: A holomorphic vector bundle E of rank $r = r(E)$ is called *stable* (or ω-stable) if for any holomorphic subbundle $F \subsetneq E$ $(F \neq 0)$ defined outside an analytic subset $\mathcal{S} \subset X$ of complex codimension 2 one has

$$\mu(F) = \frac{\deg_\omega(F)}{r(F)} < \mu(E) = \frac{\deg_\omega(E)}{r(E)}.$$

The number $\mu(E)$ is called the *slope* of E. The slope is a topological invariant.

Remark 7.10: For F and $\mathcal{S}$ as in the definition, the first Chern character $\mathrm{ch}_1(F)$ belongs a priori to $H^2(X \smallsetminus \mathcal{S};\mathbf{C})$. The condition on $\mathcal{S}$ precisely insures that $H^2(X \smallsetminus \mathcal{S};\mathbf{C}) \simeq H^2(X;\mathbf{C})$ and hence the degree $\deg_\omega(F)$ is well defined.

The usual definition of stability involves (coherent) subsheaves of the sheaf of holomorphic sections of E; the above definition makes the use of the fact that such a subsheaf is the sheaf of holomorphic sections of a subbundle defined outside a set of codimension 2, see e.g. [13] (the torsion-free hypothesis in that reference is automatically satisfied for a subsheaf of the sheaf of holomorphic sections of E).

8. The correspondence between flat bundles and Higgs bundles

This section starts by stating a generalization of the theorem of Narasimhan and Seshadri using the language of connections and pseudo-connections. The definition of a Higgs bundle is then given as well as the result of Hitchin and Simpson asserting the correspondence between flat bundles (with conditions) and Higgs bundles (with conditions). The end of this section is dedicated to that correspondence in the case of line bundles: the "classical" Hodge theory is the necessary tool for that case.

8.1. *The theorem of Narasimhan and Seshadri*

On surfaces, i.e. when $\dim_{\mathbb{C}} X = 1$, the theorem was proved by Narasimhan and Seshadri [16]. In this more general form it is due to Uhlenbeck-Yau [22] or Donaldson [4, 5, 6] (the version presented here is not the most general).

Theorem 8.1: *Let X be a compact Kähler manifold.*
Then there is a bijective correspondence between

(1) the conjugacy classes of irreducible representations

$$\rho : \pi_1(X) \longrightarrow \mathrm{U}(r),$$

(2) and the isomorphism classes of holomorphic vector bundles of rank r that are stable and with $\mathrm{ch}_1 = \mathrm{ch}_2 = 0$.

This correspondence is implemented by the map which sends a flat unitary connection to its $(0,1)$ part.

8.2. *Higgs bundles*

Definition 8.2: A *Higgs bundle* is a pair of a holomorphic vector bundle $(E, \bar{\partial}^E)$ together with a holomorphic $(1,0)$-form ϕ with coefficients in $\mathrm{End}(E)$ (i.e. $\bar{\partial}^{\mathrm{End}(E)}\phi = 0$) verifying $\phi \wedge \phi = 0$.

Remark 8.3: The conditions are equivalent to saying that $D'' = \bar{\partial}^E + \phi$ (an operator of mixed type but satisfying $D''(f\sigma) = \bar{\partial} f \sigma + f D''\sigma$) verifies the equality $(D'')^2 = 0$.

On surfaces the conditions $(\bar{\partial}^E)^2 = 0$ and $\phi \wedge \phi = 0$ are automatic since there are neither $(0,2)$-forms nor $(2,0)$-forms.

The stability condition for Higgs bundles takes the following form:

Definition 8.4: $(E, \bar{\partial}^E, \phi)$ is *stable* if for any holomorphic vector bundle $F \subsetneq E$ $(F \neq 0)$ defined outside an analytic subset of codimension 2 and which is ϕ-invariant, one has

$$\mu(F) = \frac{\deg_\omega(F)}{r(F)} < \mu(E) = \frac{\deg_\omega(E)}{r(E)}.$$

Here the meaning of ϕ-invariance is that $\phi(\sigma) \in \Omega^{1,0}(X; F)$ for any $\sigma \in \Omega^0(X; F) \subset \Omega^0(X; E)$.

8.3. *The Hitchin-Kobayashi correspondence*

It is sometimes under that name that the correspondence between flat bundles and Higgs bundles is found in the literature (see for example the book [15] and the introduction of [5]). No attempt to retrace the historical progression of the results will be made here. In particular no results of Kobayashi will be cited, however his book on vector bundles [13] is a very nice account on the theory.

Let E be a smooth bundle over a Kähler manifold. Suppose that $\mathrm{ch}_1(E) = \mathrm{ch}_2(E) = 0$. The correspondence is the following

$$\frac{\{\text{irreducible flat connections on } E\}}{\text{isomorphisms}} \leftrightarrow \frac{\{\text{stable Higgs bundle structures on } E\}}{\text{isomorphisms}}.$$

Or, taking any bundle E,

$$\frac{\{\text{irreducible representations of } \pi_1(X, x_0)\}}{\text{conjugacy}} \leftrightarrow \frac{\{\text{stable Higgs bundles } (E, \bar{\partial}^E, \phi) \text{ with } \mathrm{ch}_1(E) = \mathrm{ch}_2(E) = 0\}}{\text{equivalence}}.$$

The details of the constructions involved in the 2 directions of this correspondence are the subject of the next sections.

8.4. *From flat bundles to Higgs bundles*

Both directions involve Hermitian structures on E.

Let D be a (flat) connection on E over X and H a Hermitian structure on E.

Recall the decomposition of D:

$$\begin{aligned} D &= D_H + \psi_H \quad \text{unitary + Hermitian} \\ &= \partial_H + \bar{\partial}_H + \psi^{1,0} + \psi^{0,1} \quad \text{into types.} \end{aligned}$$

Theorem 8.5: *[Donaldson [7], Corlette [3], Labourie [14]]*

If D is a flat irreducible connection, then there exists a (unique up to a scalar multiple) Hermitian structure H on E such that $(E, \bar{\partial}_H, \psi^{1,0})$ is a Higgs bundle.

Furthermore this Higgs bundle is stable.

The explanation why the Higgs bundles arising this way are stable will be given in Section 10.3.

In fact this statement is rather a translation in terms of Higgs bundles of the theorem of Donaldson, Corlette and Labourie. Their theorem is the existence of harmonic Hermitian metric on (E, D); on a Kähler manifold, an harmonic metric leads to a Higgs bundle.

8.5. *From Higgs bundles to flat bundles*

Let $(E, \bar{\partial}^E, \phi)$ be a Higgs bundle (or more generally a pseudo-connection $\bar{\partial}^E$ and a $(1,0)$-form ϕ with coefficients in $\mathrm{End}(E)$).

Let H be a Hermitian structure on E, call ∂_H the operator of type $(1,0)$ such that $\partial_H + \bar{\partial}^E$ is the Chern connection.

Theorem 8.6: *[Hitchin [11], Simpson [18]]*

Let $(E, \bar{\partial}^E, \phi)$ be a stable Higgs bundle with $\mathrm{ch}_1(E) = \mathrm{ch}_2(E) = 0$.

Then there exists a unique (up to multiple) Hermitian metric H on E such that the connection

$$D = \partial_H + \bar{\partial}^E + \phi + \phi^{*_H}$$

is flat.

Furthermore the connection D is irreducible.

8.6. *The case of line bundles*

This case does not really shed light on the general theory. However it explains why the Hitchin-Kobayashi correspondence deserves the name of "non-abelian Hodge theory".

First some particularities for line bundles over X (or even over M) are recalled. Let L be a line bundle.

The bundle $\mathrm{End}(L)$ is always trivial:

$$\begin{aligned} X \times \mathbf{C} &\longrightarrow \mathrm{End}(L) \\ f &\longmapsto f\mathrm{id}_L. \end{aligned}$$

(Indeed this bundle is not only trivial it is *trivialized.*)

For any connection D^L on L, the induced connection $D^{\mathrm{End}(L)}$ on $\mathrm{End}(L) \simeq \mathbf{C}$ is always trivial. In view of the above isomorphism it is enough to prove that $D^{\mathrm{End}(L)}\mathrm{id}_L = 0$. In fact this holds in any rank.

Lemma 8.7: *For any bundle with connection* (E, D^E)*, one has* $D^{\mathrm{End}(E)}\mathrm{id}_E = 0$.

Proof: The connection $D^{\mathrm{End}(E)}$ is constructed so that for any section σ of E and any section ϕ of $\mathrm{End}(E)$, the following holds:

$$D^E\phi(\sigma) = (D^{\mathrm{End}(E)}\phi)\sigma + \phi D^E\sigma.$$

Applying this to $\phi = \mathrm{id}_E$ only one term remains

$$(D^{\mathrm{End}(E)}\mathrm{id}_E)\sigma = 0 \quad \forall\sigma$$

which says that $D^{\mathrm{End}(E)}\mathrm{id}_E = 0$. □

Even more is true: for any Hermitian structure H on L, the induced Hermitian structure on $\mathrm{End}(L) \simeq \mathbf{C}$ is trivial. For this we need to show that, for the induced norm on $\mathrm{End}(L)$, id_L is of norm one: let e be a section of L of norm one, the dual basis (!) e^* is a section of L^* of norm one (this follows from the definition of the norm on L^*) and hence $e \otimes e^*$ is of norm one. But $\mathrm{id}_L = e \otimes e^*$.

From this discussion one gets that:

- for any connection D on L
- any form ψ in $\Omega^1(M; \mathrm{End}(L)) = \Omega^1(M; \mathbf{C})$
- and any Hermitian structure H on L

the following holds

(1) $F(D + \psi) = F(D) + d\psi$ (since $D\psi = d\psi$ and $\psi \wedge \psi = 0$) and
(2) $\psi^{*_H} = \bar{\psi}$.

8.7. *Line bundles: From flat to Higgs*

Let now D be a flat connection on a line bundle L over a Kähler manifold X.

Let H_0 be one Hermitian metric on L. We will search for the Hermitian structure H solution to the Theorem 8.5 under the form $H = e^{2f} H_0$.

Some automatic cancellations hold in this case: $(\bar{\partial}_H)^2 = 0$ and $\phi^{1,0} \wedge \phi^{1,0} = 0$. The second is obvious since $\phi^{1,0}$ is a differential form.

For the first one, recall that

$$\begin{aligned} D &= D_H + \psi_H \quad \text{(unitary + Hermitian)} \\ D_H &= \partial_H + \bar{\partial}_H \quad \text{(into types)} \end{aligned}$$

since $0 = F(D) = F(D_H) + d\psi_H$, one gets

$$\begin{aligned} F(D_H) &= 0 \quad \text{(antihermitian part)} \\ d\psi_H &= 0 \quad \text{(Hermitian part).} \end{aligned}$$

But $F(D_H) = (D_H)^2 = \partial_H^2 + (\partial_H \bar{\partial}_H + \bar{\partial}_H \partial_H) + \bar{\partial}_H^2$ so that the $(0,2)$-part of $F(D_H)$ is $\bar{\partial}_H^2$ and $\bar{\partial}_H^2 = 0$.

For H_0, denote the above decomposition

$$\begin{aligned} D &= D_0 + \psi_0 \\ &= (\partial_0 + \bar{\partial}_0) + (\phi_0 + \bar{\phi}_0) \end{aligned}$$

where ϕ_0 is of type $(1,0)$. Decomposing $d\psi_0 = 0$ into type gives that $\partial\phi_0 = 0$, $\bar{\partial}\phi_0 = -\partial\bar{\phi}_0$ and $\bar{\partial}\bar{\phi}_0 = 0$. In particular $d\phi_0 = \bar{\partial}\phi_0$ and $\sqrt{-1}\bar{\partial}\phi_0$ is a real 2-form:

$$\overline{\sqrt{-1}\bar{\partial}\phi_0} = -\sqrt{-1}\partial\bar{\phi}_0 = \sqrt{-1}\bar{\partial}\phi_0 \quad \text{since } \bar{\partial}\phi_0 = -\partial\bar{\phi}_0.$$

Application of the classical $\partial\bar{\partial}$ lemma (this is where Hodge theory is used) implies that there is a real function f on X such that

$$\bar{\partial}\phi_0 = \bar{\partial}\partial f \quad \text{i.e. } \bar{\partial}(\phi_0 - \partial f) = 0.$$

The sought for metric is $H = e^{2f} H_0$. The key is to prove the relation

$$\psi_H = \psi_0 - df$$

so that $\psi_H^{1,0} = \phi_0 - \partial f$ is indeed holomorphic and the Higgs bundle equations are satisfied for $(L, \bar{\partial}_H, \psi_H^{1,0})$. For this, observe first that for any sections σ and σ' of L:

$$\begin{aligned} dH(\sigma,\sigma') &= H(D_H\sigma,\sigma') + H(\sigma, D_H\sigma') \text{ as } D_H \text{ is unitary} \\ &= H((D-\psi_H)\sigma,\sigma') + H(\sigma,(D-\psi_H)\sigma') \\ &= H(D\sigma,\sigma') + H(\sigma, D\sigma') - 2H(\psi_H\sigma,\sigma') \text{ as } \psi_H \text{ is } H\text{-Hermitian.} \end{aligned}$$

But also $H(\sigma,\sigma') = e^{2f}H_0(\sigma,\sigma')$ so

$$\begin{aligned} dH(\sigma,\sigma') &= 2df\, e^{2f}H_0(\sigma,\sigma') + e^{2f}dH_0(\sigma,\sigma') \\ &= 2df\, H(\sigma,\sigma') + e^{2f}(H_0(D\sigma,\sigma') + H_0(\sigma,\sigma') - 2H_0(\psi_0\sigma,\sigma')) \\ &= 2df\, H(\sigma,\sigma') + H(D\sigma,\sigma') + H(\sigma,\sigma') - 2H(\psi_0\sigma,\sigma'). \end{aligned}$$

Comparing these two formulas:

$$2H(\psi_H\sigma,\sigma') = -2df\, H(\sigma,\sigma') + 2H(\psi_0\sigma,\sigma')$$

therefore $\psi_H = -df + \psi_0$ as claimed.

8.8. *Line bundles: From Higgs to flat*

Suppose that $\mathcal{L} = (L, \bar{\partial}^L)$ is a holomorphic line bundle and ϕ is a holomorphic (1,0)-form (seen as an element of $\Omega^{1,0}(X, \mathrm{End}(\mathcal{L}))$). The first Chern class of L is supposed to vanish (the second Chern character is always 0 for line bundles). The stability condition here is empty.

The sought for Hermitian metric H on L satisfies that the associated connection:

$$D = \partial_H + \bar{\partial}^L + \phi + \bar{\phi}$$

is flat where $D_H = \partial_H + \bar{\partial}^L$ is the Chern connection for $(\mathcal{L}, H)$. The curvature of D is

$$F(D) = F(D_H) + d(\phi + \bar{\phi}).$$

As $(\bar{\partial}^L)^2 = 0$ one has $(\partial_H)^2 = 0$ (by adjunction) so that $F(D_H) = \partial_H\bar{\partial}^L + \bar{\partial}^L\partial_H$ is a form of type (1,1) which is antihermitian (here it means purely imaginary), i.e. $\overline{F(D_H)} = -F(D_H)$.

Also $\bar{\partial}\phi = 0$ and hence $\partial\bar{\phi} = 0$ which imply that $d\phi = \partial\phi$ and $d\bar{\phi} = \bar{\partial}\bar{\phi} = \overline{\partial\phi}$ are of type (2,0) and (0,2) respectively.

First the equation $L^\star F = 0$ is solved and then $F = F(D) = 0$ will be a consequence of the cancellation of the Chern class.

Let H_0 be a fixed metric on L and $H = e^{\varphi}H_0$ the sought for metric. The problem will again be reduced to finding the equation satisfied by the real function φ and Hodge theory will provide a solution for this equation.

The difference between ∂_H and ∂_{H_0} is controlled by φ:

$$\partial_H = \partial_{H_0} + \partial\varphi.$$

Exercise 8.1: Prove this.

Therefore $\partial_H + \bar{\partial}^L = \partial_{H_0} + \bar{\partial}^L + \partial\varphi$ and $F(\partial_H + \bar{\partial}^L) = F(\partial_{H_0} + \bar{\partial}^L) + d\partial\varphi$ but $d\partial\varphi = \partial^2\varphi + \bar{\partial}\partial\varphi = \bar{\partial}\partial\varphi$.

Note also that $L^\star(d(\phi + \bar{\phi})) = L^\star\partial\phi + L^\star\bar{\partial}\bar{\phi} = 0$ since $L^\star$ is zero on forms of type $(2,0)$ and $(0,2)$.

Hence $L^\star F = L^\star F_0 + L^\star\bar{\partial}\partial\varphi$ where $F_0 = F(\partial_{H_0} + \bar{\partial}^L)$. Moreover $L^\star\bar{\partial}\partial\varphi = [L^\star, \bar{\partial}]\partial\varphi = \sqrt{-1}\partial^\star\partial\varphi = \sqrt{-1}\Delta_\partial\varphi$ by the Kähler identities (see Section 2.12). The fact that F_0 is antihermitian implies that $L^\star F_0 = \sqrt{-1}g$ with g a real function. We have to find a function φ such that $g + \Delta_\partial\varphi = 0$. By classical Hodge theory this is possible if and only if $\int_X g\omega^n = 0$, thus the conclusion will follow from $\int_X L^\star F_0\omega^n = 0$.

With this lemma at hand the conclusion $F = 0$ follows easily: since $\mathrm{ch}_1(L) = 0$ one knows that $\int F_0 \wedge \omega^{n-1} = 0$ so the wanted cancellation will follow from:

Lemma 8.8: *For any 2-form β:*

$$\int L^\star\beta\omega^n = n\int \beta\omega^{n-1}.$$

Proof: By construction $L^\star\beta = \langle L^\star\beta, 1\rangle$ (the notation for scalar products introduced in Section 2.1 are still in use). So $\int L^\star\beta\omega^n = \int\langle L^\star\beta, 1\rangle\omega^n = \int\langle\beta,\omega\rangle\omega\wedge\omega^{n-1}$, the last equality follows from the definition of $L^\star$.

Replacing β by $\beta' = \beta - \langle\beta,\omega\rangle\omega$, it is therefore enough to prove that $\int\beta\omega^{n-1} = 0$ under the condition that $\langle\beta,\omega\rangle = 0$. Using a partition of the unity it is enough to work locally.

Let $\epsilon_1, \dots, \epsilon_n$ be a (local) orthonormal frame for $T^{1,0}X$, $e_1, \dots, e_n$ the dual basis of $T^{*1,0}X$ and $\bar{e}_1, \dots, \bar{e}_n$ the basis of $T^{*0,1}X$.

The form ω is:

$$\omega = \frac{1}{2\sqrt{-1}}\sum e_i \wedge \bar{e}_i.$$

Under the hypothesis $\langle\beta,\omega\rangle = 0$, it is safe to forget the factor $\frac{1}{2\sqrt{-1}}$ in the rest of this proof. Write

$$\beta = \sum a_{ij}e_i \wedge e_j + b_{ij}e_i \wedge \bar{e}_j + c_{ij}\bar{e}_i \wedge \bar{e}_j.$$

One has $\langle\beta,\omega\rangle = \sum b_{ii} = 0$. Also

$$\omega^{n-1} = (n-1)!\sum_i e_1 \wedge \bar{e}_1 \wedge e_2 \wedge \bar{e}_2 \wedge \cdots \wedge \widehat{e_i \wedge \bar{e}_i} \wedge \cdots \wedge e_n \wedge \bar{e}_n,$$

so that

$$\begin{aligned}\beta \wedge \omega^{n-1} &= (n-1)!\sum b_{ii}e_i \wedge \bar{e}_i \wedge e_1 \wedge \bar{e}_1 \wedge \cdots \wedge \widehat{e_i \wedge \bar{e}_i} \wedge \cdots \wedge e_n \wedge \bar{e}_n \\ &= \tfrac{1}{n}\left(\sum b_{ii}\right)\omega^n = \tfrac{1}{n}\langle\beta,\omega\rangle\omega^n.\end{aligned}$$

(See also Remark 10.5 for a short proof.) □

Exercise 8.2: Verify the formulas for ω^{n-1} and $\beta \wedge \omega^{n-1}$.

Remark 8.9: An equality between forms was proved and not only between their integrals. There is certainly a symplectic proof.

The last task is now to establish that $F = 0$ knowing that $L^\star F = 0$ and that $\mathrm{ch}_2(L) = F \wedge F = 0$. Write

$$F = F_H + \partial\phi + \overline{\partial\phi}$$

$\partial\phi$ is a form of type $(2,0)$ and $F_H = -\overline{F_H}$ is of type $(1,1)$ and $L^\star F_H = L^\star F = 0$.

Lemma 8.10: *Let β be a $(1,1)$ form, $\beta = -\bar{\beta}$, $\langle \beta, \omega \rangle = 0$ then $\beta \wedge \beta \wedge w^{n-2} = \frac{1}{n(n-1)} \|\beta\|^2 \omega^n$.*

Let α be a $(2,0)$ form, then $\alpha \wedge \bar{\alpha} \wedge \omega^{n-2} = \frac{1}{n(n-1)} \|\alpha\|^2 \omega^n$.

Since $F \wedge F \wedge \omega^{n-2} = F_H \wedge F_H \wedge \omega^{n-2} + 2\partial\phi \wedge \overline{\partial\phi} \wedge \omega^{n-2}$ (eliminating the terms that are not of type (n,n)) one get that $0 = n(n-1) \int F \wedge F \wedge \omega^{n-2} = \int \|F_H\|^2 \omega^n + 2 \int \|\partial\phi\|^2 \omega^n$ therefore F_H and $\partial\phi$ are identically 0, consequently F is 0.

Proof: It is a local statement. Take a frame as in the previous proof. Then $\omega = (2\sqrt{-1})^{-1} \sum_i e_i \wedge \bar{e}_i$ and

$$\omega^{n-2} = (n-2)! \sum_{i<j} \frac{e_1 \wedge \bar{e}_1}{2\sqrt{-1}} \wedge \cdots \wedge \widehat{\frac{e_i \wedge \bar{e}_i}{2\sqrt{-1}}} \wedge \cdots \wedge \widehat{\frac{e_j \wedge \bar{e}_j}{2\sqrt{-1}}} \wedge \cdots \wedge \frac{e_n \wedge \bar{e}_n}{2\sqrt{-1}}.$$

A form β of type $(1,1)$ is $\beta = \sum_{i,j} b_{ij} \frac{e_i \wedge \bar{e}_j}{2\sqrt{-1}}$. It satisfies $\bar{\beta} = -\beta$ if $\bar{b}_{ji} = -b_{ij}$ for all i, j and $\langle \beta, \omega \rangle = 0$ if $\sum b_{ii} = 0$. Its squared norm is $\|\beta\|^2 = \sum |b_{ij}|^2$.

Also

$$\beta \wedge \beta = \sum b_{ij} b_{kl} \frac{e_i \wedge \bar{e}_j}{2\sqrt{-1}} \wedge \frac{e_k \wedge \bar{e}_l}{2\sqrt{-1}},$$

wedging with ω^{n-2} gives

$$\begin{aligned}
\beta \wedge \beta \wedge \omega^{n-2} &= \tfrac{1}{n(n-1)} \textstyle\sum_{i \neq j} b_{ii} b_{jj} \omega^n - \tfrac{1}{n(n-1)} \textstyle\sum_{i \neq j} b_{ij} b_{ji} \omega^n \\
&= (-\tfrac{1}{n(n-1)} \textstyle\sum_{i,j} b_{ij} b_{ji}) \omega^n \\
&= \tfrac{1}{n(n-1)} \textstyle\sum b_{ij} \bar{b}_{ij} \omega^n = \tfrac{1}{n(n-1)} \|\beta\|^2 \omega^n
\end{aligned}$$

where the equality at the second line follows from $\sum b_{ii} = 0$.

Similarly a $(2,0)$-form α has the form $\sum_{i<j} a_{ij} e_i \wedge e_j$ and

$$\begin{aligned}\alpha \wedge \bar{\alpha} &= \textstyle\sum_{i<j,k<l} a_{ij}\bar{a}_{kl} e_i \wedge e_j \wedge \bar{e}_k \wedge \bar{e}_l \\ &= 4 \textstyle\sum a_{ij}\bar{a}_{kl} \frac{e_i\wedge\bar{e}_k}{2\sqrt{-1}} \wedge \frac{e_j\wedge\bar{e}_l}{2\sqrt{-1}}.\end{aligned}$$

Thus $n(n-1)\,\alpha \wedge \bar{\alpha} \wedge \omega^{n-2} = 4\sum_{i<j} a_{ij}\bar{a}_{ij}\omega^n = \|\alpha\|^2\omega^n$. □

9. Examples

Riemann surfaces are the places where a lot of examples can be provided. A few general facts on Riemann surfaces and on their line bundles are first recalled. The examples of Higgs bundles are first given for rank 2 vector bundles, it is also explained when the corresponding representation has image in $\mathrm{SL}_2(\mathbf{C})$ or in $\mathrm{GL}_2(\mathbf{R})$. General statements are also given for detecting in terms of the Higgs bundle when the image of the holonomy representation has determinant 1 or is real valued. The Teichmüller component, a remarkable discovery of Hitchin, is also described. Doing this other important aspects of the theory of Higgs bundles are eluded: the $\mathbf{C}^*$-action and the fact that stability is an open condition. This section ends with the illustration how to understand Higgs bundles corresponding to representations in $\mathrm{U}(p,q)$ or $\mathrm{Sp}_{2n}(\mathbf{R})$.

9.1. *Every Riemann surface is Kähler*

If Σ is a complex manifold of dimension 1 and $\langle\cdot,\cdot\rangle$ is a compatible metric then Σ is a Kähler manifold ($d\omega = 0$ since $\dim_{\mathbf{R}} \Sigma < \deg(d\omega)$). Note that in this situation J is entirely determined by $\langle\cdot,\cdot\rangle$ and the orientation of Σ: it is the rotation of angle $\pi/2$.

9.2. *Line bundles over* $\boldsymbol{\Sigma}$

For $\mathcal{L}$ a holomorphic line bundle over Σ, denote by $H^0(\mathcal{L})$ —or $H^0(\Sigma;\mathcal{L})$— the space of holomorphic sections of $\mathcal{L}$, i.e. the elements σ of $\Omega^0(\Sigma,\mathcal{L}) = \Gamma(\Sigma,\mathcal{L})$ satisfying $\bar{\partial}^{\mathcal{L}}\sigma = 0$.

Then $\dim H^0(\mathcal{L}) = 0$ if $\deg \mathcal{L} < 0$ and the Riemann-Roch theorem states that

$$\dim H^0(\mathcal{L}) - \dim H^0(\mathcal{L}^* \otimes \mathcal{K}) = \deg \mathcal{L} - (g-1)$$

where $\mathcal{K}$ is the holomorphic line bundle $T^{*1,0}\Sigma$ whose degree is $2g-2$.

Exercise 9.1: $T^{*1,0}X$ is a holomorphic bundle (any X).

Exercise 9.2: Let L and K be two (smooth) line bundles over Σ. Let $\theta : L \otimes L \to K$ be an isomorphism.

For any holomorphic structure $\bar{\partial}^K$ on K, there is a unique holomorphic structure $\bar{\partial}^L$ on L such that θ becomes a holomorphic isomorphism.

Exercise 9.3: (Corollary) If $\deg \mathcal{K}$ is even, there is a holomorphic line bundle $\mathcal{L}$ such that $\mathcal{L} \otimes \mathcal{L} = \mathcal{K}$. (You need to know the smooth classification of line bundles over Σ.)

$\mathcal{L}$ is denoted by $\mathcal{K}^{1/2}$.

9.3. *Some rank 2 Higgs bundles*

Let $\mathcal{L}$ be a line bundle over Σ and $\mathcal{E} = \mathcal{L} \oplus \mathcal{L}^{-1}$ (a negative power of a line bundle means the dual of the corresponding positive power for the tensor product). The bundle $\mathrm{End}(\mathcal{E})$ decomposes as the sum of 4 line bundles

$$\begin{array}{ll} \mathrm{Hom}(\mathcal{L}, \mathcal{L}) \simeq \mathbf{C} & \mathrm{Hom}(\mathcal{L}^{-1}, \mathcal{L}) \simeq \mathcal{L}^2 \\ \mathrm{Hom}(\mathcal{L}, \mathcal{L}^{-1}) \simeq \mathcal{L}^{-2} & \mathrm{Hom}(\mathcal{L}^{-1}, \mathcal{L}^{-1}) \simeq \mathbf{C}. \end{array}$$

A holomorphic $(1,0)$-form with coefficients in $\mathrm{End}(\mathcal{E})$ is the same thing as a holomorphic section of $\mathrm{End}(\mathcal{E}) \otimes \mathcal{K}$ (compare with Exercise 2.7). Therefore the Higgs field ϕ will be given as a matrix

$$\phi = \begin{pmatrix} a & b \\ c & d \end{pmatrix} \text{ with } \begin{cases} a \in H^0(\mathcal{K}), & b \in H^0(\mathcal{L}^2 \otimes \mathcal{K}), \\ c \in H^0(\mathcal{L}^{-2} \otimes \mathcal{K}), & d \in H^0(\mathcal{K}). \end{cases}$$

Lemma 9.1: *If $(\mathcal{E}, \phi)$ is stable then $|\deg(\mathcal{L})|$ is bounded by $g - 1$.*

Proof: If $c = 0$ then $\mathcal{L} \subset \mathcal{L} \oplus \mathcal{L}^{-1} = \mathcal{E}$ is a sub-Higgs bundle hence the stability condition says that $\mu(\mathcal{L}) = \deg(\mathcal{L}) < \mu(\mathcal{E}) = 0$.

If $c \neq 0$ then the bundle $\mathcal{L}^{-2} \otimes \mathcal{K}$ has a non-trivial holomorphic section and hence its degree is positive: $\deg(\mathcal{L}^{-2} \otimes \mathcal{K}) = -2\deg(\mathcal{L}) + 2g - 2 \geq 0$ so that $\deg(\mathcal{L}) \leq g - 1$.

In any case, $\deg(\mathcal{L}) \leq g - 1$.

Similarly, working with b, $\deg(\mathcal{L}) \geq 1 - g$. □

Lemma 9.2: *If $(\mathcal{E}, \phi)$ is stable then the corresponding representation $\rho : \pi_1(\Sigma) \to \mathrm{GL}_2(\mathbf{C})$ has image in $\mathrm{SL}_2(\mathbf{C})$ if and only if $a + d = 0$.*

Proof: One needs to convince him/herself (i.e. to prove) that the representation $\det(\rho)$ corresponds to the Higgs bundle $(\bigwedge^2 \mathcal{E}, \bigwedge^2 \phi)$. Here $\bigwedge^2 \mathcal{E} = \mathcal{L} \otimes \mathcal{L}^{-1} = \mathbf{C}$ and $\bigwedge^2 \phi = a + d$.

Since the trivial representation corresponds to the trivial Higgs bundle, one has $\det(\rho) = 1$ if and only if $a + d = 0$. □

Proposition 9.3: *If $a = d = 0$ and $(\mathcal{E}, \phi)$ is stable then the image of the corresponding representation is in* $\mathrm{SL}_2(\mathbf{R})$.

Proof: Let $g \in \mathcal{A}ut(\mathcal{E})$ given in "matrix" coordinates by

$$g = \begin{pmatrix} 1 & 0 \\ 0 & -1 \end{pmatrix},$$

then

$$\begin{aligned} g\phi g^{-1} &= \begin{pmatrix} 1 & 0 \\ 0 & -1 \end{pmatrix} \begin{pmatrix} 0 & b \\ c & 0 \end{pmatrix} \begin{pmatrix} 1 & 0 \\ 0 & -1 \end{pmatrix} \\ &= \begin{pmatrix} 0 & -b \\ -c & 0 \end{pmatrix} = -\phi. \end{aligned}$$

If H is the Yang-Mills Hermitian metric for $(\mathcal{E}, \phi)$ than H is the metric for $(\mathcal{E}, -\phi)$ and $g \cdot H$ is the metric for $(\mathcal{E}, g \cdot \phi) = (\mathcal{E}, -\phi)$. Hence, by the uniqueness of the solution, $g \cdot H = H$ and g is H-unitary.

This means that the eigenspaces of g are H-orthogonal, thus $H = H_+ \overset{\perp}{\oplus} H_-$ where $H_\pm$ are metrics on $\mathcal{L}^{\pm 1}$.

As $\bigwedge^2 H$ is the Hermitian metric for $\bigwedge^2 \phi = 0$, then $\bigwedge^2 H$ should be the constant metric on $\mathbf{C} = \mathcal{L} \otimes \mathcal{L}^{-1}$. But $\bigwedge^2 H = H_+ \otimes H_-$; this means that H_- is the metric on $\mathcal{L}^{-1}$ induced from $(\mathcal{L}, H_+)$ by duality. More precisely if $\Theta_+ : \mathcal{L} \to \mathcal{L}^{-1}$ is the conjugate-linear isomorphism deduced from H_+ (i.e. $H_+(\sigma, \sigma') = \Theta_+(\sigma) \cdot \sigma'$) and the same for Θ_- and H_-, then Θ_+ and Θ_- are inverse of each other.

Let τ be the conjugate-linear automorphism of $\mathcal{E}$ given by

$$\tau(\sigma_+, \sigma_-) = \big(\Theta_-(\sigma_-), \Theta_+(\sigma_+)\big).$$

Then $\tau^2 = \mathrm{id}_{\mathcal{E}}$ and $H(\tau v, \tau w) = H(w, v)$, this is enough to imply that $D'\tau = \tau D''$ ($D' = \partial_H + \phi^{*H}$ and $D'' = \bar{\partial} + \phi$) so that τ commutes with D.

The fixed points set of τ is a rank 2 real bundle and, as D commutes with τ, D restricts to a connection of this real bundle.

Consequently the holonomy is a representation in $\mathrm{GL}_2(\mathbf{R})$ —and also in $\mathrm{SL}_2(\mathbf{C})$ by Lemma 9.2. □

Hitchin [11] used this to give a parametrization of the set $\mathrm{Hom}(\pi_1(\Sigma), \mathrm{SL}_2(\mathbf{R}))/\sim$. He took advantage of the fact that the divisor of a (non-zero) holomorphic section of a line bundle determines completely the line bundle and the section (up to a multiple). For example if $\deg \mathcal{L} \geq 0$ then the divisor of c (belonging to $\mathrm{Sym}^{2g-2-2\deg \mathcal{L}} \Sigma$) and b (belonging to some vector space) almost completely determine the Higgs bundle.

There is a general procedure to detect real representation.

Proposition 9.4: *Let* $(E, D'' = \bar{\partial} + \phi)$ *be a Higgs bundle (stable,* $\mathrm{ch}_1 = \mathrm{ch}_2 = 0$*).*

Then the corresponding representation has image in $\mathrm{GL}_d(\mathbf{R})$ *if and only if there is a holomorphic quadratic form* Q *on* E *such that* ϕ *is* Q*-symmetric.*

Of course the conclusion is that the representation has image in a conjugate of $\mathrm{GL}_d(\mathbf{R})$.

Proof: Let H the corresponding Hermitian metric (and always $D' = \partial_H + \phi^{*_H}$). If the holonomy of D is real, there exists a conjugate-linear automorphism τ of E with $\tau^2 = \mathrm{id}_E$ and commuting with D. One can show that this implies that $H(\tau v, \tau w) = \overline{H(v, w)}$: indeed $\overline{H(\tau v, \tau w)}$ is a formula for the Hermitian metric $\tau \cdot H$ and that metric is the solution of Theorem 8.5 for the connection $\tau \cdot D$, since $\tau \cdot D = D$, by uniqueness $\tau \cdot H = H$.

In turns this implies that:

(1) $Q(v, w) = H(\tau v, w)$ is symmetric (hence a quadratic form);
(2) τ commutes with $\partial_H + \bar{\partial}$ and with $\phi + \phi^{*_H}$;
(3) ϕ is Q-symmetric ($\Leftarrow \tau$ commutes with $\phi + \phi^{*_H}$);
(4) Q is holomorphic ($\Leftarrow \tau$ commutes with $\partial_H + \bar{\partial}$).

Conversely, let Q be as in the proposition. Define $\varrho : E \to E$ conjugate-linear isomorphism by $Q(v, w) = H(\varrho v, w)$; then $\varrho D'' = D' \varrho$ (and ϱ commutes with D). But ϱ might not be an involution. However:

$$H(\varrho v, \varrho w) = H(v, \varrho^2 w) = H(\varrho^2 v, w)$$

so ϱ^2 is positive definite, self-adjoint and commutes with D, therefore $\varrho^2 = \lambda^2 \mathrm{id}_E$ with $\lambda > 0$ and $\tau = \varrho/\lambda$ is the sought for real structure. □

In fact the converse bijection can be used to prove the uniformization theorem.

Theorem 9.5: *[Hitchin [11]] Let* $(\mathcal{E} = \mathcal{K}^{-1/2} \oplus \mathcal{K}^{1/2}, \phi = \left(\begin{smallmatrix} 0 & 1 \\ 0 & 0 \end{smallmatrix}\right))$.

Then $(\mathcal{E}, \phi)$ *is stable, let* H *be the Hermitian metric solution of Theorem 8.5. Then* H *is diagonal and induces a metric on* $\mathcal{K}^{-1/2}$ *and finally on* $\mathcal{K}^{-1} = T\Sigma$.

Hence a Riemannian metric is obtained on Σ *and its curvature is equal to* -4.

9.4. *Teichmüller component*

Les $\mathcal{E} = \mathcal{K}^{-n/2} \oplus \mathcal{K}^{-n/2+1} \oplus \cdots \oplus \mathcal{K}^{n/2}$, holomorphic rank n bundle of degree 0. The bundle of endomorphisms of $\mathcal{E}$ decomposes accordingly:

$$\mathrm{End}(\mathcal{E}) = \begin{pmatrix} \mathbf{C} & \mathcal{K}^{-1} & \mathcal{K}^{-2} & \dots & \mathcal{K}^{-n+1} \\ \mathcal{K} & \mathbf{C} & \ddots & \ddots & \vdots \\ \mathcal{K}^2 & \mathcal{K} & \ddots & \ddots & \vdots \\ \vdots & \ddots & \ddots & \ddots & \mathcal{K}^{-1} \\ \mathcal{K}^{n-1} & \dots & \mathcal{K}^2 & \mathcal{K} & \mathbf{C} \end{pmatrix}.$$

For $\underline{\alpha} = (\alpha_2, \dots, \alpha_n)$, with $\alpha_i \in H^0(\mathcal{K}^i)$ define

$$\phi(\underline{\alpha}) = \begin{pmatrix} 0 & 1 & & \\ \alpha_2 & \ddots & \ddots & \\ \vdots & \ddots & \ddots & 1 \\ \alpha_n & \cdots & \alpha_2 & 0 \end{pmatrix}$$

a holomorphic section of $\mathrm{End}(\mathcal{E}) \otimes \mathcal{K}$.

The results of Hitchin [12] are:

(1) for all $\underline{\alpha}$, $(\mathcal{E}, \phi(\underline{\alpha}))$ is stable;
(2) the corresponding representation $\rho(\underline{\alpha})$ is real;
(3) the map

$$\begin{aligned} \bigoplus_{i=2}^n H^0(\mathcal{K}^i) &\longrightarrow \mathrm{Hom}(\pi_1(\Sigma), \mathrm{SL}_n(\mathbf{R}))/\sim \\ \underline{\alpha} &\longmapsto \rho(\underline{\alpha}) \end{aligned}$$

is a bijection onto a connected component of the character variety $\mathrm{Hom}(\pi_1(\Sigma), \mathrm{SL}_n(\mathbf{R}))/\sim$.

Proof: [Proof of 1] It uses 3 ingredients:

(i) $(\mathcal{E}, \phi(0))$ is stable. Let $(\mathcal{E}_2, \psi) = (\mathcal{K}^{-1/2} \oplus \mathcal{K}^{1/2}, \left(\begin{smallmatrix} 0 & 1 \\ 0 & 0 \end{smallmatrix}\right))$ be the Higgs bundle of Theorem 9.5, then the corresponding representation $\rho : \pi_1(\Sigma) \to \mathrm{SL}_2(\mathbf{R})$ is the uniformization of Σ, in particular it is Zariski dense.

The Higgs bundle $(\mathcal{E}, \phi(0))$ is the $(n-1)$-th symmetric power $\mathrm{Sym}^{n-1}(\mathcal{E}_2, \psi)$ so that the corresponding representation is $\mathrm{Sym}^{n-1}\rho$, i.e. the composition of ρ with the morphism $\mathrm{SL}_2(\mathbf{R}) \to \mathrm{SL}_n(\mathbf{R})$ given by the action of $\mathrm{SL}_2(\mathbf{R})$ on $\mathrm{Sym}^{n-1}\mathbf{R}^2 \simeq \mathbf{R}^n$. This representation is irreducible, thus the corresponding Higgs bundle is stable.

(ii) Stability is an open condition. [Sketch of proof] Let $(E, \bar{\partial}_n, \phi_n)$, $n \in \mathbf{N}$, be a sequence of Higgs bundles converging to $(E, \bar{\partial}^E, \phi)$ (in the sense that $\bar{\partial}_n - \bar{\partial}^E$ and $\phi_n - \phi$ converge to 0 in $\Omega^1(\Sigma; \mathrm{End}(E))$).

Suppose that for all n there is a subbundle $F_n \subset E$ that is stable by $\bar{\partial}_n$ and ϕ_n, i.e. such that $(\bar{\partial}_n + \phi_n)(\Omega^0(\Sigma; F_n)) \subset \Omega^1(\Sigma; F_n)$, and with $\mu(F_n) \geq \mu(E)$.

Up to extracting a subsequence, the subbundles F_n can be supposed to have the following properties:

(a) $\mathrm{rank}(F_n) = k$ is constant;
(b) F_n converges to F.

F_n can be seen as a (smooth) section of the bundle $\mathrm{Gr}_k(E)$ in Grassmannian manifolds, as the space of sections $\Gamma(\Sigma; \mathrm{Gr}_k(E))$ is compact, there is a converging subsequence.

Now if $\sigma \in \Omega^0(\Sigma; F)$ there are sections $\sigma_n \in \Omega^0(\Sigma; F_n)$ converging to σ and one finds that

$$(\bar{\partial}^E + \phi)\sigma = \lim(\bar{\partial}_n + \phi_n)\sigma_n \in \Omega^1(\Sigma; \lim F_n).$$

Hence F is a Higgs subbundle of $(E, \bar{\partial}^E, \phi)$. Furthermore $\mu(F) = \lim \mu(F_n)$ (in fact $F \simeq F_n$ for n big enough; this can be seen using a projection $E \to F$). Therefore $(E, \bar{\partial}^E, \phi)$ is not stable: non-stability is a closed condition.

(iii) Let $\underline{\alpha}$ in $\bigoplus H^0(\mathcal{K}^i)$. If $(\mathcal{E}, \phi(\underline{\alpha}))$ is not stable, then the same holds for any $(\mathcal{E}, \lambda g \cdot \phi(\underline{\alpha}))$, $\lambda \in \mathbf{C}^*$, $g : \mathcal{E} \to \mathcal{E}$ holomorphic gauge isomorphism (the action is of course $g \cdot \phi = g\phi g^{-1}$). Taking

$$g = g_\lambda = \begin{pmatrix} \lambda & & & \\ & \lambda^2 & & \\ & & \ddots & \\ & & & \lambda^n \end{pmatrix},$$

notice that

$$g \cdot \phi(\underline{\alpha}) = \begin{pmatrix} 0 & \lambda^{-1} & & \\ \lambda\alpha_2 & \ddots & \ddots & \\ \vdots & \ddots & \ddots & \lambda^{-1} \\ \lambda^{n-1}\alpha_n & \cdots & \lambda\alpha_2 & 0 \end{pmatrix}$$

so that $\lambda g \cdot \phi(\underline{\alpha}) = \phi(\lambda^2\alpha_2, \ldots, \lambda^n\alpha_n)$. In the limit $\lambda \to 0$, the bundle $(\mathcal{E}, \phi(0))$ would be obtained as a limit of non-stable Higgs bundles, contradicting (i) and (ii). $\square$

Proof: [Proof of 2] Take

$$Q = \begin{pmatrix} & & & 1 \\ & & \cdot^{\cdot^{\cdot}} & \\ & 1 & & \\ 1 & & & \end{pmatrix}$$

holomorphic quadratic form and apply Proposition 9.4. □

Proof: [Proof of 3] The map $\underline{\alpha} \mapsto (\mathcal{E}, \phi(\underline{\alpha}), Q)$ is (almost) a section of

$$\begin{aligned} \{(\mathcal{E}, \phi, Q) \text{ stable...}\} &\xrightarrow{P} \bigoplus_{i=2}^{n} H^0(\mathcal{K}^i) \\ (\mathcal{E}, \phi, Q) &\longmapsto (\operatorname{tr}(\phi^i))_{i=2,\ldots,n}. \end{aligned}$$

(Since ϕ is a holomorphic morphism $\mathcal{E} \to \mathcal{E} \otimes \mathcal{K}$, ϕ^i is a holomorphic morphism $\mathcal{E} \to \mathcal{E} \otimes \mathcal{K}^i$ and its trace is a holomorphic section of $\mathcal{K}^i$.) The above P is well defined since tr is conjugacy invariant.

For all i there is a constant $c_i \neq 0$ such that

$$\operatorname{tr}(\phi(\underline{\alpha})^i) = c_i \alpha_i + \text{polynomial}(\alpha_2, \ldots, \alpha_{i-1}).$$

This is what is meant by $\underline{\alpha} \to \phi(\underline{\alpha})$ is almost a section of P.

From this one deduces:

(a) $\underline{\alpha} \to \rho(\underline{\alpha})$ is injective, thus its image is open by dimensionality;
(b) its image is also closed: if a sequence $(\rho(\underline{\alpha}_k))_{k \in \mathbf{N}}$ converges then $P(\rho(\underline{\alpha}_k))$ also converges so that $(\underline{\alpha}_k)_{k \in \mathbf{N}}$ converges equally.

(a) and (b) prove that the image of $\underline{\alpha} \to \rho(\underline{\alpha})$ is a connected component of the representation variety $\operatorname{Hom}(\pi_1(\Sigma), \mathrm{SL}_n(\mathbf{R}))/\sim$ and that this connected component is diffeomorphic to a vector space. □

Hitchin [12] proved the existence of such a component in $\operatorname{Hom}(\pi_1(\Sigma), G)/\sim$ for any adjoint simple real Lie group that is $\mathbf{R}$-split.

9.5. *Representations in unitary or symplectic groups*

The end of this section illustrates how representations in $\mathrm{U}(p,q)$ or in $\mathrm{Sp}_{2n}(\mathbf{R})$ can be obtained.

9.5.1. *The groups*

The stabilizer of a Hermitian form of signature (p, q) is the unitary group $\mathrm{U}(p, q)$:

$$\mathrm{U}(p,q) = \left\{ g \in \mathrm{GL}_{p+q}(\mathbf{C}) \mid {}^t\bar{g}\begin{pmatrix} \mathrm{id}_p & \\ & -\mathrm{id}_q \end{pmatrix} g = \begin{pmatrix} \mathrm{id}_p & \\ & -\mathrm{id}_q \end{pmatrix} \right\}.$$

It is a reductive Lie group of real rank equal to $\min(p, q)$.

Symplectic groups are stabilizers of symplectic forms:

$$\begin{aligned} \mathrm{Sp}_{2n}(\mathbf{C}) &= \left\{ g \in \mathrm{GL}_{2n}(\mathbf{C}) \mid {}^t g \begin{pmatrix} & \mathrm{id}_n \\ -\mathrm{id}_n & \end{pmatrix} g = \begin{pmatrix} & \mathrm{id}_n \\ -\mathrm{id}_n & \end{pmatrix} \right\} \\ \mathrm{Sp}_{2n}(\mathbf{R}) &= \mathrm{Sp}_{2n}(\mathbf{C}) \cap \mathrm{GL}_{2n}(\mathbf{R}) \\ &= \left\{ g \in \mathrm{GL}_{2n}(\mathbf{R}) \mid {}^t g \begin{pmatrix} & \mathrm{id}_n \\ -\mathrm{id}_n & \end{pmatrix} g = \begin{pmatrix} & \mathrm{id}_n \\ -\mathrm{id}_n & \end{pmatrix} \right\} \\ &= \left\{ g \in \mathrm{GL}_{2n}(\mathbf{C}) \mid g = \bar{g},\ {}^t g \begin{pmatrix} & \mathrm{id}_n \\ -\mathrm{id}_n & \end{pmatrix} g = \begin{pmatrix} & \mathrm{id}_n \\ -\mathrm{id}_n & \end{pmatrix} \right\}. \end{aligned}$$

Exercise 9.4:

$$\begin{aligned} \mathrm{Sp}_{2n}(\mathbf{R}) &= \mathrm{Sp}_{2n}(\mathbf{C}) \cap \mathrm{U}(n, n) \\ &= \Big\{ g \in \mathrm{GL}_{2n}(\mathbf{C}) \mid {}^t g \begin{pmatrix} & \mathrm{id}_n \\ -\mathrm{id}_n & \end{pmatrix} g = \begin{pmatrix} & \mathrm{id}_n \\ -\mathrm{id}_n & \end{pmatrix} \text{ and} \\ &\qquad {}^t\bar{g} \begin{pmatrix} & \sqrt{-1}\mathrm{id}_n \\ -\sqrt{-1}\mathrm{id}_n & \end{pmatrix} g = \begin{pmatrix} & \sqrt{-1}\mathrm{id}_n \\ -\sqrt{-1}\mathrm{id}_n & \end{pmatrix} \Big\}. \end{aligned}$$

Exercise 9.5:

$$\begin{aligned} &\mathrm{Sp}_{2n}(\mathbf{R}) = \mathrm{GL}_{2n}(\mathbf{R}) \cap \mathrm{U}(n, n) \\ &= \left\{ g \in \mathrm{GL}_{2n}(\mathbf{C}) \mid g = \bar{g} \text{ and } {}^t\bar{g} \begin{pmatrix} & \sqrt{-1}\mathrm{id}_n \\ -\sqrt{-1}\mathrm{id}_n & \end{pmatrix} g = \begin{pmatrix} & \sqrt{-1}\mathrm{id}_n \\ -\sqrt{-1}\mathrm{id}_n & \end{pmatrix} \right\}. \end{aligned}$$

9.5.2. *The bundles*

Let $\mathcal{V}$ and $\mathcal{W}$ two holomorphic vector bundles of rank p and q respectively.

For α a form in $\Omega^{1,0}(X; \mathrm{Hom}(\mathcal{V}, \mathcal{W}))$ holomorphic and β belonging to $\Omega^{1,0}(X; \mathrm{Hom}(\mathcal{W}, \mathcal{V}))$ holomorphic, form the following Higgs bundle:

$$\left(\mathcal{E} = \mathcal{V} \oplus \mathcal{W}, \phi = \begin{pmatrix} & \beta \\ \alpha & \end{pmatrix} \right).$$

Proposition 9.6: *If $(\mathcal{E}, \phi)$ is stable and $\mathrm{ch}_1(\mathcal{E}) = \mathrm{ch}_2(\mathcal{E}) = 0$, then the corresponding representation $\rho : \pi_1(X) \to \mathrm{GL}_{p+q}(\mathbf{C})$ has image in the subgroup $\mathrm{U}(p, q)$.*

Conversely if $\rho : \pi_1(X) \to \mathrm{U}(p, q)$, one can show that the associated Higgs bundle has the above form.

Proof: Let H be the Hermitian structure provided by Theorem 8.6. The reasoning is similar to the proof of Proposition 9.3.

The gauge isomorphism $g = \begin{pmatrix} \mathrm{id}_{\mathcal{V}} & \\ & -\mathrm{id}_{\mathcal{W}} \end{pmatrix}$ is holomorphic and anti-commutes with ϕ. This implies (again by the uniqueness of H and the fact that H is also the Hermitian metric for $(\mathcal{E}, -\phi)$) that g is H-unitary, therefore its eigenspaces are H-orthogonal so that $H = H_{\mathcal{V}} \oplus H_{\mathcal{W}}$ where $H_{\mathcal{V}}$, $H_{\mathcal{W}}$ are Hermitian structure on $\mathcal{V}$ and $\mathcal{W}$.

The consequence will be that the Hermitian form $K = H_{\mathcal{V}} - H_{\mathcal{W}}$ is flat for the connection $D = \partial_H + \bar{\partial}^{\mathcal{E}} + \phi + \phi^{*_H}$, i.e. this connection is K-unitary. This will imply that the holonomy of D is in $\mathrm{U}(p,q)$.

First $\partial_H + \bar{\partial}^{\mathcal{E}}$ is K-unitary: indeed as $\bar{\partial}^{\mathcal{E}} = \begin{pmatrix} \bar{\partial}^{\mathcal{V}} & \\ & \bar{\partial}^{\mathcal{W}} \end{pmatrix}$ and $H = \begin{pmatrix} H_{\mathcal{V}} & \\ & H_{\mathcal{W}} \end{pmatrix}$ one has $\partial_H = \begin{pmatrix} \partial_{H_{\mathcal{V}}} & \\ & \partial_{H_{\mathcal{W}}} \end{pmatrix}$. Since $\partial_H + \bar{\partial}^{\mathcal{E}}$ is H-unitary, the connection $\partial_{H_{\mathcal{V}}} + \bar{\partial}^{\mathcal{V}}$ is $H_{\mathcal{V}}$-unitary and the connection $\partial_{H_{\mathcal{W}}} + \bar{\partial}^{\mathcal{W}}$ is $H_{\mathcal{W}}$-unitary. Therefore $\partial_H + \bar{\partial}^{\mathcal{E}} = \begin{pmatrix} \partial_{H_{\mathcal{V}}} + \bar{\partial}^{\mathcal{V}} & \\ & \partial_{H_{\mathcal{W}}} + \bar{\partial}^{\mathcal{W}} \end{pmatrix}$ is $(H_{\mathcal{V}} - H_{\mathcal{W}})$-unitary.

Second $\phi + \phi^{*_H}$ is K-antihermitian: it is known that $\phi + \phi^{*_H}$ is H-Hermitian and that $\phi^{*_H} = \begin{pmatrix} & \alpha^{*_H} \\ \beta^{*_H} & \end{pmatrix}$. For (v, w) and (v', w') sections of $\mathcal{E} = \mathcal{V} \oplus \mathcal{W}$, developing the equality

$$H(\phi \cdot (v, w), (v', w')) = H((v, w), \phi^{*_H}(v', w'))$$

gives

$$\begin{aligned} H_{\mathcal{V}}(\beta w, v') &= H_{\mathcal{W}}(w, \beta^{*_H} v') \\ \text{and } H_{\mathcal{W}}(\alpha v, w') &= H_{\mathcal{V}}(v, \alpha^{*_H} w'). \end{aligned}$$

But

$$\begin{aligned} K(\phi \cdot (v, w), (v', w')) &= H_{\mathcal{V}}(\beta w, v') - H_{\mathcal{W}}(\alpha v, w') \\ &= H_{\mathcal{W}}(w, \beta^{*_H}) - H_{\mathcal{V}}(v, \alpha^{*_H} w') \\ &= -K((v, w), \phi^{*_H} \cdot (v', w')), \end{aligned}$$

equally $K(\phi^{*_H} \cdot (v, w), (v', w')) = -K((v, w), \phi \cdot (v', w'))$ and $\phi + \phi^{*_H}$ is K-antihermitian. So the sum $D = \partial_H + \bar{\partial}^{\mathcal{E}} + \phi + \phi^{*_H}$ is K-unitary. □

Now suppose that $\mathcal{W} = \mathcal{V}^*$, that $\alpha \in \Omega^{1,0}(X; \mathrm{Hom}(\mathcal{V}, \mathcal{V}^*))$ is symmetric (i.e. if v, v' are sections of $\mathcal{V}$ —so that $\alpha(v)$ is a 1-form with coefficient in $\mathcal{V}^*$ thus $\alpha(v)(v')$ is a 1-form— then $\alpha(v)(v') = \alpha(v')(v)$.) and $\beta \in \Omega^{1,0}(X; \mathrm{Hom}(\mathcal{V}^*, \mathcal{V}))$ is symmetric.

Proposition 9.7: *Under these conditions, the representation corresponding to $(\mathcal{E}, \phi)$ has image into $\mathrm{Sp}_{2n}(\mathbf{R})$.*

Proof: In view of the equality $\mathrm{Sp}_{2n}(\mathbf{R}) = \mathrm{Sp}_{2n}(\mathbf{C}) \cap \mathrm{U}(n, n)$ and the preceding proposition it is enough to produce a flat symplectic form for

$D = \partial_H + \bar{\partial}^{\mathcal{E}} + \phi + \phi^{*_H}$. For this it is enough to produce a holomorphic symplectic form such that $\omega(\phi\sigma, \sigma') = -\omega(\sigma, \phi\sigma')$ for all sections σ and σ'.

Exercise 9.6: This is indeed sufficient.

Let us give a formula for ω:

$$\omega((v_1, f_1), (v_2, f_2)) = f_2(v_1) - f_1(v_1).$$

(v_i sections of $\mathcal{V}$, f_i sections of $\mathcal{V}^*$ so that (v_i, f_i) are sections of $\mathcal{E}$.) It clearly defines a holomorphic symplectic form and the property of ϕ with respect to ω is a consequence of the symmetry of α and β. □

10. On Hermitian metrics and more

This section tries to explain how to prove Theorem 8.5 (solving the "harmonic equation") and Theorem 8.6 (solving the "Hitchin equation"). The first thing is to explain why, in reality, these equation are *not* overdetermined. The necessity of the hypothesis in those theorems will be also explained and some considerations on their proofs will be given.

10.1. *Opportune cancellations: The case of flat bundle*

Recall that the terminology "opportune cancellations" was chosen for the purpose of this notes only. Let D be a flat irreducible connection on a bundle E over a (compact and connected) Kähler manifold X.

Given a metric H on E, the decomposition of D into a unitary connection and a self-adjoint 1-form is again denoted by:

$$D = D_H + \psi_H,$$

and the decomposition into "types" of those are:

$$D_H = \partial_H + \bar{\partial}_H, \quad \psi_H = \phi + \phi^{*_H}.$$

The following operators are often useful:

$$\begin{aligned} D'_H &= \partial_H + \phi^{*_H} \\ D''_H &= \bar{\partial}_H + \phi \\ \text{and } D_C &= -D'_H + D''_H. \end{aligned}$$

Proposition 10.1: $G_H = (D''_H)^2 = 0 \Leftrightarrow L^\star G_H = 0 \Leftrightarrow D^\star_H \psi_H = 0$.

The adjunction for the L^2-scalar product is denoted with the $\star$ and L is the operation of wedging with the Kähler form ω.

Remark 10.2: This is the meaning of "opportune cancellations": as soon as $L^\star G_H$ is zero (and of course under the condition $D = 0$) then G_H is also zero.

The proof below follows [19] (see also the section preceding that lemma in [19] for further historical comments).

Before the proof, recall the Kähler identities:

$$
\begin{array}{lll}
[L^\star, \bar\partial_H] = \sqrt{-1}\partial_H^\star & [L^\star, D''_H] = \sqrt{-1}D'^\star_H & [L^\star, D_C] = \sqrt{-1}D^\star \\
[L^\star, \partial_H] = -\sqrt{-1}\bar\partial_H^\star & [L^\star, D'_H] = -\sqrt{-1}D''^\star_H & [L^\star, D] = -\sqrt{-1}D_C^\star.
\end{array}
$$

Proof:

(1) $G_H = \frac{1}{4}(DD_C + D_C D)$.

Since $D^2 = 0 = D'^2_H + D''^2_H + D'_H D''_H + D''_H D'_H$ one has (separating the Hermitian and antihermitian parts)

$$D'^2_H + D''^2_H = 0 \quad \text{and} \quad D'_H D''_H + D''_H D'_H = 0. \tag{10.1}$$

Thus $D_C^2 = D'^2_H + D''^2_H - D'_H D''_H - D''_H D'_H = 0$. Therefore $G_H = D''^2_H = \left(\frac{D+D_C}{2}\right)^2 = \frac{DD_C + D_C D}{4}$.

(2) $DG_H = D_C G_H = 0$.

Only one of this identities will be proved. Write, for a moment, in superscript the bundle on which a connection (or another operator) acts: the sought for identity is $D^{\mathrm{End}(E)} G_H = 0$.

For any section σ of E, the above holds:

$$
\begin{aligned}
D^E(G_H\sigma) &= \big(D^{\mathrm{End}(E)}G_H\big)\sigma + G_H D^E\sigma \\
&= \big(D^{\mathrm{End}(E)}G_H\big)\sigma + \tfrac{1}{4}(D^E D^E_C + D^E_C D^E)D^E\sigma \\
&= \big(D^{\mathrm{End}(E)}G_H\big)\sigma + \tfrac{1}{4}D^E D^E_C D^E\sigma \quad \text{since } (D^E)^2 = 0.
\end{aligned}
$$

But

$$
\begin{aligned}
D^E(G_H\sigma) &= D^E\Big(\frac{D^E D^E_C + D^E_C D^E}{4}\Big)\sigma \\
&= \frac{1}{4}D^E D^E_C D^E\sigma \quad \text{again since } (D^E)^2 = 0.
\end{aligned}
$$

Comparing the 2 equations gives $(D^{\mathrm{End}(E)}G_H)\sigma = 0$, for all sections σ, hence $D^{\mathrm{End}(E)}G_H = 0$.

(3) $D(\phi - \phi^{*_H}) = 2G_H$.

The $(0,2)$-part and the $(1,1)$-part of the first term in equation (10.1) are respectively

$$\bar{\partial}_H^2 + \phi^{*_H} \wedge \phi^{*_H} = 0, \quad \bar{\partial}_H \phi + \partial_H \phi^{*_H} = 0,$$

also, the $(2,0)$-part and the $(0,2)$-part of the second term in the same equation gives $\partial_H \phi = 0$ and $\bar{\partial}_H \phi^{*_H} = 0$. On the other hand

$$G_H = \bar{\partial}_H^2 + \bar{\partial}_H \phi + \phi \wedge \phi.$$

And the last term to calculate is

$$\begin{aligned} D(\phi - \phi^{*_H}) &= \partial_H \phi + \bar{\partial}_H \phi - \partial_H \phi^{*_H} - \bar{\partial}_H \phi^{*_H} + 2\phi \wedge \phi - 2\phi^{*_H} \wedge \phi^{*_H} \\ &= 2\bar{\partial}_H \phi + 2\phi \wedge \phi + 2\bar{\partial}_H^2 = 2G_H. \end{aligned}$$

Exercise 10.1: Calculating $G_H\sigma = D''_H(D''_H\sigma)$ and (in 2 different ways) $D((\phi - \phi^{*_H})\sigma)$, validate the above identities for G_H and $D(\phi - \phi^{*_H})$. (Remember that the Leibniz identity involves signs.)

(4) Suppose now that $L^\star G_H = 0$, since $D_C G_H = 0$, by the Kähler identity one get $D^\star G_H = 0$. Thus

$$\begin{aligned} 2\textstyle\int \langle G_H, G_H \rangle \omega^n &= \textstyle\int \langle G_H, D(\phi - \phi^{*_H}) \rangle \omega^n \\ &= \textstyle\int \langle D^\star G_H, \phi - \phi^{*_H} \rangle \omega^n = 0, \end{aligned}$$

and hence $G_H = 0$.

(5) $D_H^\star \psi_H = -2\sqrt{-1} L^\star G_H$.

Indeed

$$\begin{aligned} D_H^\star \psi_H &= (\partial_H^\star + \bar{\partial}_H^\star)(\phi + \phi^{*_H}) \\ &= \bar{\partial}_H^\star \phi^{*_H} + \partial_H^\star \phi \quad \text{eliminating terms not of type } (0,0) \\ &= -\sqrt{-1} L^\star (\bar{\partial}_H \phi - \partial_H \phi^{*_H}) \quad \text{by the Kähler identities} \\ &= -2\sqrt{-1} L^\star \bar{\partial}_H \phi \quad \text{since } \bar{\partial}_H \phi = -\partial_H \phi^{*_H}. \end{aligned}$$

Also

$$\begin{aligned} L^\star G_H &= L^\star(-\phi^{*_H} \wedge \phi^{*_H} + \bar{\partial}_H \phi + \phi \wedge \phi) \\ &= L^\star \bar{\partial}_H \phi \quad \text{since } L^\star = 0 \text{ in degrees } (2,0) \text{ and } (0,2). \end{aligned}$$

This finishes the proof of the proposition. □

10.2. *Opportune cancellations: Higgs bundles*

Let $(E, \bar{\partial}^E, \phi)$ be a Higgs bundle such that

$$\int \mathrm{ch}_1(E)\omega^{n-1} = 0, \quad \text{and} \quad \int \mathrm{ch}_2(E)\omega^{n-2} = 0.$$

Remark 10.3: Those topological assumption are a priori weaker than those stated in Theorem 8.6. Nevertheless, only the cancellations of those 2 integrals are involved in the proof of that result. The present section illustrates why.

For an Hermitian metric H on E, construct:

$$\begin{aligned} D_H &= \partial_H + \bar\partial^E && \text{the Chern connection;} \\ \psi_H &= \phi + \phi^{*_H} && \text{sum of } \phi \text{ and its adjoint,} \end{aligned}$$

and $D'' = \bar\partial^E + \phi$, $D'_H = \partial_H + \phi^{*_H}$, $D = D'_H + D'' = D_H + \psi_H$. The curvature of D is

$$\begin{aligned} F = F(D) &= F(D_H) + D_H\psi_H + \psi_H \wedge \psi_H \\ &= D'^2_H + (D'_H D'' + D'' D'_H) + D''^2. \end{aligned}$$

Proposition 10.4: *If $L^\star F = 0$, then $F = 0$.*

Remark 10.5: The equation $L^\star F = 0$ already implies $\deg_\omega E = 0$. Indeed let β be the 2-form $\mathrm{tr}(F)$, then $L^\star\beta = \mathrm{tr}(L^\star F) = 0$. Thus:

$$\begin{aligned} \tfrac{1}{(n-1)!}\int \beta\wedge\omega^{n-1} &= \int \beta \wedge \star\omega && \text{as } \star\,\omega = \tfrac{1}{(n-1)!}\omega^{n-1} \\ &= \int \langle \beta, \omega\rangle \mathrm{vol} && \text{by the definition of } \star \\ &= \int \langle \beta, L1\rangle \mathrm{vol} \\ &= \int \langle L^\star\beta, 1\rangle \mathrm{vol} = 0. \end{aligned}$$

But $\deg_\omega E$ is proportional to $\int \beta\omega^{n-1}$.

In the proof of the proposition, the following equivalent of Lemma 8.10 will be involved.

Lemma 10.6: *Let $\beta \in \Omega^{1,1}(X; \mathrm{End}(E))$ such that $\beta^{*_H} = -\beta$ and $L^\star\beta = 0$, then $\mathrm{tr}(\beta\wedge\beta)\omega^{n-2} = \frac{1}{n(n-1)}\langle \beta, \beta\rangle\omega^n$.*
Let $\alpha \in \Omega^{2,0}(X; \mathrm{End}(E))$ then $\mathrm{tr}(\alpha \wedge \alpha^{*_H})\omega^{n-2} = \frac{1}{n(n-1)}\langle \alpha, \alpha\rangle \omega^n$.

Proof: [Proof of Proposition 10.4]
In the Hermitian/antihermitian decomposition of F:

$$\begin{aligned} F &= (F(D_H) + \psi_H \wedge \psi_H) + (D_H\psi_H) \\ &= (D'_H D'' + D'' D'_H) + (D'^2_H + D''^2), \end{aligned}$$

the antihermitian part $\beta = F(D_H) + \psi_H \wedge \psi_H$ is of type $(1,1)$ whereas the Hermitian part has the form $\alpha + \alpha^{*_H}$ with $\alpha = \partial_H \phi$ of type $(2,0)$.

This already implies that $L^\star\beta = L^\star F$ since $L^\star$ is zero on forms of degrees $(2,0)$ and $(0,2)$. By the lemma (and the hypothesis $L^\star F = 0$) there is a constant $C > 0$

$$\begin{aligned} \int \mathrm{tr}\beta \wedge \beta\omega^{n-2} &= C\|\beta\|^2_{L^2} \\ \int \mathrm{tr}\alpha \wedge \alpha^{*_H}\omega^{n-2} &= C\|\alpha\|^2_{L^2}. \end{aligned}$$

The hypothesis on E says that $\int \mathrm{tr}F \wedge F\omega^{n-2} = 0$. Developing $F = \beta + \alpha$ in this equality and canceling out the terms not of degree (n,n):

$$\begin{aligned} 0 = \int \mathrm{tr}F \wedge F\omega^{n-2} &= \int \mathrm{tr}\beta \wedge \beta\omega^{n-2} + \int \mathrm{tr}\alpha \wedge \alpha^{*_H}\omega^{n-2} + \int \mathrm{tr}\alpha^{*_H} \wedge \alpha\omega^{n-2} \\ &= C\|\beta\|^2_{L^2} + 2C\|\alpha\|^2_{L^2}. \end{aligned}$$

Consequently $\beta = \alpha = 0$ and $F = 0$. □

10.3. *Complete reducibility and poly-stability*

We illustrate in this section the necessity of the hypothesis in Theorems 8.5 and 8.6.

Proposition 10.7: *Let (E,D) be a flat bundle over a Riemannian manifold M. Suppose there exists a harmonic metric H (i.e. such that $D^\star_H\psi_H = 0$ always using the decomposition $D = D_H + \psi_H$ as the sum of a unitary connection and an Hermitian 1-form).*

The flat bundle (E,D) is then a direct orthogonal sum of irreducible connections.

One sometimes says that D is completely reducible or semisimple.

Proof: By a recurrence on $\dim E$, it is enough to prove that $F^\perp$ (orthogonal with respect to H) is D-stable as soon as the subbundle F is D-stable.

In the direct sum $E = F \oplus F^\perp$, the connection D has the form

$$D = \begin{pmatrix} D_1 & 2\beta \\ 0 & D_2 \end{pmatrix}$$

with $\beta \in \Omega^1(M; \mathrm{Hom}(F^\perp, F))$ (the factor 2 here is to avoid factors 1/2 in the subsequent formulas).

The unitary part of D is the connection:

$$D_H = \begin{pmatrix} D_{H_1} & \beta \\ -\beta^{*_H} & D_{H_2} \end{pmatrix}.$$

Let $S = \begin{pmatrix} -\mathrm{id}_F & \\ & id_{F^\perp} \end{pmatrix} \in \Omega^0(M; \mathrm{End}(E))$, then

$$0 = \int \langle D^\star_H\psi_H, S\rangle \mathrm{vol}_M = \int \langle \psi_H, D_H S\rangle \mathrm{vol}_M.$$

But

$$D_H S = \begin{pmatrix} D_{H_1} & \\ & D_{H_2} \end{pmatrix} S + \left[\begin{pmatrix} 0 & \beta \\ -\beta^{*H} & 0 \end{pmatrix}, S\right]$$
$$= 0 + 2\begin{pmatrix} 0 & \beta \\ \beta^{*H} & 0 \end{pmatrix}.$$

Also

$$\psi_H = \begin{pmatrix} \psi_1 & \beta \\ \beta^{*H} & \psi_2 \end{pmatrix}$$

so $\langle \psi_H, D_H S\rangle$ is the function equal to (local calculation in an orthonormal frame $(\epsilon_i)_{i=1,\dots,n}$ of TM are understood here):

$$\langle \psi_H, D_H S\rangle = \sum \mathrm{tr}(\psi_H(\epsilon_i)(D_H S(\epsilon_i))^{*_H})$$
$$= \sum 2\mathrm{tr}\begin{pmatrix} & \beta \\ \beta^{*H} & \end{pmatrix}^2(\epsilon_i) = 4\sum \mathrm{tr}(\beta(\epsilon_i)\beta^{*_H}(\epsilon_i))$$
$$= 4\sum\langle \beta(\epsilon_i), \beta(\epsilon_i)\rangle = 4\langle \beta, \beta\rangle.$$

The conclusion is that $\int\langle\beta,\beta\rangle\mathrm{vol}_M = 0$, i.e. $\beta = 0$ and the flat bundle splits. □

Proposition 10.8: *Let $(E, \bar\partial^E, \phi)$ be a Higgs bundle. Suppose that there is a Hermitian metric H and a scalar $\lambda \in \mathbf{C}$ such that $L^\star F = \lambda \mathrm{id}_E$ (with $F = F(D)$, $D = D'_H + D''$, etc.).*

Then any Higgs subbundle $\mathcal{F}$ (defined outside an analytic subset $\mathcal{S}$ of complex codimension 2) satisfies $\mu(\mathcal{F}) = \frac{\deg \mathcal{F}}{\mathrm{rank}\mathcal{F}} \leq \mu(\mathcal{E})$. In case of equality $\mu(\mathcal{F}) = \mu(\mathcal{E})$, the subbundle $\mathcal{F}^\perp$ is holomorphic and ϕ-invariant (i.e. it is a Higgs subbundle) (and $\mathcal{S}$ is empty).

The consequence of this proposition is that $\mathcal{E}$ is poly-stable: it is a direct sum of stable Higgs bundles of the same slope (the slope of $\mathcal{E}$ is $\mu(\mathcal{E})$).

Remark 10.9: The complex number λ is determined by the topological (or $\mathcal{C}^\infty$) vector bundle E:

$$\mathrm{rank}(E)\frac{\sqrt{-1}}{2\pi}\lambda\mathrm{vol}(X) = \deg(E).$$

Proof: (The proof is done only for a bundle $\mathcal{F}$ defined *everywhere.*)

In the orthogonal decomposition $\mathcal{E} = \mathcal{F} \oplus \mathcal{F}^\perp = \mathcal{F} \oplus \mathcal{Q}$, the operator $\bar\partial^{\mathcal{E}}$ and the form ϕ are triangular:

$$\bar\partial^{\mathcal{E}} = \begin{pmatrix} \bar\partial^{\mathcal{F}} & \alpha^{*H} \\ & \bar\partial^{\mathcal{Q}} \end{pmatrix}, \quad \phi = \begin{pmatrix} \phi_{\mathcal{F}} & \beta \\ & \phi_{\mathcal{Q}} \end{pmatrix},$$

$$\text{with } \beta \in \Omega^{1,0}(X; \mathrm{Hom}(\mathcal{Q}, \mathcal{F}))$$
$$\alpha \in \Omega^{1,0}(X; \mathrm{Hom}(\mathcal{F}, \mathcal{Q})).$$

One has

$$D = \begin{pmatrix} D_{\mathcal{F}} & \alpha^{*_H} + \beta \\ -\alpha + \beta^{*_H} & D_{\mathcal{Q}} \end{pmatrix} = \begin{pmatrix} D_{\mathcal{F}} & \\ & D_{\mathcal{Q}} \end{pmatrix} + \begin{pmatrix} & \alpha^{*_H} + \beta \\ -\alpha + \beta^{*_H} & \end{pmatrix},$$

so that

$$F(D) = \begin{pmatrix} F_{\mathcal{F}} & \\ & F_{\mathcal{Q}} \end{pmatrix} + \begin{pmatrix} & D(\alpha^{*_H} + \beta) \\ D(-\alpha + \beta^{*_H}) & \end{pmatrix}$$
$$+ \begin{pmatrix} (\alpha^{*_H} + \beta) \wedge (-\alpha + \beta^{*_H}) & \cdots \\ \cdots & \cdots \end{pmatrix}$$

($\cdots$ are for the terms that are not calculated), looking at the component in $\mathrm{End}(\mathcal{F})$ and applying $L^\star$ give:

$$\lambda \mathrm{id}_{\mathcal{F}} = L^\star F_{\mathcal{F}} + L^\star(-\alpha^{*_H} \wedge \alpha + \beta \wedge \beta^{*_H}).$$

Multiplying by $\frac{\sqrt{-1}}{2\pi}$, taking the trace, wedging with ω^{n-1} and integrating over X gives:

$$\frac{\sqrt{-1}}{2\pi}\lambda \mathrm{vol}(X)\mathrm{rank}(\mathcal{F}) - \deg(\mathcal{F})$$
$$= -\frac{\sqrt{-1}}{2\pi}\int \mathrm{tr}\alpha^{*_H} \wedge \alpha \wedge \omega^{n-1} + \frac{\sqrt{-1}}{2\pi}\int \mathrm{tr}\beta \wedge \beta^{*_H} \wedge \omega^{n-1}$$
$$= -\frac{1}{\pi}\int \|\alpha\|^2\omega^n - \frac{1}{\pi}\int \|\beta\|^2\omega^n.$$

Exercise 10.2: Prove the last equalities.

Hence $\mu(\mathcal{E}) - \mu(\mathcal{F}) \leq 0$ and if this last difference is zero, α and β are zero and the Higgs bundle splits. □

10.4. *Minimizing functionals*

The two equations under concern here (the harmonic equation or the Hitchin equation) are in fact related to minimizing functionals.

10.4.1. *The energy*

For a flat connection D on $E \to M$ and a Hermitian metric H, let us define:

$$\mathscr{E}(H) = \mathscr{E}(D, H) = \int \langle \psi_H, \psi_H \rangle \mathrm{vol}_M = \langle \psi_H, \psi_H \rangle_{L^2}.$$

The point is now to understand the minimum of $\mathscr{E}$ when H varies and D is fixed. However, for any gauge automorphism $g \in \mathcal{A}ut(E)$, the following holds:

$$\mathscr{E}(g \cdot D, g \cdot H) = \mathscr{E}(D, H).$$

Exercise 10.3: Recall the action $H \mapsto g \cdot H$ and verify this equality.

Hence, as the action of $\mathcal{A}ut(E)$ is transitive on the space of Hermitian structures on E, an equivalent point of view consists in fixing H and letting D vary in a gauge orbit. The benefits of this change are that the L^2-metrics are not varying anymore. (The definition of the L^2-metric on $\mathrm{End}(E)$ involves H.)

Lemma 10.10: *If H is a minimum for the functional $\mathscr{E}$, then H is a solution of the harmonic equation: $D_H^\star \psi_H = 0$.*

Remark 10.11: The energy functional $\mathscr{E}$ is in fact convex so that any critical point is a minimum.

Proof: A formula for the first variation of the energy $\frac{d}{dt}|_{t=0}\mathscr{E}(D, H_t)$ (with $H_0 = H$) is sought for.

As there is $g_t \in \mathcal{A}ut(E)$ with $g_0 = \mathrm{id}_E$ and $H_t = g_t \cdot H$, it is equivalent to understand

$$\mathscr{E}(g_t^{-1} \cdot D, H) = \mathscr{E}(D, g_t \cdot H)$$

up to first order.

Since $g_t^{-1} \cdot D - D = g_t^{-1} D^{\mathrm{End}(E)} g_t$, one has

$$\begin{aligned}
\tfrac{d}{dt}|_{t=0} g_t^{-1} \cdot D &= \lim_{t\to 0} \tfrac{g_t^{-1}\cdot D - D}{t} \\
&= \lim_{t\to 0} g_t^{-1} \tfrac{D^{\mathrm{End}(E)} g_t}{t} \\
&= \lim_{t\to 0} g_t^{-1} \tfrac{D^{\mathrm{End}(E)} (g_t - g_0)}{t} \qquad \text{as } Dg_0 = D\mathrm{id}_E = 0 \\
&= D\dot{g} \quad \text{with } \dot{g} = \tfrac{d}{dt}|_{t=0} g_t \\
&= D_H \dot{g} + [\psi_H, \dot{g}].
\end{aligned}$$

Let ψ_t the Hermitian part of $g_t^{-1} \cdot D$. The derivative $\frac{d}{dt}|_{t=0}\psi_t$ is the Hermitian part of $\frac{d}{dt}|_{t=0} g_t^{-1} \cdot D$. Writing $\dot{g} = A + S$ as the sum of an antihermitian and a Hermitian endomorphism and observing that the operator D_H respects these decompositions whereas the operator $[\psi_H, \cdot]$ permutes them, one obtain:

$$\frac{d}{dt}|_{t=0}\psi_t = D_H S + [\psi_H, A].$$

Consequently

$$\frac{d}{dt}|_{t=0}\mathscr{E}(D, H_t) = 2\int \langle \psi_H, D_H S + [\psi_H, A]\rangle \mathrm{vol}_M.$$

It is not difficult to establish that $\langle \psi_H, [\psi_H, A]\rangle = 0$. Hence the following formula is obtained:

$$\frac{d}{dt}|_{t=0}\mathscr{E}(D, H_t) = 2\langle D_H^{\star}\psi_H, S\rangle_{L^2}.$$

As $D_H^{\star}\psi_H$ is Hermitian, taking $S = D_H^{\star}\psi_H$ in this last equation implies that the critical values of $\mathscr{E}$ must satisfy $D_H^{\star}\psi_H = 0$. □

10.4.2. *Why is this the energy?*

To establish this, let us write $E = E_\rho = \Gamma\backslash(\widetilde{M} \times \mathbf{C}^d)$ (as in Section 4.3) for $\rho : \Gamma = \pi_1(M) \to \mathrm{GL}_d(\mathbf{C})$.

A Hermitian metric H is the same thing as a ρ-equivariant metric on $\widetilde{M} \times \mathbf{C}^d$. But the space of metrics on $\mathbf{C}^d$ is

$$N = \{S \in M_d(\mathbf{C}) \mid {}^t\overline{S} = S,\ S > 0\}$$

the space of positive definite Hermitian matrices. Hence an equivariant metric on $\widetilde{M} \times \mathbf{C}^d$ is a map

$$f : \widetilde{M} \longrightarrow N$$

satisfying the equivariance relation:

$$f(m) = {}^t\overline{\rho(\gamma)} f(\gamma \cdot m)\rho(\gamma), \quad \forall \gamma \in \Gamma,\ m \in \widetilde{M}.$$

Lemma 10.12: $\psi_H = -\frac{1}{2} f^{-1} df$.

The first observation is that this equality makes sense: $\alpha = f^{-1}df \in \Omega^1(\widetilde{M}; \mathrm{End}(\mathbf{C}^d))$ is equivariant:

$$\alpha_{\gamma \cdot m} = \rho(\gamma)\alpha_m \rho(\gamma)^{-1}$$

and hence descends to M: it "belongs" to $\Omega^1(M; \mathrm{End}(E))$.

Proof: By the definitions of d (the exterior derivative), D_H and ψ_H and since $H(\sigma, \sigma') = {}^t\bar{\sigma} f \sigma'$ for σ and σ' sections of $\widetilde{M} \times \mathbf{C}^d$, one has

$$dH(\sigma, \sigma') = H(D_H\sigma, \sigma') + H(\sigma, D_H\sigma')$$
$$d = D_H + \psi_H.$$

(Here the equivariant decomposition of the trivial connection d on $\widetilde{M} \times \mathbf{C}^d$ is introduced with the same notations as on M.) Therefore

$$\begin{aligned} dH(\sigma, \sigma') &= d({}^t\bar{\sigma} f \sigma') \\ &= {}^t\overline{d\sigma} f \sigma' + {}^t\bar{\sigma} df \sigma' + {}^t\bar{\sigma} f d\sigma' \\ &= H(d\sigma, \sigma') + {}^t\bar{\sigma} df \sigma' + H(\sigma, d\sigma') \\ &= H(D_H\sigma, \sigma') + H(\sigma, D_H\sigma') + H(\psi_H\sigma, \sigma') + H(\sigma, \psi_H\sigma') + {}^t\bar{\sigma} df \sigma' \\ &= dH(\sigma, \sigma') + 2H(\sigma, \psi_H\sigma') + {}^t\bar{\sigma} df \sigma', \end{aligned}$$

thus $2f\psi_H + df = 0$. □

The open subset $N \subset \mathrm{Herm}(\mathbf{C}^d) = \{S \in M_d(\mathbf{C}) \mid {}^t\overline{S} = S\}$ has a structure of a Riemannian manifold:

$$\text{if } X, Y \in T_A N = \mathrm{Herm}(\mathbf{C}^d), \text{ then } \quad \langle X, Y\rangle_A = \mathrm{tr} A^{-1} X A^{-1} Y.$$

Using this and the above formula in the lemma:

$$\mathscr{E}(H) = \int_M \langle \psi_H, \psi_H \rangle \mathrm{vol}_M = \frac{1}{4} \int_M \langle df, df \rangle \mathrm{vol}_M,$$

where $\langle df, df\rangle$ uses the scalar products from the Riemannian structures on M and N. One notes that $\langle df, df\rangle : \widetilde{M} \to \mathbf{R}$ is $\pi_1(M)$-invariant and hence is (or descends to) a function on M as the notation $\int_M$ already suggested.

The conclusion of this discussion is that $\mathscr{E}$ represents the energy of the function f.

10.4.3. *The heat flow*

One of the possible methods to finding a solution of the harmonic equation is to use the associated gradient flow. Here this flow is called the heat flow, and in terms of the metric H, it is given by the equation:

$$H_t^{-1} \frac{d}{dt} H_t = -D_H^{\star} \psi_H.$$

This is still a non-linear PDE, but its linearization is a parabolic linear equation. Using this (and after a significant amount of work) the existence of solutions for short period of time can be shown.

Of course the existence of H_t is equivalent to the existence of a family $(f_t)_{t \in I}$ of ρ-equivariant functions $f_t : \widetilde{M} \to N$ satisfying the corresponding heat equation.

An important observation is that the sectional curvature of N is non-positive. Then a Bochner type formula (and another significant amount of work) shows that, for any two solutions f_t and f'_t of the heat equation, the function $t \mapsto \sup_{m \in M} d_N(f_t(m), f'_t(m))$ is bounded.

This in turn implies that solutions are defined for all $t \in \mathbf{R}_+$ but it also implies that for all $\gamma \in \pi_1(M)$ there is a constant $c(\gamma)$ such that, for all $t \geq 0$, $d_N(f_t(m_0), \rho(\gamma) f_t(m_0)) \leq c(\gamma)$.

The case when the solution f_t "converges to infinity" needs to be excluded. This would means (maybe up to extraction) that the family of points $(f_t(m_0))_{t \geq 0}$ converges to some ξ in the visual boundary at infinity of N. Using the above bound, one deduces that $\rho(\gamma)\xi = \xi$ for all γ in $\pi_1(M)$. This means precisely that ρ is not irreducible.

10.4.4. *The Yang-Mills functional*

In order to solve the Hitchin equation for a Higgs bundle (E, D''), a natural approach is to try to minimize the Yang-Mills functional $\mathscr{YM}(H) = \|F(D'_H + D'')\|_{L^2} = \|L^\star F(D'_H + D'')\|_{L^2}$ (in fact the equality of the 2 last terms is only up to an additive constant depending on the topological bundle E). The gradient flow for the Yang-Mills functional leads to an evolution equation involving fourth order derivatives of H and it is not the route pursued. Instead of the Yang-Mills flow is the following evolution equation:

$$H_t^{-1} \frac{d}{dt} H_t = -L^\star F_t,$$

where F_t is the curvature of $D'_H + D''$. The justification for considering this equation is as follows (this discussion is borrowed from [20]). One can consider the map $H \mapsto L^\star F(D'_H + D'')$ as a vector field on the space $\mathrm{Herm}(E)$ of Hermitian metric on E and one searches for a zero of that vector field. A possible approach to find such a zero is to follow the gradient flow of the vector field (its opposite here). It turns out that in the present situation, the norm of the vector field is indeed decreasing along its gradient flow.

It is again a significant effort to prove the existence of the flow for all positive time and its convergence to a solution of the Hitchin equation under the stability hypothesis. In its full generality, this result is due to Simpson [18], based on an important result by Uhlenbeck and Yau [22]. Earlier results have been obtained by Donaldson [6] and by Hitchin [11].

References

1. Werner Ballmann, *Lectures on Kähler manifolds*, ESI Lectures in Mathematics and Physics, European Mathematical Society (EMS), Zürich, 2006.
2. Kenneth S. Brown, *Cohomology of groups*, Springer Verlag, New York, 1982.
3. Kevin Corlette, *Flat G-bundles with canonical metrics*, J. Differential Geom. **28** (1988), no. 3, 361–382.
4. Simon K. Donaldson, *A new proof of a theorem of Narasimhan and Seshadri*, J. Differential Geom. **18** (1983), no. 2, 269–277.
5. ____, *Anti self-dual Yang-Mills connections over complex algebraic surfaces and stable vector bundles*, Proc. London Math. Soc. (3) **50** (1985), no. 1, 1–26.
6. ____, *Infinite determinants, stable bundles and curvature*, Duke Math. J. **54** (1987), no. 1, 231–247.
7. ____, *Twisted harmonic maps and the self-duality equations*, Proc. London Math. Soc. (3) **55** (1987), no. 1, 127–131.
8. Simon K. Donaldson and Peter B. Kronheimer, *The geometry of four-manifolds*, Oxford Mathematical Monographs, The Clarendon Press Oxford University Press, New York, 1990, Oxford Science Publications.
9. Phillip Griffiths and Joseph Harris, *Principles of algebraic geometry*, Wiley Classics Library, John Wiley & Sons, Inc., New York, 1994, Reprint of the 1978 original.
10. Victor Guillemin and Shlomo Sternberg, *Symplectic techniques in physics*, second ed., Cambridge University Press, Cambridge, 1990.
11. Nigel J. Hitchin, *The self-duality equations on a Riemann surface*, Proc. London Math. Soc. (3) **55** (1987), no. 1, 59–126.
12. ____, *Lie groups and Teichmüller space*, Topology **31** (1992), no. 3, 449–473.
13. Shoshichi Kobayashi, *Differential geometry of complex vector bundles*, Publications of the Mathematical Society of Japan, vol. 15, Princeton University Press, Princeton, NJ; Iwanami Shoten, Tokyo, 1987, Kanô Memorial Lectures, 5.
14. François Labourie, *Existence d'applications harmoniques tordues à valeurs dans les variétés à courbure négative*, Proc. Amer. Math. Soc. **111** (1991), no. 3, 877–882.
15. Martin Lübke and Andrei Teleman, *The Kobayashi-Hitchin correspondence*, World Scientific Publishing Co., Inc., River Edge, NJ, 1995.
16. Mudumbai S. Narasimhan and Conjeeveram S. Seshadri, *Stable and unitary vector bundles on a compact Riemann surface*, Ann. of Math. (2) **82** (1965), 540–567.
17. Jean-Pierre Serre, *Cohomologie des groupes discrets*, Prospects in mathematics (Proc. Sympos., Princeton Univ., Princeton, N.J., 1970), Princeton Univ. Press, Princeton, N.J., 1971, pp. 77–169. Ann. of Math. Studies, No. 70.
18. Carlos T. Simpson, *Constructing variations of Hodge structure using Yang-Mills theory and applications to uniformization*, J. Amer. Math. Soc. **1** (1988), no. 4, 867–918.

19. ____, *Higgs bundles and local systems*, Inst. Hautes Études Sci. Publ. Math. (1992), no. 75, 5–95.
20. Yum Tong Siu, *Lectures on Hermitian-Einstein metrics for stable bundles and Kähler-Einstein metrics*, DMV Seminar, vol. 8, Birkhäuser Verlag, Basel, 1987.
21. Norman Steenrod, *The Topology of fibre bundles*, Princeton Mathematical Series, vol. 14, Princeton University Press, Princeton, N. J., 1951. MR 0039258 (12,522b).
22. Karen Uhlenbeck and Shing-Tung Yau, *On the existence of Hermitian-Yang-Mills connections in stable vector bundles*, Comm. Pure Appl. Math. **39** (1986), no. S, suppl., S257–S293, Frontiers of the mathematical sciences: 1985 (New York, 1985).
23. Richard Wentworth, *Higgs bundles and local systems on Riemann surfaces*, Advanced Courses in Mathematics series - CRM Barcelona, Birkhäuser, 2014, to appear.

An Introduction to Spectral Data for Higgs Bundles

Laura P. Schaposnik
Department of Mathematics
University of Illinois at Urbana-Champaign
IL 61801, USA
schapos@illinois.edu

These notes provide an introduction to Higgs bundles for complex and real Lie groups, as well as a description of their associated spectral data.

1. Introduction

The first lecture shall introduce classical Higgs bundles and the Hitchin fibration, and describe the associated spectral data in the case of principal Higgs bundles for classical complex Lie groups, following mainly Hitchin's papers [31, 32, 33, 34]. The second lecture is dedicated to the construction of Higgs bundles for real forms of classical complex Lie groups as fixed points of involutions, and the description of the corresponding spectral data, following mainly [35, 46, 47, 48]. Along the way, we shall mention different applications and open questions related to the methods introduced in both lectures.

Each lecture contains exercises of varying difficulty, whose solutions can be found in [47]. Open questions which might be tackled with methods similar to the ones introduced in the lectures appear indicated with (*) together with references which feature approaches that may be useful. Since it proves to be very difficult to give a comprehensive and exhaustive account of research in tangential areas, we shall restrain ourselves to mentioning related work only when it directly involves methods using *spectral data*. The reader should refer to references in the bibliography for further research in related topics (e.g., see references in [4, 42, 47]).

2. Spectral data for G_c-Higgs bundles

The art of doing mathematics consists in finding that special case which contains all the germs of generality.

David Hilbert

Following [31, 32, 34] we dedicate this lecture to overview classical Higgs bundles as well as G_c-Higgs bundles for the groups $G_c = SL(n, \mathbb{C})$, $Sp(n, \mathbb{C})$, $SO(2n+1, \mathbb{C})$ and $SO(2n, \mathbb{C})$. In each case we introduce the Hitchin fibration and describe the generic fibres through *spectral data*, i.e., an associated spectral curve and a line bundle on it.

2.1. G_c-*Higgs bundles*

Consider Σ a compact Riemann surface of genus $g \geq 2$ with canonical bundle $K = T^*\Sigma$. Classically, a Higgs bundle on Σ is defined as follows:

Definition 2.1: A *Higgs bundle* is a pair (E, Φ) for E a holomorphic vector bundle on Σ, and Φ, the *Higgs field*, a holomorphic section in $H^0(\Sigma, \mathrm{End}(E) \otimes K)$.

In order to understand better what Higgs bundles are and how to generalise the definition, we shall first look at the moduli space of vector bundles and then study the moduli space of classical Higgs bundles and its associated spectral data. For more details the reader should refer to [31, 32, 23, 19, 51, 40, 52].

2.1.1. *Moduli space of vector bundles*

Holomorphic vector bundles E on a compact Riemann surface Σ of genus $g \geq 2$ are topologically classified by their rank $rk(E)$ and their degree $deg(E)$.

Definition 2.2: The *slope* of a holomorphic vector bundle E is defined as $\mu(E) := deg(E)/rk(E)$ and is used to define stability conditions: A vector bundle E is said to be *stable* (*semi-stable*) if for any proper, non-zero subbundle $F \subset E$ one has $\mu(F) < \mu(E)$ ($\mu(F) \leq \mu(E)$). It is *polystable* if it is a direct sum of stable bundles whose slope is the same as E.

It is known that the space of holomorphic bundles of fixed rank and fixed degree, up to isomorphism, is not a Hausdorff space. However, through Mumford's Geometric Invariant Theory one can construct the moduli space $\mathcal{N}(n,d)$ of stable bundles of fixed rank n and degree d, which has the natural structure of an algebraic variety.

Remark 2.3: For coprime n and d, the moduli space $\mathcal{N}(n,d)$ is a smooth projective algebraic variety of dimension $n^2(g-1)+1$.

Remark 2.4: All line bundles are stable, and thus $\mathcal{N}(1,d)$ contains all line bundles of degree d, and is isomorphic to $\mathrm{Jac}^d(\Sigma)$ of Σ, an abelian variety of dimension g.

Let G_c be a complex semisimple Lie group. Following [44] one can define stability for principal G_c-bundles as follows (see [4] for a comprehensive study):

Definition 2.5: A holomorphic principal G_c-bundle P is said to be *stable* (*semi-stable*) if for every reduction $\sigma : \Sigma \to P/Q$ to maximal parabolic subgroups Q of G_c one has $\deg \sigma^* T_{rel} > 0$ (≥ 0), where T_{rel} is the relative tangent bundle for the projection $P/Q \to \Sigma$.

The notion of polystability may be carried over to principal G_c-bundles, allowing one to construct the moduli space of isomorphism classes of polystable principal G_c-bundles of fixed topological type over the compact Riemann surface Σ.

2.1.2. *Moduli space of classical Higgs bundles*

In order to define the moduli space of Higgs bundles, the following stability condition is considered:

Definition 2.6: A vector subbundle F of E for which $\Phi(F) \subset F \otimes K$ is said to be a *Φ-invariant subbundle* of E. A Higgs bundle (E,Φ) is

- *stable* (*semi-stable*) if for each proper Φ-invariant $F \subset E$ one has $\mu(F) < \mu(E)$ (*equiv.* $\leq$);
- *polystable* if $(E,\Phi) = (E_1,\Phi_1) \oplus (E_2,\Phi_2) \oplus \cdots \oplus (E_r,\Phi_r)$, where (E_i,Φ_i) is stable with $\mu(E_i) = \mu(E)$ for all i.

Example 2.7: Choose a square root $K^{1/2}$ of the canonical bundle K, and a section ω of K^2. A family of classical Higgs bundles (E,Φ_ω) may be obtained by considering the vector bundle $E = K^{\frac{1}{2}} \oplus K^{-\frac{1}{2}}$ and the Higgs bundle Φ_ω given by

$$\Phi_\omega = \begin{pmatrix} 0 & \omega \\ 1 & 0 \end{pmatrix} \in H^0(\Sigma, \mathrm{End}(E) \otimes K).$$

Problem 2.8: *Show that the pairs (E, Φ_ω) from Example 2.7 are stable.*

Problem 2.9: *Prove that if a Higgs bundle (E, Φ) is stable, then for any $\lambda \in \mathbb{C}^*$ and α a holomorphic automorphism of E, the induced Higgs bundles $(E, \lambda\Phi)$ and $(E, \alpha^*\Phi)$ are stable.*

In order to define the moduli space of classical Higgs bundles, we shall first define an appropriate equivalence relation. For this, consider a strictly semi-stable Higgs bundle (E, Φ). As it is not stable, E admits a subbundle $F \subset E$ of the same slope which is preserved by Φ. If F is a subbundle of E of least rank and same slope which is preserved by Φ, it follows that F is stable and hence the induced pair (F, Φ) is stable. Then, by induction one obtains a flag of subbundles $F_0 = 0 \subset F_1 \subset \cdots \subset F_r = E$ where $\mu(F_i/F_{i-1}) = \mu(E)$ for $1 \leq i \leq r$, and where the induced Higgs bundles $(F_i/F_{i-1}, \Phi_i)$ are stable. This is the *Jordan-Hölder filtration* of E, and it is not unique. However, the graded object $\mathrm{Gr}(E, \Phi) := \bigoplus_{i=1}^{r}(F_i/F_{i=1}, \Phi_i)$ is unique up to isomorphism.

Definition 2.10: Two semi-stable Higgs bundles (E, Φ) and (E', Φ') are said to be *S-equivalent* if $\mathrm{Gr}(E, \Phi) \cong \mathrm{Gr}(E', \Phi')$.

Problem 2.11: *If a pair (E, Φ) is strictly stable, what is the induced Jordan-Hölder filtration?*

Following [40] we let $\mathcal{M}(n, d)$ be the moduli space of S-equivalence classes of semi-stable Higgs bundles of fixed degree d and fixed rank n. The moduli space $\mathcal{M}(n, d)$ is a quasi-projective scheme, and has an open subscheme $\mathcal{M}'(n, d)$ which is the moduli scheme of stable pairs. Thus, every point is represented by either a stable or a polystable Higgs bundle. When d and n are coprime, the moduli space $\mathcal{M}(n, d)$ is smooth.

The cotangent space of $\mathcal{N}(n, d)$ over the stable locus is contained in $\mathcal{M}(n, d)$ as a Zariski open subset. The moduli space $\mathcal{M}(n, d)$ is a non-compact variety which has complex dimension $2n^2(g-1)+2$. Moreover, it is a hyperkähler manifold with natural symplectic form ω defined on the infinitesimal deformations $(\dot{A}, \dot{\Phi})$ of a Higgs bundle (E, Φ) by

$$\omega((\dot{A}_1, \dot{\Phi}_1), (\dot{A}_2, \dot{\Phi}_2)) = \int_\Sigma \mathrm{tr}(\dot{A}_1\dot{\Phi}_2 - \dot{A}_2\dot{\Phi}_1), \tag{2.1}$$

where $\dot{A} \in \Omega^{0,1}(\mathrm{End}_0 E)$ and $\dot{\Phi} \in \Omega^{1,0}(\mathrm{End}_0 E)$ (see [31, 32] for details). For simplicity, we shall fix n and d and write $\mathcal{M}$ for $\mathcal{M}(n, d)$.

2.1.3. *Moduli space of G_c-Higgs bundles*

The notion of Higgs bundle can be generalized to encompass principal G_c-bundles, for G_c a complex semi-simple Lie group. For more details, the reader should refer to [32].

Definition 2.12: A *G_c-Higgs bundle* is a pair (P, Φ) where P is a principal G_c-bundle over Σ, and the Higgs field Φ is a holomorphic section of the vector bundle $adP \otimes_{\mathbb{C}} K$, for adP the vector bundle associated to the adjoint representation.

When $G_c \subset GL(n, \mathbb{C})$, a G_c-Higgs bundle gives rise to a Higgs bundle in the classical sense, with some extra structure reflecting the definition of G_c. In particular, classical Higgs bundles are given by $GL(n, \mathbb{C})$-Higgs bundles.

Example 2.13: The Higgs bundles in Example 2.7 have traceless Higgs field, and the determinant $\Lambda^2 E$ is trivial. Hence, for each quadratic differential ω one has an $SL(2, \mathbb{C})$-Higgs bundle (E, Φ_ω).

By extending the stability definitions for principal G_c-bundles, one can define *stable*, *semi-stable* and *polystable* G_c-Higgs bundles. Since in these notes we shall be working with Higgs pairs which do not preserve any subbundle, they will be automatically stable and thus we shall not dedicate time to recall the main study of stability for principal Higgs bundles. For details about the corresponding constructions, the reader should refer for example to [12, 4]. We denote by $\mathcal{M}_{G_c}$ the moduli space of S-equivalence classes of polystable G_c-Higgs bundles.

In the remainder of this Section, following [32] and [34] we introduce the Hitchin fibration and describe the generic fibres for G_c-Higgs bundles where $G_c = GL(n, \mathbb{C})$, $SL(n, \mathbb{C})$, $Sp(2n, \mathbb{C})$, $SO(2n+1, \mathbb{C})$ and $SO(2n, \mathbb{C})$. We shall cover with more detail the initial cases, and leave as an exercise to the reader some of the results for the latter groups.

2.2. *The Hitchin fibration*

A natural way of studying $\mathcal{M}_{G_c}$ is through the Hitchin fibration, as introduced in [32]. We shall denote by p_i, for $i = 1, \ldots, k$, a homogeneous basis for the algebra of invariant polynomials of the Lie algebra $\mathfrak{g}_c$ of G_c, and let d_i be their degrees. Then, the *Hitchin fibration* is given by

$$h\ :\ \mathcal{M}_{G_c} \longrightarrow \mathcal{A}_{G_c} := \bigoplus_{i=1}^{k} H^0(\Sigma, K^{d_i}), \tag{2.2}$$

$$(E,\Phi) \ \mapsto\ (p_1(\Phi),\ldots,p_k(\Phi)). \tag{2.3}$$

The map h is referred to as the *Hitchin map*, and is a proper map for any choice of basis [32]. Furthermore, $\dim\mathcal{A}_{G_c} = \dim\mathcal{M}_{G_c}/2$, making the Higgs bundle moduli space into an integrable system.

Remark 2.14: Note that, in a local frame, a Higgs field Φ has values in a Lie algebra, and thus since this is well defined up to conjugation, evaluating the invariant polynomials is globally well defined.

Remark 2.15: Let $\mathfrak{g}^c$ be one of the classical Lie algebras $\mathfrak{sl}(n,\mathbb{C})$, $\mathfrak{sp}(2n,\mathbb{C})$, $\mathfrak{so}(2n+1,\mathbb{C})$. Then, for $\pi : \mathfrak{g}^c \to \mathfrak{gl}(V)$ a representation of $\mathfrak{g}^c$, the ring of invariant polynomials of $\mathfrak{g}^c$ is generated by $\mathrm{Tr}(\pi(X)^i)$, for $i \in \mathbb{N}$ and $X \in \mathfrak{g}^c$. Hence, a homogeneous basis of invariant polynomials for classical Higgs bundles (E,Φ) of rank n can be taken as $tr(\Phi^i)$ for $1 \leq i \leq n$.

Remark 2.16: Whilst a formal definition of the smooth locus of the Hitchin base can be given (e.g., see [20]) in these lectures we shall note that the generic fibres of the Hitchin fibration are smooth, and thus generic points in the Hitchin base are in the smooth locus.

In what follows we shall describe the spectral data associated to G_c-Higgs bundles as introduced in [32, 34].

2.2.1. *$GL(n,\mathbb{C})$-Higgs bundles*

As before, let K be the canonical bundle of Σ, and X its total space with projection $\pi : X \to \Sigma$. We shall denote by η the tautological section of the pull back π^*K on X. Abusing notation we denote with the same symbols the sections of powers K^i on Σ and their pull backs to X. The characteristic polynomial of a Higgs bundle (E,Φ) in a generic fibre $h^{-1}(a)$ defines a smooth curve $\pi : S_a \to \Sigma$ in X, the *spectral curve* of Φ, whose equation is

$$\det(\eta Id - \pi^*\Phi) = \eta^n + a_1\eta^{n-1} + a_2\eta^{n-2} + \cdots + a_{n-1}\eta + a_n = 0, \tag{2.4}$$

for $a_i \in H^0(\Sigma, K^i)$ (for simplicity, we shall write $\det(\eta - \Phi)$ for the characteristic polynomial of the Higgs field Φ, and drop the subscript a of S_a). By the adjunction formula on X (see e.g. [29]), since the canonical bundle

K has trivial cotangent bundle one has $K_S \cong \pi^* K^n$, and hence the genus of S is

$$g_S = 1 + n^2(g-1). \tag{2.5}$$

The spectral data *for classical Higgs bundles in a smooth fibre of the Hitchin fibration is given by a spectral curve S defined as in (2.4) and a line bundle $L \in Jac(S)$.*

In order to see that the smooth fibres of the Hitchin fibration are Jacobians, starting with a line bundle L on the smooth curve $\pi : S \to \Sigma$ with equation as in (2.4), we shall obtain a classical Higgs bundle by considering the direct image $\pi_* L$ of L. Recall that by definition of direct image, given an open set $\mathcal{U} \subset \Sigma$, one has $H^0(\pi^{-1}(\mathcal{U}), L) = H^0(\mathcal{U}, \pi_* L)$. Multiplication by the tautological section η induces the map

$$H^0(\pi^{-1}(\mathcal{U}), L) \xrightarrow{\eta} H^0(\pi^{-1}(\mathcal{U}), L \otimes \pi^* K),$$

which by definition of direct image can be pushed down to give

$$\Phi : \pi_* L \to \pi_* L \otimes K.$$

Then, one obtains a Higgs field $\Phi \in H^0(\Sigma, \mathrm{End}E \otimes K)$ for $E := \pi_* L$.

Problem 2.17: *Use Grothendieck-Riemann-Roch to show that the degree of E is $deg(E) = deg(L) + (n^2 - n)(1-g)$.*

Moreover, the Higgs field satisfies its characteristic equation, which by construction is given by $\eta^n + a_1\eta^{n-1} + a_2\eta^{n-2} + \cdots + a_{n-1}\eta + a_n = 0$. Furthermore, since S is irreducible, from Remark 2.21 there are no invariant subbundles of the Higgs field, making the induced Higgs bundle (E, Φ) stable.

Conversely, let (E, Φ) be a classical Higgs bundle. The characteristic polynomial is given by $\det(x - \Phi) = x^n + a_1 x^{n-1} + a_2 x^{n-2} + \cdots + a_{n-1}x + a_n$, and its coefficients define the *spectral curve* S in the total space X whose equation is (2.4).

From [11], there is a bijective correspondence between Higgs bundles (E, Φ) and the line bundles L on the spectral curve S described previously. This correspondence identifies the fibre of the Hitchin map with the Picard variety of line bundles of the appropriate degree. By tensoring the line bundles L with a chosen line bundle of degree $-\deg(L)$, one obtains a point in the Jacobian $\mathrm{Jac}(S)$, the abelian variety of line bundles of degree zero on S, which has dimension g_S as in (2.5). In particular, the Jacobian variety is the connected component of the identity in the Picard group $H^1(S, \mathcal{O}_S^*)$. Thus, the fibre of the classical Hitchin fibration $h : \mathcal{M} \to \mathcal{A}$ is isomorphic to the Jacobian of the spectral curve S. For more details, the reader should refer for example to [34].

Example 2.18: In the case of a classical rank 2 Higgs bundle (E, Φ), the characteristic polynomial of Φ defines a spectral curve $\pi : S \to \Sigma$. This is a 2-fold cover of Σ in the total space of K, and has equation $\eta^2 + a_2 = 0$, for a_2 a quadratic differential and η the tautological section of π^*K. By [11] the curve is smooth when a_2 has simple zeros, and in this case the ramification points are given by the divisor of a_2. For z a local coordinate near a ramification point, the covering is given by $z \mapsto z^2 := w$. In a neighbourhood of $z = 0$, a section of the line bundle M can be expressed as $f(w) = f_0(w) + zf_1(w)$. Since the Higgs field is obtained via multiplication by η, one has

$$\Phi(f_0(w) + zf_1(w)) = wf_1(w) + zf_0(w), \tag{2.6}$$

and thus a local form of the Higgs field Φ is given by

$$\Phi = \begin{pmatrix} 0 & w \\ 1 & 0 \end{pmatrix}.$$

Remark 2.19: When $G_c \subset GL(n, \mathbb{C})$, for the groups G_c we are considering in these notes, the spectral data of a G_c-Higgs bundle is given by the spectral data of the pair as a classical Higgs bundle, satisfying extra conditions.

Remark 2.20: For general G_c, a similar description of the fibres can be obtained though a Lie theoretic approach, by means of what is known as *Cameral covers.* The reader should refer to [21] (see also [20]) for this generic description, and note that it is equivalent to the one given in the next sections for the groups considered in these lecture notes.

Remark 2.21: The characteristic polynomial of Φ restricted to an invariant subbundle divides the characteristic polynomial of Φ.

2.2.2. *$SL(n, \mathbb{C})$-Higgs bundles*

When $G_c = SL(n, \mathbb{C})$ we apply Definition 2.12 to obtain the following:

Definition 2.22: An $SL(n, \mathbb{C})$-Higgs bundle is a classical Higgs bundle (E, Φ) where the rank n vector bundle E has trivial determinant and the Higgs field has zero trace.

A basis for the invariant polynomials on the Lie algebra $\mathfrak{sl}(n,\mathbb{C})$ is given by the coefficients of the characteristic polynomial of a trace-free matrix $A \in \mathfrak{sl}(n,\mathbb{C})$. In this case, the spectral curve $\pi : S \to \Sigma$ associated to the Higgs bundle has equation

$$\eta^n + a_2\eta^{n-2} + \cdots + a_{n-1}\eta + a_n = 0, \tag{2.7}$$

where $a_i \in H^0(\Sigma, K^i)$ are the coefficients of the characteristic polynomial of the Higgs field Φ. In particular, one may consider $a_i = Tr(\Phi^i)$, from where it is clear that in this case $a_1 = Tr(\Phi) = 0$. Generically S is a smooth curve of genus $g_S = 1 + n^2(g-1)$, and the coefficients define the corresponding Hitchin fibration

$$h \ : \ \mathcal{M}_{SL(n,\mathbb{C})} \longrightarrow \mathcal{A}_{SL(n,\mathbb{C})} := \bigoplus_{i=2}^{n} H^0(\Sigma, K^i). \tag{2.8}$$

In this case the generic fibres of the Hitchin fibration are given by the subset of $\text{Jac}(S)$ of line bundles L on S for which $\pi_* L = E$ and $\Lambda^n \pi_* L$ is trivial. These conditions in terms of L lead to the following:

The generic fibre of the $SL(n,\mathbb{C})$ Hitchin fibration is biholomorphically equivalent to the $Prym(S,\Sigma)$, for S the spectral curve defined as in (2.4).

In order to see why one has to take the Prym variety, recall that the Norm map

$$\text{Nm} : \text{Pic}(S) \to \text{Pic}(\Sigma),$$

associated to π is defined on divisor classes by $\text{Nm}(\sum n_i p_i) = \sum n_i \pi(p_i)$. In particular,

$$\text{Nm}(\pi^{-1}(x)) = \pi(\pi^{-1}(x)) = nx.$$

The kernel of the Norm map is the *Prym variety*, and is denoted by $\text{Prym}(S,\Sigma)$. From [11], the determinant bundle of L satisfies

$$\Lambda^n \pi_* L \cong \text{Nm}(L) \otimes K^{-n(n-1)/2}.$$

Thus, $\Lambda^n \pi_* L$ is trivial if and only if

$$\text{Nm}(L) \cong K^{n(n-1)/2}. \tag{2.9}$$

Equivalently, since $\text{Nm}(\sum n_i \pi^{-1}(p_i)) = n \sum n_i p_i$, the determinant bundle $\Lambda^n \pi_* L$ is trivial if the line bundle $M := L \otimes \pi^* K^{-(n-1)/2}$ is in the Prym variety.

Remark 2.23: In the case of even rank, equation (2.9) implies a choice of a square root of K (see [31] and [34] for more details).

2.2.3. *$Sp(2n,\mathbb{C})$-Higgs bundles*

Let $G_c = Sp(2n,\mathbb{C})$, and let V be a $2n$-dimensional vector space with a non-degenerate skew-symmetric form $<,>$. For v_i, v_j eigenvectors of $A \in \mathfrak{sp}(2n,\mathbb{C})$ with eigenvalues λ_i and λ_j,

$$\begin{aligned}\lambda_i < v_i, v_j > &= < \lambda_i v_i, v_j > && (2.10)\\ &= < A v_i, v_j > && (2.11)\\ &= - < v_i, A v_j > && (2.12)\\ &= - < v_i, \lambda_j v_j > = -\lambda_j < v_i, v_j > .\end{aligned}$$

From the above, $< v_i, v_j >= 0$ unless $\lambda_i = -\lambda_j$. Since $< v_j, v_j >= 0$, from the non-degeneracy of the symplectic inner product it follows that if λ_i is an eigenvalue so is $-\lambda_i$. Thus, distinct eigenvalues of A must occur in $\pm\lambda_i$ pairs, and the corresponding eigenspaces are paired by the symplectic form. The characteristic polynomial of A must therefore be of the form

$$\det(x - A) = x^{2n} + a_1 x^{2n-2} + \cdots + a_{n-1} x^2 + a_n,$$

and a basis for the invariant polynomials on the Lie algebra $\mathfrak{sp}(2n,\mathbb{C})$ is given by $a_1, \ldots, a_n$.

Definition 2.24: An $Sp(2n,\mathbb{C})$-Higgs bundle is a pair (E,Φ) for E a rank $2n$ vector bundle with a symplectic form $\omega(\ ,\ )$, and the Higgs field $\Phi \in H^0(\Sigma, \mathrm{End}(E) \otimes K)$ satisfying

$$\omega(\Phi v, w) = -\omega(v, \Phi w).$$

The volume form ω^n trivialises the determinant bundle $\Lambda^{2n} E^*$. The characteristic polynomial $\det(\eta - \Phi)$ defines a spectral curve $\pi : S \to \Sigma$ in X with equation

$$\eta^{2n} + a_1 \eta^{2n-1} + \cdots + a_{n-1} \eta^2 + a_n = 0, \qquad (2.13)$$

whose genus is $g_S := 1 + 4n^2(g-1)$. The curve S has a natural involution $\sigma(\eta) = -\eta$ and thus one can define the quotient curve $\bar{\pi} : \overline{S} = S/\sigma \to \Sigma$, of which S is a 2-fold cover $p : S \to \overline{S}$. Note that the Norm map associated to p satisfies $p^*\mathrm{Nm}(x) = x + \sigma x$, and thus the Prym variety $\mathrm{Prym}(S, \overline{S})$ is given by the line bundles $M \in \mathrm{Jac}(S)$ for which $\sigma^* M \cong M^*$.

As in the case of classical Higgs bundles, the characteristic polynomial of a Higgs field Φ gives the Hitchin fibration

$$h \,:\, \mathcal{M}_{Sp(2n,\mathbb{C})} \longrightarrow \mathcal{A}_{Sp(2n,\mathbb{C})} := \bigoplus_{i=1}^{n} H^0(\Sigma, K^{2i}), \tag{2.14}$$

and one has the following:

The generic fibres $h^{-1}(a)$ of the Hitchin fibration for $Sp(2n,\mathbb{C})$-Higgs bundles is given by Prym varieties $Prym(S,\overline{S})$, where S and its quotient $\bar{S}$ are the curves defined by a as above.

The spectral data described above for an $Sp(2n,\mathbb{C})$-Higgs bundle (E,Φ) can be obtained by looking at the extra conditions needed on $L \in \text{Jac}(S)$ associated to the corresponding classical Higgs pair for which $\pi_* L = E$. In order to understand this, note that for $\mathcal{V} \subset S$ an open set, we have $\mathcal{V} \subset \pi^{-1}(\pi(\mathcal{V}))$ and hence a natural restriction map $H^0(\pi^{-1}(\pi(\mathcal{V})), L) \to H^0(\mathcal{V}, L)$, which gives the evaluation map $ev:\ \pi^*\pi_* L \to L$. Multiplication by η commutes with this linear map and so the action of $\pi^*\Phi$ on the dual of the vector bundle $\pi^*\pi_* L$ preserves a one-dimensional subspace. Hence L^* is an eigenspace of $\pi^*\Phi^t$, with eigenvalue η. Equivalently, L is the cokernel of $\pi^*\Phi - \eta$ acting on $\pi^* E \otimes \pi^* K^*$. By means of the Norm map for π, this correspondence can be seen on the curve S via the exact sequence

$$0 \to L \otimes \pi^* K^{1-2n} \to \pi^* E \xrightarrow{\pi^*\Phi - \eta} \pi^*(E \otimes K^*) \xrightarrow{ev} L \otimes \pi^* K \to 0, \tag{2.15}$$

and its dualised sequence

$$0 \to L^* \otimes \pi^* K^* \to \pi^*(E^* \otimes K^*) \to \pi^* E^* \to L^* \otimes \pi^* K^{2n-1} \to 0. \tag{2.16}$$

In particular, from the relative duality theorem one has that

$$\pi_*(L)^* \cong \pi_*(K_S \otimes \pi^* K^{-1} \otimes L^*), \tag{2.17}$$

and thus E^* is the direct image sheaf $\pi_*(L^* \otimes \pi^* K^{2n-1})$.

Given an $Sp(2n,\mathbb{C})$-Higgs bundle (E,Φ), one has $\Phi^t = -\Phi$ and an eigenspace L of Φ with eigenvalue η is transformed to $\sigma^* L$ for the eigenvalue $-\eta$. Moreover, since the line bundle L is the cokernel of $\pi^*\Phi - \eta$ acting on $\pi^*(E \otimes K^*)$, one can consider the corresponding exact sequences (2.15) and its dualised sequence, which identify L^* with $L \otimes \pi^* K^{1-2n}$, or equivalently, $L^2 = \pi^* K^{2n-1}$. By choosing a square root $K^{1/2}$ one has a line bundle $M := L \otimes \pi^* K^{-n+1/2}$ for which $\sigma^* M \cong M^*$, i.e., which is in the Prym variety $\text{Prym}(S,\overline{S})$.

Conversely, an $Sp(2p,\mathbb{C})$-Higgs bundle can be recovered from a line bundle $M \in \text{Prym}(S,\overline{S})$, for S a smooth curve with equation (2.13) and $\bar{S}$ its quotient curve. Indeed, by Bertini's theorem, such a smooth curve S with equation (2.13) always exists. Letting $E := \pi_* L$ for $L = M \otimes \pi^* K^{n-1/2}$, one has the exact sequences (2.15) and its dualised on the curve S. Moreover, since $L^2 \cong \pi^* K^{2n-1}$, there is an isomorphism $E \cong E^*$ which induces the symplectic structure on E. Hence, the generic fibres of the corresponding Hitchin fibration can be identified with the Prym variety $\mathrm{Prym}(S, \overline{S})$.

2.2.4. $SO(2n+1, \mathbb{C})$-*Higgs bundles*

We shall now consider the special orthogonal group $G_c = SO(2n+1, \mathbb{C})$ and the corresponding Higgs bundles. Following a similar analysis as in the previous case, one can see that for a generic matrix $A \in \mathfrak{so}(2n+1, \mathbb{C})$, its distinct eigenvalues occur in $\pm\lambda_i$ pairs, and necessarily A has a zero eigenvalue. Thus, the characteristic polynomial of A must be of the form

$$\det(x - A) = x(x^{2n} + a_1 x^{2n-2} + \cdots + a_{n-1} x^2 + a_n), \qquad (2.18)$$

where the coefficients $a_1, \ldots, a_n$ give a basis for the invariant polynomials on $\mathfrak{so}(2n+1, \mathbb{C})$.

Definition 2.25: An $SO(2n+1, \mathbb{C})$-Higgs bundle is a pair (E, Φ) for E a holomorphic vector bundle of rank $2n+1$ with a non-degenerate symmetric bilinear form (v, w), and Φ a Higgs field in $H^0(\Sigma, \mathrm{End}_0(E) \otimes K)$ which satisfies $(\Phi v, w) = -(v, \Phi w)$.

The moduli space $\mathcal{M}_{SO(2n+1,\mathbb{C})}$ has two connected components, characterised by a class $w_2 \in H^2(\Sigma, \mathbb{Z}_2) \cong \mathbb{Z}_2$, depending on whether E has a lift to a spin bundle or not. The spectral curve induced by the characteristic polynomial in (2.18) is a reducible curve: an $SO(2n+1, \mathbb{C})$-Higgs field Φ always has a zero eigenvalue, and from [34] the zero eigenspace E_0 is given by $E_0 \cong K^{-n}$.

From (2.18), the characteristic polynomial $\det(\eta - \Phi)$ defines a component of the spectral curve, which we shall denote by $\pi : S \to \Sigma$, and whose equation is $\eta^{2n} + a_1 \eta^{2n-2} + \cdots + a_{n-1}\eta^2 + a_n = 0$, where $a_i \in H^0(\Sigma, K^{2i})$. This is a $2n$-fold cover of Σ, with genus $g_S = 1 + 4n^2(g-1)$. The Hitchin fibration in this case is given by the map

$$h : \mathcal{M}_{SO(2n+1,\mathbb{C})} \longrightarrow \mathcal{A}_{SO(2n+1,\mathbb{C})} := \bigoplus_{i=1}^{n} H^0(\Sigma, K^{2i}), \qquad (2.19)$$

which sends each pair (E, Φ) to the coefficients of $\det(\eta - \Phi)$. As in the case of $Sp(2n, \mathbb{C})$, the curve S has an involution σ which acts as $\sigma(\eta) = -\eta$.

Thus, we may consider the quotient curve $\overline{S} = S/\sigma$ in the total space of K^2, for which S is a double cover $p : S \to \overline{S}$. In this case the regular fibres can be described as follows:

The regular fibres $h^{-1}(a)$ of the $SO(2n+1,\mathbb{C})$ Hitchin fibration are given by Prym varieties Prym$(S,\bar{S})$ together with a trivialization of each $M \in$ Prym$(S,\bar{S})$ over the zeros of a_n defining S as in (2.18).

Following [34], the symmetric bilinear form (v, w) canonically defines a skew form $(\Phi v, w)$ on E/E_0 with values in K. Moreover, choosing a square root $K^{1/2}$ one can define

$$V = E/E_0 \otimes K^{-1/2},$$

on which the corresponding skew form is non-degenerate. The Higgs field Φ induces a transformation Φ' on V which has characteristic polynomial

$$\det(x - \Phi') = x^{2n} + a_1 x^{2n-2} + \cdots + a_{n-1}x^2 + a_n.$$

Note that this is exactly the case of $Sp(2n,\mathbb{C})$ described in Section 2.2.3, and thus we may describe the above with a choice of a line bundle M_0 in the Prym variety Prym$(S,\overline{S})$. In particular, S corresponds to the smooth spectral curve of an $Sp(2n,\mathbb{C})$-Higgs bundle.

When reconstructing the vector bundle E with an $SO(2n+1,\mathbb{C})$ structure from an $Sp(2n,\mathbb{C})$-Higgs bundle (V,Φ') as in [34], there is a mod 2 invariant associated to each zero of the coefficient a_n of the characteristic polynomial $\det(\eta - \Phi')$. This data comes from choosing a trivialisation of $M \in$ Prym$(S,\overline{S})$ over the zeros of a_n, and defines a covering P' of the Prym variety Prym$(S,\overline{S})$. The covering has two components corresponding to the spin and non-spin lifts of the vector bundle. The identity component of P', which corresponds to the spin case, is isomorphic to the dual of the symplectic Prym variety, and this is the generic fibre of the $SO(2n+1,\mathbb{C})$ Hitchin map - the reader should refer to Hitchin's work [32] and [33] for a thorough explanation of how the above description is obtained.

2.2.5. *$SO(2n,\mathbb{C})$-Higgs bundles*

Lastly, we consider $G_c = SO(2n,\mathbb{C})$. As in previous cases, the distinct eigenvalues of a matrix $A \in \mathfrak{so}(2n,\mathbb{C})$ occur in pairs $\pm\lambda_i$, and thus the characteristic polynomial of A is of the form $\det(x - A) = x^{2n} + a_1x^{2n-2} + \cdots + a_{n-1}x^2 + a_n$. In this case the coefficient a_n is the square of a polynomial

p_n, the Pfaffian, of degree n. A basis for the invariant polynomials on the Lie algebra $\mathfrak{so}(2n,\mathbb{C})$ is $a_1, a_2, \ldots, a_{n-1}, p_n$, (the reader should refer, for example, to [5] and references therein for further details).

Definition 2.26: An $SO(2n,\mathbb{C})$-Higgs bundle is a pair (E,Φ), for E a holomorphic vector bundle of rank $2n$ with a non-degenerate symmetric bilinear form $(\ ,\ )$, and the Higgs field $\Phi \in H^0(\Sigma, \mathrm{End}_0(E) \otimes K)$ satisfying $(\Phi v, w) = -(v, \Phi w)$.

Considering the characteristic polynomial $\det(\eta - \Phi)$ of a Higgs bundle (E,Φ) one obtains a $2n$-fold cover $\pi : S \to \Sigma$ whose equation is given by

$$\det(\eta - \Phi) = \eta^{2n} + a_1\eta^{2n-2} + \cdots + a_{n-1}\eta^2 + p_n^2,$$

for $a_i \in H^0(\Sigma, K^{2i})$ and $p_n \in H^0(\Sigma, K^n)$. Note that this curve has always singularities, which are given by $\eta = 0$. The curve S has a natural involution $\sigma(\eta) = -\eta$, whose fixed points in this case are the singularities of S. The *virtual* genus of S can be obtained via the adjunction formula, giving $g_S = 1 + 4n^2(g-1)$.

In order to define the spectral data, one may consider its non-singular model $\hat{\pi} : \hat{S} \to \Sigma$, whose genus is given by

$$\begin{aligned} g_{\hat{S}} &= g_S - \#\text{singularities} \\ &= 1 + 4n^2(g-1) - 2n(g-1) \\ &= 1 + 2n(2n-1)(g-1). \end{aligned}$$

As the fixed points of σ are double points, the involution extends to an involution $\hat{\sigma}$ on $\hat{S}$ which does not have fixed points. Considering the associated basis of invariant polynomials for each Higgs field Φ, one may define the Hitchin fibration

$$h \ : \ \mathcal{M}_{SO(2n,\mathbb{C})} \longrightarrow \mathcal{A}_{SO(2n,\mathbb{C})} := H^0(\Sigma, K^n) \oplus \bigoplus_{i=1}^{n-1} H^0(\Sigma, K^{2i}). \quad (2.20)$$

In this case the line bundle associated to an $SO(2n,\mathbb{C})$-Higgs bundle is defined on the desingularisation $\hat{S}$ of S:

The smooth fibres of the $SO(2n,\mathbb{C})$ Hitchin fibration are given by Prym$(\hat{S}, \hat{S}/\hat{\sigma})$, for $\hat{S}$ the desingularisation of the curve S associated to the regular base point a.

Starting with an $SO(2n,\mathbb{C})$-Higgs bundle, since $\hat{S}$ is smooth we obtain an eigenspace bundle $L \subset \ker(\eta - \Phi)$ inside the vector bundle E pulled back to $\hat{S}$. In particular, this line bundle satisfies $\hat{\sigma}^*L \cong L^* \otimes (K_{\hat{S}} \otimes \pi^*K^*)^{-1}$, thus defining a point in $\mathrm{Prym}(\hat{S}, \hat{S}/\hat{\sigma})$ given by

$$M := L \otimes (K_{\hat{S}} \otimes \pi^*K^*)^{1/2}.$$

Conversely, a Higgs bundle (E, Φ) may be recovered from a curve S with has equation $\eta^{2n} + a_1\eta^{2n-2} + \cdots + a_{n-1}\eta^2 + p_n^2 = 0$, and a line bundle L on its desingularisation $\hat{S}$. Note that given the sections

$$s = \eta^{2n} + a_1\eta^{2n-2} + \cdots + a_{n-1}\eta^2 + p_n^2$$

for fixed p_n with simple zeros, one has a linear system whose only base points are when $\eta = 0$ and $p_n = 0$. Hence, by Bertini's theorem the generic divisor of the linear system defined by the sections s has those base points as its only singularities. Moreover, as p_n is a section of K^n, in general there are $2n(g-1)$ singularities which are generically ordinary double points. A generic divisor of the above linear system defines a curve S which has an involution $\sigma(\eta) = -\eta$ whose only fixed points are the base points.

The involution σ induces an involution $\hat{\sigma}$ on the desingularisation $\hat{S}$ of S which has no fixed points, and thus we may consider the quotient $\hat{S}/\hat{\sigma}$ and the corresponding Prym variety $\mathrm{Prym}(\hat{S}, \hat{S}/\hat{\sigma})$. Following a similar procedure as before, a line bundle $M \in \mathrm{Prym}(\hat{S}, \hat{S}/\hat{\sigma})$ induces a Higgs bundle (E, Φ) where E is the direct image sheaf of $L = M \otimes (K_{\hat{S}} \otimes \pi^*K^*)^{-1/2}$. It is thus the Prym variety of $\hat{S}$ which is a generic fibre of the corresponding Hitchin fibration.

Problem 2.27: *Show that the genus $g_{\hat{S}/\hat{\sigma}}$ of $\hat{S}/\hat{\sigma}$ is $n(2n-1)(g-1)$.*

2.3. *Spectral data for complex Higgs bundles*

Considering S a spectral curve, $\hat{S}$ a normalized spectral curve, and $\overline{S}$ and $\overline{\hat{S}}$ the quotients of S and $\hat{S}$ by the involution $\eta \mapsto -\eta$. Moreover, let D be a sub-divisor of $[a_n]$. Then, the spectral data described in this lecture can be summarised as follows:

Table 1. Spectral data for complex Higgs bundles.

Group	Spectral curve	Generic fibre	Ref.
$GL(n, \mathbb{C})$	$\eta^n + a_1\eta^{n-1} + \cdots + a_{n-1}\eta + a_n$	$\mathrm{Jac}(S)$	[32], [34]
$SL(n, \mathbb{C})$	$\eta^n + a_2\eta^{n-2} + \cdots + a_{n-1}\eta + a_n$	$\mathrm{Prym}(S/\Sigma)$	[32], [34]
$Sp(n, \mathbb{C})$	$\eta^{2n} + a_1\eta^{2n-2} + \cdots + a_{n-1}\eta^2 + a_n$	$\mathrm{Prym}(S/\overline{S})$	[32], [34]
$SO(2n+1, \mathbb{C})$	$\eta^{2n} + a_1\eta^{2n-2} + \cdots + a_{n-1}\eta^2 + a_n$	$\mathrm{Prym}(S/\overline{S})$ + D	[32], [34]
$SO(2n, \mathbb{C})$	$\eta^{2n} + a_1\eta^{2n-2} + \cdots + a_{n-1}\eta^2 + p_n^2$	$\mathrm{Prym}(\hat{S}/\overline{\hat{S}})$	[32]

3. Spectral data for G-Higgs bundles

But most of all a good example is a thing of beauty. It shines and convinces. It gives insight and understanding. It provides the bedrock of belief.

Sir Michael Atiyah

Higgs bundles for real forms were first studied by N. Hitchin in [31], and the results for $SL(2,\mathbb{R})$ were generalised in [33], where Hitchin studied the case of $G = SL(n,\mathbb{R})$. Using Higgs bundles he counted the number of connected components and, in the case of split real forms, he identified a component homeomorphic to $\mathbb{R}^{\dim G(2g-2)}$ and which naturally contains a copy of a Teichmüller space. The aim of this Lecture is to introduce principal Higgs bundles for real forms and their corresponding spectral data as studied in [47] and further developed in [35, 36].

3.1. *G-Higgs bundles*

We shall begin by reviewing definitions and properties related to real forms of Lie algebras and Lie groups (see e.g., [24, 37, 41, 39, 45]), and then define G-Higgs bundles for real forms G of classical semisimple complex Lie groups G_c, or of $GL(n,\mathbb{C})$. Through the approach of [33], we describe these Higgs bundles as the fixed points of a certain involution on the moduli space of G_c-Higgs bundles. In later sections we study G-Higgs bundles for non-compact real forms G and in each case give an overview of the corresponding spectral data when available.

3.1.1. *Real forms*

Let $\mathfrak{g}_c$ be a complex Lie algebra with complex structure i, whose Lie group is G_c.

Definition 3.1: A *real form of* $\mathfrak{g}^c$ *is a real Lie algebra which satisfies* $\mathfrak{g}^c = \mathfrak{g} \oplus i\mathfrak{g}$.

Given a real form $\mathfrak{g}$ of $\mathfrak{g}^c$, an element $Z \in \mathfrak{g}^c$ in the Lie algebra may be written as $Z = X + iY$ for $X, Y \in \mathfrak{g}$. The mapping $X + iY \mapsto X - iY$ is called the *conjugation* with respect to $\mathfrak{g}$.

Remark 3.2: Any real form $\mathfrak{g}$ of $\mathfrak{g}^c$ is given by the fixed points set of an antilinear involution τ on $\mathfrak{g}^c$. In particular the conjugation with respect to $\mathfrak{g}$ satisfies these properties.

Definition 3.3: A real form of a complex Lie group G_c is an antiholomorphic Lie group automorphism $\tau : G_c \to G_c$ of order two, i.e., $\tau^2 = 1$.

Every $X \in \mathfrak{g}^c$ defines an endomorphism ad$X$ of the Lie algebra $\mathfrak{g}^c$ given by $\mathrm{ad}X(Y) = [X, Y]$ for $Y \in \mathfrak{g}^c$. For Tr the trace of a vector space endomorphism, $B(X,Y) = \mathrm{Tr}(\mathrm{ad}X\mathrm{ad}Y)$ is a the bilinear form on $\mathfrak{g}^c \times \mathfrak{g}^c$ called the *Killing form* of $\mathfrak{g}^c$.

Definition 3.4: A real Lie algebra $\mathfrak{g}$ is called *compact* if the Killing form is negative definite on it. The corresponding Lie group G is a compact Lie group.

Definition 3.5: Let $\mathfrak{g}$ be a real form of a complex simple Lie algebra $\mathfrak{g}^c$, given by the fixed points of an antilinear involution τ. Then, if there is a Cartan subalgebra invariant under τ on which the Killing form is negative definite, the real form $\mathfrak{g}$ is called a *compact real form*. Such a compact real form of $\mathfrak{g}^c$ corresponds to a compact real form G of G_c; if there is an invariant Cartan subalgebra on which the Killing form is positive definite, the form is called a *split (or normal) real form*. The corresponding Lie group G is the split real form of G_c.

Any complex semisimple Lie algebra $\mathfrak{g}^c$ has a compact and a split real form which are unique up to conjugation via $\mathrm{Aut}_{\mathbb{C}}\mathfrak{g}^c$ (e.g., for $\mathfrak{sl}(\mathfrak{n}, \mathbb{C})$ these are $\mathfrak{su}(\mathfrak{n})$ and $\mathfrak{sl}(n, \mathbb{R})$ respectively).

Remark 3.6: Recall that all Cartan subalgebras $\mathfrak{h}$ of a finite dimensional Lie algebra $\mathfrak{g}$ have the same dimension. The rank of $\mathfrak{g}$ is defined to be this dimension, and a real form $\mathfrak{g}$ of a complex Lie algebra $\mathfrak{g}^c$ is split if and only if the real rank of $\mathfrak{g}$ equals the complex rank of $\mathfrak{g}^c$.

An involution θ of a real semisimple Lie algebra $\mathfrak{g}$ such that the symmetric bilinear form $B_\theta(X, Y) = -B(X, \theta Y)$ is positive definite is called a *Cartan involution*. Any real semisimple Lie algebra has a Cartan involution, and any two Cartan involutions θ_1, θ_2 of $\mathfrak{g}$ are conjugate via an automorphism of $\mathfrak{g}$, i.e., there is a map φ in $\mathrm{Aut}\mathfrak{g}$ such that $\varphi\theta_1\varphi^{-1} = \theta_2$. The decomposition of $\mathfrak{g}$ into eigenspaces of a Cartan involution θ is called the *Cartan decomposition* of $\mathfrak{g}$.

Proposition 3.7: [39] *Let $\mathfrak{g}^c$ be a complex semisimple Lie algebra, and ρ the conjugation with respect to a compact real form $\mathfrak{u}$ of $\mathfrak{g}^c$. Then, ρ is a Cartan involution.*

Proposition 3.8: [37] *Any non-compact real form $\mathfrak{g}$ of a complex simple Lie algebra $\mathfrak{g}^c$ can be obtained from a pair $(\mathfrak{u}, \theta)$, for $\mathfrak{u}$ the compact real form of $\mathfrak{g}^c$ and θ an involution on $\mathfrak{u}$.*

For completion, we shall recall here the construction of real forms from [37]. Let $\mathfrak{h}$ be the $+1$-eigenspaces of θ and $i\mathfrak{m}$ the -1-eigenspace of θ acting on $\mathfrak{u}$, thus having

$$\mathfrak{u} = \mathfrak{h} \oplus i\mathfrak{m}. \tag{3.1}$$

Since $\mathfrak{g}^c = \mathfrak{h} \oplus \mathfrak{m} \oplus i(\mathfrak{h} \oplus \mathfrak{m})$, there is a natural non-compact real form $\mathfrak{g}$ of $\mathfrak{g}^c$ given by

$$\mathfrak{g} = \mathfrak{h} \oplus \mathfrak{m}. \tag{3.2}$$

Moreover, if a linear isomorphism θ_0 induces the decomposition as in (3.2), then θ_0 is a Cartan involution of $\mathfrak{g}$ and $\mathfrak{h}$ is the maximal compact subalgebra of $\mathfrak{g}$.

Following the notation of Proposition 3.8, let ρ be the antilinear involution defining the compact form $\mathfrak{u}$ of a complex simple Lie algebra $\mathfrak{g}^c$ whose decomposition via an involution θ is given by equation (3.1). Moreover, let τ be an antilinear involution which defines the corresponding non-compact real form $\mathfrak{g} = \mathfrak{h} \oplus \mathfrak{m}$ of $\mathfrak{g}^c$. Considering the action of the two antilinear involutions ρ and τ on $\mathfrak{g}^c$, we may decompose the Lie algebra $\mathfrak{g}^c$ into eigenspaces

$$\mathfrak{g}^c = \mathfrak{h}^{(+,+)} \oplus \mathfrak{m}^{(-,+)} \oplus (i\mathfrak{m})^{(+,-)} \oplus (i\mathfrak{h})^{(-,-)}, \tag{3.3}$$

where the upper index $(\cdot, \cdot)$ represents the $\pm$-eigenvalue of ρ and τ respectively. From the decomposition (3.3), the involution θ on the compact real form $\mathfrak{u}$ giving a non-compact real form $\mathfrak{g}$ of $\mathfrak{g}^c$ can be seen as acting on $\mathfrak{g}^c$ as $\sigma := \rho\tau$. Moreover, this induces an involution on the corresponding Lie group $\sigma := G_c \to G_c$.

Remark 3.9: The fixed point set $\mathfrak{g}^\sigma$ of σ is given by $\mathfrak{g}^\sigma = \mathfrak{h} \oplus i\mathfrak{h}$, and thus it is the complexification of the maximal compact subalgebra $\mathfrak{h}$ of $\mathfrak{g}$. Equivalently, the anti-invariant set under the involution σ is given by $\mathfrak{m}^{\mathbb{C}}$.

3.1.2. *G-Higgs bundles through involutions*

As mentioned previously, non-abelian Hodge theory on the compact Riemann surface Σ gives a correspondence between the moduli space of reductive representations of $\pi_1(\Sigma)$ in a complex Lie group G_c and the moduli space of G_c-Higgs bundles. The anti-holomorphic operation of conjugating by a real form τ of G_c in the moduli space of representations can be seen via this correspondence as a holomorphic involution Θ of the moduli space of G_c-Higgs bundles.

Following [32], in order to obtain a G-Higgs bundle, for A the connection which solves Hitchin equations, one requires the flat $GL(n,\mathbb{C})$ connection

$$\nabla = \nabla_A + \Phi + \Phi^* \tag{3.4}$$

to have holonomy in a non-compact real form G of $GL(n,\mathbb{C})$, whose real structure is τ and Lie algebra is $\mathfrak{g}$. More generally, for a complex Lie group G_c with non-compact real form G and real structure τ, one requires

$$\nabla = \nabla_A + \Phi - \rho(\Phi) \tag{3.5}$$

to have holonomy in G, where ρ is the compact real structure of G_c. Since A has holonomy in the compact real form of G_c, we have $\rho(\nabla_A) = \nabla_A$. Hence, requiring $\nabla = \tau(\nabla)$ is equivalent to requiring $\nabla_A = \tau(\nabla_A)$ and $\Phi - \rho(\Phi) = \tau(\Phi - \rho(\Phi))$. In terms of $\sigma = \rho\tau$, these two equalities are given by $\sigma(\nabla_A) = \nabla_A$ and $\Phi - \rho(\Phi) = \tau(\Phi - \rho(\Phi)) = \tau(\Phi) - \sigma(\Phi) = \sigma(\rho(\Phi) - \Phi)$. Hence, ∇ has holonomy in the real form G if ∇_A is invariant under σ, and Φ anti-invariant. In terms of a G_c-Higgs bundle (P,Φ), one has that for $\mathcal{U}$ and $\mathcal{V}$ two trivialising open sets in the compact Riemann surface Σ, the involution σ induces an action on the transition functions $g_{uv} : \mathcal{U} \cap \mathcal{V} \to G_c$ given by $g_{uv} \mapsto \sigma(g_{uv})$, and on the Higgs field by sending $\Phi \mapsto -\sigma(\Phi)$.

Concretely, for G a real form of a complex semisimple lie group G_c, we may construct G-Higgs bundles as follows. For H the maximal compact subgroup of G, we have seen that the Cartan decomposition of $\mathfrak{g}$ is given by $\mathfrak{g} = \mathfrak{h} \oplus \mathfrak{m}$, for $\mathfrak{h}$ the Lie algebra of H, and $\mathfrak{m}$ its orthogonal complement. This induces the following decomposition of the Lie algebra $\mathfrak{g}^c$ of G_c in terms of the eigenspaces of the corresponding involution σ as defined before: $\mathfrak{g}^c = \mathfrak{h}^{\mathbb{C}} \oplus \mathfrak{m}^{\mathbb{C}}$. Note that the Lie algebras satisfy $[\mathfrak{h},\mathfrak{h}] \subset \mathfrak{h}$, $[\mathfrak{h},\mathfrak{m}] \subset \mathfrak{m}$, and $[\mathfrak{m},\mathfrak{m}] \subset \mathfrak{h}$. Hence there is an induced isotropy representation given by $\mathrm{Ad}|_{H^{\mathbb{C}}} : H^{\mathbb{C}} \to GL(\mathfrak{m}^{\mathbb{C}})$. Then, Definition 2.12 generalises to the following (see e.g. [27]):

Definition 3.10: A principal G-*Higgs bundle* is a pair (P, Φ) where P is a holomorphic principal $H^{\mathbb{C}}$-bundle on Σ, and Φ is a holomorphic section of $P \times_{Ad} \mathfrak{m}^{\mathbb{C}} \otimes K$.

Example 3.11: For a compact real form G, one has $G = H$ and $\mathfrak{m} = \{0\}$, and thus σ is the identity and the Higgs field must vanish: a G-Higgs bundle becomes a principal G_c-bundle.

In terms of involutions, following [33] and recalling the previous analysis leading to Remark 3.9, we have the following:

Proposition 3.12: *Let G be a real form of a complex semi-simple Lie group G_c, whose real structure is τ. Then,* G-Higgs bundles *are given by the fixed points in $\mathcal{M}_{G_c}$ of the involution Θ_G acting by*

$$\Theta_G : \ (P, \Phi) \mapsto (\sigma(P), -\sigma(\Phi)),$$

where $\sigma = \rho\tau$, for ρ the compact real form of G_c.

Similarly to the case of G_c-Higgs bundles, there is a notion of stability, semi stability and polystability for G-Higgs bundles. Following [14] and [15], one can see that the polystability of a G-Higgs bundle for $G \subset GL(n, \mathbb{C})$ is equivalent to the polystability of the corresponding $GL(n, \mathbb{C})$-Higgs bundle. However, a G-Higgs bundle can be stable as a G-Higgs bundle but not as a $GL(n, \mathbb{C})$-Higgs bundle. We shall denote by $\mathcal{M}_G$ the moduli space of polystable G-Higgs bundles on Σ.

Problem 3.13: *(*) Considering the notion of "strong real form" from [2], describe the corresponding Higgs bundles and give a definition of Θ_G for which one does not have the problem described in the above paragraph. The reader might find useful the notes in [1, 2] for a concise definition.*

One should note that a fixed point of Θ_G in $\mathcal{M}_{G^c}$ gives a representation of $\pi_1(\Sigma)$ into the real form G up to the equivalence of conjugation by the normalizer of G in G^c. This may be bigger than G itself, and thus two distinct classes in $\mathcal{M}_G$ could be isomorphic in $\mathcal{M}_{G^c}$ via a complex map. Hence, although there is a map from $\mathcal{M}_G$ to the fixed point subvarieties in $\mathcal{M}_{G^c}$, this might not be an embedding. The reader may refer to [25] for the Hitchin-Kobayashi type correspondence for real forms.

Remark 3.14: A description of the above phenomena in the case of rank 2 Higgs bundles is given in [46], where one can see how the $SL(2, \mathbb{R})$-Higgs bundles which have different topological invariants lie in the same connected component as $SL(2, \mathbb{C})$-Higgs bundles.

Remark 3.15: As mentioned previously, the study of real Higgs bundles as fixed point sets of involutions was initiated by Hitchin in [33] in the case of split real forms, and developed for other real forms in [47]. Moreover, this approach has been taken in several papers recently (see, among others, [7, 8, 10, 13, 18, 49]), and continues to be used (see, among others, [9]).

Remark 3.16: The point of view of Proposition 3.12, which is considered throughout [47], fits into a more global picture where Θ_G is one of three natural involutions acting on the moduli space of Higgs bundles [7, 8], giving three families of (B, A, A), (A, B, A) and (A, A, B) branes in $\mathcal{M}_{G_c}$ as the fixed point sets. One should note that the fixed point sets of these involutions are of great importance when studying the relation of Langlands duality with Higgs bundles, as initiated in [30, 38] and [34].

3.2. *Spectral data for G-Higgs bundles*

As mentioned in the first Lecture, the moduli spaces $\mathcal{M}_{G_c}$ have a natural symplectic structure, which we denoted by ω. Moreover, following [32], the involutions Θ_G send $\omega \mapsto -\omega$. Thus, at a smooth point, the fixed point set must be Lagrangian and so the expected dimension of $\mathcal{M}_G$ is half the dimension of $\mathcal{M}_{G_c}$. In order to describe the spectral data for real G-Higgs bundles, one considers the moduli space $\mathcal{M}_G$ sitting inside $\mathcal{M}_{G_c}$ as fixed points of Θ_G in the Hitchin base $\mathcal{A}_{G_c}$ and the corresponding preserved fibres.

By considering Cartan's classification of classical semisimple Lie algebras, we shall now describe G-Higgs bundles and their spectral data for non-compact real forms of a classical semisimple complex Lie algebra $\mathfrak{g}^c$. For I_n the unit matrix of order n, we denote by $I_{p,q}$, J_n and $K_{p,q}$ the matrices

$$I_{p,q} = \begin{pmatrix} -I_p & 0 \\ 0 & I_q \end{pmatrix}, \quad J_n = \begin{pmatrix} 0 & I_n \\ -I_n & 0 \end{pmatrix}, \quad K_{p,q} = \begin{pmatrix} -I_p & 0 & 0 & 0 \\ 0 & I_q & 0 & 0 \\ 0 & 0 & -I_p & 0 \\ 0 & 0 & 0 & I_q \end{pmatrix}.$$

Following Proposition 3.8, we study each complex Lie algebra $\mathfrak{g}_c$ and compact form $\mathfrak{u}$ with different involutions θ which give decompositions $\mathfrak{u} = \mathfrak{h} \oplus i\mathfrak{m}$. Then the corresponding natural non-compact real form is $\mathfrak{g} = \mathfrak{h} \oplus \mathfrak{m}$, and to make sense of Proposition 3.12 we consider the following Lie algebras, Lie groups, real forms, and holomorphic and anti holomorphic involutions:

Table 2. Compact forms $\mathfrak{u}$ of classical Lie algebras.

$\mathfrak{g}_c$	Lie group G_c	Split form	Compact form $\mathfrak{u}$	ρ fixing $\mathfrak{u}$
$\mathfrak{a}_n$	$SL(n,\mathbb{C})$	$\mathfrak{sl}(n,\mathbb{R})$	$\mathfrak{su}(n)$	$\rho(X) = -\overline{X}^t$
$\mathfrak{b}_n$	$SO(2n+1,\mathbb{C})$	$\mathfrak{so}(n,n+1)$	$\mathfrak{so}(2n+1)$	$\rho(X) = \overline{X}$
$\mathfrak{c}_n$	$Sp(2n,\mathbb{C})$	$\mathfrak{sp}(2n,\mathbb{R})$	$\mathfrak{sp}(n)$	$\rho(X) = J_n\overline{X}J_n^{-1}$
$\mathfrak{d}_n$	$SO(2n,\mathbb{C})$	$\mathfrak{so}(n,n)$	$\mathfrak{so}(2n)$	$\rho(X) = \overline{X}$

Table 3. Non-compact forms G of classical Lie algebras G_c.

$\mathfrak{g}_c$	Real form G	τ fixing G	Involution θ on $\mathfrak{u}$
$\mathfrak{a}_n$	$SL(n,\mathbb{R})$	$\rho(X) = -\overline{X}^t$	$\theta(X) = \overline{X}$.
	$SU^*(2m)$	$\tau(X) = J_m\overline{X}J_m^{-1}$	$\theta(X) = J_m\overline{X}J_m^{-1}$
	$SU(p,q)$	$\tau(X) = -I_{p,q}\overline{X}^tI_{p,q}$	$\theta(X) = I_{p,q}XI_{p,q}$
$\mathfrak{b}_n$	$SO(p,q)$	$\tau(X) = I_{p,q}\overline{X}I_{p,q}$.	$\theta(X) = I_{p,q}XI_{p,q}$
$\mathfrak{c}_n$	$Sp(2n,\mathbb{R})$	$\tau(X) = \overline{X}$	$\theta(X) = \overline{X}$
	$Sp(2p,2q)$	$\tau(X) = -K_{p,q}X^*K_{p,q}$.	$\theta(X) = K_{p,q}XK_{p,q}$,
$\mathfrak{d}_n$	$SO(p,q)$	$\tau(X) = I_{p,q}\overline{X}I_{p,q}$.	$\theta(X) = I_{p,q}XI_{p,q}$
	$SO^*(2m)$	$\tau(X) = J_m\overline{X}J_m^{-1}$.	$\theta(X) = J_m\overline{X}J_m^{-1}$

In the case of split real forms, following the methods of [33] one obtains a description of real Higgs bundles which we shall use in subsequent sections:

Theorem 3.17: [47] *For G the split real form of G_c, the fixed points of Θ_G in the smooth fibres of the Hitchin fibration for G_c-Higgs bundles are given by points of order two.*

3.2.1. $SL(n,\mathbb{R})$-*Higgs bundles*

Higgs bundles for $SL(n,\mathbb{R})$ were first considered in [33], where Hitchin studied a copy of Teichmüller space inside the moduli space of Higgs bundles for split real forms. Following Definition 3.10, an $SL(n,\mathbb{R})$-Higgs bundle is a pair (E,Φ) where E is a rank n orthogonal vector bundle and the Higgs field $\Phi : E \to E \otimes K$ is a symmetric and traceless holomorphic map.

Proposition 3.18: *$SL(n,\mathbb{R})$-Higgs bundles are given by the fixed points of*

$$\Theta_{SL(n,\mathbb{R})} : (E,\Phi) \mapsto (E^*,\Phi^t)$$

in $\mathcal{M}_{SL(n,\mathbb{C})}$ corresponding to automorphisms $f : E \to E^*$ giving a symmetric form on E.

Problem 3.19: *Find the decomposition of $\mathfrak{u} = \mathfrak{su}(n)$ induced by the corresponding θ in Table 5, and use this to deduce Proposition 3.18.*

Recalling that the trace is invariant under transposition, one has that the ring of invariant polynomials of $\mathfrak{g}^c = \mathfrak{sl}(n, \mathbb{C})$ is acted on trivially by the involution $-\sigma$, and thus the Hitchin base is preserved by $\Theta_{SL(n,\mathbb{R})}$. In order to find the spectral data for $SL(n, \mathbb{R})$-Higgs bundles, following Theorem 3.17 we look at elements of order two in the fibres of the Hitchin fibration for $SL(n, \mathbb{C})$-Higgs bundles:

Over a smooth point in the Hitchin base $\mathcal{A}_{SL(n,\mathbb{C})}$, Higgs bundles for $SL(n, \mathbb{R})$ correspond to line bundles $L \in Prym(S, \Sigma)$ such that $L^2 \cong \mathcal{O}$.

In the case of $n = 2$, the $SL(2, \mathbb{C})$-spectral curve S given as in (2.7) has a natural involution $\sigma : \eta \mapsto -\eta$ and $\mathrm{Prym}(S, \Sigma) = \{L \in \mathrm{Jac}(S) : \sigma^* L \cong L^*\}$. Hence, points in the smooth fibres corresponding to $SL(2, \mathbb{R})$-Higgs bundles are given by line bundles $L \in \mathrm{Jac}(S)$ such that $\sigma^* L \cong L$.

Problem 3.20: *Let $L \in Prym(S, \Sigma)$ be a line bundle of order two. Then, its direct image is a rank 2 bundle on Σ which decomposes into the sum of two line bundles $V \oplus V^*$. How can the Lefschetz fixed point formula (which relates the action of an involution on a line bundle, and the dimension of the spaces of invariant and anti-invariant sections of a line bundle) from* [6] *be used to relate the degree of V and the action of σ on L in the spirit of* [48]*?*

The topological invariant associated to $SL(n, \mathbb{R})$-Higgs bundles is the characteristic class $\omega_2 \in \mathbb{Z}_2$ which is the obstruction to lifting the orthogonal bundle to a spin bundle, and its study was carried through in [36].

Problem 3.21: *For $n = 2$, use the approach of* [36] *to relate ω_2 to the invariants in Problem 3.20.*

The spectral data of $SL(n, \mathbb{R})$-Higgs bundles gives a finite cover of the smooth locus of the Hitchin fibration. For $n = 2$, an explicit description of the monodromy action whose orbits are the connected components of $\mathcal{M}_{SL(2,\mathbb{R})}$ is given in [46].

Problem 3.22: *(*) How can the methods in* [46] *be extended to study monodromy for $SL(n, \mathbb{R})$-Higgs bundles for $n \geq 3$?*

3.2.2. *$SU^*(2m)$-Higgs bundles*

The group $SU^*(2m)$ is the subgroup of $SL(2m, \mathbb{C})$ which commutes with an antilinear automorphism J of $\mathbb{C}^{2m}$ such that $J^2 = -1$. At the level of the Lie algebras we have that the involution θ decomposes $\mathfrak{u} = \mathfrak{h} \oplus i\mathfrak{m}$ where $\mathfrak{h} = \mathfrak{sp}(m)$. The induced non-compact real form $\mathfrak{g} = \mathfrak{h} \oplus \mathfrak{m}$ is

$$\mathfrak{g} = \mathfrak{su}^*(2m) = \left\{ \begin{pmatrix} Z_1 & Z_2 \\ -\overline{Z}_2 & \overline{Z}_1 \end{pmatrix} \middle| \begin{array}{c} Z_1, Z_2 \ m \times m \quad \text{complex matrices,} \\ \mathrm{Tr} Z_1 + \mathrm{Tr} \overline{Z}_1 = 0 \end{array} \right\}.$$

Definition 3.23: An $SU^*(2m)$ Higgs bundle on Σ is a pair (E, Φ) for E a rank $2m$ vector bundle with a symplectic form ω, and the Higgs field $\Phi \in H^0(\Sigma, \mathrm{End}(E) \otimes K)$ traceless and symmetric with respect to ω.

These Higgs bundles are the first example considered in this notes for which one has nonabelian spectral data. It was first studied in [35], providing what one may call the nonabelianization of the Hitching fibration. In what follows we shall describe the nonabelian spectral data, and also do so for other cases which behave similarly, which have been studied in [35, 47].

Proposition 3.24: *Isomorphism classes of $SU^*(2m)$-Higgs bundles are given by fixed points of the involution*

$$\Theta_{SU^*} : (E, \Phi) \mapsto (E^*, \Phi^t)$$

on $SL(2m, \mathbb{C})$-Higgs bundles corresponding to pairs which have an automorphism $f : E \to E^*$ endowing it with a symplectic structure, and which trivialises its determinant bundle.

As the trace is invariant under conjugation and transposition, one has that the involution $-\sigma(X) = J_m X^t J_m^{-1}$ acts trivially on the ring of invariant polynomials of $\mathfrak{sl}(2m, \mathbb{C})$, and thus preserves the Hitchin base. The spectral data associated to $SU^*(2m)$-Higgs bundles (E, Φ) was studied in [35], and we shall describe here its main features.

The characteristic polynomial of an $SU^*(2m)$-Higgs bundle (E, Φ) can be seen to be the square of a Pfaffian, $\det(\eta - \Phi) = p(\eta)^2$ and thus all fixed points of Θ_{SU^*} lie over singular points of the $SL(2m, \mathbb{C})$ Hitchin fibration. With a slight abuse of notation, we denote by S the spectral curve in the total space of K defined by

$$p(\eta) = \eta^m + a_2 \eta^{m-2} + \cdots + a_m = 0$$

where the coefficients $a_i \in H^0(\Sigma, K^i)$. It is a ramified m-fold cover of Σ whose ramification points are the zeros of a_m. As in the case of complex groups, we interpret $p(\eta) = 0$ as the vanishing of a section of $\pi^* K^m$ over the total space of the canonical bundle $\pi : K \to \Sigma$, where η is the tautological section of $\pi^* K$, and Bertini's theorem assures us that for generic a_i the curve is nonsingular.

Problem 3.25: *What is the genus g_S of S?*

On the spectral curve S, the cokernel of $(\eta - \Phi)$ is a rank two holomorphic vector bundle V on S. Then, following [11] (and using $p(\Phi) = 0$ instead of the Cayley-Hamilton theorem), we can identify E with the direct image $\pi_* V$ and Φ as the direct image of $\eta : V \to V \otimes \pi^* K$. From [35] one has a description of the spectral data:

The fixed point set of $\Theta_{SU^*(2m)}$ in a smooth fibre of the $SL(2m, \mathbb{C})$-Hitchin fibration is the moduli space of semi-stable rank 2 vector bundles on S with fixed determinant $\pi^* K^{m-1}$.

Problem 3.26: *Use Remark 2.21 together with Grothendieck-Riemann-Roch to show that semi-stability of V implies semi-stability of (E, Φ).*

Problem 3.27: *Follow the approach of $SL(n, \mathbb{C})$-Higgs bundles to note that by fixing the determinant of V one obtains a trivialization of the determinant of $\pi_* V$ on Σ.*

3.2.3. *$SU(p, q)$-Higgs bundles*

Definition 3.28: An $SU(p, q)$-Higgs bundle over Σ is a pair (E, Φ) where $E = W_1 \oplus W_2$ for W_1, W_2 vector bundles over Σ of rank p and q such that $\Lambda^p W_1 \cong \Lambda^q W_2^*$, and the Higgs field Φ is given by $\Phi = \begin{pmatrix} 0 & \beta \\ \gamma & 0 \end{pmatrix}$, for $\beta : W_2 \to W_1 \otimes K$ and $\gamma : W_1 \to W_2 \otimes K$.

Problem 3.29: *Find the decomposition $\mathfrak{u} = \mathfrak{h} \oplus i\mathfrak{m}$ via the action of θ in Table 2 and deduce that $\theta\rho$ is the anti-holomorphic involution fixing the non-compact real form $\mathfrak{u}(p, q)$.*

Proposition 3.30: *$SU(p, q)$-Higgs bundles are fixed points of $\Theta_{SU(p,q)}$: $(E, \Phi) \mapsto (E, -\Phi)$ on $SL(p+q, \mathbb{C})$-Higgs bundles corresponding to bundles E which have an automorphism conjugate to $I_{p,q}$ sending Φ to $-\Phi$, and whose ± 1 eigenspaces have dimensions p and q.*

The involution $-\sigma$ acts trivially on the polynomials of even degree. Whilst the spectral data is not known for $p \neq q$, in the case of $p = q$ it has been described in [47] and [48] by looking at $U(p, p)$-Higgs bundles $(W_1 \oplus W_2, \Phi)$, which when satisfying $\Lambda^p W_1 \cong \Lambda^q W_2^*$ correspond to $SU(p, p)$-Higgs bundles. In this case, the characteristic polynomial defines a spectral curve $\pi : S \to \Sigma$ through the equation $\det(\eta - \Phi) = \eta^{2p} + a_2 \eta^{2p-2} +$

$\cdots + a_{2p-2}\eta^2 + a_{2p} = 0$, where η is the tautological section of $\pi^* K$ and $a_i \in H^0(\Sigma, K^i)$. This is a $2p$-fold cover of Σ, ramified over the $4p(g-1)$ zeros of a_{2p}, and has a natural involution $\eta \mapsto -\eta$ which has as fixed points the ramification points of the cover, and which by abuse of notation, we shall call σ.

The involution σ plays an important role when constructing the spectral data as described in [48]. A line bundle L on S which defines a classical Higgs bundle induces a $U(p,p)$-Higgs bundle if and only if $\sigma^* L \cong L$. In this case, at a fixed point $x \in S$ of the involution, there is a linear action of σ on the fibre L_x given by scalar multiplication of ± 1. This description of the spectral data can be then seen in terms of Jacobians through [48]:

The fixed point set of $\Theta_{U(p,p)}$ in a smooth fibre of the classical Hitchin fibration can be seen in terms of pull backs of Jac(S/σ) on a symmetric product of Σ to a point of the Hitching base.

As described in [48], the topological invariants associated to a $U(p,p)$-Higgs bundle $(W_1 \oplus W_2, \Phi)$ are the degrees $\deg(W_1)$ and $\deg(W_2)$, and can be seen in terms of the degree of the line bundle L on S and the number of ramification points of S over which the linear action of σ on the fibre of L is -1.

Problem 3.31: *Use the Lefschetz fixed point formula in [6] to see that the parity of the degree of L and the number of points over which σ acts as -1 needs to be the same.*

Problem 3.32: *Following [14], a $U(p,p)$-Higgs bundles has an associated invariant $\tau(deg(W_1), deg(W_2)) := deg(W_1) - deg(W_2)$, known as the Toledo invariant. Use Problem 3.31 to express the invariant in terms of fixed points of σ and obtain natural bounds.*

In the case of $SU(p,p)$-Higgs bundles, for maximal Toledo invariant (i.e., when the invariant in the above problem achieves the bounds), the fixed point set of $\Theta_{SU(p,p)}$ in a smooth fibre of the $SL(2p,\mathbb{C})$-Hitchin fibration is given by a covering of Prym($S/\sigma, \Sigma$), the Prym variety of the quotient curve S/σ. For $SU(p,p+1)$-Higgs bundles, the methods and arguments of [20] can be adapted and used to obtain the spectral data as seen in [42].

Problem 3.33: *(*) How can the methods from [48] together with the approach of [53] be used to obtain the spectral data for $SU(p,1)$-Higgs bundles?*

Remark 3.34: The moduli space of real Higgs bundles is a brane in the moduli space of complex Higgs bundles, and as such it has a dual space, a brane in the moduli space of complex Higgs bundles for the Langlands dual group. Properties of this dual space have been conjectured in [8], and it is interesting to note that the spectral data for $SU(p,p)$-Higgs bundles from [48] is used to conjecture a dual space to $\mathcal{M}_{SU(p,p)}$ through Langlands duality in [36].

3.2.4. *$SO(p,q)$-Higgs bundles*

In this case, if $p+q$ is even, $\mathfrak{g}$ is a split real form if and only if $p=q$; if $p+q$ is odd, $\mathfrak{g}$ is a split real form if and only if $p=q+1$. Whilst we shall give some details on the construction of $SO(p,q)$-Higgs bundles, for a more detailed description of the approach needed to understand groups with signature the reader should refer to the following section on $Sp(2p,2q)$-Higgs bundles.

The vector space V associated to the standard representation of $\mathfrak{h}^{\mathbb{C}}$ can be decomposed into $V = V_p \oplus V_q$, for V_p and V_q complex vector spaces of dimension p and q respectively, with orthogonal structures. The maximal compact subalgebra of $\mathfrak{so}(p,q)$ is $\mathfrak{h} = \mathfrak{so}(p) \times \mathfrak{so}(q)$ and the Cartan decomposition of $\mathfrak{so}(p+q,\mathbb{C})$ is given by $(\mathfrak{so}(p,\mathbb{C}) \oplus \mathfrak{so}(q,\mathbb{C})) \oplus \mathfrak{m}^{\mathbb{C}}$, for

$$\mathfrak{m} = \left\{ \begin{pmatrix} 0 & X_2 \\ X_2^t & 0 \end{pmatrix} \middle| X_2 \text{ real } p\times q \text{ matrix} \right\}.$$

Definition 3.35: An $SO(p,q)$ Higgs bundle is a pair (E,Φ) where the vector bundle is $E = V_p \oplus V_q$ for V_p and V_q complex vector spaces of dimension p and q respectively, with orthogonal structures, and the Higgs field is a section in $H^0(\Sigma, (\mathrm{Hom}(V_q,V_p) \oplus \mathrm{Hom}(V_p,V_q)) \otimes K)$ given by

$$\Phi = \begin{pmatrix} 0 & \beta \\ \gamma & 0 \end{pmatrix} \quad \text{for } \gamma \equiv -\beta^{\mathrm{T}}, \text{ and } \beta^{\mathrm{T}} \text{the orthogonal transpose of } \beta.$$

Proposition 3.36: *$SO(p,q)$ Higgs bundles are fixed points of*

$$\Theta_{SO(p,q)} : (E,\Phi) \mapsto (E,-\Phi)$$

on the moduli space of $SO(p+q,\mathbb{C})$ corresponding to vector bundles E which have an automorphism f conjugate to $I_{p,q}$ sending Φ to $-\Phi$ and whose ±1 eigenspaces have dimensions p and q.

Problem 3.37: *The involution θ from the Table 2 decomposes $\mathfrak{u} = \mathfrak{h} \oplus i\mathfrak{m}$. Give an explicit description of $\mathfrak{m}$ and $\mathfrak{h}$ and of the real form $\mathfrak{g} = \mathfrak{h} \oplus \mathfrak{m}$.*

Since the ring of invariant polynomials of $\mathfrak{g}^c = \mathfrak{so}(2m+1,\mathbb{C})$ is generated by $\mathrm{Tr}(X^i)$ for $X \in \mathfrak{g}^c$, for $p+q=2m+1$ one has that the induced action of the involution $\Theta_{SO(p,q)}$ is trivial on the ring of invariant polynomials of the Lie algebra $\mathfrak{so}(2m+1,\mathbb{C})$, i.e., when p and q have different parity.

Problem 3.38: *In the case of $\mathfrak{so}(2m,\mathbb{C})$, for $2m = p+q$, the ring of invariant polynomials is generated by $Tr(X^i)$ for $X \in \mathfrak{g}^c$ and $i < 2m$, together with the Pfaffian p_m, which is of degree m. Under which conditions on p and q is the induced action of $\Theta_{SO(p,q)}$ trivial on the ring of invariant polynomials?*

The spectral data for $SO(p,q)$-Higgs bundles when $p=q$ or $p=q+1$ can be seen through Theorem 3.17 from [47] as points of order two in the smooth fibres of the $SO(p+q,\mathbb{C})$-Hitchin fibration.

In both cases a key ingredient is the double cover $p : S \to S/\sigma$ given by the spectral curve (the desingularised curve in the case of $SO(2n,\mathbb{C})$) over the quotient curve, which through K-theoretic methods allows one to express the topological invariants involved in terms of the action of σ [9].

3.2.5. *$SO^*(2m)$-Higgs bundles*

The action of θ in Table 2 decomposes the compact form $\mathfrak{u} = \mathfrak{h} \oplus i\mathfrak{m}$ for $\mathfrak{h} = \mathfrak{u}(m) \cong \mathfrak{so}(2m) \cap \mathfrak{sp}(m)$, and

$$i\mathfrak{m} = \left\{ \begin{pmatrix} X_1 & X_2 \\ X_2 & -X_1 \end{pmatrix} \middle| X_1,\ X_2 \in \mathfrak{so}(m) \right\}, \tag{3.6}$$

and the induced non-compact real form $\mathfrak{g} = \mathfrak{h} \oplus \mathfrak{m}$ is

$$\mathfrak{g} = \mathfrak{so}^*(2m) = \left\{ \begin{pmatrix} Z_1 & Z_2 \\ -\overline{Z}_2 & \overline{Z}_1 \end{pmatrix} \middle| \begin{matrix} Z_1, Z_2\ m \times m \quad \text{complex matrices} \\ Z_1\ \text{ skew symmetric, } Z_2 \text{ Hermitian} \end{matrix} \right\}.$$

The vector space associated to the standard representation of $\mathfrak{h}^{\mathbb{C}}$ has an orthogonal and symplectic structure J. Since $J^{-1} = J^t$ and $J^2 = -1$, the vector space may be expressed in terms of the $\pm i$ eigenspaces of J as $V \oplus V^*$, for V a rank m vector space. Thus, we have the following definition:

Definition 3.39: An $SO^*(2m)$-Higgs bundle is given by a pair (E,Φ) where $E = V \oplus V^*$ for V a rank m holomorphic vector bundle, and where the Higgs field Φ is given by

$$\Phi = \begin{pmatrix} 0 & \beta \\ \gamma & 0 \end{pmatrix} \quad \text{for} \quad \begin{cases} \gamma : \ V \to V^* \otimes K & \text{satisfying } \gamma = -\gamma^t \\ \beta : \ V^* \to V \otimes K & \text{satisfying } \beta = -\beta^t \end{cases}.$$

In terms of involutions, these Higgs bundles may be seen as follows:

Proposition 3.40: *$SO^*(2m)$-Higgs bundles are fixed points of the involution*

$$\Theta_{SO^*(2m)}: \ (E,\Phi) \mapsto (E,-\Phi)$$

on the moduli space of $SO(2m,\mathbb{C})$-Higgs bundles corresponding to vector bundles E which have an orthogonal automorphism f conjugate to J_m, sending Φ to $-\Phi$ and which squares to -1, equipping E with a symplectic structure.

As in the previous case, the involution induced action of $\Theta_{SO^*(2m)}$ is trivial on the ring of invariant polynomials of $\mathfrak{g}^c$. The spectral data for these Higgs bundles is studied in [35], and we shall give a short description below.

In order to understand the associated spectral data, one notes that $SO^*(2m)$-Higgs bundles (E,Φ) may be regarded as $SU^*(2m)$-Higgs bundle with extra conditions. Hence, one may define a natural m cover of the Riemann surface $\pi : S \to \Sigma$ by taking

$$\sqrt{\mathrm{char}(\Phi)} = \eta^m + a_2\eta^{2m-2} + \cdots + a_m,$$

and a rank 2 vector bundle V on S whose direct image on Σ gives E. Since in this case the equation of the spectral curve only has even coefficients, there is a natural involution $\sigma : \eta \to -\eta$ and one may consider the induced action of σ on V and on its determinant bundle. In particular, from [35] the vector bundle V gives an $SO^*(2m)$-Higgs bundle if and only if it is preserved by the involution and the induced action on it satisfies some conditions:

The fixed point set of $\Theta_{SO^*(2m)}$ in a smooth fibre of the $SO(2m,\mathbb{C})$-Hitchin fibration is given by the moduli space of semi-stable rank 2 vector bundles V on S with fixed determinant π^*K^{2m-1}, whose induced action by σ on the determinant bundle is trivial.

Problem 3.41: *The relative duality theorem gives*

$$(\pi_*(V))^* \cong \pi_*(V^* \otimes K_S) \otimes K^*.$$

Use this to see that in order to have $E \cong E^*$ through a skew form, the action of σ needs to be trivial on the determinant bundle of V for $\pi_*V = E$.

Problem 3.42: *(*) Describe how the vector bundles of rank 2 in [35] appear in the description of the connected components of $\mathcal{M}_{SO^*(2m)}$ in [17].*

3.2.6. *$Sp(2n,\mathbb{R})$-Higgs bundles*

In this section and the one which follows we consider the non-compact real forms of the complex Lie group $Sp(2n,\mathbb{C})$. For this, recall that the symplectic Lie algebra $\mathfrak{sp}(2n,\mathbb{C})$ is given by the set of $2n \times 2n$ complex matrices X that satisfy $J_n X + X^t J_n = 0$ or equivalently, $X = -J_n^{-1} X^t J_n$.

Let $\mathfrak{u}$ be the compact real form $\mathfrak{u} = \mathfrak{sp}(n)$ and $\theta(X) = \overline{X} = J_n \overline{X} J_n^{-1}$. The Lie algebra $\mathfrak{sp}(n)$ is given by the quaternionic skew-Hermitian matrices; that is, the set of $n \times n$ quaternionic matrices X which satisfy $X = -\overline{X}^t$. The compact form is $\mathfrak{u} = \mathfrak{h} \oplus i\mathfrak{m}$, for $\mathfrak{h} = \mathfrak{u}(n) \cong \mathfrak{so}(2n) \cap \mathfrak{sp}(n)$, which leads to the split real form $\mathfrak{g} = \mathfrak{h} \oplus \mathfrak{m}$ defined by

$$\mathfrak{g} = \mathfrak{sp}(2n,\mathbb{R}) = \left\{ \begin{pmatrix} X_1 & X_2 \\ X_3 & -X_1^t \end{pmatrix} \middle| \begin{matrix} X_i \text{ real } n \times n \quad \text{matrices} \\ X_2, X_3 \text{ symmetric} \end{matrix} \right\}.$$

Definition 3.43: An $Sp(2n,\mathbb{R})$-Higgs bundle is given by a pair (E,Φ) where $E = V \oplus V^*$ for V a rank n holomorphic vector bundle, and for Φ the Higgs field given by

$$\Phi = \begin{pmatrix} 0 & \beta \\ \gamma & 0 \end{pmatrix} \quad \text{for} \quad \begin{cases} \gamma : \ V \to V^* \otimes K & \text{satisfying } \gamma = \gamma^t \\ \beta : \ V^* \to V \otimes K & \text{satisfying } \beta = \beta^t \end{cases}.$$

Proposition 3.44: *$Sp(2n,\mathbb{C})$ Higgs bundles, and $Sp(2n,\mathbb{R})$-Higgs bundles are given by the fixed points of*

$$\Theta_{Sp(2n,\mathbb{R})} : \ (E,\Phi) \mapsto (E,-\Phi)$$

on $Sp(2n,\mathbb{C})$-Higgs bundles corresponding to vector bundles E which have a symplectic isomorphism sending Φ to $-\Phi$, and whose square is the identity, endowing E with an orthogonal structure.

The invariant polynomials of $\mathfrak{g}^c$ are of even degree, and hence the involution $-\sigma$ acts trivially on them, making $\Theta_{Sp(2n,\mathbb{R})}$ preserve the whole Hitchin base $\mathcal{A}_{Sp(2n,\mathbb{C})}$. In the case of rank 4 Higgs bundles, the spectral data was first consider in P. Gothen's thesis [27, 28], and through Theorem 3.17 and the spectral data for complex Higgs bundles one has the following:

The fixed points of $\Theta_{Sp(2n,\mathbb{R})}$ in the smooth fibres of the $Sp(2n,\mathbb{C})$-Hitchin fibration are given by line bundles $L \in Prym(S,S/\sigma)$ such that $L^2 \cong \mathcal{O}$.

In particular, since S is a ramified double cover of S/σ, one has that $L \in \mathrm{Prym}(S, S/\sigma)$ if and only if $\sigma^* L \cong L^*$. Hence, by considering points of order two one has that $\sigma^* L \cong L$ and thus there is a natural induced action of σ on the line bundle L. The topological invariants associated to these Higgs bundles were studied in [36] through the natural action of σ.

Problem 3.45: *Compare the calculations in [27] which lead to Milnor-Wood type inequalities for $Sp(2n, \mathbb{R})$-Higgs bundles, with the inequalities one obtains by using the involution σ as in [36].*

Problem 3.46: *(*) Express the invariants from [27] in terms of different choices of the natural involution σ on S as well as in terms of the action of a second natural involution appearing in some situations on S/σ.*

3.2.7. *Sp(2p, 2q)-Higgs bundles*

The induced non-compact real form $\mathfrak{g} = \mathfrak{h} \oplus \mathfrak{m}$ is

$$\mathfrak{sp}(2p, 2q) = \left\{ \begin{pmatrix} Z_{11} & Z_{12} & Z_{13} & Z_{14} \\ \overline{Z}_{12}^t & Z_{22} & Z_{14}^t & Z_{24} \\ -\overline{Z}_{13} & \overline{Z}_{14} & \overline{Z}_{11} & -\overline{Z}_{12} \\ \overline{Z}_{14}^t & -\overline{Z}_{24} & -Z_{12}^t & \overline{Z}_{22} \end{pmatrix} \middle| \begin{array}{c} Z_{i,j} \quad \text{complex matrices,} \\ Z_{11},\ Z_{13} \ \text{order}\ p, \\ Z_{12},\ Z_{14}\ p \times q \ \text{matrices,} \\ Z_{11},\ Z_{22}\ \text{skew Hermitian,} \\ Z_{13},\ Z_{24}\ \text{symmetric.} \end{array} \right\}.$$

Problem 3.47: *Show that $\mathfrak{m}^{\mathbb{C}}$ can be expressed as subset of certain off-diagonal matrices.*

Definition 3.48: An $Sp(2p, 2q)$-Higgs bundle is given by a pair (E, Φ) for $E = V_{2p} \oplus V_{2q}$ is a direct sum of symplectic vector spaces of rank $2p$ and $2q$, and

$$\Phi = \begin{pmatrix} 0 & -\gamma^{\mathrm{T}} \\ \gamma & 0 \end{pmatrix} \quad \text{for} \quad \begin{cases} \gamma & : V_{2p} \to V_{2q} \otimes K \\ -\gamma^{\mathrm{T}} & : V_{2q} \to V_{2p} \otimes K \end{cases},$$

for γ^{T} the symplectic transpose of γ.

Proposition 3.49: *$Sp(2p, 2q)$-Higgs bundles are the fixed points of*

$$\Theta_{Sp(2p,2q)} : \ (E, \Phi) \mapsto (E, -\Phi^T)$$

on the moduli space of $Sp(2p + 2q, \mathbb{C})$-Higgs bundles corresponding to symplectic vector bundles E which have an endomorphism $f : E \to E$ conjugate to $\tilde{K}_{p,q}$, sending Φ to the symplectic transpose $-\Phi^T$, and whose ± 1 eigenspaces are of dimension $2p$ and $2q$.

As the trace is invariant under conjugation and transposition, the induced action of $\Theta_{Sp(2p,2q)}$ is trivial on the ring of invariant polynomials of $\mathfrak{g}^c = \mathfrak{sp}(2(p+q),\mathbb{C})$. In the case of $p = q$, one can see that $Sp(2p,2p)$-Higgs bundles are a particular case of $SU^*(2p)$-Higgs bundles, and thus one needs to understand which extra conditions to the spectral data for $SU^*(2p)$-Higgs bundles needs to be added in order to have the Higgs bundles for the symplectic real form.

From the previous section, when $p = q$ the corresponding spectral curve is a $2p$-fold cover of the Riemann surface Σ whose equation is given by the square root of the characteristic polynomial of the Higgs field. Moreover, it has a natural involution σ whose action determines the associated spectral data. More precisely, the following is shown in [47] and [35]:

The fixed point set of $\Theta_{Sp(2p,2p)}$ in a smooth fibre of the $Sp(4p,\mathbb{C})$-Hitchin fibration is given by the moduli space of semi-stable rank 2 vector bundles V on S with fixed determinant π^*K^{2p-1}, whose induced action by σ on $\Lambda^2 V$ is -1.

Since the action on $\Lambda^2 V$ is -1, the involution $\sigma : S \to S$ acts with different eigenvalues ± 1 on the fibres of V over the ramification points of S, and thus through [3], the spectral data relates to the moduli space of admissible parabolic rank 2 bundles on S/σ as seen in [47].

Problem 3.50: *(*) Nonabelianization can also be seen through Cameral covers* [42]. *Realise the action of σ in terms of Cameral covers.*

4. Spectral data for real Higgs bundles

From the above sections, we have seen that spectral data can be been defined for G-Higgs bundles, and this has been done for several groups G. A summary of the state of the art in this direction is given as follows (to the best of the author's knowledge), where the notation is as in the previous sections (for the precise objects in each case, the reader should refer to the previous sections)[a]:

[a]For the groups missing, no spectral data has yet been defined.

Table 4. Spectral data for real Higgs bundles.

Group	Spectral curve	generic fibre	Ref.
$GL(n,\mathbb{R})$	$\eta^n + a_1\eta^{n-1} + \cdots + a_{n-1}\eta + a_n$	Jac$(S)[2]$	[47]
$U(p,p)$	$\eta^{2p} + a_1\eta^{2p-2} + \cdots + a_{p-1}\eta^2 + a_p$	Jac$(\overline{S})$ + D + m	[47, 48]
$SL(n,\mathbb{R})$	$\eta^n + a_2\eta^{n-2} + \cdots + a_{n-1}\eta + a_n$	Prym$(S,\Sigma)[2]$	[47, 46]
$SU^*(2m)$	$\eta^{2p} + a_1\eta^{2p-2} + \cdots + a_{p-1}\eta^2 + a_p$	Subspace of $\mathcal{N}_2$(S)	[35]
$SU(p,p)$	$\eta^{2p} + a_1\eta^{2p-2} + \cdots + a_{p-1}\eta^2 + a_p$	Prym$(S,\overline{S}) + D + m$	[47, 48]
$SU(p,p+1)$	$\eta^{2p} + a_1\eta^{2p-2} + \cdots + a_{p-1}\eta^2 + a_p$	Jac$(\overline{S})$ + f	[43, 47]
$SO(p,p+1)$	$\eta^{2p} + a_1\eta^{2p-2} + \cdots + a_{p-1}\eta^2 + a_p$	Prym$(S,\overline{S})[2]$ +D	[47, 50, 9]
$Sp(2p,\mathbb{R})$	$\eta^{2p} + a_1\eta^{2p-2} + \cdots + a_{p-1}\eta^2 + a_p$	Prym$(S,\overline{S})[2]$	[36, 47]
$Sp(2p,2p)$	$\eta^{2p} + a_1\eta^{2p-2} + \cdots + a_{p-1}\eta^2 + a_p$	Subspace of $\mathcal{N}_2$(S)	[35]
$Sp(2p,2p+2q)$	$\eta^{2q}(\eta^{4p} + a_1\eta^{4p-2} + \cdots + a_{2p-1}\eta^2 + a_{2p})$	Abelian & non-abelian	[9]
$SO(p,p)$	$\eta^{2p} + a_1\eta^{2p-2} + \cdots + a_{p-1}\eta^2 + a_p^2$	Prym$(\hat{S},\overline{\hat{S}})[2]$	[47]
$SO(p,p+q)$	$\eta^q(\eta^{2p} + a_1\eta^{2p-2} + \cdots + a_{p-1}\eta^2 + a_p)$	Abelian & non-abelian	[9]
$SO^*(2p)$	$\eta^{2p} + a_1\eta^{2p-2} + \cdots + a_{p-1}\eta^2 + a_p$	Subspace of $\mathcal{N}_2$(S)	[35, 47]

In the above table, we have considered the following notation, following the previous sections:

- S stands for a spectral curve and $\hat{S}$ for the normalized curve;
- $\overline{S}$ and $\overline{\hat{S}}$ denote the quotients of S and $\hat{S}$ by the involution $\eta \mapsto -\eta$;
- D denotes a positive divisor and f an extension class;
- $\mathcal{N}_2(S)$ is the moduli space of semi-stable rank 2 vector bundles on the spectral curve S.

Although Morse theoretic approaches (following [31], for a partial list, see [14, 15, 17] and references therein) are usually considered to study connectivity of the moduli space of G-Higgs bundles, spectral data may also be used to calculate the number of connected components of the moduli spaces of G-Higgs bundles. This approach was taken for the following groups[b]:

[b] In the table we give references for work done through spectral data, and in Remark 4 we mention the original sources of those results, when done previously with other methods.

Table 5. Connectivity for real Higgs bundles.

Group	restriction	components	Ref.
$PGL_i(n,\mathbb{R})$	$n=2$ and $i=0$	$2^{2g}+g-1$	[10]
	$n=2$ and $i=1$	$2^{2g}+g-2$	[10]
$PSL_i(n,\mathbb{R})$	$n=2$ and $i=0$	$2g-1$	[10]
	$n=2$ and $i=1$	$2g-2$	[10]
$GL(n,\mathbb{R})$	$n=2$	$3\cdot 2^{2g}+g-3$	[10]
$SL(n,\mathbb{R})$	$n=2$	$2\cdot 2^{2g}+2g-3$	[10, 46, 47]
$SU^*(2m)$	-	1	[35]
$SU(p,q)$	$p=q$, maximal, over generic loci	2^{2g}	[46, 47]
$Sp(2n,\mathbb{R})$	$n=2$, maximal	$3\cdot 2^{2g}+2g-4$	[10]
$Sp(2p,2q)$	$p=q$	1	[35, 47]
$SO(p,p)$	$p=2$	$3\cdot 6^{2g}+4g^2-6g-3$	[10]
$SO^*(2m)$	-	1	[35]

Some of the above connectivity results have been obtained before with other methods, some of which do not require the restrictions in the table:

- The number of components for $Sp(4,\mathbb{R})$ was obtained originally by Gothen in [27].
- The number of components $2.2^{2g}+2g-3$ for $SL(2,\mathbb{R})$ and $4g-3$ for $PSL(2,\mathbb{R})$ were shown by Goldman in [26]. Xia [54, 55] showed that the number of components of the space of homomorphisms $Hom(\pi_1(\Sigma), PSL(2,\mathbb{R}))$ is $2.2^{2g}+4g-5$. This number is different to the number $2.2^{2g}+2g-3$ of components of $PGL(2,\mathbb{R})$ because upon taking the quotient of the conjugation action of $PGL(2,\mathbb{R})$, certain pairs of components are identified.
- Connectivity for $SU(p,p)$-Higgs bundles was first studied in [14], and for $SU^*(2m)$ in [16] and for $SO^*(2m)$ in [17].

The remaining results mentioned in the above table have not, as far as we are aware, been obtained elsewhere.

Acknowledgments

The author would like to thank IMS at the National University of Singapore and the organizers of the program *The Geometry, Topology and Physics of Moduli Spaces of Higgs Bundles* for an ideal environment for research and collaborations, as well as Steve Bradlow for most useful comments on several drafts of the manuscript. The author also acknowledges support from U.S. National Science Foundation grants DMS 1107452, 1107263, 1107367 "sRNMS: GEometric structures And Representation varieties" (the GEAR

Network). Her research is also partially supported by the NSF grant DMS-1509693.

References

1. J. Adams, "Real Forms and Strong Real Forms Quick Reference Guide", Atlas Meeting, Palo Alto, July 2005.
2. J. Adams, "Strong real forms and the Kac classification", *At. of Lie Groups and Rep*, 2005.
3. J.E. Andersen, J. Grove, "Automorphism fixed points in the moduli space of semi-stable bundles", *Q. J. Math*, vol. 57, no. 1, pp. 1-35, 2006.
4. M.A. Arroyo, "The Geometry of $SO(p,q)$-Higgs Bundles", *Ph.D., Univ. de Salamanca,* 2009.
5. H. Aslaksen, "$SO(2)$ invariants of a set of 2×2 matrices", *MATH. SCAND.*, vol. 65, pp. 59-66, 1989.
6. M.F. Atiyah, R. Bott, "A Lefschetz Fixed Point Formula for Elliptic Complexes: II. Applications", *Ann. Math.*, vol. 88, no. 3, pp. 451-491, 1968.
7. D. Baraglia, L.P. Schaposnik, "Higgs bundles and (A,B,A)-branes", *Communications in Mathematical Physics*, vol. 331, pp. 1271-1300, 2014.
8. D. Baraglia, L.P. Schaposnik, "Real structures on moduli spaces of Higgs bundles", *Advances in Theoretical and Mathematical Physics*, vol. 20, no. 3, pp. 525–531, 2014.
9. D. Baraglia, L.P. Schaposnik (2015), "Cayley and Langlands type correspondences for orthogonal Higgs bundles", arXiv:1708.08828.
10. D. Baraglia, L.P. Schaposnik (2015), "Monodromy of rank 2 twisted Hitchin systems and real character varieties", *Trans. Amer. Math. Soc.*, DOI: https://doi.org/10.1090/tran/7144, Arxiv. 1506.00372.
11. A. Beauville, M.S. Narasimhan and S. Ramanan, "Spectral curves and the generalised theta divisor", *J. reine angew. Math.*, vol. 398, pp. 169-179, 1989.
12. I. Biswas, T.L. Gómez, "Connections and Higgs fields on a principal bundle", *Ann. Global Anal. Geom.*, vol. 33, no. 1, pp. 19-46, 2008.
13. I. Biswas, G. Wilkin (2015), "Anti-holomorphic involutive isometry of hyper-Kähler manifolds and branes", Journal of Geometry and Physics, 88, 52-55.
14. S. Bradlow, O. García-Prada, P.G. Gothen, "Surface group representations and $U(p,q)$-bundles", *J. Diff. Geom.*, vol. 64, pp. 111-170, 2003.
15. S. Bradlow, O. García-Prada, P.G. Gothen, "Maximal surface group representations in isometry groups of classical Hermitian symmetric spaces", *Geom. Ded.*, vol. 122, pp. 185-213, 2006.
16. O. García-Prada, A. Oliveira (2010), "Higgs bundles for the non-compact dual of the unitary group", Illinois journal of mathematics, 04/2010; 55(3).
17. S. Bradlow, O. García-Prada, P.G. Gothen, "Higgs bundles for the non-compact dual of the special orthogonal group", 2013. arXiv:math/1303.1058.
18. S.B. Bradlow, L.P. Schaposnik, "Higgs bundles and exceptional isogenies", *Res. Math. Sci.* (2016) 3:14; DOI:10.1186/s40687-016-0062-0, Arxiv. 1508.02650.

19. K. Corlette, "Flat G-bundles with canonical metrics", *J. Differential Geom.*, vol. 28, pp. 361-382, 1988.
20. R. Donagi, "Spectral Covers", *MSRI Series*, vol. 28, 1995.
21. R. Donagi, E. Markman, "Spectral covers, algebraically completely integrable, Hamiltonian systems, and moduli of bundles", *Integrable Systems and Quantum Groups*, 1993.
22. R. Donagi, T. Pantev, "Langlands duality for Hitchin systems", *Invent. Math.*, 2012.
23. S.K. Donaldson, "Twisted harmonic maps and the self-duality equations", *Proc. London Math. Soc.*, vol. 55, no. 3, pp. 127-131, 1987.
24. J. Fuchs, C. Schweigert, "Symmetries, Lie Algebras and Representations: A Graduate Course for Physicists", *Cambridge Monographs on Mathematical Physics*, Camb. Univ. Press, 2003.
25. O. García-Prada, P.B. Gothen, I. Mundet I. Riera, "The Hitchin-Kobayashi correspondence, Higgs pairs and surface group representations", 2009. arXiv:math/0909.4487.
26. W.M. Goldman, Topological components of spaces of representations. *Invent. Math.* **93** (1988), no. 3, 557-607.
27. P.B. Gothen, "The topology of Higgs bundle moduli spaces", *Ph.D. thesis*, University of Warwick, 1995.
28. P.B. Gothen, "Components of spaces of representations and stable triples", *Topology*, vol. 40, pp. 823-850, 2001.
29. P. Griffiths, J. Harris, "Principles of Algebraic Geometry", *Wiley Class. Library*, vol. 52, 2011.
30. T. Hausel and M. Thaddeus, "Mirror symmetry, Langlands duality, and the Hitchin system", *Invent. Math.*, vol. 153, pp. 197-229, 2003.
31. N.J. Hitchin, "The self-duality equations on a Riemann surface", *Proc. LMS*, vol. 55, no. 3, pp. 59-126, 1987.
32. N.J. Hitchin, "Stable bundles and integrable systems", *Duke Math.*, J. vol. 54, no. 1, pp. 91-114, 1987.
33. N.J. Hitchin, "Lie Groups and Teichmüller Space", *Topology*, vol. 31, no. 3, pp. 449-473, 1992.
34. N.J. Hitchin, "Langlands duality and $G2$ spectral curves", *Q. J. Math.*, vol. 58, pp. 319-344, 2007.
35. N.J. Hitchin, L.P. Schaposnik, "Nonabelianization of Higgs bundles", *J. Differential Geometry*, vol. 97, pp. 79-89, 2014.
36. N.J. Hitchin, "Higgs bundles and characteristic classes", 2013. arXiv:math/1308.4603.
37. S. Helgason, "Differential geometry, Lie groups, and symmetric spaces", *Graduate Studies in Mathematics, AMS*, vol. 34, 2001.
38. A. Kapustin, E. Witten, "Electric-magnetic duality and the geometric Langlands program", *Commun. Number Theory Phys.*, vol. 1, no. 1, pp. 1-236, 2007.
39. A.W. Knapp, "Lie groups beyond an introduction", *Progress in mathematics*, vol. 140, 2002.

40. N. Nitsure, "Moduli spaces of semistable pairs on a curve", *Proc. Lon. Math. Sot.*, vol. 62, pp. 275-300, 1991.
41. A.L. Onishchik, E.B. Vinberg, "Lie groups and Lie algebras III: structure of Lie groups and Lie algebras", *Encyclopaedia of mathematical sciences: Lie groups and Lie algebras*, Springer, vol. 3, 1994.
42. A. Peón, "Higgs bundles, real forms and the Hitchin fibration", *PhD Thesis*, UAM, 2013.
43. A. Peón (2015), Cameral data for $SU(p+1,p)$-Higgs bundles, arXiv:1506.01318.
44. A. Ramanathan, "Stable principal bundles on a compact Riemann surface", *Math. Ann.*, vol. 213, pp. 129-152, 1975.
45. A.A. Sagle, R.E. Walde, "Introduction to Lie groups and Lie algebras", *Pure and Applied Mathematics*, Academic Press, vol. 51, 1973.
46. L.P. Schaposnik, "Monodromy of the SL_2 Hitchin fibration", *Int. Math. Journal*, vol. 24, no. 2, 2013.
47. L.P. Schaposnik, "Spectral data for G-Higgs bundles", *PhD Thesis*, University of Oxford, 2013. arXiv:1301.1981.
48. L.P. Schaposnik, "Spectral data for U(m,m)-Higgs bundles", *Int. Math. Research Notices*, vol. 2015, no. 11, pp. 3486–3498, 2015, arXiv:1307.4419.
49. L.P. Schaposnik (2015), "Higgs bundles and applications", Oberwolfach Reports, Report No. 31, 31-33 (DOI: 10.4171/OWR/2015/3), arXiv:1603.06691.
50. L.P. Schaposnik, A geometric approach to orthogonal Higgs bundles, *Eur. J. Math.* (2017), https://doi.org/10.1007/s40879-017-0206-9, arXiv:1608.00300.
51. C.T. Simpson, "Constructing variations of Hodge structure using Yang-Mills theory and applications to uniformisation", *Journal of the AMS*, vol. 1, pp. 867-918, 1988.
52. C.T. Simpson, "Higgs bundles and local systems", *Publ. Math. I.H.E.S.*, vol. 75, pp. 5-95, 1992.
53. E.Z. Xia, "The moduli of flat $U(p,1)$ structures on Riemann surfaces", *Geometriae Dedicata*, vol. 97, pp. 33-43, 2003.
54. E.Z. Xia, Components of $Hom(\pi_1, PGL(2,\mathbb{R}))$. *Topology* **36** (1997), no. 2, 481-499.
55. E.Z. Xia, The moduli of flat $PGL(2,\mathbb{R})$ connections on Riemann surfaces. *Comm. Math. Phys.* **203** (1999), no. 3, 531-549.

Lectures on the Topological Recursion for Higgs Bundles and Quantum Curves

Olivia Dumitrescu
Department of Mathematics, Central Michigan University
Mount Pleasant, MI 48859, USA
and
Simion Stoilow Institute of Mathematics, Romanian Academy
21 Calea Grivitei Street, 010702 Bucharest, Romania
dumit1om@cmich.edu

Motohico Mulase
Department of Mathematics, University of California
Davis, CA 95616–8633, USA
and
Kavli Institute for Physics and Mathematics of the Universe
The University of Tokyo, Kashiwa, Japan
mulase@math.ucdavis.edu

This chapter aims at giving an introduction to the notion of *quantum curves*. The main purpose is to describe the new discovery of the relation between the following two disparate subjects: one is the topological recursion, that has its origin in random matrix theory and has been effectively applied to many enumerative geometry problems; and the other is the quantization of Hitchin spectral curves associated with Higgs bundles. Our emphasis is on explaining the motivation and examples. Concrete examples of the direct relation between Hitchin spectral curves and enumeration problems are given. A general geometric framework of quantum curves is also discussed.[a,b]

[a] *2010 Mathematics Subject Classification.* Primary: 14H15, 14N35, 81T45; Secondary: 14F10, 14J26, 33C05, 33C10, 33C15, 34M60, 53D37.
[b] *Keywords and phrases.* Topological quantum field theory; topological recursion; quantum curves; Hitchin spectral curves, Higgs bundles.

Contents

We're not going to tell you the story
the way it happened.
We're going to tell it
the way we remember it.

1. Introduction

Mathematicians often keep their childhood dream for a long time. When you saw a perfect rainbow as a child, you might have wondered what awaited you when you went over the arch. In a lucky situation, you might have seen the double, or even triple, rainbow arches spanning above the brightest one, with the reversing color patterns on the higher arches. Yet we see nothing underneath the brightest arch.

One of the purposes of these lectures is to offer you a vision: **on the other side of the rainbow, you see quantum invariants.** This statement describes only the tip of the iceberg. We believe something like the following is happening: Let C be a smooth projective curve over $\mathbb{C}$, and

$$\begin{array}{ccc}\Sigma & \xrightarrow{\;i\;} & \overline{T^*C} \\ & {\scriptstyle \pi}\searrow & \downarrow{\scriptstyle \pi} \\ & & C\end{array} \tag{1.1}$$

be an arbitrary Hitchin *spectral curve* associated with a particular *meromorphic* Higgs bundle (E, ϕ) on C. Then the asymptotic expansion at an essential singularity of a solution (flat section) of the $\hbar$-connection on C, that is the image of the **quantization** applied to Σ, carries the information of quantum invariants of a totally different geometric structure, which should be considered as the **mirror** to the geometric context (1.1).

In this introduction, we are going to tell you a story of an example to this mirror correspondence using the rainbow integral of Airy. The Hitchin spectral curve is a singular compactification of a parabola $x = y^2$ in a Hirzebruch surface. The corresponding quantum invariants, the ones hidden underneath the rainbow, are the cotangent class intersection numbers of the moduli space $\overline{\mathcal{M}}_{g,n}$. These numbers then determine the coefficients of the tautological relations among the generators of the tautological rings of $\mathcal{M}_g$ and $\overline{\mathcal{M}}_{g,n}$. The uniqueness of the asymptotic expansion relates the WKB analysis of the quantization of the parabola at infinity to the intersection numbers on $\overline{\mathcal{M}}_{g,n}$, through a combinatorial estimate of the *volume* of the moduli space $\mathcal{M}_{g,n}$.

The story begins in 1838.

1.1. *On the other side of the rainbow*

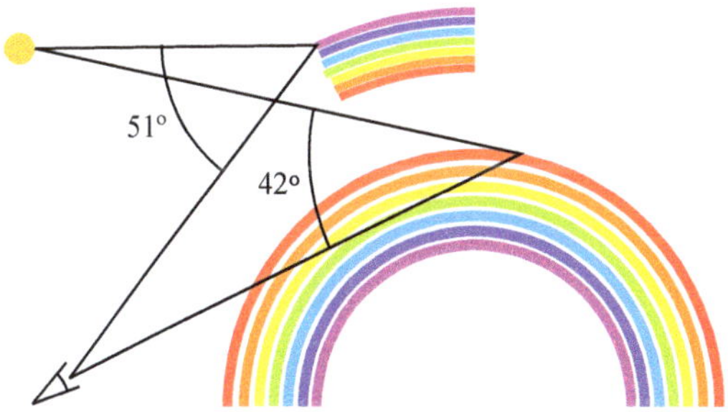

Fig. 1.1. Rainbow arches

Sir George Biddel Airy devised a simple formula, which he called the *rainbow integral*

$$Ai(x) = \frac{1}{2\pi}\int_{-\infty}^{\infty} e^{ipx} e^{i\frac{p^3}{3}}\, dp \tag{1.2}$$

and now carries his name, in his attempt of explaining the rainbow phenomena in terms of wave optics [2]. The angle between the sun and the observer measured at the brightest arch is always about 42°. The higher arches also have definite angles, independent of the rainbow. Airy tried to explain these angles and the brightness of the rainbow arches by the peak points of the rainbow integral.

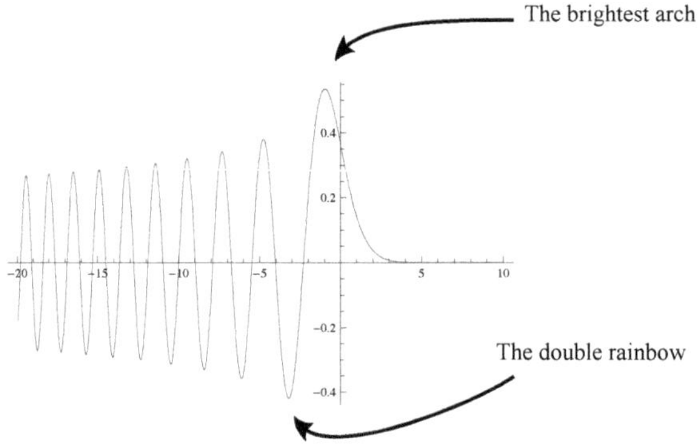

Fig. 1.2. The Airy function

We note that (1.2) is an *oscillatory integral*, and determines a real analytic function in $x \in \mathbb{R}$. It is easy to see, by integration by parts and taking care of the boundary contributions in oscillatory integral, that

$$\begin{aligned}\frac{d^2}{dx^2}Ai(x) &= \frac{1}{2\pi}\int_{-\infty}^{\infty}(-p^2)e^{ipx}e^{i\frac{p^3}{3}}\,dp = \frac{1}{2\pi}\int_{-\infty}^{\infty} e^{ipx}\left(i\frac{d}{dp}e^{i\frac{p^3}{3}}\right)dp \\ &= -\frac{1}{2\pi}\int_{-\infty}^{\infty}\left(i\frac{d}{dp}e^{ipx}\right)e^{i\frac{p^3}{3}}\,dp = \frac{1}{2\pi}\,x\int_{-\infty}^{\infty} e^{ipx}e^{i\frac{p^3}{3}}\,dp.\end{aligned}$$

Thus the Airy function satisfies a second order differential equation (known as the Airy differential equation)

$$\left(\frac{d^2}{dx^2} - x\right) Ai(x) = 0. \tag{1.3}$$

Now we consider $x \in \mathbb{C}$ as a complex variable. Since the coefficients of (1.3) are entire functions (they are just 1 and x), any solution of this differential equation is automatically entire, and has a convergent power series expansion

$$Ai(x) = \sum_{n=0}^{\infty} a_n x^n \tag{1.4}$$

at the origin with the radius of convergence ∞. Plugging (1.4) into (1.3), we obtain a recursion formula

$$a_{n+2} = \frac{1}{(n+2)(n+1)} a_{n-1}, \qquad n \geq 0, \tag{1.5}$$

with the initial condition $a_{-1} = 0$. Thus we find

$$a_{3n} = a_0 \cdot \frac{\prod_{j=1}^{n}(3j-2)}{(3n)!}, \qquad a_{3n+1} = a_1 \cdot \frac{\prod_{j=1}^{n}(3j-1)}{(3n+1)!}, \qquad a_{3n+2} = 0.$$

Here, a_0 and a_1 are arbitrary constants, so that the Airy differential equation has a two-dimensional space of solutions. These coefficients do not seem to be particularly interesting. The oscillatory integral (1.2) tells us that as $x \to +\infty$ on the real axis, the Airy function defined by the rainbow integral vanishes, because $e^{ipx+ip^3/3}$ oscillates so much that the integral cancels. More precisely, $Ai(x)$ satisfies a limiting formula

$$\lim_{x \to +\infty} \frac{Ai(x)}{\frac{1}{2\sqrt{\pi}} \cdot \frac{1}{\sqrt[4]{x}} \exp\left(-\frac{2}{3}x^{\frac{3}{2}}\right)} = 1. \tag{1.6}$$

Hence it exponentially decays, as $x \to +\infty$, $x \in \mathbb{R}$. Among the Taylor series solutions (1.4), there is *only one* solution that satisfies this exponential decay property, which is given by the following initial condition for (1.5):

$$a_0 = \frac{1}{3^{\frac{2}{3}}\Gamma(\frac{2}{3})}, \qquad a_1 = -\frac{1}{3^{\frac{1}{3}}\Gamma(\frac{1}{3})}.$$

The exponential decay on the positive real axis explains why we do not see any rainbows under the brightest arch. Then what do we really see underneath the rainbow? Or on the *other side* of the rainbow?

The differential equation (1.3) tells us that the Airy function has an essential singularity at $x = \infty$. Otherwise, the solution would be a polynomial in x, but (1.5) does not terminate at a finite n. How do we analyze the behavior of a holomorphic function at its essential singularity? And what kind of information does it tell us?

Definition 1.1: Asymptotic expansion. Let $f(z)$ be a holomorphic function defined on an open domain Ω of the complex plane $\mathbb{C}$ having the origin 0 on its boundary. A formal power series

$$\sum_{n=0}^{\infty} a_n z^n$$

is an *asymptotic expansion* of $f(z)$ on Ω at $z = 0$ if

$$\lim_{\substack{z\to 0\\ z\in\Omega}} \frac{1}{z^{m+1}}\left(f(z) - \sum_{n=0}^{m} a_n z^n\right) = a_{m+1} \tag{1.7}$$

holds for every $m \geq 0$.

The asymptotic expansion is a domain specific notion. For example, $f(z) = e^{-1/z}$ is holomorphic on $\mathbb{C}^* = \mathbb{C} \setminus \{0\}$, but it does not have any asymptotic expansion on all of $\mathbb{C}^*$. However, it has an asymptotic expansion

$$e^{-1/z} \sim 0$$

on

$$\Omega = \left\{ z \in \mathbb{C}^* \;\middle|\; |\mathrm{Arg}(z)| < \frac{\pi}{2} - \epsilon \right\}, \qquad \epsilon > 0.$$

If there is an asymptotic expansion of f on a domain Ω, then it is unique, and if $\Omega' \subset \Omega$ with $0 \in \partial\Omega'$, then obviously the same asymptotic expansion holds on Ω'.

The Taylor expansion (1.4) shows that $Ai(x)$ is real valued on the real axis, and from (1.6), we see that the value is positive for $x > 0$. Therefore, $\log Ai(x)$ is a holomorphic function on $Re(x) > 0$.

Theorem 1.2: *Asymptotic expansion of the Airy function. Define*

$$S_0(x) = -\frac{2}{3}x^{\frac{3}{2}}, \qquad S_1(x) = -\frac{1}{4}\log x - \log(2\sqrt{\pi}). \tag{1.8}$$

Then $\log Ai(x) - S_0(x) - S_1(x)$ *has the following asymptotic expansion on* $Re(x) > 0$.

$$\log Ai(x) - S_0(x) - S_1(x) = \sum_{m=2}^{\infty} S_m(x), \tag{1.9}$$

$$S_m(x) := x^{-\frac{3}{2}(m-1)} \cdot \frac{1}{2^{m-1}} \sum_{\substack{g\geq 0, n>0\\ 2g-2+n=m-1}} \frac{(-1)^n}{n!} \sum_{\substack{d_1+\cdots+d_n\\ =3g-3+n}} \langle \tau_{d_1}\cdots\tau_{d_n}\rangle_{g,n} \prod_{i=1}^{n} |2d_i - 1|!! \tag{1.10}$$

for $m \geq 2$. *The key coefficients are defined by*

$$\langle \tau_{d_1} \cdots \tau_{d_n} \rangle_{g,n} := \int_{\overline{\mathcal{M}}_{g,n}} \psi_1^{d_1} \cdots \psi_n^{d_n}, \tag{1.11}$$

where, $\overline{\mathcal{M}}_{g,n}$ *is the moduli space of stable curves of genus* g *with* n *smooth marked points. Let* $[C, (p_1, \ldots, p_n)] \in \overline{\mathcal{M}}_{g,n}$ *be a point of the moduli space. We can construct a line bundle* $\mathbb{L}_i$ *on the smooth Deligne-Mumford stack* $\overline{\mathcal{M}}_{g,n}$ *by attaching the cotangent line* $T^*_{p_i}C$ *at the point* $[C, (p_1, \ldots, p_n)]$ *of the moduli space. The symbol*

$$\psi_i = c_1(\mathbb{L}_i) \in H^2(\overline{\mathcal{M}}_{g,n}, \mathbb{Q})$$

denotes its first Chern class. Since $\overline{\mathcal{M}}_{g,n}$ *has dimension* $3g - 3 + n$, *the integral* (1.11) *is automatically* 0 *unless* $d_1 + \cdots + d_n = 3g - 3 + n$.

Surprisingly, on the other side of the rainbow, i.e., when $x > 0$, we see the intersection numbers (1.11)!

Remark 1.3: The relation between the Airy function and the intersection numbers was discovered by Kontsevich [55]. He replaces the variables x and p in (1.2) by *Hermitian matrices.* It is a general property that a large class of Hermitian matrix integrals satisfy an integrable system of KdV and KP type (see, for example, [62]). Because of the cubic polynomial in the integrand, the matrix Airy function of [55] counts trivalent ribbon graphs through the Feynman diagram expansion, which represent open dense subsets of $\overline{\mathcal{M}}_{g,n}$. By identifying the intersection numbers and the Euclidean volume of these open subsets, Kontsevich proves the Witten conjecture [86]. Our formulas (1.9) and (1.10) are a consequence of the results reported in [15,27,65,68]. We will explain the relation more concretely in these lectures.

Remark 1.4: The numerical value of the asymptotic expansion (1.9) is given by

$$\begin{aligned}
\log Ai(x) &= -\frac{2}{3}x^{\frac{3}{2}} - \frac{1}{4}\log x - \log(2\sqrt{\pi}) \\
&- \frac{5}{48}x^{-\frac{3}{2}} + \frac{5}{64}x^{-3} - \frac{1105}{9216}x^{-\frac{9}{2}} + \frac{565}{2048}x^{-6} - \frac{82825}{98304}x^{-\frac{15}{2}} + \frac{19675}{6144}x^{-9} \\
&- \frac{1282031525}{88080384}x^{-\frac{21}{2}} + \frac{80727925}{1048576}x^{-12} - \frac{1683480621875}{3623878656}x^{-\frac{27}{2}} + \cdots .
\end{aligned} \tag{1.12}$$

This follows from the WKB analysis of the Airy differential equation, which will be explained in this introduction.

Remark 1.5: Although the asymptotic expansion is not *equal* to the holomorphic function itself, we use the equality sign in these lectures to avoid further cumbersome notations.

The Airy differential equation appears in many different places, showing the feature of a *universal object* in the WKB analysis. It reflects the fact that the intersection numbers (1.11) are the most fundamental objects in Gromov-Witten theory. In contrast to the Airy differential equation, the gamma function is a universal object in the context of *difference equations.* We recall that

$$\Gamma(z+1) = z\Gamma(z),$$

and its asymptotic expansion for $Re(z) > 0$ is given by

$$\log \Gamma(z) = z \log z - z - \frac{1}{2} \log z + \frac{1}{2} \log(2\pi) + \sum_{m=1}^{\infty} \frac{B_{2m}}{2m(2m-1)} z^{-(2m-1)}, \tag{1.13}$$

where B_{2m} is the $(2m)$-th Bernoulli number defined by the generating function

$$\frac{x}{e^x - 1} = \sum_{n=0}^{\infty} \frac{B_n}{n!} x^n.$$

This is called **Stirling's formula**, and its main part gives the well-known approximation of the factorial:

$$n! \sim \sqrt{2\pi n}\, \frac{n^n}{e^n}.$$

The asymptotic expansion of the gamma function is deeply related to the moduli theory of algebraic curves. For example, from Harer and Zagier [45] we learn that the orbifold Euler characteristic of the moduli space of smooth algebraic curves is given by the formula

$$\chi(\mathcal{M}_{g,n}) = (-1)^{n-1} \frac{(2g-3+n)!}{(2g-2)!n!} \zeta(1-2g). \tag{1.14}$$

Here, the special value of the Riemann zeta function is the Bernoulli number

$$\zeta(1-2g) = -\frac{B_{2g}}{2g}.$$

The expression (1.14) is valid for $g = 0, n \geq 3$ if we use the gamma function for $(2g-2)!$. Stirling's formula (1.13) is much simpler than the expansion of $\log Ai(x)$. As the prime factorization of one of the coefficients

$$\frac{1683480621875}{3623878656} = \frac{5^5 \cdot 13 \cdot 17 \cdot 2437619}{2^{27} \cdot 3^3}$$

shows, we do not expect any simple closed formula for the coefficients of (1.12), like Bernoulli numbers. Amazingly, still there is a close relation between these two asymptotic expansions (1.12) and (1.13) through the work on *tautological relations* of Chow classes on the moduli space $\mathcal{M}_g$ by Mumford [70], followed by recent exciting developments on the Faber-Zagier conjecture [36, 49, 77]. In Theorem 2.7, we will see yet another close relationship between the Euler characteristic of $\mathcal{M}_{g,n}$ and the intersection numbers on $\overline{\mathcal{M}}_{g,n}$, through two special values of the *same* function.

The asymptotic expansion of the Airy function $Ai(x)$ itself for $Re(x) > 0$ has actually a rather simple expression:

$$\begin{aligned} Ai(x) &= \frac{e^{-\frac{2}{3}x^{\frac{3}{2}}}}{2\sqrt{\pi}x^{\frac{1}{4}}} \sum_{m=0}^{\infty} \frac{\left(-\frac{3}{4}\right)^m \Gamma\left(m+\frac{5}{6}\right)\Gamma\left(m+\frac{1}{6}\right)}{2\pi m!} x^{-\frac{3}{2}m} \\ &= \frac{e^{-\frac{2}{3}x^{\frac{3}{2}}}}{2\sqrt{\pi}x^{\frac{1}{4}}} \sum_{m=0}^{\infty} (-1)^m \left(\frac{1}{576}\right)^m \frac{(6m)!}{(2m)!(3m)!} x^{-\frac{3}{2}m}. \end{aligned} \tag{1.15}$$

The expansion in terms of the gamma function values of the first line of (1.15) naturally arises from a hypergeometric function. The first line is equal to the second line because

$$\Gamma(z)\Gamma(1-z) = \frac{\pi}{\sin \pi z},$$

and induction on m. Since the $m = 0$ term in the summation is 1, we can apply the formal logarithm expansion

$$\log(1-X) = -\sum_{j=1}^{\infty} \frac{1}{j} X^j$$

to (1.15) with

$$X = -\sum_{m=1}^{\infty} (-1)^m \left(\frac{1}{576}\right)^m \frac{(6m)!}{(2m)!(3m)!} x^{-\frac{3}{2}m},$$

and obtain

$$\begin{aligned} \log Ai(x) &= -\frac{2}{3}x^{\frac{3}{2}} - \frac{1}{4}\log x - \log(2\sqrt{\pi}) \\ &\quad - \frac{1}{576}\frac{6!}{2!3!}x^{-\frac{3}{2}} + \left(\frac{1}{576}\right)^2 \left(\frac{12!}{4!6!} - \frac{1}{2}\left(\frac{6!}{2!3!}\right)^2\right) x^{-3} \\ &\quad - \left(\frac{1}{576}\right)^3 \left(\frac{18!}{6!9!} - \frac{12!}{4!6!}\cdot\frac{6!}{2!3!} + \frac{1}{3}\left(\frac{6!}{2!3!}\right)^3\right) x^{-\frac{9}{2}} + \cdots. \end{aligned} \tag{1.16}$$

In general, for $m \geq 1$, we have

$$S_{m+1}(x) = (-1)^m x^{-\frac{3}{2}m} \left(\frac{1}{576}\right)^m \sum_{\lambda \vdash m} (-1)^{\ell(\lambda)-1} \frac{(\ell(\lambda)-1)!}{|\mathrm{Aut}(\lambda)|} \prod_{i=1}^{\ell(\lambda)} \frac{(6\lambda_i)!}{(2\lambda_i)!(3\lambda_i)!}, \tag{1.17}$$

where λ is a partition of m, $\ell(\lambda)$ its length, and $\mathrm{Aut}(\lambda)$ is the group of permutations of the parts of λ of equal length.

Comparing (1.10) and (1.17), we establish concrete relations among the intersection numbers.

Theorem 1.6: *Rainbow formula. The cotangent class intersection numbers of $\overline{\mathcal{M}}_{g,n}$ satisfy the following relation for every $m \geq 1$:*

$$\begin{aligned} &\sum_{\substack{g \geq 0, n > 0 \\ 2g-2+n=m}} \frac{1}{n!} \sum_{\substack{d_1+\cdots+d_n \\ =3g-3+n}} \langle \tau_{d_1} \cdots \tau_{d_n} \rangle_{g,n} \prod_{i=1}^{n} |2d_i - 1|!! \\ &= \left(\frac{1}{288}\right)^m \sum_{\lambda \vdash m} (-1)^{\ell(\lambda)-1} \frac{(\ell(\lambda)-1)!}{|Aut(\lambda)|} \prod_{i=1}^{\ell(\lambda)} \frac{(6\lambda_i)!}{(2\lambda_i)!(3\lambda_i)!}. \end{aligned} \tag{1.18}$$

Here, we use the fact that $(-1)^n = (-1)^m$ if $2g-2+n = m$. For example, for $m = 2$, we have

$$\frac{1}{6}\langle \tau_0^3 \tau_1 \rangle_{0,4} + \frac{1}{2}\langle \tau_1^2 \rangle_{1,2} + 3\langle \tau_0 \tau_2 \rangle_{1,2} = \left(\frac{1}{288}\right)^2 \left(\frac{12!}{4!6!} - \frac{1}{2}\left(\frac{6!}{2!3!}\right)^2\right) = \frac{5}{16}.$$

This can be verified by evaluating

$$\langle \tau_0^3 \tau_1 \rangle_{0,4} = 1, \qquad \langle \tau_1^2 \rangle_{1,2} = \langle \tau_0 \tau_2 \rangle_{1,2} = \frac{1}{24}.$$

Remark 1.7: The main purpose of these lectures is to relate the topological recursion of [34] and quantization of Hitchin spectral curves. The left-hand side of (1.18) represents the topological recursion in this example, since the intersection numbers can be computed through this mechanism, as explained below. Actually, this is an important example that leads to the universal structure of the topological recursion. The right-hand side is the asymptotic expansion of a function that is coming from the geometry of the Hitchin spectral curve of a Higgs bundle.

The structure of the cohomology ring (or more fundamental Chow ring) of the moduli space $\overline{\mathcal{M}}_{g,n}$, and its open part $\mathcal{M}_{g,n}$ consisting of smooth curves, attracted much attention since the publication of the inspiring paper by Mumford [70] mentioned above. Let us focus on a simple situation

$$\pi : \mathcal{M}_{g,1} \longrightarrow \mathcal{M}_g,$$

which *forgets* the marked point on a smooth curve. By gluing the canonical line bundle of the fiber of each point on the base $\mathcal{M}_g$, which is the curve represented by the point on the moduli, we obtain the relative dualizing sheaf ω on $\mathcal{M}_{g,1}$. Its first Chern class, considered as a divisor on $\mathcal{M}_{g,1}$ and an element of the Chow group $A^1(\mathcal{M}_{g,1})$, is denoted by ψ. In the notation of (1.11), this is the same as ψ_1. Mumford defines *tautological* classes

$$\kappa_a := \pi_*(\psi^{a+1}) \in A^a(\mathcal{M}_g).$$

One of the statements of the Faber-Zagier conjecture of [36], now a theorem due to Ionel [49], says the following.

Conjecture 1.8: A part of the Faber-Zagier Conjecture [36] define rational numbers $a_j \in \mathbb{Q}$ by

$$\sum_{j=1}^{\infty} a_j t^j = -\log\left(\sum_{m=0}^{\infty} \frac{(6m)!}{(2m)!(3m)!} t^m\right). \tag{1.19}$$

Then the coefficient of t^ℓ of the expression

$$\exp\left(\sum_{j=1}^{\infty} a_j \kappa_j t^j\right) \in \big(\mathbb{Q}[\kappa_1, \kappa_2, \dots]\big)[[t]]$$

for each $\ell \geq 1$ gives the unique codimension ℓ tautological relation among the κ-classes on the moduli space $\mathcal{M}_{3\ell-1}$.

We see from (1.18), these coefficients a_j's are given by the intersection numbers (1.11), by a change of the variables

$$t = -\frac{1}{576}\, x^{-\frac{3}{2}}.$$

Indeed, we have

$$a_j = -288^j \sum_{\substack{g\geq 0, n>0\\ 2g-2+n=j}} \frac{1}{n!} \sum_{\substack{d_1+\cdots+d_n\\ =3g-3+n}} \langle \tau_{d_1}\cdots\tau_{d_n}\rangle_{g,n} \prod_{i=1}^{n} |2d_i - 1|!!\,. \tag{1.20}$$

These tautological relations are generalized for the moduli spaces $\overline{\mathcal{M}}_{g,n}$, and are proved in [77]. Amazingly, still the rainbow integral (1.2) plays the essential role in determining the tautological relations in this generalized work.

Remark 1.9: As we have seen above, the asymptotic expansions of the gamma function and the Airy function carry information of quantum invariants, in particular, certain topological information of $\mathcal{M}_{g,n}$ and $\overline{\mathcal{M}}_{g,n}$. We note that these quantum invariants are stored in the **insignificant** part of the asymptotic expansions.

Here come questions.

Question 1.10: The Airy function is a one-variable function. It cannot be a generating function of all the intersection numbers (1.11). Then how do we obtain all intersection numbers from the Airy function, or the Airy differential equation, alone?

Question 1.11: The relations between the Airy function, the gamma function, and intersection numbers are all great. But then how does this relation have anything to do with **Higgs bundles**?

Question 1.12: As we remarked, the information of the quantum invariants is stored in the insignificant part of the asymptotic expansion. In the Airy example, they correspond to $S_m(x)$ of (1.10) for $m \geq 2$. Then what does the main part of the asymptotic behavior of the function, i.e., those functions in (1.8), determine?

As we have remarked earlier, Kontsevich [55] utilized matrix integral techniques to answer Question 1.10. The key idea is to replace the variables in (1.2) by Hermitian matrices, and then use the asymptotic expansion on the result. Through the Feynman diagram expansion, he was able to obtain a generating function of all the intersection numbers.

What we explain in these lectures is the concept of **topological recursion** of [34]. Without going into matrix integrals, we can directly obtain (a totally different set of) generating functions of the intersection numbers from the Airy function. Here, the Airy differential equation is identified as a **quantum curve**, and application of the **semi-classical limit** and the topological recursion enable us to calculate generating functions of the intersection numbers.

But before going into detail, let us briefly answer Question 1.11 below. The point is that the geometry of the Airy function is a special example of Higgs bundles.

For the example of the Airy differential equation, the topological recursion is exactly the same as the *Virasoro constraint conditions* for the intersection numbers (1.11), and the semi-classical limit recovers the **Hitchin spectral curve** of the corresponding Higgs bundle. The information stored in the main part of the asymptotic expansion (1.8), as asked in Question 1.12, actually determines the spectral curve and its geometry. We can turn the story in the other way around: we will see that the functions $S_0(x)$ and $S_1(x)$ corresponding to (1.8) in the general context are indeed determined by the geometry of the Hitchin spectral curve of an appropriate Higgs bundle.

The stage setting is the following. As a base curve, we have $\mathbb{P}^1$. On this curve we have a vector bundle

$$E = \mathcal{O}_{\mathbb{P}^1}(-1) \oplus \mathcal{O}_{\mathbb{P}^1}(1) = K_{\mathbb{P}^1}^{\frac{1}{2}} \oplus K_{\mathbb{P}^1}^{-\frac{1}{2}} \tag{1.21}$$

of rank 2. The main character of the Second Act is a *meromorphic* Higgs field

$$\phi = \begin{bmatrix} & x(dx)^2 \\ 1 & \end{bmatrix} : E \longrightarrow K_{\mathbb{P}^1}(5) \otimes E. \tag{1.22}$$

Here, x is an affine coordinate of $\mathbb{A}^1 \subset \mathbb{P}^1$, 1 on the $(2,1)$-component of ϕ is the natural morphism

$$1 : K_{\mathbb{P}^1}^{\frac{1}{2}} \xrightarrow{=} K_{\mathbb{P}^1}^{\frac{1}{2}} \longrightarrow K_{\mathbb{P}^1}^{\frac{1}{2}} \otimes \mathcal{O}_{\mathbb{P}^1}(5),$$

and

$$x(dx)^2 \in H^0\big(\mathbb{P}^1, K_{\mathbb{P}^1}^{\otimes 2}(5-1)\big) \cong \mathbb{C}$$

is the unique (up to a constant factor) meromorphic *quadratic differential* on $\mathbb{P}^1$ that has one zero at $x = 0$ and a pole of order 5 at $x = \infty$. We use K_C to denote the canonical sheaf on a projective algebraic curve C. The data (E, ϕ) is called a *Higgs pair*. Although ϕ contains a quadratic differential in its component, because of the shape of the vector bundle E, we see that

$$E = K_{\mathbb{P}^1}^{\frac{1}{2}} \oplus K_{\mathbb{P}^1}^{-\frac{1}{2}} \xrightarrow{\phi} \left(K_{\mathbb{P}^1}^{\frac{3}{2}} \oplus K_{\mathbb{P}^1}^{\frac{1}{2}}\right) \otimes \mathcal{O}_{\mathbb{P}^1}(5) = K_{\mathbb{P}^1}(5) \otimes \left(K_{\mathbb{P}^1}^{\frac{1}{2}} \oplus K_{\mathbb{P}^1}^{-\frac{1}{2}}\right),$$

hence

$$\phi \in H^0\big(\mathbb{P}^1, K_{\mathbb{P}^1}(5) \otimes \mathrm{End}(E)\big)$$

is indeed an $\mathrm{End}(E)$-valued meromorphic 1-form on $\mathbb{P}^1$.

The cotangent bungle

$$\pi : T^*\mathbb{P}^1 \longrightarrow \mathbb{P}^1$$

is the total space of $K_{\mathbb{P}^1}$. Therefore, the pull-back bundle $\pi^* K_{\mathbb{P}^1}$ has a tautological section $\eta \in H^0(T^*\mathbb{P}^1, \pi^* K_{\mathbb{P}^1})$, which is a globally defined holomorphic 1-form on $T^*\mathbb{P}^1$. The global holomorphic 2-form $-d\eta$ gives the holomorphic symplectic structure, hence a hyper-Kähler structure, on $T^*\mathbb{P}^1$. If we trivialize the cotangent bundle on the affine neighborhood of $\mathbb{P}^1$ with a coordinate x, and use a fiber coordinate y, then $\eta = ydx$. We wish to define the spectral curve of this Higgs pair. Due to the fact that ϕ is singular at $x = \infty$, we cannot capture the whole story within the cotangent bundle. We note that the cotangent bundle $T^*\mathbb{P}^1$ has a natural compactification

$$\overline{T^*\mathbb{P}^1} := \mathbb{P}(K_{\mathbb{P}^1} \oplus \mathcal{O}_{\mathbb{P}^1}) = \mathbb{F}_2 \xrightarrow{\pi} \mathbb{P}^1,$$

which is known as a Hirzebruch surface. The holomorphic 1-form η extends to $\overline{T^*\mathbb{P}^1}$ as a meromorphic 1-form with simple poles along the divisor at infinity.

Now we can consider the characteristic polynomial

$$\det(\eta - \pi^*\phi) \in H^0\left(\overline{T^*\mathbb{P}^1}, \pi^*\left(K_{\mathbb{P}^1}^{\otimes 2}(5)\right)\right)$$

as a meromorphic section of the line bundle $\pi^* K_{\mathbb{P}^1}^{\otimes 2}$ on the compact space $\overline{T^*\mathbb{P}^1}$. It defines the Hitchin spectral curve

$$\Sigma = \left(\det(\eta - \pi^*\phi)\right)_0 \subset \overline{T^*\mathbb{P}^1} \tag{1.23}$$

as a divisor. Again in terms of the local coordinate (x, y) of $\overline{T^*\mathbb{P}^1}$, the spectral curve Σ is simply given by

$$x = y^2. \tag{1.24}$$

It is a perfect parabola in the (x, y)-plane. But our Σ is in the Hirzebruch surface, not in the projective plane. Choose the coordinate $(u, w) \in \mathbb{F}_2$ around $(x, y) = (\infty, \infty)$ defined by

$$\begin{cases} u = \frac{1}{x} \\ \frac{1}{w} du = ydx \end{cases} . \tag{1.25}$$

Then the local expression of Σ around $(u, w) = (0, 0)$ becomes a quintic cusp

$$w^2 = u^5. \tag{1.26}$$

So the spectral curve Σ is indeed highly singular at infinity!

1.2. *Quantum curves, semi-classical limit, and the WKB analysis*

At this stage we have come to the point of introducing the notion of *quantization.* We wish to quantize the spectral curve Σ of (1.23). In terms of the affine coordinate (x, y), the **quantum curve** of $y^2 - x = 0$ should be the Airy differential equation

$$\left(\left(\hbar\frac{d}{dx}\right)^2 - x\right)\Psi(x,\hbar) = 0. \tag{1.27}$$

This is the Weyl quantization, in which we change the commutative algebra $\mathbb{C}[x, y]$ to a Weyl algebra $\mathbb{C}[\hbar]\langle x, y\rangle$ defined by the commutation relation

$$[x, y] = -\hbar. \tag{1.28}$$

We consider $x \in \mathbb{C}[\hbar]\langle x, y\rangle$ as the multiplication operator by the coordinate x, and $y = \hbar\frac{d}{dx}$ as a differential operator.

How do we know that (1.27) is the right quantization of the spectral curve (1.24)? Apparently, the limit $\hbar \to 0$ of the differential operator does not reproduce the spectral curve. Let us now recall the WKB method for analyzing differential equations like (1.27). This is a method that relates classical mechanics and quantum mechanics. As we see below, the WKB method is not for finding a convergent analytic solution to the differential equation. Since the equation we wish to solve is considered to be a quantum equation, the corresponding classical problem, if it exists, should be recovered by taking $\hbar \to 0$. We denote by an unknown function $S_0(x)$ the "solution" to the corresponding classical problem. To emphasize the classical behavior at the $\hbar \to 0$ limit, we expand the solution to the quantum equation as

$$\Psi(x,\hbar) = \exp\left(\sum_{m=0}^{\infty}\hbar^{m-1}S_m(x)\right) := \exp\left(\frac{1}{\hbar}S_0(x)\right)\cdot\exp\left(\sum_{m=1}^{\infty}\hbar^{m-1}S_m(x)\right). \tag{1.29}$$

The idea is that as $\hbar \to 0$, the effect of $S_0(x)$ is magnified. But as a series in $\hbar$, (1.29) is ill defined because the coefficient of each power of $\hbar$ is an infinite sum. It is also clear that $\hbar \to 0$ does not make sense for $\Psi(x, \hbar)$. Instead of expanding (1.29) immediately in $\hbar$ and take its 0 limit, we use the following standard procedure. First we note that (1.27) is equivalent to

$$\left[\exp\left(-\frac{1}{\hbar}S_0(x)\right)\cdot\left(\left(\hbar\frac{d}{dx}\right)^2 - x\right)\cdot\exp\left(\frac{1}{\hbar}S_0(x)\right)\right]\exp\left(\sum_{m=1}^{\infty}\hbar^{m-1}S_m(x)\right) = 0.$$

Since the conjugate differential operator

$$\exp\left(-\frac{1}{\hbar}S_0(x)\right)\cdot\left(\left(\hbar\frac{d}{dx}\right)^2 - x\right)\cdot\exp\left(\frac{1}{\hbar}S_0(x)\right)$$
$$= \left(\hbar\frac{d}{dx}\right)^2 + 2\hbar S_0'(x)\frac{d}{dx} + \left(S_0'(x)\right)^2 - x + \hbar S_0''(x)$$

is a well-defined differential operator, *and* its limit $\hbar \to 0$ makes sense, we interpret (1.27) as the following differential equation:

$$\left[\left(\hbar\frac{d}{dx}\right)^2 + 2\hbar S_0'(x)\frac{d}{dx} + \left(S_0'(x)\right)^2 - x + \hbar S_0''(x)\right]\exp\left(\sum_{m=1}^{\infty}\hbar^{m-1}S_m(x)\right) = 0. \tag{1.30}$$

Here, $'$ indicates the x-derivative. This equation is equivalent to

$$\left(\sum_{m=0}^{\infty}\hbar^m S_m'(x)\right)^2 + \sum_{m=0}^{\infty}\hbar^{m+1}S_m''(x) - x = 0 \tag{1.31}$$

for every $m \geq 0$. The coefficient of the $\hbar^0$, or the $\hbar \to 0$ limit of (1.31), then gives

$$\left(S_0'(x)\right)^2 - x = 0, \tag{1.32}$$

and that of $\hbar^1$ gives

$$S_0''(x) + 2S_0'(x)S_1'(x) = 0. \tag{1.33}$$

The $\hbar^0$ term is what we call the **semi-classical limit** of the differential equation (1.27). From (1.32) we obtain

$$S_0(x) = \pm\frac{2}{3}x^{\frac{3}{2}} + c_0, \tag{1.34}$$

with a constant of integration c_0. Then plugging $S_0(x)$ into (1.33) we obtain

$$S_1(x) = -\frac{1}{4}\log x - \log(2\sqrt{\pi}) + c_1,$$

again with a constant of integration c_1. Note that these solutions are consistent with (1.8). For $m \geq 1$, the coefficient of $\hbar^{m+1}$ gives

$$S_{m+1}'(x) = -\frac{1}{2S_0'(x)}\left(S_m''(x) + \sum_{a=1}^{m}S_a'(x)S_{m+1-a}'(x)\right), \tag{1.35}$$

which can be solved recursively, term by term from $S_0(x)$. This mechanism is the method of Wentzel-Kramers-Brillouin (WKB) approximation.

We can ignore the constants of integration when solving (1.35) because it is easy to restore them, if necessary, just by adding c_m to each $S_m(x)$ in (1.29). The solution then simply changes to another one

$$\left(\exp\left(\frac{1}{\hbar}c_0\right)\exp\left(\frac{1}{\hbar}S_0(x)\right)\right)\cdot\exp\left(\sum_{m=1}^{\infty}\hbar^{m-1}c_m\right)\exp\left(\sum_{m=1}^{\infty}\hbar^{m-1}S_m(x)\right).$$

In terms of the main variable x, the above solution is a constant multiple of the original one. The two choices of the sign in (1.34) lead to two linearly independent solutions of (1.27). If we impose

$$\lim_{x\to\infty} S_m(x) = 0, \qquad m \geq 2, \tag{1.36}$$

then the differential equation (1.35) uniquely determines all terms $S_m(x)$. Thus, with the choice of the negative sign in (1.34) and imposing $c_0 = c_1 = 0$ and (1.36), we obtain the unique exponentially decaying solution for $x \to \infty$ along the real line. This solution agrees with (1.10) and (1.17). Thus we obtain the second line of the Rainbow formula (1.18).

We also see from the semi-classical limit (1.32) that if we put $y = S_0'(x)$, then we recover the Hitchin spectral curve $x = y^2$. The functions $S_m(x)$ actually live on the spectral curve rather than the base $\mathbb{P}^1$, because of the appearance of $\sqrt{x}$ in (1.10).

Remark 1.13: One can ask a question: **Does** (1.29) **give a convergent solution?** The answer is a flat **No!** Suppose we solve the Airy differential equation with the WKB method explained above, and define a "solution" by (1.29). Expand the second exponential factor as a power series in $\hbar$, and write the solution as

$$\Psi(x,\hbar) = \exp\left(\frac{1}{\hbar}S_0(x)\right)\sum_{n=0}^{\infty} f_n(x)\hbar^n.$$

Then for any compact subset $K \subset \mathbb{C} \setminus \{0\}$, there is a constant C_K such that

$$\sup_{x\in K} |f_n(x)| \leq C_K^n n!.$$

Therefore, unless we are in an extremely special case, the second exponential factor in the expression (1.29) does not converge as a power series in $\hbar$ at all! Since

$$\Psi(x,\hbar) := Ai\left(x/\hbar^{\frac{2}{3}}\right)$$

is a solution that is entire in x and any $\hbar$ for which $1/\hbar^{\frac{2}{3}}$ makes sense, the WKB method around $x \in \mathbb{C} \setminus \{0\}$ is the same as the asymptotic expansion of the Airy function $Ai(x)$ given in (1.15). There we see the factorial growth of the coefficients. Thus the WKB method is **not** for finding a convergent analytic solution.

1.3. *The topological recursion as quantization*

Then what is good about the WKB method and the purely formal solution (1.29)? Let us examine (1.10). We note that $S_m(x)$s are one variable functions, and different values of g and n are summed in its definition. Therefore, knowing the solution $\Psi(x, \hbar)$ of (1.27) that decays exponentially as $x \to \infty$ along the real axis, assuming $\hbar > 0$, does not seem to possibly recover intersection numbers $\langle \tau_{d_1} \cdots \tau_{d_n} \rangle_{g,n}$ for all values of $(d_1, \dots, d_n)$ and (g, n). Then how much information does the quantum curve (1.27) really have?

Here comes the idea of **topological recursion** of Eynard and Orantin [34]. This mechanism gives a refined expression of each $S_m(x)$, and computes all intersection numbers. The solution Ψ of (1.29) is never holomorphic, and it makes sense only as the asymptotic expansion of a holomorphic solution at its essential singularity. The expansion of the Airy function $Ai(x)$ at a holomorphic point does not carry any interesting information. The function's key information is concentrated in the expansion at the essential singularity. The topological recursion is for obtaining this hidden information when applied at the essential singularity of the solution, by giving an explicit formula for the WKB expansion. And the WKB analysis is indeed a method that determines the relation between the quantum behavior and the classical behavior of a given system, i.e., the process of quantization.

As we have seen above, the quantum curve (1.27) recovers the spectral curve (1.24) by the procedure of semi-classical limit. We recall that the spectral curve lives in the Hirzebruch surface $\mathbb{F}_2$, and it has a quintic cusp singularity (1.26) at $(x, y) = (\infty, \infty)$. It requires two blow-ups of $\mathbb{F}_2$ to resolve the singularity of Σ. Let us denote this minimal resolution by $\widetilde{\Sigma}$. The proper transform is a smooth curve of genus 0, hence it is a $\mathbb{P}^1$. Let $B \cong \mathbb{P}^1$ be the 0-section, and $F \cong \mathbb{P}^1$ a fiber, of $\mathbb{F}_2$. Then after two blow-ups, $\widetilde{\Sigma} \subset Bl(\mathbb{F}_2)$ is identified as a divisor by the equation

$$\widetilde{\Sigma} = 2B + 5F - 4E_2 - 2E_1 \in \text{Pic}\big(Bl(\mathbb{F}_2)\big),$$

where E_i is the exceptional divisor introduced at the i-th blow-up [25, Section 5].

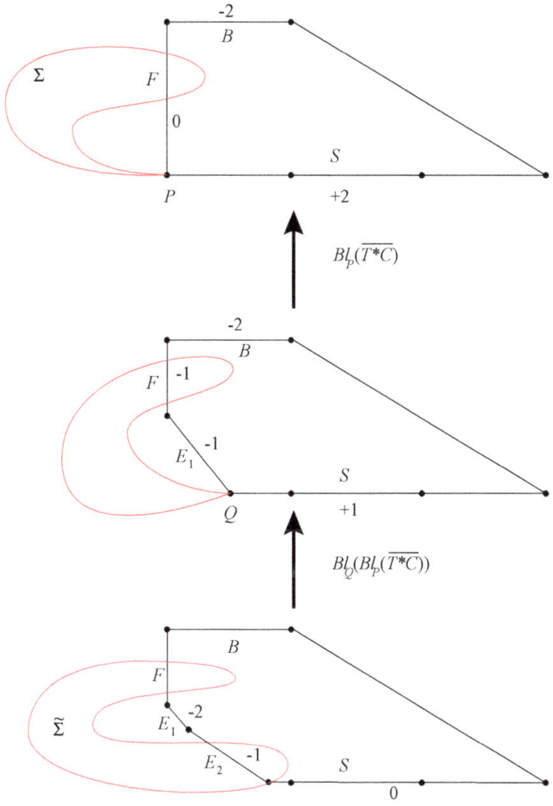

Fig. 1.3. Blowing up $\mathbb{F}_2$ twice. The parabola in $\mathbb{F}_2$ has a quintic cusp singularity at infinity (top). After the second blow-up, the proper transform becomes non-singular (bottom). Since all fibers of $\mathbb{F}_2 \longrightarrow \mathbb{P}^1$ are equivalent, the toric picture has only one representative of the fiber $F = \mathbb{P}^1$. It is a stretch to place Σ and $\widetilde{\Sigma}$ in this diagram, because the spectral curve is a double cover of the base B, rather than a degree 4 covering that the picture may suggest.

Since the desingularization $\widetilde{\Sigma}$ is just a copy of a $\mathbb{P}^1$, we can choose a **normalization coordinate** t on it so that the map $\tilde{\pi} : \widetilde{\Sigma} \longrightarrow \mathbb{P}^1$ to the base curve $\mathbb{P}^1$ is given by

$$\begin{cases} x = \frac{4}{t^2} \\ y = -\frac{2}{t} \end{cases}, \qquad \begin{cases} u = \frac{t^2}{4} \\ w = \frac{t^5}{32} \end{cases}. \tag{1.37}$$

With respect to the normalization coordinate, define a homogeneous polynomial of degree $6g-6+3n$ for $2g-2+n>0$ by

$$F^A_{g,n}(t_1,\dots,t_n) := \frac{(-1)^n}{2^{2g-2+n}} \sum_{\substack{d_1+\cdots+d_n\\=3g-3+n}} \langle \tau_{d_1}\cdots\tau_{d_n}\rangle_{g,n} \prod_{i=1}^n |2d_i-1|!! \left(\frac{t_i}{2}\right)^{2d_i+1} \tag{1.38}$$

as a function on $(\widetilde{\Sigma})^n$, and an n-linear differential form

$$W^A_{g,n}(t_1,\dots,t_n) := d_{t_1}\cdots d_{t_n} F^A_{g,n}(t_1,\dots,t_n). \tag{1.39}$$

For *unstable geometries* $(g,n)=(0,1)$ and $(0,2)$, we need to define differential forms separately:

$$W^A_{0,1}(t) := \eta = y\,dx = \frac{16}{t^4}dt, \tag{1.40}$$

$$W^A_{0,2}(t_1,t_2) := \frac{dt_1\cdot dt_2}{(t_1-t_2)^2}. \tag{1.41}$$

The definition of $W^A_{0,1}$ encodes the geometry of the singular spectral curve $\Sigma\subset\mathbb{F}_2$ embedded in the Hirzebruch surface, and $W^A_{0,2}$ depends only on the intrinsic geometry of the normalization $\widetilde{\Sigma}$. Then we have

Theorem 1.14: *Topological recursion for the intersection numbers, [27]*

$$\begin{aligned} &W^A_{g,n}(t_1,\dots,t_n)\\ &= -\frac{1}{2\pi i}\int_\gamma \left(\frac{1}{t+t_1}+\frac{1}{t-t_1}\right)\frac{t^4}{64}\frac{1}{dt}\,dt_1 \Bigg[W^A_{g-1,n+1}(t,-t,t_2,\dots,t_n)\\ &\qquad + \sum_{\substack{g_1+g_2=g\\ I\sqcup J=\{2,\dots,n\}}}^{\text{no }(0,1)\text{ terms}} W^A_{g_1,|I|+1}(t,t_I)W^A_{g_2,|J|+1}(-t,t_J)\Bigg], \end{aligned} \tag{1.42}$$

where the integral is taken with respect to the contour of Figure 1.4, and the sum is over all partitions of g and set partitions of $\{2,3,\dots,n\}$ without including $g_1=0$ and $I=\emptyset$, or $g_2=0$ and $J=\emptyset$. The notation $\frac{1}{dt}$ represents the ratio operation, which acts on a differential 1-form to produce a global meromorphic function. When acted on the quadratic differential, $\frac{1}{dt}$ yields a 1-form.

This is an example of the axiomatic mechanism of [34] called the *topological recursion*. We give a more geometric interpretation of the formula in the later sections. The derivation of the topological recursion (1.42) from (1.38) is the subject of Section 2 where the origin of the recursion formula

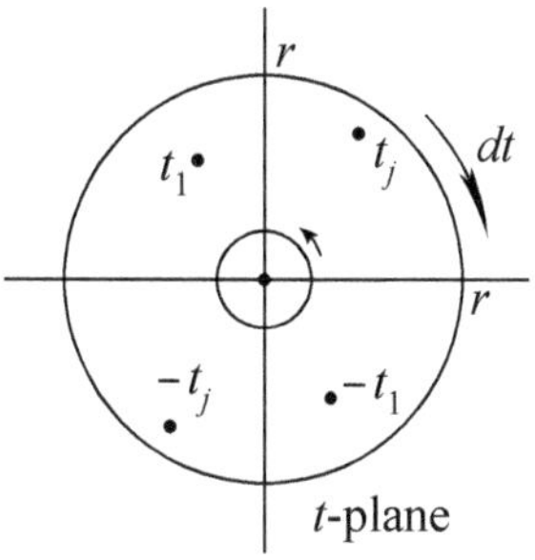

Fig. 1.4. The integration contour γ. This contour encloses an annulus bounded by two concentric circles centered at the origin. The outer one is a small circle around $\infty \in \widetilde{\Sigma} = \mathbb{P}^1$, and the inner one a small circle around the origin. Both circles are positively oriented on $\widetilde{\Sigma}$. Using the t-plane as the affine chart, the small loop around ∞ becomes a circle of a large radius $r >> \infty$ with the opposite orientation, as in the figure. Geometrically, $t = 0$ and $t = \infty$ correspond to the two simple ramification points of the Galois covering $\tilde{\pi} : \widetilde{\Sigma} \longrightarrow \mathbb{P}^1$, and the contour γ consists of two small loops around these ramification points.

is identified via graph enumeration. The power of this formula is that all $W^A_{g,n}$ for $2g-2+n > 0$ are calculated from the initial values (1.40) and (1.41). For example,

$$\begin{aligned} W^A_{1,1}(t_1) &= -\left[\frac{1}{2\pi i}\int_\gamma \left(\frac{1}{t+t_1} + \frac{1}{t-t_1}\right)\frac{t^4}{64}\frac{1}{dt}W^A_{0,2}(t,-t)\right] dt_1 \\ &= -\left[\frac{1}{2\pi i}\int_\gamma \left(\frac{1}{t+t_1} + \frac{1}{t-t_1}\right)\frac{t^4}{64}\frac{(-dt)}{4t^2}\right] dt_1 \\ &= -\frac{1}{128}t_1^2 dt_1 \\ &= -\frac{3}{16}\langle\tau_1\rangle_{1,1} t_1^2 dt_1. \end{aligned}$$

Thus we find $\langle\tau_1\rangle_{1,1} = \frac{1}{24}$.

The functions $F^A_{g,n}$ for $2g-2+n > 0$ can be calculated by integration:

$$F^A_{g,n}(t_1,\ldots,t_n) = \int_0^{t_1}\cdots\int_0^{t_n} W^A_{g,n}(t_1,\ldots,t_n). \tag{1.43}$$

Note that

$$t \to 0 \Longleftrightarrow x \to \infty.$$

Therefore, we are considering the expansion of quantities at the essential singularity of $Ai(x)$. It is surprising to see that the topological recursion indeed determines all intersection numbers (1.11)! Now define

$$S_m(x) = \sum_{2g-2+n=m-1} \frac{1}{n!} F^A_{g,n}\big(t(x), \dots, t(x)\big), \tag{1.44}$$

where we choose a branch of $\pi : \Sigma \longrightarrow \mathbb{P}^1$, and consider $t = t(x)$ as a function in x. It coincides with (1.10). Since the *moduli* spaces $\overline{\mathcal{M}}_{0,1}$ and $\overline{\mathcal{M}}_{0,2}$ do not exist, we do not have an expression (1.38) for these unstable geometries. So let us formally apply (1.43) to (1.40):

$$F^A_{0,1}(t) := \int_0^t \frac{16}{t^4} dt = -\frac{16}{3} t^{-3} = -\frac{2}{3} x^{\frac{3}{2}} = S_0(x). \tag{1.45}$$

We do not have this type of integration procedure to produce $S_1(x)$ from $W^A_{0,2}$. So we simply define $S_1(x)$ by solving (1.33). Then the residues in (1.42) can be concretely computed, and produce a system of recursive partial differential equations among the $F^A_{g,n}$s. Their **principal specialization** (1.44) produces (1.35)! Therefore, we obtain (1.18).

In this context, the topological recursion is the process of quantization, because it actually constructs the function $\Psi(x, \hbar)$ by giving a closed formula for the WKB analysis, and hence the differential operator (1.27) that annihilates it, all from the classical curve $x = y^2$.

1.4. *Non-Abelian Hodge correspondence and quantum curves*

Then what is the quantum curve? Since it is a second order differential equation with a deformation parameter $\hbar$, and its semi-classical limit is the spectral curve of a Higgs bundle, which is a rank 2 bundle in our example, we can easily *guess* that it should be the result of the **non-Abelian Hodge correspondence**. Since the quantization procedure is a *holomorphic* correspondence, while the non-Abelian Hodge correspondence is *not* holomorphic as a map, we do not expect that these two are the same.

To have a glimpse of the geometric effect of quantization, let us start with a Higgs bundle (E, ϕ) of (1.21) and (1.22). The transition function of the vector bundle

$$E = \mathcal{O}_{\mathbb{P}^1}(-1) \oplus \mathcal{O}_{\mathbb{P}^1}(1)$$

on $\mathbb{P}^1 = U_\infty \cup U_0$ defined on $\mathbb{C}^* = U_\infty \cap U_0$ is given by $\begin{bmatrix} x & \\ & \frac{1}{x} \end{bmatrix}$, where $U_0 = \mathbb{P}^1 \setminus \{\infty\}$ and $U_\infty = \mathbb{A}^1 = \mathbb{P}^1 \setminus \{0\}$. The trivial extension

$$0 \longrightarrow \mathcal{O}_{\mathbb{P}^1}(-1) \longrightarrow E \longrightarrow \mathcal{O}_{\mathbb{P}^1}(1) \longrightarrow 0$$

has a unique 1-parameter family of deformations as the extension of $\mathcal{O}_{\mathbb{P}^1}(1)$ by $\mathcal{O}_{\mathbb{P}^1}(-1)$:

$$0 \longrightarrow \mathcal{O}_{\mathbb{P}^1}(-1) \longrightarrow E_\hbar \longrightarrow \mathcal{O}_{\mathbb{P}^1}(1) \longrightarrow 0,$$

where $\hbar \in \mathrm{Ext}^1\left(\mathcal{O}_{\mathbb{P}^1}(1), \mathcal{O}_{\mathbb{P}^1}(-1)\right) \cong H^1\left(\mathbb{P}^1, K_{\mathbb{P}^1}\right) \cong \mathbb{C}$, and the transition function of $E_\hbar$ is given by

$$\begin{bmatrix} x & \hbar \\ & \frac{1}{x} \end{bmatrix}.$$

Since

$$\begin{bmatrix} 1 & \\ -\frac{1}{\hbar x} & 1 \end{bmatrix} \begin{bmatrix} x & \hbar \\ & \frac{1}{x} \end{bmatrix} \begin{bmatrix} & -\hbar \\ \frac{1}{\hbar} & x \end{bmatrix} = \begin{bmatrix} 1 & \\ & 1 \end{bmatrix},$$

$$E_\hbar \cong \begin{cases} \mathcal{O}_{\mathbb{P}^1}(-1) \oplus \mathcal{O}_{\mathbb{P}^1}(1) & \hbar = 0 \\ \mathcal{O}_{\mathbb{P}^1} \oplus \mathcal{O}_{\mathbb{P}^1} & \hbar \neq 0. \end{cases} \tag{1.46}$$

The Higgs field (1.22) satisfies the transition relation

$$-\begin{bmatrix} & \frac{1}{u^5} \\ 1 & \end{bmatrix} du = \begin{bmatrix} x & \\ & \frac{1}{x} \end{bmatrix} \begin{bmatrix} & x \\ 1 & \end{bmatrix} dx \begin{bmatrix} x & \\ & \frac{1}{x} \end{bmatrix}^{-1},$$

where $u = 1/x$ is a coordinate on U_∞. Because of the relation

$$-\begin{bmatrix} & \frac{1}{u^5} \\ 1 & \end{bmatrix} du = \begin{bmatrix} x & \hbar \\ & \frac{1}{x} \end{bmatrix} \begin{bmatrix} & x \\ 1 & \end{bmatrix} dx \begin{bmatrix} x & \hbar \\ & \frac{1}{x} \end{bmatrix}^{-1} - d\begin{bmatrix} x & \hbar \\ & \frac{1}{x} \end{bmatrix} \begin{bmatrix} x & \hbar \\ & \frac{1}{x} \end{bmatrix}^{-1},$$

somewhat miraculously, $\nabla^\hbar = \hbar d + \phi$ with the same Higgs fields as a connection matrix becomes an $\hbar$-connection in $E_\hbar$:

$$\nabla^\hbar = \hbar d + \phi : E_\hbar \longrightarrow K_{\mathbb{P}^1}(5) \otimes E_\hbar. \tag{1.47}$$

It gives the Higgs field at $\hbar = 0$. A flat section with respect to $\nabla^\hbar$ for $\hbar \neq 0$ can be obtained by solving

$$\left(\hbar \frac{d}{dx} + \begin{bmatrix} & x \\ 1 & \end{bmatrix}\right) \begin{bmatrix} -\hbar \Psi(x,\hbar)' \\ \Psi(x,\hbar) \end{bmatrix} = \begin{bmatrix} 0 \\ 0 \end{bmatrix}, \tag{1.48}$$

where $\Psi(x,\hbar)'$ is the x-derivative. Clearly, (1.48) is equivalent to (1.27).

For $\hbar = 1$, (1.47) is a holomorphic flat connection on $E_1|_{\mathbb{A}^1} = \mathcal{O}_{\mathbb{A}^1}^{\oplus 2}$, the restriction of the vector bundle E_1 on the affine coordinate neighborhood. Therefore, $\left(E_\hbar, \nabla^\hbar\right)\big|_{\hbar=1}$ defines a $\mathcal{D}$-module on $E_1|_{\mathbb{A}^1}$. As a holomorphic $\mathcal{D}$-module over $\mathbb{A}^1$, we have an isomorphism

$$\left(\mathcal{O}_{\mathbb{A}^1}^{\oplus 2}, \nabla^{\hbar=1}\right)\big|_{\mathbb{A}^1} \cong \mathcal{D}/\left(\mathcal{D} \cdot P(x,1)\right),$$

where

$$P(x,\hbar) := \left(\hbar\frac{d}{dx}\right)^2 - x, \tag{1.49}$$

and $\mathcal{D}$ denotes the sheaf of linear differential operators with holomorphic coefficients on $\mathbb{A}^1$. The above consideration indicates how we construct a quantum curve as a $\mathcal{D}$-module from a given particular Higgs bundle.

Hitchin's original idea [46] of constructing *stable* Higgs bundles is to solve a system of differential equations, now known as *Hitchin's equations.* The stability condition for the Higgs bundles can be translated into a system of nonlinear elliptic partial differential equations defined on a **Hermitian** vector bundle $E \longrightarrow C$ over a compact Riemann surface C. It takes the following form:

$$\begin{aligned} F(D) + [\phi, \phi^{\dagger_h}] &= 0 \\ D^{0,1}\phi &= 0. \end{aligned} \tag{1.50}$$

Here, h is a Hermitian metric in E, D is a unitary connection in E with respect to h, $D^{0,1}$ is the $(0,1)$-component of the covariant differentiation with respect to the complex structure of the curve C, $F(D)$ is the curvature of D, $\dagger_h$ is the Hermitian conjugation with respect h, and ϕ is a differentiable Higgs field on E. Solving Hitchin's equation is equivalent to constructing a 1-parameter family of flat connections of the form

$$D(\zeta) = \frac{\phi}{\zeta} + D + \phi^{\dagger_h}\zeta, \qquad \zeta \in \mathbb{C}^* \tag{1.51}$$

(see [38, 71, 80]). The non-Abelian Hodge correspondence is the association of $\left(\widetilde{E}, D(1)^{1,0}\right)$ to the given stable holomorphic Higgs bundle (E,ϕ), i.e., a solution to Hitchin's equations. Here, $D(1) = \phi + D + \phi^{\dagger_h}$, and $\widetilde{E}$ is a holomorphic vector bundle with the complex structure given by the flat connection $D(1)^{0,1}$. The connection $D(1)^{1,0}$ is then a holomorphic connection in $\widetilde{E}$.

A new idea that relates the non-Abelian Hodge correspondence and *opers* is emerging [23]. The role of the topological recursion in this context, and the identification of opers as globally defined quantum curves, are being developed. Since it is beyond our scope of the current lecture notes, it will be discussed elsewhere.

So far, we have considered $\Psi(x,\hbar)$ as a formal "function." What is it indeed? Since our vector bundle E is of the form (1.21), the shape of the equation (1.48) suggests that

$$\Psi(x,\hbar) \in \widehat{H}^0\left(\mathbb{P}^1, K_{\mathbb{P}^1}^{-\frac{1}{2}}\right), \tag{1.52}$$

where the hat sign indicates that we really do not have any good space to store the formal wave function $\Psi(x, \hbar)$. The reason for the appearance of $K_{\mathbb{P}^1}^{-\frac{1}{2}}$ as its home is understood as follows. Since $W_{g,n}^A$ for $2g-2+n$ is an n-linear differential form, (1.43) tells us that $F_{g,n}^A$ is just a number. Therefore, $S_m(x)$ that is determined by (1.44) for $m \geq 2$ is also just a number. Here, by a "number" we mean a genuine function in x. The differential equation (1.32) should be written

$$(dS_0(x))^{\otimes 2} = x(dx)^2 = q(x) \tag{1.53}$$

as an equation of quadratic differentials. Here,

$$q(x) \in K_{\mathbb{P}^1}^{\otimes 2}$$

can be actually any meromorphic quadratic differential on $\mathbb{P}^1$, including $x(dx)^2$, so that

$$\phi = \begin{bmatrix} & 1 \\ q & \end{bmatrix} \in K_{\mathbb{P}^1} \otimes \mathrm{End}(E)$$

is a meromorphic Higgs field on the vector bundle E in the same way. Similarly, (1.33) should be interpreted as

$$S_1(x) = -\frac{1}{2} \int d \log(dS_0) = -\frac{1}{2} \log \sqrt{q(x)} = -\frac{1}{4} \log q(x). \tag{1.54}$$

Now recall the conjugate differential equation (1.30). Its solution takes the form

$$\frac{1}{\sqrt[4]{q(x)}} \exp\left(\sum_{m=2}^{\infty} S_m(x) \hbar^{m-1}\right), \tag{1.55}$$

and as we have noted, the exponential factor is just a number. Therefore, the geometric behavior of this solution is determined by the factor

$$\frac{1}{\sqrt[4]{q(x)}} \in K_{\mathbb{P}^1}^{-\frac{1}{2}},$$

which is a meromorphic section of the negative half-canonical sheaf.

We recall that $\Psi(x, \hbar)$ has another factor $\exp\big(S_0(x)/\hbar\big)$. It should be understood as a "number" defined on the spectral curve Σ, because $dS_0(x) = \eta$ is the tautological 1-form on $T^*\mathbb{P}^1$ restricted to the spectral curve, and $S_0(x)$ is its integral on the spectral curve. Therefore, this factor tells us that the equation (1.27) should be considered on Σ. Yet its local x-dependence is indeed determined by $K_{\mathbb{P}^1}^{-\frac{1}{2}}$.

We have thus a good answer for Question 1.12 now. The main part of the asymptotic expansion (1.8) tells us what geometry we should consider. It tells us what the Hitchin spectral curve should be, and it also includes the information of Higgs bundle (E, ϕ) itself.

Remark 1.15: The Airy example, and another example we consider in the next section, are in many ways special, in the sense that the $S_2(x)$-term of the WKB expansion is given by integrating the solutions $W_{1,1}$ and $W_{0,3}$ of the topological recursion (1.42). In general, the topological recursion mechanism of computing $W_{1,1}$ and $W_{0,3}$ from $W_{0,2}$ does not correspond to the WKB equation for S_2. As discovered in [24], the topological recursion in its PDE form is equivalent to the WKB equations for all $S_m(x)$ in the range of $m \geq 3$. But the PDE recursion, which we discuss in detail in the later sections, does not determine S_2. It requires a new way of viewing the topological recursion in its differential equation formulation: **we consider $F_{1,1}$ and $F_{0,3}$ as the initial condition for topological recursion**, rather than $W_{0,1}$ and $W_{0,2}$, which have been more commonly considered as the starting point for the topological recursion.

1.5. *The Lax operator for Witten-Kontsevich KdV equation*

Surprisingly, the operator $P(x,1)$ of (1.49) at $\hbar = 1$ is the initial value of the *Lax operator* for the KdV equations that appears in the work of Witten [86] and Kontsevich [55]. Witten considered a different generating function of the intersection numbers (1.11) given by

$$\begin{aligned} F(s_0, s_1, s_2, \ldots) &= \left\langle \exp\left(\sum_{d=0}^{\infty} s_d \tau_d\right)\right\rangle \\ &= \sum_{k_0,k_1,k_2,\cdots=0}^{\infty} \left\langle \tau_0^{k_0}\tau_1^{k_1}\tau_2^{k_2}\cdots\right\rangle \prod_{j=0}^{\infty} \frac{s_j^{k_j}}{k_j!} \\ &= \sum_{g=0}^{\infty}\sum_{n=1}^{\infty}\frac{1}{n!}\sum_{\substack{d_1+\cdots+d_n\\=3g-3+n}} \langle \tau_{d_1}\cdots\tau_{d_n}\rangle_{g,n} s_{d_1}\cdots s_{d_n} \\ &= \langle \tau_0^3\rangle_{0,3}\frac{s_0^3}{3!} + \langle \tau_1\rangle_{1,1} s_1 + \langle\tau_0^4\rangle_{0,4}\frac{s_0^4}{4!} + \cdots. \end{aligned} \tag{1.56}$$

Define

$$t_{2j+1} := \frac{s_j}{(2j+1)!!},$$

and

$$u(t_1, t_3, t_5, \dots) := \left(\frac{\partial}{\partial s_0}\right)^2 F(s_0, s_1, s_2, \dots). \tag{1.57}$$

Then $u(t_1, t_3, t_5, \dots)$ satisfies the system of KdV equations, whose first equation is

$$u_{t_3} = \frac{1}{4} u_{t_1 t_1 t_1} + 3 u u_{t_1}. \tag{1.58}$$

The system of KdV equations are the deformation equation for the universal iso-spectral family of second order ordinary differential operators of the form

$$L(X, t) := \left(\frac{d}{dX}\right)^2 + 2u(X + t_1, t_3, t_5, \dots) \tag{1.59}$$

in X and deformation parameters $t = (t_1, t_3, t_5, \dots)$. The operator is often referred to as the *Lax operator* for the KdV equations. The expression

$$\sqrt{L} = \frac{d}{dX} + u \cdot \left(\frac{d}{dX}\right)^{-1} - \frac{1}{2} u' \cdot \left(\frac{d}{dX}\right)^{-2} + \cdots$$

makes sense in the ring of pseudo-differential operators, where $'$ denotes the X-derivative. The KdV equations are the system of *Lax equations*

$$\frac{\partial L}{\partial t_{2m+1}} = \left[\left(\sqrt{L}^{2m+1}\right)_+, L\right], \qquad m \geq 0,$$

where $_+$ denotes the differential operator part of a pseudo-differential operator. The commutator on the right-hand side explains the invariance of the eigenvalues of the Lax operator L with respect to the deformation parameter t_{2m+1}. The t_1-deformation is the translation $u(X) \longmapsto u(X + t_1)$, and the t_3-deformation is given by the KdV equation (1.58).

For the particular function (1.57) defined by the intersection numbers (1.56), the *initial* value of the Lax operator is

$$\left(\frac{d}{dX}\right)^2 + 2u(X + t_1, t_3, t_5, \dots)\Big|_{t_1 = t_3 = t_5 = \cdots = 0} = \left(\frac{d}{dX}\right)^2 + 2X.$$

In terms of yet another variable $x = -\frac{2}{\sqrt[3]{4}} X$, the initial value becomes

$$\frac{1}{\sqrt[3]{4}} \left(\left(\frac{d}{dX}\right)^2 + 2X\right) = \left(\frac{d}{dx}\right)^2 - x = P(x, \hbar)\big|_{\hbar = 1},$$

which is the Airy differential operator.

Kontsevich [55] used the matrix Airy function to obtain all intersection numbers. The topological recursion replaces the asymptotic analysis of matrix integrations with a series of residue calculations on the spectral curve $x = y^2$.

1.6. *All things considered*

What we have in front of us is an interesting example of a theory yet to be constructed.

- We have generating functions of quantum invariants, such as Gromov-Witten invariants. They are symmetric functions.
- We take the *principal specialization* of these functions, and form a generating function of the specialized generating functions.
- This function then solves a 1-dimensional stationary Schrödinger equation. The equation is what we call a *quantum curve*.
- From this Schrödinger equation (or a quantum curve), we construct an algebraic curve, a *spectral curve* in the sense of Hitchin, through the process of *semi-classical limit*.
- The differential version of the *topological recursion* [24,25] applied to the spectral curve then recovers the starting quantum invariants.
- The spectral curve can be also expressed as the Hitchin spectral curve of a particular meromorphic Higgs bundle.
- Then the quantum curve is equivalent to the $\hbar$-connection in the $\hbar$-deformed vector bundle on the base curve, on which the initial Higgs bundle is defined.
- The topological recursion therefore constructs a flat section, although formal, of the $\hbar$-connection from the Hitchin spectral curve. At least locally, the $\hbar$-connection itself is thus constructed by the topological recursion.

We do not have a general theory yet. In particular, we do not have a global definition of quantum curves. As mentioned above, right at this moment, the notion of *opers* is emerging as a mathematical definition of quantum curves, at least for the case of smooth spectral covers in the cotangent bundle T^*C of a smooth curve C of genus greater than 1. We will report our finding in this exciting direction elsewhere. Here, we present what we know as of now.

The idea of topological recursion was devised for a totally different context. In the authors' work [24], for the first time the formalism of Eynard and Orantin was placed in the Higgs bundle context. The formalism depends purely on the geometry of the Hitchin spectral curve. Therefore, quantities that the topological recursion computes should represent the geometric information. Then in [25], we have shown through examples that quantization of singular spectral curves are related to certain enumerative geometry problems, when the quantum curve is analyzed at its singular point and the function $\Psi(x,\hbar)$, which should be actually considered as a formal section of $K_C^{-\frac{1}{2}}$, is expanded at its essential singularity. In the example described above, the corresponding counting problem is computing the intersection numbers of certain cohomology classes on $\overline{\mathcal{M}}_{g,n}$. The original topological recursion of [34] is generalized to singular spectral curves in [25] for this purpose.

The question we still do not know its answer is how to directly connect the Higgs bundle information with the geometric structure whose quantum invariants are captured by the topological recursion.

Since the time of inception of the topological recursion [16, 34], numerous papers have been produced, in both mathematics and physics. It is far beyond the authors' ability to make any meaningful comments on this vast body of literature in the present chapter. Luckily, interested readers can find useful information in Eynard's talk at the ICM 2014 [32]. Instead of attempting the impossible, we review here a glimpse of **geometric developments** inspired by the topological recursion that have taken place in the last few years.

The geometry community's keen attention was triggered when a concrete **remodeling conjecture** was formulated by string theorists, first by Mariño [59], and then in a more precise and generalized framework by Bouchard, Klemm, Mariño and Pasquetti [11,12]. The conjecture states that open Gromov-Witten invariants of an arbitrary toric Calabi-Yau orbifold of dimension 3 can be calculated by the topological recursion formulated on the *mirror curve*. A physical argument for the correctness of the conjecture is pointed out in [76]. Bouchard and Mariño [13] then derived a new conjectural formula for simple Hurwitz numbers from the remodeling conjecture. The correctness of the Hurwitz number conjecture can be easily checked by a computer for many concrete examples. At the same time, it was clear that the conjectural formula was totally different from the combinatorial formula known as the **cut-and-join equation** of [42, 43, 84].

After many computer experiments, one of the authors noticed that the conjectural formula of Bouchard and Mariño was exactly the *Laplace transform* of a particular variant of the cut-and-join equation. Once the precise relation between the knowns and unknowns is identified, often the rest is straightforward, even though a technical difficulty still remains. The conjecture for simple Hurwitz numbers of [13] was solved in [33,69]. Its generalization to the orbifold Hurwitz numbers is then established in [10]. In each case, the Laplace transform plays the role of the mirror symmetry, changing the combinatorial problem on the A-model side to a complex analysis problem on the B-model side. The first case of the remodeling conjecture for $\mathbb{C}^3$ was solved, using the same idea, shortly afterwords in [90]. The remodeling conjecture in its full generality is recently solved in its final form by Fang, Liu, and Zong (announced in [37]), based on an earlier work of [35].

Independent of these developments, the relation between the topological recursion and combinatorics of enumeration of various graphs drawn on oriented topological surfaces has been studied by the Melbourne group of mathematicians, including Do, Leigh, Manesco, Norbury, and Scott (see, for example, [22,72–74]). The authors' earlier papers [15,27,65] are inspired by their work. A surprising observation of the Melbourne group, formulated in a conjectural formula, is that the Gromov-Witten invariants of $\mathbb{P}^1$ themselves should satisfy the topological recursion. Since $\mathbb{P}^1$ is not a Calabi-Yau manifold, this conjecture does not follow from the remodeling conjecture.

The $GW(\mathbb{P}^1)$ conjecture of [74] is solved by the Amsterdam group of mathematicians, consisting of Dunin-Barkowski, Shadrin, and Spitz, in collaboration with Orantin [30]. Their discovery, that the topological recursion on a disjoint union of open discs as its spectral curve is equivalent to *cohomological field theory*, has become a key technique of many later works [3, 28, 37].

The technical difficulty of the topological recursion lies in the evaluation of residue calculations involved in the formula. When the spectral curve is an open disc, this difficulty does not occur. But if the global structure of the spectral curve has to be considered, then one needs a totally different idea. The work of [10, 33, 69] has overcome the complex analysis difficulty in dealing with simple and orbifold Hurwitz numbers. There, the key idea is the use of the (piecewise) *polynomiality* of these numbers through the ELSV formula [31] and its orbifold generalization [51]. The paper [28] proves the converse: they first prove the polynomiality of the Hurwitz numbers without assuming the relation to the intersection numbers over $\overline{\mathcal{M}}_{g,n}$, and then establish the ELSV formula from the topological recursion, utilizing a technique of [69].

The topological recursion is a byproduct of the study of random matrix theory/matrix models [16, 34]. A recursion of the same nature appeared earlier in the work of Mirzakhani [60, 61] on the Weil-Petersson volume of the moduli space of bordered hyperbolic surfaces. The Laplace transform of the Mirzakhani recursion is an example of the topological recursion. Its spectral curve, the sine curve, was first identified in [66] as the intertwiner of two representations of the Virasoro algebra.

The notion of quantum curves goes back to [1]. It is further developed in the physics community by Dijkgraaf, Hollands, Sułkowski, and Vafa [18, 19, 48]. The geometry community was piqued by [17, 44], which speculated on the relation between the topological recursion, quantization of the $SL(2,\mathbb{C})$-character variety of the fundamental group of a knot complement in S^3, the AJ-conjecture due to [39, 40], and the K_2-group of algebraic K-theory. Although it is tantalizingly interesting, so far no mathematical results have been established in this direction for hyperbolic knots, or even it may be impossible (see for example, [9]), mainly due to the reducible nature of the spectral curve in this particular context. For torus knots, see also [14]. A rigorous construction of quantization of spectral curves was established for a few examples in [68] for the first time, but without any relation to knot invariants. One of these examples will be treated in these lectures below. By now, we have many more mathematical examples of quantum curves [10, 22, 29, 67, 91, 92]. We note that in many of these examples, the spectral curves have a global parameter, even though the curves are not necessarily the rational projective curve. Therefore, the situation is in some sense still the "genus 0 case." The difficulty of quantization lies in dealing with complicated entire functions, and the fact that the quantum curves are **difference equations**, rather than differential equations of finite order.

We are thus led to another question.

Question 1.16: What can we do when we have a different situation, where the spectral curve of the theory is global, compact, and of a high genus?

It comes as a surprise that there is a system of recursive partial differential equations, resembling the residue calculation formula for the topological recursion, when the spectral curve of the topological recursion is precisely a Hitchin spectral curve associated with a rank 2 Higgs bundle. The result of the calculation then leads us to a construction of a quantum curve. In this way a connection between quantization of Hitchin spectral curves and the topological recursion is discovered in [24].

If we regard the topological recursion as a method of calculation of quantum invariants, then we need to allow singular spectral curves, as we have seen earlier. The simplest quantization of the singular Hitchin spectral curve is then obtained by the topological recursion again, but this time, it has to be applied to the normalization of the singular spectral curve constructed by a particular way of using blow-ups of the ambient compactified cotangent bundle. This is the content of [25], and we obtain a **Rees $\mathcal{D}$-module** as the result of quantization.

We note here that what people call by quantization is not unique. Depending on the purpose, one needs to use a different guideline for quantization. The result would be a different differential equation, but still having the same semi-classical limit.

For example, it has been rigorously proved in [50] that surprisingly the quantization procedure of [24, 25], including the desingularization of the spectral curve, *automatically* leads to an iso-monodromic deformation family, for the case of the Painlevé I equation. Their global parameter is essentially the normalization coordinate of the singular elliptic curve. In their work [50], they ask what one obtains if the straightforward topological recursion is applied for the quantization of a singular elliptic curve with a prescribed parameter in a particular way. They find that the quantum curve is a Schrödinger equation whose coefficients have nontrivial dependence on $\hbar$, yet it is an iso-spectral family with respect to the parameter. This work, and also the numerous mathematical examples of quantum curves that have been already constructed, suggest that the idea of using $\mathcal{D}$-modules for the definition of quantum curves ([18, 25]) is not the final word. Differential operators of an infinite order, or difference operators mixed with differential operators, also have to be considered.

For the mirror curves of toric Calabi-Yau orbifolds of dimension 3 appearing in the context of the remodeling conjecture [11, 12, 37, 59], the conjectural quantum curves acquire a very different nature. It has deep connections to number theory and quantum dilogarithm functions [53].

A suggestion of using deformation quantization modules for quantum curves is made by Kontsevich in late 2013 in a private communication to the authors. An interesting work toward this direction is proposed by Petit [78].

The relation of the quantization discussed in these lectures with the **non-Abelian Hodge correspondence** and **opers** is being investigated as of now [23]. A coordinate-free global definition of quantum curves is emerging, and a direct relationship among quantum curves, non-Abelian Hodge correspondence, and opers is being developed.

The story is expanding its horizon. We have come to the other side of the rainbow. And there we find ourselves on Newton's seashore. So far we have found only a few smoother pebbles or prettier shells, whilst...

2. From Catalan numbers to the topological recursion

Let us consider a function $f(X)$ in one variable, where X is an $N \times N$ Hermitian matrix. One of the main problems of matrix integration theory is to calculate the *expectation value*

$$\langle \mathrm{tr} f(X)\rangle := \frac{\int_{H_{N\times N}} \mathrm{tr} f(X) e^{V(X)} dX}{\int_{H_{N\times N}} e^{V(X)} dX}, \tag{2.1}$$

where the *potential function* $V(X)$ is given, such as the Gaussian potential

$$V(X) = -\frac{1}{2}\mathrm{tr}(X^2), \tag{2.2}$$

so that

$$C_N = \int_{H_{N\times N}} e^{V(X)} dX$$

is finite. The integration measure dX is the standard $U(N)$-invariant Lebesgue measure of the space $H_{N\times N} = \mathbb{R}^{N^2}$ of Hermitian matrices of size N. When a Gaussian potential (2.2) is chosen, $e^{V(X)}dX$ is a probability measure after an appropriate normalization, and $\langle \mathrm{tr}(X^m)\rangle$ is the m-th *moment.* If we know all the moments, then we can calculate the expectation value of any polynomial function $f(X)$. Therefore, the problem changes into calculating a generating function of the moments

$$\frac{1}{C_N}\sum_{m=0}^{\infty} \frac{1}{z^{m+1}} \int_{H_{N\times N}} \mathrm{tr}(X^m) e^{V(X)} dX. \tag{2.3}$$

For a norm bounded matrix X, we have

$$tr\left(\frac{1}{z-X}\right) = \sum_{m=0}^{\infty} \frac{1}{z^{m+1}} \mathrm{tr}(X^m).$$

Therefore, the *resolvent* of a random matrix X,

$$\left\langle \mathrm{tr}\left(\frac{1}{z-X}\right)\right\rangle = \frac{1}{C_N}\int_{H_{N\times N}} \mathrm{tr}\left(\frac{1}{z-X}\right) e^{V(X)} dX, \tag{2.4}$$

looks the same as (2.3). But they are not the same. For example, let us consider the $N = 1$ case and write $X = x \in \mathbb{R}$. Then the formula

$$\frac{1}{\sqrt{2\pi}}\int_{-\infty}^{\infty} \frac{1}{z-x}\, e^{-\frac{1}{2}x^2} dx \overset{?}{=} \sum_{m=0}^{\infty} \frac{(2m-1)!!}{z^{2m+1}} = \frac{1}{\sqrt{2\pi}}\sum_{m=0}^{\infty} \frac{1}{z^{m+1}} \int_{-\infty}^{\infty} x^m\, e^{-\frac{1}{2}x^2} dx$$

is valid only as the asymptotic expansion of the analytic function

$$\frac{1}{\sqrt{2\pi}}\int_{-\infty}^{\infty} \frac{1}{z-x}\, e^{-\frac{1}{2}x^2} dx$$

in z for $Im(z) \neq 0$ and $Re(z) \to +\infty$. Still we can see that the information of the generating function of the moment (2.3) can be extracted from the resolvent (2.4) if we apply the technique of asymptotic expansion of a holomorphic function at an essential singularity. The asymptotic method of matrix integrals leads to many interesting formulas, such as calculating the orbifold Euler characteristic $\chi(\mathcal{M}_{g,n})$ of the moduli space of smooth pointed curves [45]. We refer to [63, 64] for introductory materials of these topics.

More generally, we can consider *multi-resolvent correlation functions*

$$\left\langle \prod_{i=1}^{n} \mathrm{tr}\left(\frac{1}{z_i - X}\right)\right\rangle = \frac{1}{C_N}\int_{H_{N\times N}} \prod_{i=1}^{n} \mathrm{tr}\left(\frac{1}{z_i - X}\right) e^{V(X)} dX. \tag{2.5}$$

When we say "calculating" the expectation value (2.5), we wish to identify it as a *holomorphic* function in all the parameters, i.e., $(z_1, \ldots, z_n)$, the coefficients of the potential V, and the matrix size N. In particular, the analytic dependence on the parameter N is an important feature we wish to determine.

It is quite an involved problem in analysis, and we do not attempt to follow this route in these lecture notes. The collaborative effort of the random matrix community has devised a recursive method of solving this analysis problem (see, for example, [16, 32, 34]), which is now known as the **topological recursion**. One thing we can easily expect here is that since (2.5) is an analytic function in $(z_1, \ldots, z_n)$, there must be an obvious relation between the topological recursion and algebraic geometry.

What is amazing is that the exact same recursion formula happens to appear in the context of many different enumerative geometry problems, again and again. Even though the counting problems are different, the topological recursion always takes the same general formalism. Therefore, to understand the nature of this formalism, it suffices to give the simplest non-trivial example. This is what we wish to accomplish in this section.

2.1. *Counting graphs on a surface*

Let us start with the following problem:

Problem 2.1: *Find the number of distinct cell-decompositions of a given closed oriented topological surface of genus g, with the specified number of* 0*-cells, and the number of* 1*-cells that are incident to each* 0*-cell.*

Denote by C_g a compact, 2-dimensional, oriented topological manifold of genus g without boundary. Two cell-decompositions of C_g are identified if there is an orientation-preserving homeomorphism of C_g onto itself that brings one to the other. If there is such a map for the same cell-decomposition, then it is an automorphism of the data. The 1-*skeleton* of a cell-decomposition, which we denote by γ, is a *graph* drawn on C_g. We call a 0-cell a *vertex*, a 1-cell an *edge*, and a 2-cell a *face*. The midpoint of an edge separates the edge into two *half-edges* joined together at the midpoint. The *degree* of a vertex is the number of half-edges incident to it. For the purpose of counting, we label all vertices. To be more precise, we give a total ordering to the set of vertices. Most of the time we simply use $[n] = \{1, \ldots, n\}$ to label the set of n vertices.

The 1-skeleton γ is usually called a *ribbon graph*, which is a graph with a cyclic order assigned to incident half-edges at each vertex. The face-labeled ribbon graphs describe an orbifold cell-decomposition of $\mathcal{M}_{g,n} \times \mathbb{R}_+^n$. Since we label vertices of γ, there is a slight difference as to what the graph represents. It is the *dual* graph of a ribbon graph, and its vertices are labeled. To emphasize the dual nature, we call γ a **cell graph**.

Most cell graphs do not have any non-trivial automorphisms. If there is one, then it induces a cyclic permutation of half-edges at a vertex, since we label all vertices. Therefore, if we pick one of the incident half-edges at each vertex, assign an outgoing arrow to it, and require that an automorphism also fixes the arrowed half-edges, then the graph has no non-trivial automorphisms. For a counting problem, no automorphism is a desirable situation because the bijective counting method works better there. Let us call such a graph an **arrowed cell graph**. Now the refined problem:

Problem 2.2: *Let $\vec{\Gamma}_{g,n}(\mu_1, \ldots, \mu_n)$ denote the set of arrowed cell graphs drawn on a closed, connected, oriented surface of genus g with n labeled vertices of degrees $\mu_1, \ldots, \mu_n$. Find its cardinality*

$$C_{g,n}(\mu_1, \ldots, \mu_n) := |\vec{\Gamma}_{g,n}(\mu_1, \ldots, \mu_n)|. \tag{2.6}$$

When $\gamma \in \vec{\Gamma}_{g,n}(\mu_1, \ldots, \mu_n)$, we say γ has *type* (g, n). Denote by $c_\alpha(\gamma)$ the number of α-cells of the cell-decomposition associated with γ. Then we have

$$2 - 2g = c_0(\gamma) - c_1(\gamma) + c_2(\gamma), \quad c_0(\gamma) = n, \quad 2c_1(\gamma) = \mu_1 + \cdots + \mu_n.$$

Therefore, $\vec{\Gamma}_{g,n}(\mu_1, \ldots, \mu_n)$ is a finite set.

Example 2.1: An arrowed cell graph of type $(0, 1)$ is a collection of loops drawn on a plane as in Figure 2.2. If we assign a pair of parenthesis to

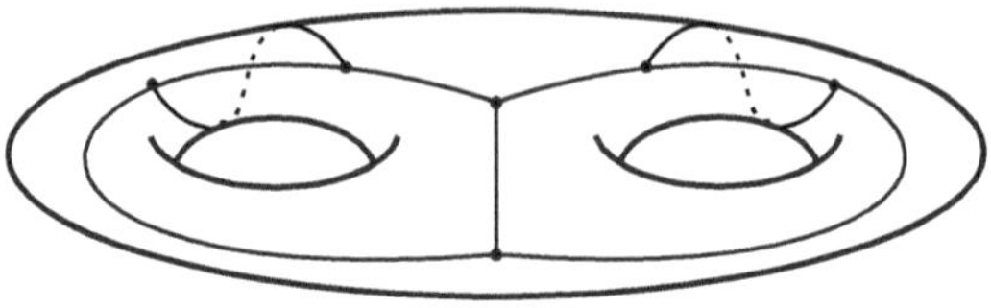

Fig. 2.1. A cell graph of type $(2,6)$.

each loop, starting from the arrowed one as (, and go around the unique vertex counter-clock wise, then we obtain the parentheses pattern ((( ))). Therefore,

$$C_{0,1}(2m) = C_m = \frac{1}{m+1}\binom{2m}{m} \tag{2.7}$$

is the m-th Catalan number. Thus it makes sense to call $C_{g,n}(\mu_1, \cdots, \mu_n)$ **generalized Catalan numbers**. Note that it is a symmetric function in n integer variables.

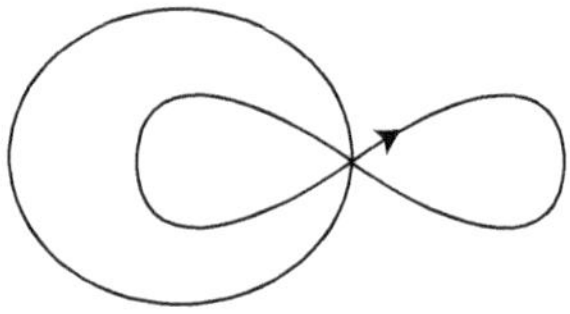

Fig. 2.2. A cell graph of type $(0,1)$ with a vertex of degree 6.

Theorem 2.3: *Catalan Recursion, [27, 85]. The generalized Catalan numbers satisfy the following equation.*

$$\begin{aligned} C_{g,n}(\mu_1, \dots, \mu_n) &= \sum_{j=2}^{n} \mu_j C_{g,n-1}(\mu_1 + \mu_j - 2, \mu_2, \dots, \widehat{\mu_j}, \dots, \mu_n) \\ &\quad + \sum_{\alpha+\beta=\mu_1-2} \Bigg[C_{g-1,n+1}(\alpha, \beta, \mu_2, \cdots, \mu_n) \\ &\quad + \sum_{\substack{g_1+g_2=g \\ I \sqcup J = \{2,\dots,n\}}} C_{g_1,|I|+1}(\alpha, \mu_I) C_{g_2,|J|+1}(\beta, \mu_J) \Bigg], \end{aligned} \tag{2.8}$$

where $\mu_I = (\mu_i)_{i \in I}$ for an index set $I \subset [n]$, $|I|$ denotes the cardinality of I, and the third sum in the formula is for all partitions of g and set partitions of $\{2, \dots, n\}$.

Proof: Let γ be an arrowed cell graph counted by the left-hand side of (2.8). Since all vertices of γ are labeled, we write the vertex set by $\{p_1, \ldots, p_n\}$. We take a look at the half-edge incident to p_1 that carries an arrow.

Case 1: The arrowed half-edge extends to an edge E that connects p_1 and p_j for some $j > 1$.

In this case, we contract the edge and join the two vertices p_1 and p_j together. By this process we create a new vertex of degree $\mu_1 + \mu_j - 2$. To make the counting bijective, we need to be able to go back from the contracted graph to the original, provided that we know μ_1 and μ_j. Thus we place an arrow to the half-edge next to E around p_1 with respect to the counter-clockwise cyclic order that comes from the orientation of the surface. In this process we have μ_j different arrowed graphs that produce the same result, because we must remove the arrow placed around the vertex p_j in the original graph. This gives the first line of the right-hand side of (2.8). See Figure 2.3.

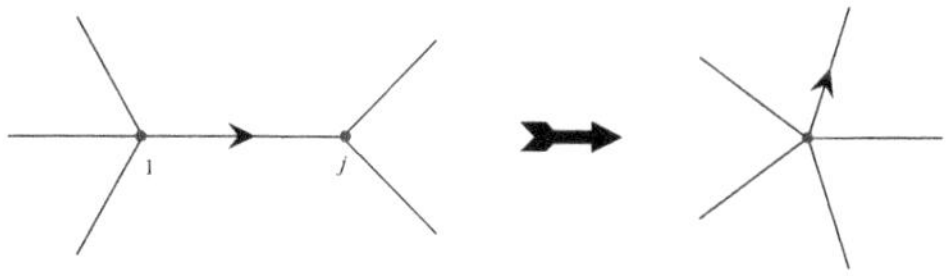

Fig. 2.3. The process of contracting the arrowed edge E that connects vertices p_1 and p_j, $j > 1$.

Case 2: The arrowed half-edge at p_1 is a loop E that goes out from, and comes back to, p_1.

The process we apply is again contracting the loop E. The loop E separates all other incident half-edges at p_1 into two groups, one consisting of α of them placed on one side of the loop, and the other consisting of β half-edges placed on the other side. It can happen that $\alpha = 0$ or $\beta = 0$. Contracting a loop on a surface causes pinching. Instead of creating a pinched (i.e., singular) surface, we separate the double point into two new vertices of degrees α and β. Here again we need to remember the place of the loop E. Thus we put an arrow to the half-edge next to the loop in each group. See Figure 2.4.

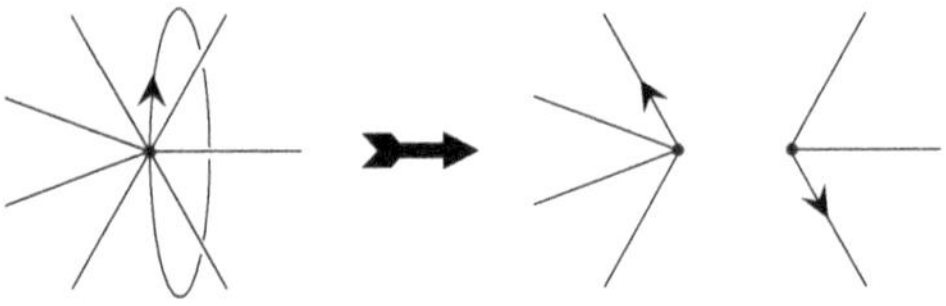

Fig. 2.4. The process of contracting the arrowed loop E that is attached to p_1.

After the pinching and separating the double point, the original surface of genus g with n vertices $\{p_1, \ldots, p_n\}$ may change its topology. It may have genus $g-1$, or it splits into two pieces of genus g_1 and g_2. The second line of (2.8) records all such cases. Normally we would have a factor $\frac{1}{2}$ in front of the second line of the formula. We do not have it here because the arrow on the loop could be in two different directions. Placing the arrow on the other half-edge of the loop is equivalent to interchanging α and β.

This completes the proof. □

Remark 2.4: For $(g, n) = (0, 1)$, the above formula reduces to

$$C_{0,1}(\mu_1) = \sum_{\alpha+\beta=\mu_1-2} C_{0,1}(\alpha)C_{0,1}(\beta). \tag{2.9}$$

Since the degree of the unique vertex is always even for type $(0, 1)$ graphs, by defining $C_{0,1}(0) = 1$, (2.9) gives the Catalan recursion. Only for $(g, n) = (0, 1)$, this irregular case of $\mu_1 = 0$ happens, because a degree 0 single vertex is *connected*, and gives a cell-decomposition of S^2. All other cases, if one of the vertices has degree 0, then the Catalan number $C_{g,n}(\mu_1, \ldots, \mu_n)$ is simply 0 because there is no corresponding *connected* cell decomposition.

Remark 2.5: Equation (2.8) is a recursion with respect to

$$2g - 2 + n + \sum_{i=1}^{n} \mu_i.$$

The values are therefore completely determined by the initial value $C_{0,1}(0) = 1$. The formula does not give a recursion of a *function* $C_{g,n}(\mu_1, \ldots, \mu_n)$, because the same type (g, n) appears on the right-hand side.

The classical Catalan recursion (2.9) determines all values of $C_{0,1}(2m)$, but the closed formula (2.7) requires a different strategy. Let us introduce a generating function

$$z = z(x) = \sum_{m=0}^{\infty} C_{0,1}(2m)x^{-2m-1}. \tag{2.10}$$

From (2.9) we have

$$z^2 = \sum_{m=0}^{\infty} \left(\sum_{a+b=m} C_{0,1}(2a)C_{0,1}(2b) \right) x^{-2m-2} = \sum_{m=0}^{\infty} C_{0,1}(2m+2)x^{-2m-2}.$$

Since

$$xz = \sum_{m=-1}^{\infty} C_{0,1}(2m+2)x^{-2m-2},$$

we obtain an equation $xz = z^2 + 1$, or

$$x = z + \frac{1}{z}. \tag{2.11}$$

This is the inverse function of the complicated-looking generating function $z(x)$ at the branch $z \to 0$ as $x \to +\infty$! Thus the curve (2.11) knows everything about the Catalan numbers. For example, we can prove the closed formula (2.7). The solution of (2.11) as a quadratic equation for z that gives the above branch is given by

$$z(x) = \frac{x - \sqrt{x^2 - 4}}{2} = \frac{x}{2}\left(1 - \sqrt{1 - \frac{4}{x^2}}\right).$$

The binomial expansion of the square root

$$\sqrt{1+X} = \sum_{m=0}^{\infty} \binom{\frac{1}{2}}{m} X^m,$$

$$\binom{\frac{1}{2}}{m} := \frac{\frac{1}{2}(\frac{1}{2}-1)(\frac{1}{2}-2)\cdots(\frac{1}{2}-m+1)}{m!} = (-1)^{m-1}\frac{(2m-3)!!}{2^m m!}$$
$$= \frac{(-1)^{m-1}}{4^m}\frac{1}{2m-1}\binom{2m}{m},$$

then gives the closed formula (2.7) for the Catalan numbers:

$$\begin{aligned} z(x) &= \sum_{m=0}^{\infty} C_{0,1}(2m)x^{-2m-1} \\ &= \frac{x}{2}\left(1 - \sqrt{1 - \frac{4}{x^2}}\right) \end{aligned}$$

$$= \frac{x}{2}\left(1 - \left(\sum_{m=0}^{\infty} \frac{(-1)^{m-1}}{4^m} \frac{1}{2m-1}\binom{2m}{m}(-1)^m\left(\frac{4}{x^2}\right)^m\right)\right)$$
$$= \frac{1}{2}\sum_{m=1}^{\infty} \frac{1}{2m-1}\binom{2m}{m} x^{-2m+1}$$
$$= \sum_{m=0}^{\infty} \frac{1}{m+1}\binom{2m}{m} x^{-2m-1}.$$

Another piece of information we obtain from (2.11) is the radius of convergence of the infinite series $z(x)$ of (2.10). Since

$$dx = \left(1 - \frac{1}{z^2}\right) dz,$$

the map (2.11) is critical (or ramified) at $z = \pm 1$. The critical values are $x = \pm 2$. On the branch we are considering, (2.10) is the inverse function of (2.11) for all values of $|x| > 2$. This means the series $z(x)$ is absolutely convergent on the same domain.

Remark 2.6: In combinatorics, we often consider a generating function of interesting quantities only as a formal power series. The idea of topological recursion tells us that we should consider the **Riemann surface** of the Catalan number generating function $z = z(x)$. We then recognize that there is a global algebraic curve hidden in the scene, which is the curve of Figure 2.5. The topological recursion mechanism then tells us how to calculate all $C_{g,n}(\vec{\mu})$ from this curve alone, known as the **spectral curve**.

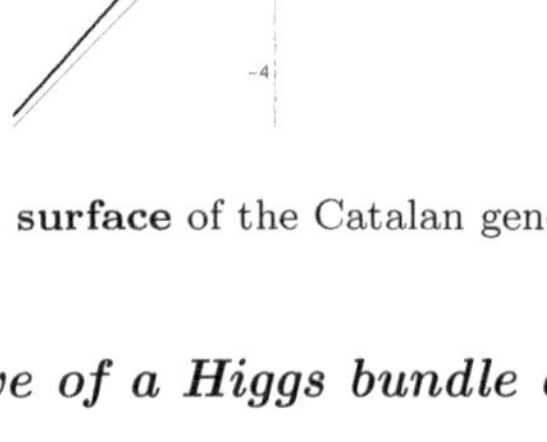

Fig. 2.5. The **Riemann surface** of the Catalan generating function $z = z(x)$.

2.2. *The spectral curve of a Higgs bundle and its desingularization*

To consider the quantization of the curve (2.11), we need to place it into a cotangent bundle. Here again, we use the base curve $\mathbb{P}^1$ and the same vector bundle E of (1.21) on it. As a Higgs field, we use

$$\phi := \begin{bmatrix} -xdx & -(dx)^2 \\ 1 & \end{bmatrix} : E \longrightarrow K_{\mathbb{P}^1}(4) \otimes E. \tag{2.12}$$

Here, $(dx)^2 \in H^0\big(\mathbb{P}^1, K_{\mathbb{P}^1}^{\otimes 2}(4)\big)$ is the unique (up to a constant factor) quadratic differential with an order 4 pole at infinity, and $xdx \in H^0\big(\mathbb{P}^1, K_{\mathbb{P}^1}(2)\big)$ is the unique meromorphic differential with a zero at $x = 0$ and a pole of order 3 at $x = \infty$. In the affine coordinate (x, y) of the Hirzebruch surface $\mathbb{F}^2$ as before, the spectral curve Σ is given by

$$\det\big(\eta - \pi^*(\phi)\big) = (y^2 + xy + 1)(dx)^2 = 0, \tag{2.13}$$

where $\pi : \mathbb{F}_2 \longrightarrow \mathbb{P}^1$ is the projection. Therefore, the generating function $z(x)$ of (2.10) gives a parametrization of the spectral curve

$$\begin{cases} x = z(x) + \frac{1}{z(x)} \\ y = -z(x). \end{cases} \tag{2.14}$$

In the other affine coordinate (u, w) of (1.25), the spectral curve is singular at $(u, w) = (0, 0)$:

$$u^4 - uw + w^2 = 0. \tag{2.15}$$

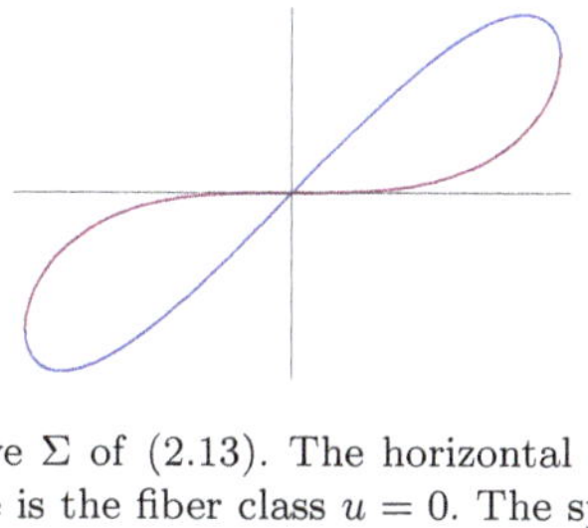

Fig. 2.6. The spectral curve Σ of (2.13). The horizontal line is the divisor $w = 0$ at infinity, and the vertical line is the fiber class $u = 0$. The spectral curve intersects with $w = 0$ four times. One of the two curve germ components is given by $w = u$, and the other by $w = u^3$.

Blow up $\mathbb{F}_2 = \overline{T^*\mathbb{P}^1}$ once at the nodal singularity $(u, w) = (0, 0)$ of the spectral curve Σ, and let $\widetilde{\Sigma} \longrightarrow \Sigma$ be the proper transform of Σ.

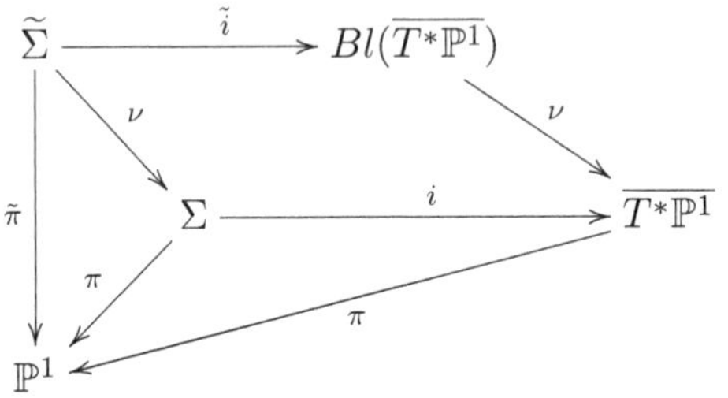

(2.16)

In terms of the coordinate w_1 defined by $w = w_1 u$, (2.15) becomes

$$u^2 + \left(w_1 - \frac{1}{2}\right)^2 = \frac{1}{4}, \tag{2.17}$$

which is the defining equation for $\widetilde{\Sigma}$. From this equation we see that its geometric genus is 0, hence it is just another $\mathbb{P}^1$. The covering $\tilde{\pi} : \widetilde{\Sigma} \longrightarrow \mathbb{P}^1$ is ramified at two points, corresponding to the original ramification points $(x, y) = (\pm 2, \mp 1)$ of $\pi : \Sigma \longrightarrow \mathbb{P}^1$. The rational parametrization of (2.17) is given by

$$\begin{cases} u = \frac{1}{2} \cdot \frac{t^2-1}{t^2+1} \\ w_1 = \frac{1}{2} - \frac{t}{t^2+1}, \end{cases} \tag{2.18}$$

where t is the affine coordinate of $\widetilde{\Sigma}$ so that $t = \pm 1$ gives $(u, w) = (0, 0)$. Indeed, the parameter t is a **normalization coordinate** of the spectral curve Σ:

$$\begin{cases} x = 2 + \frac{4}{t^2-1} \\ y = -\frac{t+1}{t-1}, \end{cases} \qquad \begin{cases} u = \frac{1}{2} \cdot \frac{t^2-1}{t^2+1} \\ w = \frac{1}{4} \cdot \frac{(t-1)^3(t+1)}{(t^2+1)^2}. \end{cases} \tag{2.19}$$

Although the expression of x and y in terms of the normalization coordinate is more complicated than (2.14), it is important to note that the spectral curve $\widetilde{\Sigma}$ is now non-singular.

2.3. *The generating function, or the Laplace transform*

For all (g, n) except for $(0, 1)$ and $(0, 2)$, let us introduce the generating function of (2.6) as follows:

$$F^C_{g,n}(x_1, \ldots, x_n) := \sum_{\mu_1 \geq 1, \ldots, \mu_n \geq 1} \frac{C_{g,n}(\mu_1, \ldots, \mu_n)}{\mu_1 \cdots \mu_n} x_1^{-\mu_1} \cdots x_n^{-\mu_n}. \tag{2.20}$$

If we consider $x_i = e^{w_i}$, then the above sum is just the discrete Laplace transform of the function in n integer variables:

$$\frac{C_{g,n}(\mu_1,\ldots,\mu_n)}{\mu_1\cdots\mu_n}.$$

Our immediate goal is to compute the Laplace transform as a holomorphic function. Since the only information we have now is the generalized Catalan recursion (2.8), how much can we say about this function? Actually, the following theorem is proved, all from using the recursion alone!

Theorem 2.7: *Let us consider the generating function* (2.20) *as a function in the normalization coordinates* $(t_1,\ldots,t_n)$ *satisfying*

$$x_i = 2 + \frac{4}{t_i^2 - 1}, \qquad i = 1,2,\ldots,n,$$

as identified in (2.19), *and by abuse of notation, we simply write it as* $F_{g,n}^C(t_1,\ldots,t_n)$. *For the range of* (g,n) *with the stability condition* $2g-2+n>0$, *we have the following.*

- *The generating function* $F_{g,n}^C(t_1,\ldots,t_n)$ *is a* ***Laurent polynomial*** *in the* t_i*-variables of the total degree* $3(2g-2+n)$.
- *The reciprocity relation holds:*

$$F_{g,n}^C(1/t_1,\ldots,1/t_n) = F_{g,n}^C(t_1,\ldots,t_n). \tag{2.21}$$

- *The special values at* $t_i = -1$ *are given by*

$$F_{g,n}^C(t_1,\ldots,t_n)\big|_{t_i=-1} = 0 \tag{2.22}$$

  for each i.
- *The diagonal value at* $t_i = 1$ *gives the orbifold Euler characteristic of the moduli space* $\mathcal{M}_{g,n}$*:*

$$F_{g,n}^C(1,\ldots,1) = (-1)^n\chi(\mathcal{M}_{g,n}). \tag{2.23}$$

- *The restriction of the Laurent polynomial* $F_{g,n}^C(t_1,\ldots,t_n)$ *to its highest degree terms gives a homogeneous polynomial defined by*

$$F_{g,n}^{C,\ highest}(t_1,\ldots,t_n) = \frac{(-1)^n}{2^{2g-2+n}} \sum_{\substack{d_1+\cdots+d_n\\=3g-3+n}} \langle\tau_{d_1}\cdots\tau_{d_n}\rangle_{g,n} \prod_{i=1}^n |2d_i-1|!! \left(\frac{t_i}{2}\right)^{2d_i+1}. \tag{2.24}$$

Thus the function $F^C_{g,n}(t_1,\dots,t_n)$ knows the orbifold Euler characteristic of $\mathcal{M}_{g,n}$, and all the cotangent class intersection numbers (1.11) for all values of (g,n) in the stable range! It is also striking that it is actually a Laurent polynomial, while the definition (2.20) is given only as a formal Laurent series. The reciprocity (2.21) is the reflection of the invariance of the spectral curve Σ under the rotation

$$(x,y)\longmapsto(-x,-y).$$

This surprising theorem is a consequence of the Laplace transform of the Catalan recursion itself.

Theorem 2.8: *Differential recursion, [93]. The Laplace transform $F^C_{g,n}(t_1,\dots,t_n)$ satisfies the following differential recursion equation for every (g,n) subject to $2g-2+n\geq 2$.*

$$\begin{aligned}
&\frac{\partial}{\partial t_1}F^C_{g,n}(t_1,\dots,t_n)\\
&\quad=-\frac{1}{16}\sum_{j=2}^n\left[\frac{t_j}{t_1^2-t_j^2}\left(\frac{(t_1^2-1)^3}{t_1^2}\frac{\partial}{\partial t_1}F^C_{g,n-1}(t_1,\dots,\widehat{t_j},\dots,t_n)\right.\right.\\
&\qquad\qquad\left.\left.-\frac{(t_j^2-1)^3}{t_j^2}\frac{\partial}{\partial t_j}F^C_{g,n-1}(t_2,\dots,t_n)\right)\right]\\
&\qquad-\frac{1}{16}\sum_{j=2}^n\frac{(t_1^2-1)^2}{t_1^2}\frac{\partial}{\partial t_1}F^C_{g,n-1}(t_1,\dots,\widehat{t_j},\dots,t_n)\\
&\quad-\frac{1}{32}\frac{(t_1^2-1)^3}{t_1^2}\left[\frac{\partial^2}{\partial u_1\partial u_2}F^C_{g-1,n+1}(u_1,u_2,t_2,t_3,\dots,t_n)\right]\Bigg|_{u_1=u_2=t_1}\\
&\quad-\frac{1}{32}\frac{(t_1^2-1)^3}{t_1^2}\sum_{\substack{g_1+g_2=g\\ I\sqcup J=\{2,3,\dots,n\}}}^{\text{stable}}\frac{\partial}{\partial t_1}F^C_{g_1,|I|+1}(t_1,t_I)\frac{\partial}{\partial t_1}F^C_{g_2,|J|+1}(t_1,t_J).
\end{aligned}\tag{2.25}$$

For a subset $I\subset\{1,2,\dots,n\}$, we denote $t_I=(t_i)_{i\in I}$. The "stable" summation means $2g_1+|I|-1>0$ and $2g_2+|J|-1>0$.

The differential recursion uniquely determines all $F^C_{g,n}(t_1,\dots,t_n)$ by integrating the right-hand side of (2.25) from -1 to t_1 with respect to the variable t_1. The initial conditions are

$$F^C_{1,1}(t)=-\frac{1}{384}\frac{(t+1)^4}{t^2}\left(t-4+\frac{1}{t}\right)\tag{2.26}$$

and

$$F^C_{0,3}(t_1,t_2,t_3) = -\frac{1}{16}(t_1+1)(t_2+1)(t_3+1)\left(1+\frac{1}{t_1\,t_2\,t_3}\right). \tag{2.27}$$

Remark 2.9: Theorem 2.8 is proved by Mincheng Zhou in his senior thesis [93]. It is the result of a rather difficult calculation of the Laplace transform of the Catalan recursion (2.8).

Remark 2.10: Theorem 2.7 has never been stated in this format before. Its proof follows from the results of [25,27,65,68], based on induction using (2.25). The essential point of the discovery of Theorem 2.7 is the use of the normalization coordinate t of (2.19). The authors almost accidentally found the coordinate transformation

$$z(x) = \frac{t+1}{t-1}$$

in [27]. Then in [25], we noticed that this coordinate was exactly the normalization coordinate that was naturally obtained in the blow-up process (2.16).

The uniqueness of the solution of (2.25) allows us to identify the solution $F^C_{g,n}$ with the Laplace transform of the number of lattice points in $\mathcal{M}_{g,n}$, as we see later in this section. Through this identification, (2.22) and (2.23) become obvious. The asymptotic behavior (2.24) follows from the lattice point approximation of the Euclidean volume of polytopes, and Kontsevich's theorem that identifies the Euclidean volume of $\mathcal{M}_{g,n}$ and the intersection numbers (1.11) on $\overline{\mathcal{M}}_{g,n}$.

2.4. *The unstable geometries and the initial value of the topological recursion*

The actual computation of the Laplace transform equation (2.25) from (2.8) requires the evaluation of the Laplace transforms of $C_{0,1}(\mu_1)$ and $C_{0,2}(\mu_1,\mu_2)$. It is done as follows.

Since a degree 0 vertex is allowed for the $(g,n)=(0,1)$ unstable geometry, we do not have the corresponding $F^C_{0,1}(x)$ in (2.20). Since

$$d_{x_1}\cdots d_{x_n}F^C_{g,n} = \sum_{\mu_1\ge 1,\dots,\mu_n\ge 1}(-1)^n C_{g,n}(\mu_1,\dots,\mu_n)x_1^{-\mu_1-1}\cdots x_n^{-\mu_n-1}dx_1\cdots dx_n,$$

we could choose

$$d_xF^C_{0,1}(x) = -\big(z(x)-x^{-1}\big)dx$$

as a defining equation for $F^C_{0,1}$.

In the light of (2.14), $ydx = -zdx$ is a natural global holomorphic 1-form on $T^*\mathbb{P}^1$, called the *tautological* 1-form. Its exterior differentiation $d(ydx) = dy \wedge dx$ defines the canonical holomorphic symplectic structure on $T^*\mathbb{P}^1$. Since we are interested in the *quantization* of the Hitchin spectral curve, we need a symplectic structure here, which is readily available for our use from ydx.

Therefore, it is reasonable for us to define the 'function' $F_{0,1}$ by

$$dF_{0,1} = ydx, \qquad \text{or} \qquad F_{0,1} = \int ydx. \tag{2.28}$$

Although this equation does not determine the constant term, it does not play any role for our purposes. Here, we choose the constant of integration to be 0. Since the symplectic structure on $T^*\mathbb{P}^1$ is non-trivial, we need to interpret the equation (2.28) being defined on the spectral curve, and be prepared that we may not obtain any meromorphic function on the spectral curve. For the Catalan case, we have to use the spectral curve coordinate z or t as a variable of $F^C_{g,n}$, instead of x. Since

$$-zdx = -zdz + \frac{dz}{z},$$

we conclude

$$\begin{aligned} F^C_{0,1}(z) &:= -\frac{1}{2}z^2 + \log z \\ &= -\frac{1}{2}\left(\frac{t+1}{t-1}\right)^2 + \log\left(\frac{t+1}{t-1}\right). \end{aligned} \tag{2.29}$$

We encounter a non-algebraic function here, indeed.

For the computation of the Laplace transform $F^C_{0,2}$, we need an explicit formula for $C_{0,2}(\mu_1, \mu_2)$. Luckily, such computation has been carried out in [54], fully utilizing the technique of the *dispersionless Toda lattice hierarchy*. It is surprising to see how much integrable system consideration is involved in computing such a simple quantity as $C_{0,2}(\mu_1, \mu_2)$! The result is the following.

Theorem 2.11: *Calculation of the 2-point Catalan numbers, [54]. For every $\mu_1, \mu_2 > 0$, the genus 0, 2-point Catalan numbers are given by*

$$C_{0,2}(\mu_1, \mu_2) = 2\frac{\lfloor\frac{\mu_1+1}{2}\rfloor\lfloor\frac{\mu_2+1}{2}\rfloor}{\mu_1+\mu_2}\binom{\mu_1}{\lfloor\frac{\mu_1}{2}\rfloor}\binom{\mu_2}{\lfloor\frac{\mu_2}{2}\rfloor}. \tag{2.30}$$

We refer to [27] for the derivation of $F^C_{0,2}$. The result is

$$\begin{aligned} F^C_{0,2}(z_1,z_2) &= -\log(1-z_1z_2) \\ &= -\log\big(-2(t_1+t_2)\big)+\log(t_1-1)+\log(t_2-1). \end{aligned} \tag{2.31}$$

For the purpose of later use, we note that

$$\begin{aligned} d_{t_1}d_{t_2}F^C_{0,2}(t_1,t_2) &= \frac{dt_1\cdot dt_2}{(t_1+t_2)^2} \\ &= \frac{dt_1\cdot dt_2}{(t_1-t_2)^2} - (\tilde{\pi}\times\tilde{\pi})^*\frac{dx_1\cdot dx_2}{(x_1-x_2)^2}, \end{aligned} \tag{2.32}$$

where $\tilde{\pi}:\widetilde{\Sigma}\longrightarrow\mathbb{P}^1$ is the projection of (2.16), i.e., the variable transformation (2.19).

2.5. *Geometry of the topological recursion*

Computation of a multi-resolvent (2.5) is one thing. What we have in front of us is a different story. We wish to compute an asymptotic expansion of a solution to the differential equation that is defined on the base curve C and gives the quantization of the Hitchin spectral curve of a meromorphic Higgs bundle (E,ϕ). The expansion has to be done at the essential singularity of the solution.

Question 2.12: Is there a mathematical framework suitable for such problems?

The discovery of [24, 25] gives an answer: **Generalize the formalism of Eynard and Orantin of [34] to the geometric situation of meromorphic Higgs bundles. Then this generalized topological recursion computes the asymptotic expansion in question.**

We are now ready to present the topological recursion, continuing our investigation of the particular example of Catalan numbers. The point here is that **the topological recursion is a universal formula depending only on geometry**. Therefore, we can arrive at the general formula from any example. *One example rules them all!*

Theorem 2.13: *The topological recursion for the generalized Catalan numbers, [27]. Define symmetric n-linear differential forms on $(\widetilde{\Sigma})^n$ for $2g-2+n>0$ by*

$$W^C_{g,n}(t_1,\ldots,t_n) := d_{t_1}\cdots d_{t_n}F^C_{g,n}(t_1,\ldots,t_n), \tag{2.33}$$

and for $(g,n)=(0,2)$ *by*

$$W^C_{0,2}(t_1,t_2) := \frac{dt_1\cdot dt_2}{(t_1-t_2)^2}. \tag{2.34}$$

Then these differential forms satisfy the following integral recursion equation, called the ***topological recursion****.*

$$\begin{aligned} W^C_{g,n}(t_1,\dots,t_n) &= -\frac{1}{64}\frac{1}{2\pi i}\int_\gamma \left(\frac{1}{t+t_1}+\frac{1}{t-t_1}\right)\frac{(t^2-1)^3}{t^2}\cdot\frac{1}{dt}\cdot dt_1 \\ &\times \Bigg[\sum_{j=2}^n \Big(W^C_{0,2}(t,t_j)W^C_{g,n-1}(-t,t_2,\dots,\widehat{t_j},\dots,t_n) \\ &\qquad + W^C_{0,2}(-t,t_j)W^C_{g,n-1}(t,t_2,\dots,\widehat{t_j},\dots,t_n)\Big) \\ &+ W^C_{g-1,n+1}(t,-t,t_2,\dots,t_n) + \sum^{stable}_{\substack{g_1+g_2=g\\ I\sqcup J=\{2,3,\dots,n\}}} W^C_{g_1,|I|+1}(t,t_I)W^C_{g_2,|J|+1}(-t,t_J)\Bigg]. \end{aligned} \tag{2.35}$$

The last sum is restricted to the stable geometries, i.e., the partitions should satisfy $2g_1-1+|I|>0$ *and* $2g_2-1+|J|$*, as in* (2.25)*. The contour integral is taken with respect to* t *on the exactly the same cycle defined by Figure 1.4 as before, where* t *is the normalization coordinate of* (2.19)*. Note that the second and the third lines of* (2.35) *is a quadratic differential in the variable* t*.*

Remark 2.14: The notation $\frac{1}{dt}$ requires a justification. We note that the ratio of two global meromorphic sections of the same line bundle is a global meromorphic function. Here we are taking the ratio of two meromorphic 1-forms on the factor $\widetilde{\Sigma}$ corresponding to the t-variable. Thus after taking this ratio, the integrand becomes a meromorphic 1-form in $(-t)$-variable, which is integrated along the cycle γ.

Remark 2.15: The recursion (2.35) is a genuine induction formula with respect to $2g-2+n$. Thus from $W^C_{0,2}$, we can calculate all $W^C_{g,n}$s one by one. This is a big difference between (2.35) and (2.8). The latter relation contains terms with $C_{g,n}$ in the right-hand side as well, therefore, $C_{g,n}$ is not determined as a function by an induction procedure.

Remark 2.16: If we apply (2.33) to $F^C_{0,2}$ of (2.31), then we obtain (2.32), not (2.34). The difference is the pull-back of the 2-form $\frac{dx_1\cdot dx_2}{(x_1-x_2)^2}$. This difference does not affect the recursion formula (2.35) for $2g-2+n>1$.

The only case affected is $(g,n) = (1,1)$. The above recursion allows us to calculate $W^C_{1,1}$ from $W^C_{0,2}$ as we see below, but we cannot use (2.32) in place of $W^C_{0,2}$ for this case because the specialization $t_1 = t, t_2 = -t$ does not make sense in $d_{t_1} d_{t_2} F_{0,2}(t_1, t_2)$.

Remark 2.17: In Subsection 3.2, we will formulate the topological recursion as a universal formalism for the context of Hitchin spectral curves in a coordinate-free manner depending only on the geometric setting. There we will explain the meaning of $W_{0,2}$, and the formula for the topological recursion. At this moment, we note that the origin of the topological recursion is the edge-contraction mechanism of the Catalan recursion (2.8). There is a surprising relation between the edge-contraction operations and two dimensional topological quantum field theories. We refer to [26] for more detail.

Remark 2.18: The integral recursion (2.35) and the PDE recursion (2.25) are equivalent in the range of $2g-2+n \geq 2$, since if we know $W^C_{g,n}$ from the integral recursion, then we can calculate $F^C_{g,n}$ by the integration

$$F^C_{g,n}(t_1, \ldots, t_n) = \int_{-1}^{t_1} \cdots \int_{-1}^{t_n} W^C_{g,n}.$$

But the differential recursion does not provide any mechanism to calculate $F^C_{1,1}$ and $F^C_{0,3}$.

To see how the topological recursion works, let us compute $W^C_{1,1}(t_1)$ from (2.35).

$$\begin{aligned}
W^C_{1,1}(t_1) &= -\frac{1}{64}\frac{1}{2\pi i}\left[\int_\gamma \left(\frac{1}{t+t_1} + \frac{1}{t-t_1}\right)\frac{(t^2-1)^3}{t^2} \cdot \frac{1}{dt} \cdot W^C_{0,2}(t,-t)\right] dt_1 \\
&= -\frac{1}{64}\frac{1}{2\pi i}\left[\int_\gamma \left(\frac{1}{t+t_1} + \frac{1}{t-t_1}\right)\frac{(t^2-1)^3}{t^2} \cdot \frac{1}{dt} \cdot \left(\frac{-(dt)^2}{(2t)^2}\right)\right] dt_1 \\
&= \frac{1}{256}\frac{1}{2\pi i}\left[\int_\gamma \left(\frac{1}{t+t_1} + \frac{1}{t-t_1}\right)\frac{(t^2-1)^3}{t^2} \cdot \frac{1}{t^2} \cdot dt\right] dt_1 \\
&= -\frac{1}{128}\frac{(t_1^2-1)^3}{t_1^4}.
\end{aligned}$$

Here, we changed the contour integral to the negative of the residue calculations at $t = \pm t_1$, as indicated in Figure 1.4. From (2.26), we find that indeed

$$F^C_{1,1}(t_1) = \int_{-1}^{t_1} W^C_{1,1}.$$

As explained in [27], we can calculate $W^C_{0,3}$ from $W^C_{0,2}$ as well, which recovers the initial condition (3.39).

To understand the geometry behind the topological recursion, we need to identify each term of the formula. First, recall the normalization morphism

$$\tilde{\pi} : \widetilde{\Sigma} \longrightarrow \mathbb{P}^1$$

of (2.16). We note that the transformation $t \longmapsto -t$ appearing in the recursion formula is the **Galois conjugation** of the global Galois covering $\tilde{\pi} : \widetilde{\Sigma} \longrightarrow \mathbb{P}^1$. From (2.19), we see that this transformation is induced by the involution $y \longmapsto -y$ of $T^*\mathbb{P}^1$. The fixed point set of the Galois conjugation is the set of ramification points of the covering $\tilde{\pi}$, and the residue integration of (2.35) is taken around the two ramification points.

We claim that $W^C_{0,2}(t_1, t_2)$ is the *Cauchy differentiation kernel* on $\widetilde{\Sigma}$. This comes from the intrinsic geometry of the curve $\widetilde{\Sigma}$. The Cauchy differentiation kernel on $\mathbb{P}^1$ is the unique meromorphic symmetric bilinear differential form

$$c(t_1, t_2) := \frac{dt_1 \cdot dt_2}{(t_1 - t_2)^2} \tag{2.36}$$

on $\mathbb{P}^1 \times \mathbb{P}^1$ such that

$$df(t_2) = q_*\big(c(t_1, t_2) p^* f(t_1)\big) \tag{2.37}$$

for every rational function f on $\mathbb{P}^1$. Here, p and q are the projection maps

$$\begin{array}{ccc} & \mathbb{P}^1 \times \mathbb{P}^1 & \\ {}^{p}\swarrow & & \searrow^{q} \\ \mathbb{P}^1 & & \mathbb{P}^1 \end{array} \tag{2.38}$$

to the first and second factors. The push-forward q_* is defined by the integral

$$q_*\big(c(t_1, t_2) p^* f(t_1)\big) = \frac{1}{2\pi i} \oint c(t_1, t_2) f(t_1)$$

along a small loop in the fiber $q^*(t_2)$ that is centered at its intersection with the diagonal of $\Delta \subset \mathbb{P}^1 \times \mathbb{P}^1$. In terms of the other affine coordinate $u_i = 1/t_i$ of $\mathbb{P}^1$, we have

$$\frac{dt_1 \cdot dt_2}{(t_1 - t_2)^2} = \frac{du_1 \cdot du_2}{(u_1 - u_2)^2} = -\frac{dt_1 \cdot du_2}{(t_1 u_2 - 1)^2} = -\frac{du_1 \cdot dt_2}{(u_1 t_2 - 1)^2}.$$

Therefore, $c(t_1, t_2)$ is a globally defined bilinear meromorphic form on $\mathbb{P}^1 \times \mathbb{P}^1$ with poles only along the diagonal Δ.

Remark 2.19: The Cauchy integration formula

$$f(w) = \frac{1}{2\pi i} \oint_{|z-w|<\epsilon} f(z) \frac{dz}{z-w}$$

is the most useful formula in complex analysis. The Cauchy integration kernel $dz/(z-w)$ is a globally defined meromorphic 1-form on $\mathbb{C}$ that has only one pole, and its order is 1. It has to be noted that on a compact Riemann surface Σ, we do not have such a form. The best we can so is the meromorphic 1-form $\omega_\Sigma^{a-b}(t)$ with the following properties:

- It is a globally defined meromorphic 1-form with a pole of order 1 and residue 1 at a and a pole of order 1 and residue -1 at b for some pair (a,b) of distinct points of Σ.
- It is holomorphic in t except for $t=a$ and $t=b$.

Since we can add any holomorphic 1-form to $\omega_\Sigma^{a-b}(t)$ without violating the characteristic properties, the ambiguity of this form is exactly $H^0(\Sigma, K_\Sigma)$. Therefore, it is unique on $\Sigma = \mathbb{P}^1$, and is given by

$$\omega_{\mathbb{P}^1}^{a-b}(t) = \left(\frac{1}{t-a} - \frac{1}{t-b}\right) dt. \tag{2.39}$$

Note that

$$d_{t_2}\omega_{\mathbb{P}^1}^{t_2-b}(t_1) = c(t_1,t_2)$$

for any b.

Let us go back to the topological recursion for the Catalan numbers (2.35). Since all $W_{g,n}^C$ are determined by $W_{0,2}^C$ using the recursion, the remaining quantity we need to identify is the *integration kernel.* Recall the tautological 1-form

$$\eta_{\mathbb{P}^1} := y dx$$

on the tangent bundle $T^*\mathbb{P}^1$. We can see from (2.16) that its pull-back to the normalized spectral curve $\widetilde{\Sigma}$ is given by

$$\begin{aligned}\eta(t) &= -\left(\frac{t+1}{t-1}\right) d\left(\frac{4}{t^2-1}\right) \\ &= \frac{8t}{(t+1)(t-1)^3}\, dt.\end{aligned} \tag{2.40}$$

The factor

$$-\frac{1}{64}\left(\frac{1}{t+t_1} + \frac{1}{t-t_1}\right) \frac{(t^2-1)^3}{t^2} \cdot \frac{1}{dt} \cdot dt_1$$

in (2.35) is called the integration kernel, which is equal to

$$\frac{1}{2}\frac{\omega_{\mathbb{P}^1}^{(-t)-t}(t_1)}{\eta(-t)-\eta(t)}. \tag{2.41}$$

The integration kernel appears in this form in more general situations.

2.6. *The quantum curve for Catalan numbers*

There is yet another important role the differential recursion (2.25) plays. This is the derivation of the quantum curve equation for the Catalan case. In terms of the coordinates (x, y) of (2.13), the spectral curve is $y^2+xy+1=0$. The generating function $z(x)$ of (2.10), the normalization coordinate t of (2.19), and the base curve coordinate x on $\mathbb{P}^1$ are related by

$$t = t(x) := \frac{z(x)+1}{z(x)-1}, \tag{2.42}$$

which we consider as a function of x that gives the branch of $\widetilde{\Sigma}$ determined by

$$\lim_{x\to+\infty} t(x) = -1.$$

Theorem 2.20: *Quantum curve for generalized Catalan numbers, [68]. Define*

$$\Psi(t,\hbar) := \exp\left(\sum_{g=0}^{\infty}\sum_{n=1}^{\infty}\frac{1}{n!}\hbar^{2g-2+n}F_{g,n}^{C}(t,t,\ldots,t)\right), \tag{2.43}$$

incorporating (2.25), (2.29), *and* (2.31). *Then as a function in x through* (2.42), $\Psi\big(t(x),\hbar\big)$ *satisfies the following differential equation*

$$\left[\left(\hbar\frac{d}{dx}\right)^2 + x\left(\hbar\frac{d}{dx}\right)+1\right]\Psi\big(t(x),\hbar\big) = 0. \tag{2.44}$$

The semi-classical limit of (2.44) *using $S_0(x) = F_{0,1}^{C}\big(t(x)\big)$ coincides with the spectral curve*

$$y^2 + xy + 1 = 0$$

of (2.13).

Remark 2.21: Since $t = t(x)$ is a complicated function, it is surprising to see that $\Psi(t,\hbar)$ satisfies such a simple equation as (2.44).

Remark 2.22: The definition (2.43) and the meaning of (2.44) is the same as in the situation we have explained in the Introduction. First, we define

$$S_m(x) := \sum_{2g-2+n=m-1} \frac{1}{n!} F^C_{g,n}\big(t(x), t(x), \ldots, t(x)\big) \tag{2.45}$$

for every $m \geq 0$. We then interpret (2.44) as

$$\left[e^{-\frac{1}{\hbar}S_0(x)}\left(\left(\hbar\frac{d}{dx}\right)^2 + x\left(\hbar\frac{d}{dx}\right) + 1\right)e^{\frac{1}{\hbar}S_0(x)}\right]\exp\left(\sum_{m=1}^{\infty}\hbar^{m-1}S_m(x)\right) = 0. \tag{2.46}$$

To prove Theorem 2.20, let us recall a lemma from [68]:

Lemma 2.23: *Consider an open coordinate neighborhood $U \subset \Sigma$ of a projective algebraic curve Σ with a coordinate t. Let $f(t_1, \ldots, t_n)$ be a meromorphic symmetric function in n variables defined on Σ^n. Then on the coordinate neighborhood U^n, we have*

$$\begin{aligned}
\frac{d}{dt}f(t,t,\ldots,t) &= n\left[\frac{\partial}{\partial u}f(u,t,\ldots,t)\right]\Bigg|_{u=t};\\
\frac{d^2}{dt^2}f(t,t,\ldots,t) &= n\left[\frac{\partial^2}{\partial u^2}f(u,t,\ldots,t)\right]\Bigg|_{u=t}\\
&\quad + n(n-1)\left[\frac{\partial^2}{\partial u_1 \partial u_2}f(u_1,u_2,t,\ldots,t)\right]\Bigg|_{u_1=u_2=t}.
\end{aligned} \tag{2.47}$$

For a meromorphic function in one variable $f(t)$ on Σ, we have

$$\lim_{t_2\to t_1}\left[\omega_\Sigma^{t_2-b}(t_1)\big(f(t_1)-f(t_2)\big)\right] = d_1 f(t_1), \tag{2.48}$$

where $\omega_\Sigma^{t_2-b}(t_1)$ is the 1-form of Remark 2.19.

Proof: [Proof of Theorem 2.20] Differential forms can be restricted to a subvariety, but partial differential equations cannot be restricted to a subvariety in general. Therefore, it is non-trivial that (2.25) has any meaningful restriction to the diagonal of $(\mathbb{P}^1)^n$. Our strategy is the following. First, we expand (2.46) with respect to powers of $\hbar$, and derive an infinite system of ordinary differential equations for a finite collection of $S_m(x)$'s. (This method is known as the WKB analysis.) We then prove that these ODEs are exactly the **principal specialization** of (2.25), using (2.47).

Let

$$F(x,\hbar) := \sum_{m=0}^{\infty}\hbar^{m-1}S_m(x). \tag{2.49}$$

Unlike the ill-defined expression (2.43), $F(x,\hbar)$ is just a generating function of $S_m(x)$'s. Then (2.44), interpreted as (2.46), is equivalent to

$$\hbar^2 \frac{d^2}{dx^2}F + \hbar^2 \frac{dF}{dx}\frac{dF}{dx} + x\hbar\frac{dF}{dx} + 1 = 0. \tag{2.50}$$

The $\hbar$-expansion of (2.50) gives

$$\hbar^0\text{-terms}: \quad (S_0'(x))^2 + xS_0'(x) + 1 = 0, \tag{2.51}$$

$$\hbar^1\text{-terms}: \quad 2S_0'(x)S_1'(x) + S_0''(x) + xS_1'(x) = 0, \tag{2.52}$$

$$\hbar^{m+1}\text{-terms}: \quad S_m''(x) + \sum_{a+b=m+1} S_a'(x)S_b'(x) + xS_{m+1}'(x) = 0, \quad m \geq 1, \tag{2.53}$$

where $'$ denotes the x-derivative. The WKB method is to solve these equations iteratively and find $S_m(x)$ for all $m \geq 0$. Here, (2.51) is the **semi-classical limit** of (2.44), and (2.52) is the *consistency condition* we need for solving the WKB expansion. Since $dS_0(x) = ydx$, we have $y = S_0'(x)$. Thus (2.51) follows from the spectral curve equation (2.13) and the definition of $F_{0,1}^C(t)$ of (2.29).

Recalling $x = z + 1/z$, we obtain

$$\frac{d}{dx} = \frac{z^2}{z^2-1}\frac{d}{dz}.$$

In z-coordinate, $F_{0,2}^C(z,z) = -\log(1-z^2)$, which follows from (2.31). Therefore,

$$S_1'(x) = -\frac{1}{2}\frac{z^2}{z^2-1}\frac{d}{dz}\log(1-z^2) = -\frac{z^3}{(z^2-1)^2}.$$

We can then calculate

$$\begin{aligned} 2S_0'(x)S_1'(x)+S_0''(x)+xS_1'(x) &= 2z\frac{z^3}{(z^2-1)^2} - \frac{z^2}{z^2-1} - \left(z+\frac{1}{z}\right)\frac{z^3}{(z^2-1)^2} \\ &= \frac{1}{(z^2-1)^2}\left(2z^4 - z^2(z^2-1) - z^4 - z^2\right) \\ &= 0. \end{aligned}$$

Therefore, (2.52) holds. We refer to [25] for the proof of (2.53) from (2.25) and (2.23). □

Remark 2.24: The solution $\Psi(t(x),\hbar)$ is a formal section of $K_{\mathbb{P}^1}^{-\frac{1}{2}}$, as before. Note that the quantum curve (2.44) has an irregular singularity at

$x = \infty$, and hence its solution has an essential singularity at infinity. The expression (2.43) gives the asymptotic expansion of a solution around

$$t \to -1 \Longleftrightarrow x \to \infty.$$

The expansion in $1/x$ for $\hbar > 0$ is given in (3.54) below, using the Tricomi confluent hypergeometric function.

2.7. *Counting lattice points on the moduli space $\mathcal{M}_{g,n}$*

The topological recursion (2.35) is a consequence of (2.25), and the PDE recursion (2.25) is essentially the Laplace transform of the combinatorial formula (2.8). Then from where does the relation to the intersection numbers (2.24) arise?

To see this relation, we need to consider the dual of the cell graphs. They are commonly known as Grothendieck's dessins d'enfants (see for example, [79]), or ribbon graphs, as mentioned earlier. Recall that a ribbon graph has unlabeled vertices and edges, but faces are labeled. A *metric* ribbon graph is a ribbon graph with a positive real number (the length) assigned to each edge. For a given ribbon graph Γ with $e = e(\Gamma)$ edges, the space of metric ribbon graphs is $\mathbb{R}_+^{e(\Gamma)}/\mathrm{Aut}(\Gamma)$, where the automorphism group acts by permutations of edges (see [64, Section 1]). We restrict ourselves to the case that $\mathrm{Aut}(\Gamma)$ fixes each 2-cell of the cell-decomposition. We also require that every vertex of a ribbon graph has degree (i.e., valence) 3 or more. Using the canonical holomorphic coordinate systems on a topological surface of [64, Section 4] and the Strebel differentials [82], we have an isomorphism of topological orbifolds

$$\mathcal{M}_{g,n} \times \mathbb{R}_+^n \cong RG_{g,n}. \tag{2.54}$$

Here

$$RG_{g,n} = \coprod_{\substack{\Gamma \text{ ribbon graph} \\ \text{of type } (g,n)}} \frac{\mathbb{R}_+^{e(\Gamma)}}{\mathrm{Aut}(\Gamma)}$$

is the orbifold consisting of metric ribbon graphs of a given topological type (g, n). The gluing of orbi-cells is done by making the length of a non-loop edge tend to 0. The space $RG_{g,n}$ is a smooth orbifold (see [64, Section 3] and [81]). We denote by $\pi : RG_{g,n} \longrightarrow \mathbb{R}_+^n$ the natural projection via (2.54), which is the assignment of the collection of perimeter lengths of each boundary to a given metric ribbon graph.

Consider a ribbon graph Γ whose faces are labeled by $[n] = \{1, 2, \ldots, n\}$. For the moment let us give a label to each edge of Γ by an index set $[e] = \{1, 2, \ldots, e\}$. The edge-face incidence matrix is then defined by

$$\begin{aligned} A_\Gamma &= \left[a_{i\eta}\right]_{i\in[n],\ \eta\in[e]}; \\ a_{i\eta} &= \text{ the number of times edge } \eta \text{ appears in face } i. \end{aligned}$$

Thus $a_{i\eta} = 0, 1$, or 2, and the sum of the entries in each column is always 2. The Γ contribution of the space $\pi^{-1}(p_1, \ldots, p_n) = RG_{g,n}(\mathbf{p})$ of metric ribbon graphs with a prescribed perimeter $\mathbf{p} = (p_1, \ldots, p_n)$ is the orbifold polytope

$$P_\Gamma(\mathbf{p})/\mathrm{Aut}(\Gamma), \qquad P_\Gamma(\mathbf{p}) = \{\mathbf{x} \in \mathbb{R}_+^e \mid A_\Gamma \mathbf{x} = \mathbf{p}\},$$

where $\mathbf{x} = (\ell_1, \ldots, \ell_e)$ is the collection of edge lengths of the metric ribbon graph Γ. We have

$$\sum_{i\in[n]} p_i = \sum_{i\in[n]} \sum_{\eta\in[e]} a_{i\eta}\ell_\eta = 2 \sum_{\eta\in[e]} \ell_\eta.$$

We recall the topological recursion for the number of metric ribbon graphs $RG_{g,n}^{\mathbb{Z}_+}$ whose edges have integer lengths, following [15]. We call such a ribbon graph an *integral ribbon graph*. We can interpret an integral ribbon graph as Grothendieck's *dessin d'enfant* by considering an edge of integer length as a chain of edges of length one connected by bivalent vertices, and reinterpreting the notion of $\mathrm{Aut}(\Gamma)$ suitably. Since we do not go into the number theoretic aspects of dessins, we stick to the more geometric notion of integral ribbon graphs.

Definition 2.25: The weighted number $\left|RG_{g,n}^{\mathbb{Z}_+}(\mathbf{p})\right|$ of integral ribbon graphs with prescribed perimeter lengths $\mathbf{p} \in \mathbb{Z}_+^n$ is defined by

$$N_{g,n}(\mathbf{p}) = \left|RG_{g,n}^{\mathbb{Z}_+}(\mathbf{p})\right| = \sum_{\substack{\Gamma \text{ ribbon graph} \\ \text{of type } (g,n)}} \frac{\left|\{\mathbf{x} \in \mathbb{Z}_+^{e(\Gamma)} \mid A_\Gamma \mathbf{x} = \mathbf{p}\}\right|}{|\mathrm{Aut}(\Gamma)|}. \tag{2.55}$$

Since the finite set $\{\mathbf{x} \in \mathbb{Z}_+^{e(\Gamma)} \mid A_\Gamma \mathbf{x} = \mathbf{p}\}$ is a collection of lattice points in the polytope $P_\Gamma(\mathbf{p})$ with respect to the canonical integral structure $\mathbb{Z} \subset \mathbb{R}$ of the real numbers, $N_{g,n}(\mathbf{p})$ can be thought of counting the number of *lattice points* in $RG_{g,n}(\mathbf{p})$ with a weight factor $1/|\mathrm{Aut}(\Gamma)|$ for each ribbon graph. The function $N_{g,n}(\mathbf{p})$ is a symmetric function in $\mathbf{p} = (p_1, \ldots, p_n)$ because the summation runs over all ribbon graphs of topological type (g, n).

Remark 2.26: Since the integral vector $\mathbf{x}$ is restricted to take strictly positive values, we would have $N_{g,n}(\mathbf{p}) = 0$ if we were to substitute $\mathbf{p} = 0$. This normalization is natural from the point of view of lattice point counting and Grothendieck's *dessins d'enphants*. However, we do not make such a substitution in these lectures because we consider $\mathbf{p}$ as a strictly positive integer vector. This situation is similar to Hurwitz theory [33, 69], where a partition μ is a strictly positive integer vector that plays the role of our $\mathbf{p}$. We note that a different assignment of values was suggested in [72, 73].

For brevity of notation, we denote by $p_I = (p_i)_{i\in I}$ for a subset $I \in [n] = \{1, 2\ldots, n\}$. The cardinality of I is denoted by $|I|$. The following topological recursion formula was proved in [15] using the idea of ciliation of a ribbon graph.

Theorem 2.27: *[15] The number of integral ribbon graphs with prescribed boundary lengthes satisfies the topological recursion formula*

$$\begin{aligned}
p_1 N_{g,n}(p_{[n]}) &= \frac{1}{2}\sum_{j=2}^{n}\Bigg[\sum_{q=0}^{p_1+p_j} q(p_1+p_j-q)N_{g,n-1}(q, p_{[n]\setminus\{1,j\}}) \\
&\quad + H(p_1-p_j)\sum_{q=0}^{p_1-p_j} q(p_1-p_j-q)N_{g,n-1}(q, p_{[n]\setminus\{1,j\}}) \\
&\quad - H(p_j-p_1)\sum_{q=0}^{p_j-p_1} q(p_j-p_1-q)N_{g,n-1}(q, p_{[n]\setminus\{1,j\}})\Bigg] \\
&\quad + \frac{1}{2}\sum_{0\le q_1+q_2\le p_1} q_1q_2(p_1-q_1-q_2)\Bigg[N_{g-1,n+1}(q_1, q_2, p_{[n]\setminus\{1\}}) \\
&\quad + \sum_{\substack{g_1+g_2=g\\ I\sqcup J=[n]\setminus\{1\}}}^{\text{stable}} N_{g_1,|I|+1}(q_1, p_I)N_{g_2,|J|+1}(q_2, p_J)\Bigg]. \qquad (2.56)
\end{aligned}$$

Here

$$H(x) = \begin{cases} 1 & x > 0 \\ 0 & x \le 0 \end{cases}$$

is the Heaviside function, and the last sum is taken for all partitions $g = g_1 + g_2$ and $I \sqcup J = \{2, 3, \ldots, n\}$ subject to the stability conditions $2g_1 - 1 + I > 0$ and $2g_2 - 1 + |J| > 0$.

For a fixed (g, n) in the stable range, *i.e.*, $2g - 2 + n > 0$, we choose n variables $t_1, t_2, \ldots, t_n$, and define the function

$$z(t_i, t_j) = \frac{(t_i + 1)(t_j + 1)}{2(t_i + t_j)}. \tag{2.57}$$

An edge η of a ribbon graph Γ bounds two faces, say i_η and j_η. These two faces may be actually the same. Now we define the *Poincaré polynomial* [65] of $RG_{g,n}$ in the z-variables by

$$F^P_{g,n}(t_1, \ldots, t_n) = \sum_{\substack{\Gamma \text{ ribbon graph} \\ \text{of type } (g,n)}} \frac{(-1)^{e(\Gamma)}}{|\mathrm{Aut}(\Gamma)|} \prod_{\substack{\eta \text{ edge} \\ \text{of } \Gamma}} z(t_{i_\eta}, t_{j_\eta}), \tag{2.58}$$

which is a polynomial in $z(t_i, t_j)$ but actually a symmetric rational function in $t_1, \ldots, t_n$.

Let us consider the *Laplace transform*

$$L_{g,n}(w_1, \ldots, w_n) \stackrel{\text{def}}{=} \sum_{\mathbf{p} \in \mathbb{Z}^n_+} N_{g,n}(\mathbf{p}) e^{-\langle \mathbf{p}, w \rangle} \tag{2.59}$$

of the number of integral ribbon graphs $N_{g,n}(\mathbf{p})$, where $\langle \mathbf{p}, w \rangle = p_1 w_1 + \cdots + p_n w_n$, and the summation is taken over all integer vectors $\mathbf{p} \in \mathbb{Z}^n_+$ of strictly positive entries. We shall prove that after the coordinate change of [15] from the w-coordinates to the t-coordinates defined by

$$e^{-w_j} = \frac{t_j + 1}{t_j - 1}, \qquad j = 1, 2, \ldots, n, \tag{2.60}$$

the Laplace transform $L_{g,n}(w_{[n]})$ becomes the Poincaré polynomial

$$F^P_{g,n}(t_1, \ldots, t_n) = L_{g,n}\big(w_1(t), \ldots, w_n(t)\big).$$

The Laplace transform $L_{g,n}(w_N)$ can be evaluated using the definition of the number of integral ribbon graphs (2.55). Let a_η be the η-th column of the incidence matrix A_Γ so that

$$A_\Gamma = \big[a_1 \big| a_2 \big| \cdots \big| a_{e(\Gamma)}\big].$$

Then

$$\begin{aligned} L_{g,n}(w_{[n]}) &= \sum_{\mathbf{p} \in \mathbb{Z}^n_+} N_{g,n}(\mathbf{p}) e^{-\langle \mathbf{p}, w \rangle} \\ &= \sum_{\substack{\Gamma \text{ ribbon graph} \\ \text{of type } (g,n)}} \sum_{\mathbf{p} \in \mathbb{Z}^n_+} \frac{1}{|\mathrm{Aut}(\Gamma)|} \Big| \{ \mathbf{x} \in \mathbb{Z}^{e(\Gamma)}_+ \mid A_\Gamma \mathbf{x} = \mathbf{p} \} \Big| e^{-\langle \mathbf{p}, w \rangle} \end{aligned}$$

$$= \sum_{\substack{\Gamma \text{ ribbon graph} \\ \text{of type } (g,n)}} \frac{1}{|\mathrm{Aut}(\Gamma)|} \sum_{\mathbf{x}\in\mathbb{Z}_+^{e(\Gamma)}} e^{-\langle A_\Gamma \mathbf{x}, w\rangle}$$

$$= \sum_{\substack{\Gamma \text{ ribbon graph} \\ \text{of type } (g,n)}} \frac{1}{|\mathrm{Aut}(\Gamma)|} \prod_{\substack{\eta \text{ edge} \\ \text{of } \Gamma}} \sum_{\ell_\eta=1}^{\infty} e^{-\langle a_\eta, w\rangle \ell_\eta}$$

$$= \sum_{\substack{\Gamma \text{ ribbon graph} \\ \text{of type } (g,n)}} \frac{1}{|\mathrm{Aut}(\Gamma)|} \prod_{\substack{\eta \text{ edge} \\ \text{of } \Gamma}} \frac{e^{-\langle a_\eta, w\rangle}}{1-e^{-\langle a_\eta, w\rangle}}. \tag{2.61}$$

Every edge η bounds two faces, which we call face i_η^+ and face i_η^-. When $a_{i\eta} = 2$, these faces are the same. We then calculate

$$\frac{e^{-\langle a_\eta, w\rangle}}{1-e^{-\langle a_\eta, w\rangle}} = -z\left(t_{i_\eta^+}, t_{i_\eta^-}\right). \tag{2.62}$$

This follows from (2.60) and

$$\frac{e^{-(w_i+w_j)}}{1-e^{-(w_i+w_j)}} = \frac{\frac{(t_i+1)(t_j+1)}{(t_i-1)(t_j-1)}}{1-\frac{(t_i+1)(t_j+1)}{(t_i-1)(t_j-1)}} = -\frac{(t_i+1)(t_j+1)}{2(t_i+t_j)} = -z(t_i,t_j),$$

$$\frac{e^{-2w_i}}{1-e^{-2w_i}} = -\frac{(t_i+1)^2}{4t_i} = -z(t_i,t_i).$$

Note that since $z(t_i,t_j)$ is a symmetric function, which face is named i_η^+ or i_η^- does not matter. From (2.61) and (2.62), we have established

Theorem 2.28: *The Poincaré polynomials and the Laplace transform, [65]. The Laplace transform $L_{g,n}(w_{[n]})$ in terms of the t-coordinates* (2.60) *is the Poincaré polynomial*

$$F^P_{g,n}(t_{[n]}) = \sum_{\substack{\Gamma \textit{ ribbon graph} \\ \textit{of type } (g,n)}} \frac{(-1)^{e(\Gamma)}}{|Aut(\Gamma)|} \prod_{\substack{\eta \textit{ edge} \\ \textit{of } \Gamma}} z\left(t_{i_\eta^+}, t_{i_\eta^-}\right). \tag{2.63}$$

Corollary 2.29: *The evaluation of $F^P_{g,n}(t_{[n]})$ at $t_1 = \cdots = t_n = 1$ gives the Euler characteristic of $RG_{g,n}$*

$$F^P_{g,n}(1,1\ldots,1) = \chi(RG_{g,n}) = (-1)^n \chi(\mathcal{M}_{g,n}). \tag{2.64}$$

Furthermore, if we evaluate at $t_j = -1$ for any j, then we have

$$F^P_{g,n}(t_1,t_2\ldots,t_n)\big|_{t_j=-1} = 0 \tag{2.65}$$

as a function in the rest of the variables $t_{[n]\setminus\{j\}}$.

Proof: The Euler characteristic calculation immediately follows from $z(1,1)=1$.

Consider a ribbon graph Γ of type (g,n). Its j-th face has at least one edge on its boundary. Therefore,

$$\prod_{\eta \text{ edge of } \Gamma} z\left(t_{i_\eta^+}, t_{i_\eta^-}\right)$$

has a factor (t_j+1) by (2.57). It holds for every ribbon graph Γ in the summation of (2.63). Therefore, (2.65) follows. □

The following theorem is established in [65]. Its proof is the computation of the Laplace transform of the lattice point recursion (2.56).

Theorem 2.30: *Differential recursion for the Poincaré polynomials, [65]. The Poincaré polynomials $F^P_{g,n}(t_1,\dots,t_n)$ satisfy exactly the same differential recursion* (2.25) *with the same initial values of $F^P_{1,1}$ and $F^P_{0,3}$.*

Since the recursion uniquely determines all $F_{g,n}$'s for $2g-2+n>0$, we have the following:

Corollary 2.31: *For every (g,n) in the stable range $2g-2+n>0$, the two functions are identical:*

$$F^C_{g,n}(t_1,\dots,t_n) = F^P_{g,n}(t_1,\dots,t_n). \tag{2.66}$$

Because of this identification, we see that (2.64) implies (2.23), and (2.22) follows from (2.65). We can also see how (2.24) holds.

The limit $t_i \to \infty$ corresponds to $x_i \to 2$ through the normalization coordinate of (2.19), and $x_i = 2$ corresponds to a branch point of $\tilde{\pi} : \widetilde{\Sigma} \longrightarrow \mathbb{P}^1$ of (2.16). The defining equation (2.20) of $F^C_{g,n}$ does not tell us what we obtain by taking the limit $x_i \to 2$. The geometric meaning of the $t_i \to \infty$ limit becomes crystal clear by the equality (2.66). Let us take a look at the definition of the Poincaré polynomial (2.58). The fact that it is a Laurent polynomial follows from the recursion (2.25) by induction. If $|t_i| > |t_j| >> 1$, then

$$z(t_i,t_j) = \frac{(t_i+1)(t_j+1)}{2(t_i+t_j)} \sim \frac{1}{2}t_j.$$

Therefore, the highest degree part of $F^P_{g,n}$ comes from the graphs of type (g,n) with the largest number of edges. Denoting the number of vertices of a ribbon graph Γ by $v(\Gamma)$, we have

$$2-2g-n = v(\Gamma) - e(\Gamma).$$

To maximize $e(\Gamma)$, we need to maximize $v(\Gamma)$, which is achieved by taking a trivalent graph (since we do not allow degree 1 and 2 vertices in our ribbon graph). By counting the number of half-edges of a trivalent graph, we obtain $2e(\Gamma) = 3v(\Gamma)$. Hence we have

$$e(\Gamma) = 6g - 6 + 3n. \tag{2.67}$$

This is the degree of $F_{g,n}^P$, which agrees with the degree of (2.24), and also consistent with the dimension of (2.54).

Fix a point $\mathbf{p} \in \mathbb{Z}_+^n$, and scale it by a large integer $\lambda >> 1$. Then from (2.55) we see that the number of lattice points in the polytope $P_\Gamma(\lambda\mathbf{p})/\mathrm{Aut}(\Gamma)$ that is counted as a part of $N_{g,n}(\lambda\mathbf{p})$ is the same as the number of scaled lattice points $\mathbf{x} \in \frac{1}{\lambda}\mathbb{Z}_+^n$ in $P_\Gamma(\mathbf{p})/\mathrm{Aut}(\Gamma)$. As $\lambda \to \infty$, the number of lattice points can be approximated by the Euclidean volume of the polytope (cf. theory of Ehrhart polynomials).

For a fixed $(w_1, \ldots, w_n)$ with $Re(w_j) > 0$, the contribution from large $\mathbf{p}$'s in the Laplace transform $L_{g,n}(w_1, \ldots, w_n)$ of (2.59) is small. The asymptotic behavior of $L_{g,n}$ as $w \to 0$ picks up the large perimeter contribution of $N_{g,n}(\mathbf{p})$, or the counting of the lattice points of smaller and smaller mesh size. Since

$$t_j \to \infty \Longleftrightarrow w_j \to 0,$$

the large t_j behavior of $F_{g,n}^P$, which is a homogeneous polynomial of degree $6g-6+3n$, reflects the information of the volume of $\mathcal{M}_{g,n}$ in its coefficients. From Kontsevich [55], we learn that the volume is exactly the intersection number appearing in (2.24).

The topological recursion for the Airy case (1.42) is the $t \to \infty$ limit of the Catalan topological recursion (2.35), as we see from the limit

$$\frac{(t^2-1)^3}{t^2} \longrightarrow t^4.$$

Since the integrand of (1.42) has no poles at $t = 0$, the small circle of the contour γ does not contribute any residue. Thus we have derived the Airy topological recursion from the Catalan topological recursion.

3. Quantization of spectral curves

Quantum curves assemble information of quantum invariants in a compact manner. The global nature of quantum curves is not well understood at this moment of writing. In this section, we focus on explaining the relation between the **PDE version** of topological recursion discovered in [24, 25], and the *local* expression of quantum curves, suggested for example, in [44]. We give the precise definition of the PDE topological recursion in a geometric and coordinate-free manner. The discovery of [24, 25] is to connect the ideas from topological recursion with the Higgs bundle theory for the first time. Global definition of the quantum curves is being established by the authors, based on a recent work [23], and will be reported elsewhere.

As we have seen in the previous sections, for the examples of topological recursion such as the Catalan numbers and the Airy function case, the integral topological recursion is always a consequence of a corresponding recursion of free energies $F_{g,n}$ in the form of partial differential equations. Although we do not discuss them in these lectures, the situation is also true for the case of various Hurwitz numbers [10, 33, 69]. Since quantum curve is a differential equation, it is more natural to expect that the PDE recursion is directly related to quantum curves than the integral topological recursion.

This consideration motivates the authors' discovery of PDE topological recursion [24, 25]. We find that the most straightforward quantization of Hitchin spectral curves is obtained from the PDE recursion. Here, it has to be remarked that if one uses the integral topological recursion for Hitchin spectral curves, that is also introduced in [24, 25], then the quantization process produces a differential equation whose coefficients depend on all powers of $\hbar$, and thus the result is totally different from what we achieve. This shows that the integral topological recursion, which is closer to the original idea of [34], and the PDE topological recursion of [24,25] are inequivalent for the case of Hitchin spectral curves of genus greater than 0.

From a geometric point of view, our quantization is a natural notion. Therefore, we believe the introduction of PDE topological recursion is crucial for building a theory of quantum curves. It is also consistent from a physics point of view. Teschner [83] relates quantization of Hitchin moduli spaces with the quantization of Hitchin spectral curves in the way we do here.

3.1. *Geometry of non-singular Hitchin spectral curves of rank 2*

We wish to transplant the idea of [34] to Hitchin spectral curves. Our first task is to determine the differential form $W_{0,2}$ that gives the initial value of the topological recursion. It can reflect many aspects of the Hitchin spectral curve. In these lectures, we choose the Cauchy differentiation kernel as $W_{0,2}$, which depends only on the intrinsic geometry of the Hitchin spectral curve, for our purpose. Of course one can make other choices for different purposes, such as the kernel associated with a connection of a spectral line bundle on the Hitchin spectral curve.

Even for the Cauchy differentiation kernel, there is no canonical choice. It depends on a symplectic basis for the homology of the spectral curve. We give one particular choice in this subsection.

In this subsection, we consider a smooth projective curve C of genus $g(C) \geq 2$ defined over $\mathbb{C}$. As before, K_C is the canonical bundle on C. The cotangent bundle $\pi : T^*C \longrightarrow C$ is the total space of K_C, and there is the tautological section

$$\eta \in H^0(T^*C, \pi^*K_C), \tag{3.1}$$

which is a globally defined holomorphic 1-form on T^*C.

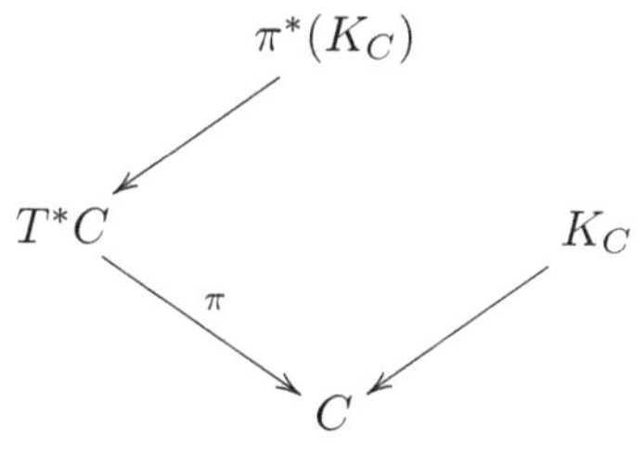

(3.2)

Choose a *generic* quadratic differential $s \in H^0\big(C, K_C^{\otimes 2}\big)$, so that the spectral curve

$$\Sigma_s \subset T^*C \tag{3.3}$$

that is defined by

$$\eta^{\otimes 2} + \pi^* s = 0 \tag{3.4}$$

is non-singular. Our spectral curve $\Sigma = \Sigma_s$ is a double sheeted ramified covering of C defined by (3.4). The genus of the spectral curve is $g(\Sigma) = 4g(C) - 3$. This is because a generic s has $\deg(K_C^{\otimes 2}) = 4g(C) - 4$ simple zeros, which correspond to branch points of the covering $\pi : \Sigma \longrightarrow C$. Thus the genus is calculated by the Riemann Hurwitz formula

$$2\big(2 - 2g(C) - \big(4g(C) - 4\big)\big) = 2 - 2g(\Sigma) - \big(4g(C) - 4\big).$$

The cotangent bundle T^*C has a natural involution

$$\sigma : T^*C \supset T_x^*C \ni (x, y) \longmapsto (x, -y) \in T_x^*C \subset T^*C, \tag{3.5}$$

which preserves the spectral curve Σ. The action of this involution is the deck-transformation. Indeed, the covering $\pi : \Sigma \longrightarrow C$ is a **Galois covering** with the Galois group $\mathbb{Z}/2\mathbb{Z} = \langle \sigma \rangle$.

Let $R \subset \Sigma$ denote the ramification divisor of this covering. Because Σ is non-singular, R is supported at $4g(C) - 4$ distinct points that are determined by $s = 0$ on C. We consider C as the 0-section of T^*C. Thus both C and Σ are divisors of T^*C. Hence R is also defined as $C \cdot \Sigma$ in T^*C supported on $C \cap \Sigma$. Note that η vanishes only along $C \subset T^*C$. As a holomorphic 1-form on Σ, $\iota^*\eta$ has $2g(\Sigma) - 2 = 8g(C) - 8$ zeros on Σ. Thus it has a degree 2 zero at each point of $\text{supp}(R)$.

As explained in the earlier sections using examples, we wish to choose a differential form $W_{0,2}$ for our geometric situation. For this purpose, let us recall the differential form ω_Σ^{a-b} of Remark 2.19. Since we can add any holomorphic 1-form to ω_Σ^{a-b}, we need to impose $g(\Sigma)$ independent conditions to make it unique. If we have a principal polarization of the period matrix for Σ, then one obvious choice would be to impose

$$\oint_{a_j} \omega_\Sigma^{a-b} = 0 \tag{3.6}$$

for all "A-cycles" a_j, $j = 1, 2, \ldots, g(\Sigma)$, following Riemann himself. The reason for this canonical choice is that we can make it for a family of spectral curves

$$\{\Sigma_s\}_{s \in U},$$

where $U \subset H^0\big(C, K_C^{\otimes 2}\big)$ is a contractible open neighborhood of s. Since the base curve C is fixed, we can choose and fix a symplectic basis for $H_1(C, \mathbb{Z})$:

$$\langle A_1, \ldots, A_g; B_1, \ldots, B_g \rangle = H_1(C, \mathbb{Z}). \tag{3.7}$$

From this choice, we construct a canonical symplectic basis for $H^1(\Sigma, \mathbb{Z})$ as follows. Let us label points of $R = \{p_1, p_2, \ldots, p_{4g-4}\}$, where $g = g(C)$. We can connect p_{2i} and p_{2i+1}, $i = 1, \ldots, 2g-3$, with a simple path on Σ that is mutually non-intersecting so that $\pi^*(\overline{p_{2i}p_{2i+1}})$, $i = 1, \ldots, 2g-3$, form a part of the basis for $H_1(\Sigma, \mathbb{Z})$. We denote these cycles by $\alpha_1, \ldots, \alpha_{2g-3}$. Since π is locally homeomorphic away from R, we have g cycles $a_1, \ldots, a_g$ on Σ_s so that $\pi_*(a_j) = A_j$ for $j = 1, \ldots, g$, where A_j's are the A-cycles of C chosen as (3.7). We define the A-cycles of Σ to be the set

$$\{a_1, \ldots, a_g, \sigma_*(a_1), \ldots, \sigma_*(a_g), \alpha_1, \ldots, \alpha_{2g-3}\} \subset H_1(\Sigma, \mathbb{Z}), \tag{3.8}$$

where σ is the Galois conjugation (see Figure 3.1). Clearly, this set can be extended into a symplectic basis for $H_1(\Sigma, \mathbb{Z})$. This choice of the symplectic basis trivializes the homology bundle

$$\{H_1(\Sigma_s, \mathbb{Z})\}_{s\in U} \longrightarrow U \subset H^0(C, K_C^{\otimes 2})$$

locally on the contractible neighborhood U.

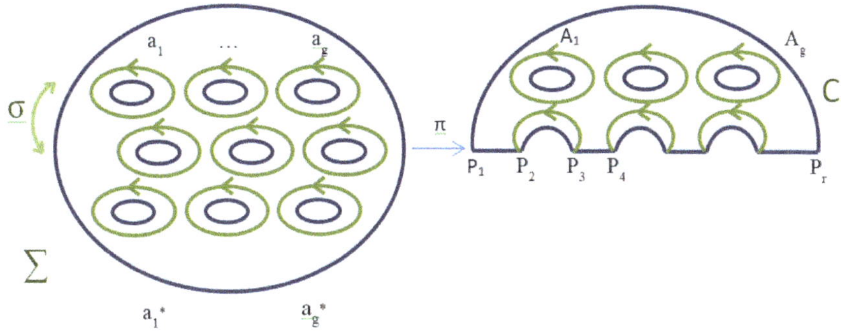

Fig. 3.1. The choice of a symplectic basis for $H_1(\Sigma, \mathbb{Z})$ from that of $H_1(C, \mathbb{Z})$.

3.2. *The generalized topological recursion for the Hitchin spectral curves*

The topological recursion of [11, 34], that has initiated the explosive developments on this subject in recent years, is restricted to non-singular spectral curves in $\mathbb{C}^2$ or $(\mathbb{C}^*)^2$. A systematic generalization of the formalism to the case of Hitchin spectral curves is given for the first time in [24, 25]. First, we gave the definition for non-singular Hitchin spectral curves associated with holomorphic Higgs bundles in [24]. We then extended the consideration to meromorphic Higgs bundles and singular spectral curves in [25]. In each case, however, the actual evaluation of the generalized topological recursion for the purpose of quantization of the Hitchin spectral curves is limited to Higgs bundles of rank 2. This is due to many technical difficulties, and at this point we still do not have a better understanding of the theory in its full generality.

The purpose of this subsection is thus to present the theory in the way we know as of now, with the scope limited to what seems to work. Many aspects of the story can be immediately generalized. Mainly for the sake of simplicity of presentation, we concentrate on the case of rank 2 Higgs bundles.

The geometric setup is the following. We have a smooth projective curve C defined over $\mathbb{C}$ of an arbitrary genus, and a meromorphic Higgs bundle (E, ϕ). Here, the vector bundle E of rank 2 is a special one of the form

$$E = K_C^{\frac{1}{2}} \oplus K_C^{-\frac{1}{2}}. \tag{3.9}$$

As a meromorphic Higgs field, we use

$$\phi = \begin{bmatrix} -s_1 & s_2 \\ 1 & \end{bmatrix} : E \longrightarrow K_C(D) \otimes E \tag{3.10}$$

with poles at an effective divisor D of C, where

$$s_1 \in H^0\big(C, K_C(D)\big),$$
$$s_2 \in H^0\big(C, K_C^{\otimes 2}(D)\big).$$

Although ϕ involves a quadratic differential in its coefficient, since

$$\begin{bmatrix} -s_1 & s_2 \\ 1 & \end{bmatrix} : K_C^{\frac{1}{2}} \oplus K_C^{-\frac{1}{2}} \longrightarrow \left(K_C^{\frac{3}{2}} \oplus K_C^{\frac{1}{2}}\right) \otimes \mathcal{O}_C(D) \longrightarrow K_C(D) \otimes \left(K_C^{\frac{1}{2}} \oplus K_C^{-\frac{1}{2}}\right),$$

we have $\phi \in H^0\big(C, K_C(D) \otimes \mathrm{End}(E)\big)$.

Remark 3.1: We use a particular section (E, ϕ) of the *Hitchin fibration* given by the form of (3.10). This section is often called a *Hitchin section.* This choice is suitable for the WKB analysis explained below. The result of our quantization through WKB constructs a **formal** section of $K_C^{-\frac{1}{2}}$, and the relation to an $\hbar$-connection on C makes our choice necessary. The theory being developed as of now (relation to opers [23]) also requires the choice of a Hitchin section.

Denote by

$$\overline{T^*C} := \mathbb{P}(K_C \oplus \mathcal{O}_C) \stackrel{\pi}{\longrightarrow} C \tag{3.11}$$

the compactified cotangent bundle of C (see [5,57]), which is a ruled surface on the base C. The Hitchin spectral curve

$$\begin{array}{ccc} \Sigma & \stackrel{i}{\longrightarrow} & \overline{T^*C} \\ & \underset{\pi}{\searrow} & \downarrow{\scriptstyle \pi} \\ & & C \end{array} \tag{3.12}$$

for a meromorphic Higgs bundle is defined as the divisor of zeros on $\overline{T^*C}$ of the characteristic polynomial of ϕ:

$$\Sigma = \left(\det(\eta - \pi^*\phi)\right)_0, \tag{3.13}$$

where $\eta \in H^0(T^*C, \pi^*K_C)$ is the tautological 1-form on T^*C extended as a meromorphic 1-form on the compactification $\overline{T^*C}$.

Definition 3.2: Integral topological recursion for a degree 2 covering. Let $\tilde{\pi} : \widetilde{\Sigma} \longrightarrow C$ be a degree 2 covering of C by a non-singular curve $\widetilde{\Sigma}$. We denote by R the ramification divisor of $\tilde{\pi}$. In this case the covering $\tilde{\pi}$ is a Galois covering with the Galois group $\mathbb{Z}/2\mathbb{Z} = \langle \tilde{\sigma} \rangle$, and R is the fixed-point divisor of the involution $\tilde{\sigma}$. The **integral topological recursion** is an inductive mechanism of constructing meromorphic differential forms $W_{g,n}$ on the Hilbert scheme $\widetilde{\Sigma}^{[n]}$ of n-points on $\widetilde{\Sigma}$ for all $g \geq 0$ and $n \geq 1$ in the *stable range* $2g - 2 + n > 0$, from given initial data $W_{0,1}$ and $W_{0,2}$.

- $W_{0,1}$ is a meromorphic 1-form on $\widetilde{\Sigma}$.
- $W_{0,2}$ is defined to be

$$W_{0,2}(z_1, z_2) = d_1 d_2 \log E_{\widetilde{\Sigma}}(z_1, z_2), \tag{3.14}$$

  where $E_{\widetilde{\Sigma}}(z_1, z_2)$ is the normalized Riemann prime form on $\widetilde{\Sigma} \times \widetilde{\Sigma}$ (see [24, Section 2]).

Let $\omega^{a-b}(z)$ be a normalized Cauchy kernel on $\widetilde{\Sigma}$ of Remark 2.19, which has simple poles at $z = a$ of residue 1 and at $z = b$ of residue -1. Then

$$d_1 \omega^{z_1 - b}(z_2) = W_{0,2}(z_1, z_2).$$

Define

$$\Omega := \tilde{\sigma}^* W_{0,1} - W_{0,1}. \tag{3.15}$$

Then $\tilde{\sigma}^* \Omega = -\Omega$, hence $\text{supp}(R) \subset \text{supp}(\Omega)$, where $\text{supp}(\Omega)$ denotes the support of both zero and pole divisors of Ω. The inductive formula of the topological recursion is then given by the following:

$$W_{g,n}(z_1, \ldots, z_n) = \frac{1}{2} \frac{1}{2\pi\sqrt{-1}} \sum_{p \in \text{supp}(\Omega)} \oint_{\gamma_p} \omega^{\tilde{z}-z}(z_1)$$
$$\times \frac{1}{\Omega(z)} \left[W_{g-1,n+1}(z, \tilde{z}, z_2, \ldots, z_n) + \sum_{\substack{g_1+g_2=g \\ I \sqcup J = \{2,\ldots,n\}}}^{\text{No }(0,1)} W_{g_1,|I|+1}(z, z_I) W_{g_2,|J|+1}(\tilde{z}, z_J) \right]. \tag{3.16}$$

Here,

- γ_p is a positively oriented small loop around a point $p \in \text{supp}(\Omega)$;
- the integration is taken with respect to $z \in \gamma_p$ for each $p \in \text{supp}(\Omega)$;

- $\tilde{z} = \tilde{\sigma}(z)$ is the Galois conjugate of $z \in \widetilde{\Sigma}$;
- the ratio of two global meromorphic 1-forms on the same curve makes sense as a global meromorphic function. The operation $1/\Omega$ applied to a meromorphic 1-form produces this ratio;
- "No (0, 1)" means that $g_1 = 0$ and $I = \emptyset$, or $g_2 = 0$ and $J = \emptyset$, are excluded in the summation;
- the sum runs over all partitions of g and set partitions of $\{2, \ldots, n\}$, other than those containing the $(0, 1)$ geometry;
- $|I|$ is the cardinality of the subset $I \subset \{2, \ldots, n\}$; and
- $z_I = (z_i)_{i \in I}$.

The main idea of [25] for dealing with singular spectral curve is the following.

- The integral topological recursion of [24, 34] is extended to the curve Σ of (3.13), as (3.16). For this purpose, we blow up $\overline{T^*C}$ several times as in (3.19) below to construct the normalization $\widetilde{\Sigma}$. The blown-up $Bl(\overline{T^*C})$ is the minimal resolution of the support $\Sigma \cup C_\infty$ of the *total* divisor

$$\Sigma - 2C_\infty = (\det(\eta - \pi^*\phi))_0 - (\det(\eta - \pi^*\phi))_\infty \tag{3.17}$$

  of the characteristic polynomial, where

$$C_\infty := \mathbb{P}(K_C \oplus \{0\}) = \overline{T^*C} \setminus T^*C \tag{3.18}$$

  is the divisor at infinity. Therefore, in $Bl(\overline{T^*C})$, the proper transform $\widetilde{\Sigma}$ of Σ is smooth and does not intersect with the proper transform of C_∞.

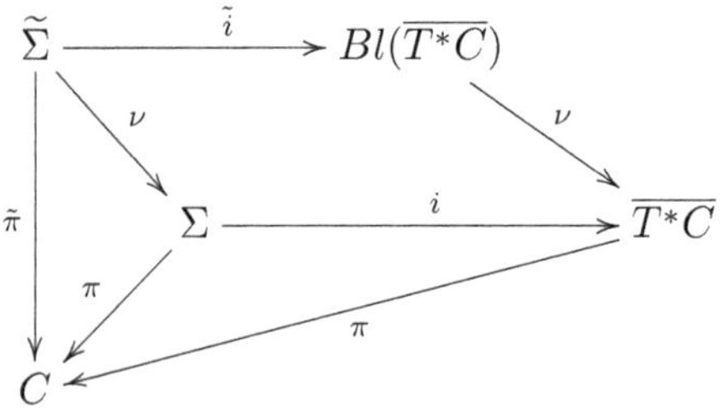

(3.19)

- The genus of the normalization $\widetilde{\Sigma}$ is given by

$$g(\widetilde{\Sigma}) = 2g(C) - 1 + \frac{1}{2}\delta,$$

  where δ is the sum of the number of cusp singularities of Σ and the ramification points of $\pi : \Sigma \longrightarrow C$.

- The generalized topological recursion Definition 3.2 requires a globally defined meromorphic 1-form $W_{0,1}$ on $\widetilde{\Sigma}$ and a symmetric meromorphic 2-form $W_{0,2}$ on the product $\widetilde{\Sigma} \times \widetilde{\Sigma}$ as the initial data. We choose

$$\begin{cases} W_{0,1} = \tilde{i}^* \nu^* \eta \\ W_{0,2} = d_1 d_2 \log E_{\widetilde{\Sigma}}, \end{cases} \tag{3.20}$$

  where $E_{\widetilde{\Sigma}}$ is a normalized Riemann prime form on $\widetilde{\Sigma}$. The form $W_{0,2}$ depends only on the intrinsic geometry of the smooth curve $\widetilde{\Sigma}$. The geometry of (3.19) is encoded in $W_{0,1}$. The integral topological recursion produces a symmetric meromorphic n-linear differential form $W_{g,n}(z_1, \ldots, z_n)$ on $\widetilde{\Sigma}$ for every (g, n) subject to $2g-2+n > 0$ from the initial data (3.20).

The key discovery of [24, (4.7)] is that we should use a partial differential equation version of the topological recursion, instead of (3.16), to construct a quantization of Σ.

3.3. *Quantization of Hitchin spectral curves*

The passage from the geometry of Hitchin spectral curve Σ of (3.13) to the quantum curve

$$\left(\left(\hbar \frac{d}{dx} \right)^2 - \operatorname{tr} \phi(x) \left(\hbar \frac{d}{dx} \right) + \det \phi(x) \right) \Psi(x, \hbar) = 0 \tag{3.21}$$

is a system of PDE recursion replacing the integration formula (3.16).

Definition 3.3: Free energies. The **free energy** of type (g, n) is a function $F_{g,n}(z_1, \ldots, z_n)$ defined on the universal covering $\mathcal{U}^n$ of $\widetilde{\Sigma}^n$ such that

$$d_1 \cdots d_n F_{g,n} = W_{g,n}.$$

Remark 3.4: The free energies may contain logarithmic singularities, since it is an integral of a meromorphic function. For example, $F_{0,2}$ is the Riemann prime form itself considered as a function on $\mathcal{U}^2$, which has logarithmic singularities along the diagonal [24, Section 2].

Definition 3.5: Differential recursion for a degree 2 covering. The **differential recursion** is the following partial differential equation for all (g, n) subject to $2g - 2 + n \geq 2$:

$$d_1 F_{g,n}(z_1, \ldots, z_n)$$

$$= \sum_{j=2}^{n} \left[\frac{\omega^{z_j-\sigma(z_j)}(z_1)}{\Omega(z_1)} \cdot d_1 F_{g,n-1}\big(z_{[\hat{j}]}\big) - \frac{\omega^{z_j-\sigma(z_j)}(z_1)}{\Omega(z_j)} \cdot d_j F_{g,n-1}\big(z_{[\hat{1}]}\big) \right] + \frac{1}{\Omega(z_1)} d_{u_1} d_{u_2} \left[F_{g-1,n+1}\big(u_1, u_2, z_{[\hat{1}]}\big) + \sum_{\substack{g_1+g_2=g \\ I \sqcup J = [\hat{1}]}}^{\text{stable}} F_{g_1,|I|+1}(u_1, z_I) F_{g_2,|J|+1}(u_2, z_J) \right] \Bigg|_{\substack{u_1=z_1 \\ u_2=z_1}}. \tag{3.22}$$

Here, $1/\Omega$ is again the ratio operation, and the index subset $[\hat{j}]$ denotes the exclusion of $j \in \{1, 2, \ldots, n\}$. The Cauchy integration kernel $\omega^{a-b}(z)$ on the spectral curve $\widetilde{\Sigma}$ is normalized differently than the A-cycle normalization we did earlier. This time we impose that

$$\lim_{a \to b} \omega^{a-b}(z) = 0. \tag{3.23}$$

Remark 3.6: As pointed out in [24, Remark 4.8], (3.22) is a globally defined coordinate-free equation, written in terms of exterior differentiations and the ratio operation, on $\widetilde{\Sigma}$.

Theorem 3.7: *The relation between the differential recursion and the integral recursion, [25]. Suppose that $F_{g,n}$ for $2g-2+n>0$ are globally meromorphic on $\widetilde{\Sigma}^{[n]}$ with poles located only along the divisor of $\widetilde{\Sigma}^{[n]}$ when one of the factors lies in the zeros of Ω. Define $W_{g,n} := d_1 \cdots d_n F_{g,n}$ for $2g-2+n>0$, and use* (3.14) *and* (3.15) *for (g,n) in the unstable range. If $F_{g,n}$s are symmetric and satisfy the differential recursion* (3.22)*, and if $W_{1,1}$ and $W_{0,3}$ satisfy the initial equations of the integral topological recursion* (3.16)*, then $W_{g,n}$s for all valued of (g,n) satisfy the integral topological recursion.*

Remark 3.8: The assumption of the theorem holds for $g(C)=0$, and therefore, for all the examples we discuss in these lectures. But we are not establishing the general **equivalence** of the integral topological recursion (3.16) and the PDE recursion (3.22). They are never equivalent when $g(\widetilde{\Sigma})>0$. Actually, what is assumed in the above theorem is that $W_{g,n}$ is exact, i.e., integrable by definition. In particular, this implies that $W_{g,n}$ has 0-period for every topological cycle. This does not happen if we start with (3.16) in general. Therefore, the above theorem serves only as a heuristic motivation for our discovery of (3.22) in [24, 25].

Proof: Although the context of the statement is slightly different, the proof is essentially the same as that of [24, Theorem 4.7]. The crucial assumption we have made is that $\tilde{\pi} : \widetilde{\Sigma} \longrightarrow C$ is a Galois covering. Therefore, the Galois conjugation $\tilde{\sigma} : \widetilde{\Sigma} \longrightarrow \widetilde{\Sigma}$ is a globally defined holomorphic mapping. To calculate the residues in the integral recursion (3.16), we need the global analysis of

$$\omega^{\tilde{z}-z}(z_1) \in H^0\big(\widetilde{\Sigma} \times \widetilde{\Sigma}, q_2^* K_{\widetilde{\Sigma}} \boxtimes q_2^* \mathcal{O}_{\widetilde{\Sigma}}(\tilde{z} + z)\big),$$

where q_1 and q_2 are projections

$$\begin{array}{ccccc} & & \widetilde{\Sigma} \times \widetilde{\Sigma} & & \\ & \swarrow^{q_1} & & \searrow^{q_2} & \\ z \in \widetilde{\Sigma} & & & & \widetilde{\Sigma} \ni z_1. \end{array} \tag{3.24}$$

The residue integration is done at each point $p \in \text{supp}(\Omega)$. The poles of the integrand of (3.16) that are enclosed in the union

$$\gamma = \bigcup_{p \in \text{supp}(\Omega)} \gamma_p$$

of the contours on the complement of $\text{supp}(\Omega)$ are located at

(1) $z = z_1, z = \tilde{z_1}$ from $\omega^{\tilde{z}-z}(z_1)$; and
(2) $z = z_j, z = \tilde{z}_j$, $j = 2, \ldots, n$, from $W_{0,2}(z, z_j)$ and $W_{0,2}(\tilde{z}, z_j)$ that appear in the second line of (3.16).

The integrand has other poles at $\text{supp}(\Omega)$ that includes the ramification divisor R, but they are not enclosed in γ. The local behavior of $\omega^{\tilde{z}-z}(z_1)$ at $z = z_1, z = \tilde{z_1}$ is well understood, and residues of the integrand of (3.16) are simply the evaluation of the differential form at $z = z_1, z = \tilde{z_1}$. The double poles coming from $W_{0,2}(z, z_j)$ and $W_{0,2}(\tilde{z}, z_j)$ contribute as differentiation of the factor it is multiplied to. Adding all contributions from the poles, we obtain (3.22). $\square$

Now let us consider a spectral curve $\Sigma \subset \overline{T^*C}$ of (3.12) defined by a pair of meromorphic sections $s_1 = -\text{tr}\phi$ of K_C and $s_2 = \det\phi$ of $K_C^{\otimes 2}$. Let $\widetilde{\Sigma}$ be the desingularization of Σ in (2.16). We apply the topological recursion (3.16) to the covering $\tilde{\pi} : \widetilde{\Sigma} \longrightarrow C$. The geometry of the spectral curve Σ provides us with a canonical choice of the initial differential forms (3.20). At this point we pay special attention to the fact that the topological recursions (3.16) and (3.22) are both defined on the spectral curve $\widetilde{\Sigma}$, while we wish to construct a differential equation on C. Since the free energies are defined on the universal covering of $\widetilde{\Sigma}$, we need to have a mechanism to relate a coordinate on the desingularized spectral curve and that of the base curve C.

To analyze the singularity structure of Σ, let us consider the discriminant of the defining equation (3.4) of the spectral curve.

Definition 3.9: Discriminant divisor. The **discriminant divisor** of the spectral curve

$$\eta^{\otimes 2} + \pi^* s_1 \eta + \pi^* s_2 = 0 \tag{3.25}$$

is a divisor on C defined by

$$\Delta := \left(\frac{1}{4}s_1^2 - s_2\right) = \Delta_0 - \Delta_\infty. \tag{3.26}$$

Here,

$$\Delta_0 = \sum_{i=1}^{m} m_i q_i, \quad m_i > 0, \quad q_i \in C, \tag{3.27}$$

is the divisor of zeros, and

$$\Delta_\infty = \sum_{j=1}^{n} n_j p_j, \quad n_j > 0, \quad p_j \in C, \tag{3.28}$$

is the divisor of ∞.

Since $\frac{1}{4}s_1^2 - s_2$ is a meromorphic section of $K_C^{\otimes 2}$, we have

$$\deg \Delta = \sum_{i=1}^{m} m_i - \sum_{j=1}^{n} n_j = 4g - 4. \tag{3.29}$$

Theorem 3.10: *Geometric genus formula, [25]. Let us define an invariant of the discriminant divisor by*

$$\delta = |\{i \mid m_i \equiv 1 \mod 2\}| + |\{j \mid n_j \equiv 1 \mod 2\}|. \tag{3.30}$$

Then the geometric genus of the spectral curve Σ *of* (3.25) *is given by*

$$g(\widetilde{\Sigma}) := p_g(\Sigma) = 2g - 1 + \frac{1}{2}\delta. \tag{3.31}$$

We note that (3.29) *implies* $\delta \equiv 0 \mod 2$.

Take an arbitrary point $p \in C \setminus \mathrm{supp}(\Delta)$, and a local coordinate x around p. By choosing a small disc V around p, we can make the inverse image of $\tilde{\pi} : \widetilde{\Sigma} \longrightarrow C$ consist of two isomorphic discs. Since V is away from the critical values of $\tilde{\pi}$, the inverse image consists of two discs in the original spectral curve Σ. Note that we choose an eigenvalue α of ϕ on V in our main construction. We are thus specifying one of the inverse image discs here. Let us name the disc V_α that corresponds to α.

At this point apply the WKB analysis to the differential equation (3.21) with the WKB expansion of the solution

$$\Psi^\alpha(x,\hbar) = \exp\left(\sum_{m=0}^{\infty} \hbar^{m-1} S_m\big(x(z)\big)\right) = \exp F^\alpha(x,\hbar), \tag{3.32}$$

where we choose a coordinate z of V_α so that the function $x = x(z)$ represents the projection $\pi : V_\alpha \longrightarrow V$. The equation $P\Psi^\alpha = Pe^{F^\alpha} = 0$ reads

$$\hbar^2 \frac{d^2}{dx^2} F^\alpha + \hbar^2 \frac{dF^\alpha}{dx}\frac{dF^\alpha}{dx} + s_1 \hbar \frac{dF^\alpha}{dx} + s_2 = 0. \tag{3.33}$$

The $\hbar$-expansion of (3.33) gives

$$\hbar^0\text{-terms}: \quad (S_0'(x))^2 + s_1 S_0'(x) + s_2 = 0, \tag{3.34}$$

$$\hbar^1\text{-terms}: \quad 2S_0'(x)S_1'(x) + S_0''(x) + s_1 S_1'(x) = 0, \tag{3.35}$$

$$\hbar^{m+1}\text{-terms}: \quad S_m''(x) + \sum_{a+b=m+1} S_a'(x)S_b'(x) + s_1 S_{m+1}'(x) = 0, \quad m \geq 1, \tag{3.36}$$

where $'$ denotes the x-derivative. The WKB method is to solve these equations iteratively and find $S_m(x)$ for all $m \geq 0$. Here, (3.34) is the **semi-classical limit** of (3.21), and (3.35) is the *consistency condition* we need to solve the WKB expansion, the same as before. Since the 1-form $dS_0(x)$ is a local section of T^*C, we identify $y = S_0'(x)$. Then (3.34) is the local expression of the spectral curve equation (3.4). This expression is the same everywhere for $p \in C \setminus \text{supp}(\Delta)$. We note s_1 and s_2 are globally defined. Therefore, we recover the spectral curve Σ from the differential operator of (3.21).

Theorem 3.11: *Main theorem. The differential topological recursion provides a formula for each $S_m(x)$, $m \geq 2$, and constructs a formal solution to the quantum curve* (3.21).

- *The quantum curve associated with the Hitchin spectral curve Σ is defined as a differential equation on C. On each coordinate neighborhood $U \subset C$ with coordinate x, a generator of the quantum curve is given by*

$$P(x,\hbar) = \left(\hbar\frac{d}{dx}\right)^2 - tr\phi(x)\left(\hbar\frac{d}{dx}\right) + \det\phi(x).$$

*In particular, the **semi-classical limit** of the quantum curve recovers the singular spectral curve* Σ*, not its normalization* $\widetilde{\Sigma}$*.*

- *The **all-order WKB expansion***

$$\Psi(x,\hbar) = \exp\left(\sum_{m=0}^{\infty} \hbar^{m-1} S_m(x)\right) \tag{3.37}$$

of a solution to the Schrödinger equation

$$\left(\left(\hbar\frac{d}{dx}\right)^2 - tr\,\phi(x)\left(\hbar\frac{d}{dx}\right) + \det\phi(x))\right)\Psi(x,\hbar) = 0,$$

near each critical value of $\pi : \Sigma \longrightarrow C$*, can be obtained by the **principal specialization** of the differential recursion* (3.22)*, after determining the first three terms. The procedure is the following. We determine* S_0, S_1*, and* S_2 *by solving*

$$\begin{aligned} \left(S_0'(x)\right)^2 - tr\phi(x)S_0'(x) + \det\phi(x) = 0,\\ 2S_0'(x)S_1'(x) + S_0''(x) - tr\phi(x)S_1'(x) = 0,\\ S_1''(x) + \sum_{a+b=2} S_a'(x)S_b'(x) - tr\phi(x)S_2'(x) = 0. \end{aligned} \tag{3.38}$$

Then find $F_{1,1}(z)$ *and* $F_{0,3}(z_1,z_2,z_3)$ *so that*

$$S_2(x) = F_{1,1}\big(z(x)\big) + \frac{1}{6}F_{0,3}\big(z(x),z(x),z(x)\big).$$

This can be achieved as follows. First integrate $W_{1,1}(z)$ *of* (3.16) *to construct* $F_{1,1}(z)$*. We do the same for the solution* $W_{0,3}(z_1,z_2,z_3)$ *of* (3.16)*. We now define*

$$F_{0,3}(z_1,z_2,z_3) = \int\!\!\int\!\!\int W_{0,3}(z_1,z_2,z_3) + 2\left(f(z_1)+f(z_2)+f(z_3)\right), \tag{3.39}$$

where

$$f(z) := \widetilde{S}_2(z) - \left(F_{1,1}(z) + \frac{1}{6}\int^z\!\!\int^z\!\!\int^z W_{0,3}(z_1,z_2,z_3)\right), \tag{3.40}$$

and $\widetilde{S}_2(z)$ *is the lift of* $S_2(x)$ *to* $\widetilde{\Sigma}$*.*

- *Suppose we have symmetric meromorphic functions* $F_{g,n}(z_1,\dots,z_n)$ *that solve the differential recursion* (3.22) *on the universal covering* $\varpi : \mathcal{U} \longrightarrow \widetilde{\Sigma}$ *with these* $F_{1,1}$ *and* $F_{0,3}$ *as initial values.*

- *Let*

$$S_m(x) = \sum_{2g-2+n=m-1} \frac{1}{n!} F_{g,n}\big(z(x)\big), \qquad m \geq 3, \tag{3.41}$$

where $F_{g,n}\big(z(x)\big)$ *is the principal specialization of* $F_{g,n}(z_1, \ldots, z_n)$ *evaluated at a local section* $z = z(x)$ *of* $\tilde{\pi} : \widetilde{\Sigma} \longrightarrow C$*. Then the* ***wave function*** $\Psi(x, \hbar)$*, a formal section of the line bundle* $K_C^{-\frac{1}{2}}$ *on* C*, solves* (3.21).
- *The canonical ordering of the quantization of the local functions on* T^*C *is automatically chosen in* (3.22) *and the principal specialization* (3.41)*. This selects the canonical ordering in* (3.21).

Remark 3.12: We do not have a closed formula for $F_{1,1}$ and $F_{0,3}$ from the given geometric data. Except for the case of $g(C) = 0$, they are **not** given by integrating $W_{1,1}$ and $W_{0,3}$ of the integral topological recursion.

Remark 3.13: The differential recursion (3.22) assumes $F_{1,1}$ and $F_{0,3}$ as the initial values. The equation itself does not give any condition for them. The discovery of [24, 25] is that the WKB equations for $S_m(x)$ are consequences of (3.22) for all $m \geq 2$. We note that there is an alternative way of constructing a quantization of the spectral curve. From the geometric data, first choose $W_{0,1}$ and $W_{0,2}$ as in (3.20), and solve the integral topological recursion (3.16). Then define a set of alternative free energies by

$$F_{g,n}^{\mathrm{alt}}(z_1, \ldots, z_n) = \int \cdots \int W_{g,n}(z_1, \ldots, z_n) \tag{3.42}$$

for all values of (g, n). Then use the same (3.41) and (3.37) to define a wave function $\Psi^{\mathrm{alt}}(x, \hbar)$. The differential equation that annihilates this alternative wave function is another quantum curve. We emphasize that $\Psi^{\mathrm{alt}}(x, \hbar)$ **does not satisfy** our quantum curve equation (3.21). It is obvious because our definition (3.40) of S_2 is different. The alternative quantum curve is a second order differential equation, but it cannot be given by a closed formula, unlike (3.21). It is also noted that every coefficient of this alternative differential operator contains terms depending on all orders of $\hbar$. Therefore, the mechanism described in these lecture notes provides a totally different notion of quantum curves. We have shown that the differential recursion (3.22) is the passage from the starting spectral curve to the quantum curve (3.21). This picture is consistent with the construction of opers in [23] and a physics point of view [83].

3.4. *Classical differential equations*

If quantum curves are natural objects, then where do we see them in classical mathematics? Indeed, they appear as classical differential equations. Riemann and Poincaré found the interplay between algebraic geometry of curves in a ruled surface and the asymptotic expansion of an analytic solution to a differential equation defined on the base curve of the ruled surface. We look at these classical subjects from a new point of view. Let us now recall the definition of regular and irregular singular points of a second order differential equation.

Definition 3.14: Let

$$\left(\frac{d^2}{dx^2} + s_1(x)\frac{d}{dx} + s_2(x)\right)\Psi(x) = 0 \tag{3.43}$$

be a second order differential equation defined around a neighborhood of $x = 0$ on a small disc $|x| < \epsilon$ with meromorphic coefficients $s_1(x)$ and $s_2(x)$ with poles at $x = 0$. Denote by k (reps. ℓ) the order of the pole of $s_1(x)$ (resp. $s_2(x)$) at $x = 0$. If $k \le 1$ and $\ell \le 2$, then (3.43) has a **regular singular point** at $x = 0$. Otherwise, consider the *Newton polygon* of the order of poles of the coefficients of (3.43). It is the upper part of the convex hull of three points $(0,0), (1,k), (2,\ell)$. As a convention, if $s_j(x)$ is identically 0, then we assign $-\infty$ as its pole order. Let $(1,r)$ be the intersection point of the Newton polygon and the line $x = 1$. Thus

$$r = \begin{cases} k & 2k \ge \ell, \\ \frac{\ell}{2} & 2k \le \ell. \end{cases} \tag{3.44}$$

The differential equation (3.43) has an **irregular singular point of class** $r - 1$ at $x = 0$ if $r > 1$.

To illustrate the scope of interrelations among the geometry of meromorphic Higgs bundles, their spectral curves, the singularities of quantum curves, $\hbar$-connections, and the quantum invariants, let us tabulate five examples here (see Table 1). The differential operators of these equations are listed in the third column. In the first three rows, the quantum curves are examples of classical differential equations known as Airy, Hermite, the Gauß hypergeometric equations. The fourth and the fifth rows are added to show that it is *not* the singularity of the spectral curve that determines the singularity of the quantum curve. In each example, the Higgs bundle (E, ϕ) we are considering consists of the base curve $C = \mathbb{P}^1$ and the rank

2 vector bundle E on $\mathbb{P}^1$ of (1.21). For this situation, the two topological recursions (3.16) and (3.22) are equivalent.

Table 1. Examples of quantum curves.

Higgs Field	Spectral Curve	Quantum Curve
$\begin{bmatrix} & x(dx)^2 \\ 1 & \end{bmatrix}$	$y^2 - x = 0$ $w^2 - u^5 = 0$ $\Sigma = 2C_0 + 5F$ $p_a = 2, p_g = 0$	Airy $\left(\hbar \frac{d}{dx}\right)^2 - x$ Class $\frac{3}{2}$ irregular singularity at ∞
$\begin{bmatrix} -xdx & -(dx)^2 \\ 1 & \end{bmatrix}$	$y^2 + xy + 1 = 0$ $w^2 - uw + u^4 = 0$ $\Sigma = 2C_0 + 4F$ $p_a = 1, p_g = 0$	Hermite $\left(\hbar \frac{d}{dx}\right)^2 + x\hbar \frac{d}{dx} + 1$ Class 2 irregular singularity at ∞
$\begin{bmatrix} \frac{2x-1}{x(1-x)}dx & \frac{(dx)^2}{4(1-x)} \\ \frac{1}{x} & \end{bmatrix}$	$y^2 + \frac{2x-1}{x(x-1)}y + \frac{1}{4x(x-1)} = 0$ $w^2 + 4(u-2)uw$ $-4u^2(u-1) = 0$ $\Sigma = 2C_0 + 4F$ $p_a = 1, p_g = 0$	Gauß Hypergeometric $\left(\hbar \frac{d}{dx}\right)^2 + \frac{2x-1}{x(x-1)}\hbar \frac{d}{dx} + \frac{1}{4x(x-1)}$ Regular singular points at $x = 0, 1, \infty$
$\begin{bmatrix} -dx & -\frac{(dx)^2}{x+1} \\ 1 & \end{bmatrix}$	$y^2 + y + \frac{1}{x+1} = 0$ $w^2 - u(u+1)w$ $+u^3(u+1) = 0$ $\Sigma = 2C_0 + 4F$ $p_a = 1, p_g = 0$	$\left(\hbar \frac{d}{dx}\right)^2 + \hbar \frac{d}{dx} + \frac{1}{x+1}$ Regular singular point at $x = -1$ and a class 1 irregular singularity at $x = \infty$
$\begin{bmatrix} -\frac{2x^2}{x^2-1}dx & \frac{(dx)^2}{x^2-1} \\ 1 & \end{bmatrix}$	$(x^2-1)y^2 + 2x^2y - 1 = 0$ non-singular $\Sigma = 2C_0 + 4F$ $p_a = p_g = 1$	$\left(\hbar \frac{d}{dx}\right)^2 + 2\frac{x^2}{x^2-1}\hbar \frac{d}{dx} - \frac{1}{x^2-1}$ Regular singular points at $x = \pm 1$ and a class 1 irregular singularity at $x = \infty$

The first column of the table shows the Higgs field $\phi : E \longrightarrow K_{\mathbb{P}^1}(5) \otimes E$. Here, x is the affine coordinate of $\mathbb{P}^1 \setminus \{\infty\}$. Since our vector bundle is a specific direct sum of line bundles, the quantization is simple in each case, due to the fact that the $\hbar$-deformation $E_\hbar$ of E satisfies the condition as described in (1.46). Thus our quantum curves are equivalent to $\hbar$-connections in the trivial bundle. Except for the Gauß hypergeometric case, the connections are given by

$$\nabla^\hbar = \hbar d - \phi, \tag{3.45}$$

where d is the exterior differentiation operator acting on the trivial bundle $E_\hbar$, $\hbar \neq 0$.

For the third example of a Gauß hypergeometric equation, we use a particular choice of parameters so that the $\hbar$-connection becomes an $\hbar$-deformed Gauß-Manin connection of (3.47). More precisely, for every $x \in \mathcal{M}_{0,4}$, we consider the elliptic curve $E(x)$ ramified over $\mathbb{P}^1$ at four points $\{0, 1, x, \infty\}$, and its two periods given by the elliptic integrals [58]

$$\omega_1(x) = \int_1^{\infty} \frac{ds}{\sqrt{s(s-1)(s-x)}}, \qquad \omega_2(x) = \int_x^1 \frac{ds}{\sqrt{s(s-1)(s-x)}}. \tag{3.46}$$

The quantum curve in this case is an $\hbar$ **-deformed meromorphic Gauß-Manin connection**

$$\nabla_{GM}^{\hbar} = \hbar d - \begin{bmatrix} \left(-\frac{2x-1}{x(x-1)} + \frac{\hbar}{x}\right) dx & -\frac{(dx)^2}{4(x-1)} \\ \frac{1}{x} & \end{bmatrix} \tag{3.47}$$

in the $\hbar$-deformed vector bundle $K_{\overline{\mathcal{M}}_{0,4}}^{\frac{1}{2}} \oplus K_{\overline{\mathcal{M}}_{0,4}}^{-\frac{1}{2}}$ of rank 2 over $\overline{\mathcal{M}}_{0,4}$. Here, d again denotes the exterior differentiation acting on this trivial vector bundle. The restriction ∇_{GM}^{1} of the connection at $\hbar = 1$ is equivalent to the Gauß-Manin connection that characterizes the two periods of (3.46), and the Higgs field is the classical limit of the connection matrix at $\hbar \to 0$:

$$\phi = \begin{bmatrix} -\frac{2x-1}{x(x-1)} dx & -\frac{(dx)^2}{4(x-1)} \\ \frac{1}{x} & \end{bmatrix}. \tag{3.48}$$

The spectral curve $\Sigma \subset \overline{T^*\overline{\mathcal{M}}_{0,4}}$ as a moduli space consists of the data $\big(E(x), \alpha_1(x), \alpha_2(x)\big)$, where $\alpha_1(x)$ and $\alpha_2(x)$ are the two eigenvalues of the Higgs field ϕ. The spectral curve $\Sigma \subset \overline{T^*\overline{\mathcal{M}}_{0,4}} = \mathbb{F}^2$ as a divisor in the Hirzebruch surface is determined by the characteristic equation

$$y^2 + \frac{2x-1}{x(x-1)} y + \frac{1}{4x(x-1)} = 0 \tag{3.49}$$

of the Higgs field. Geometrically, Σ is a singular rational curve with one ordinary double point at $x = \infty$. The quantum curve is a **quantization** of the characteristic equation (3.49) for the eigenvalues $\alpha_1(x)$ and $\alpha_2(x)$ of $\phi(x)$. It is an $\hbar$-deformed **Picard-Fuchs equation**

$$\left(\left(\hbar \frac{d}{dx}\right)^2 + \frac{2x-1}{x(x-1)} \left(\hbar \frac{d}{dx}\right) + \frac{1}{4x(x-1)}\right) \omega_i(x, \hbar) = 0,$$

and its semi-classical limit agrees with the singular spectral curve Σ. As a second order differential equation, the quantum curve has two independent solutions corresponding to the two eigenvalues. At $\hbar = 1$, these solutions are exactly the two periods $\omega_1(x)$ and $\omega_2(x)$ of the Legendre family of elliptic curves $E(x)$. The topological recursion produces asymptotic expansions of these periods as functions in $x \in \overline{\mathcal{M}}_{0,4}$, at which the elliptic curve $E(x)$ degenerates to a nodal rational curve.

This is a singular connection with simple poles at $0, 1, \infty$, and has an explicit $\hbar$-dependence in the connection matrix. The Gauß-Manin connection ∇^1_{GM} at $\hbar = 1$ is equivalent to the Picard-Fuchs equation that characterizes the periods (3.46) of the Legendre family of elliptic curves $E(x)$ defined by the cubic equation

$$t^2 = s(s-1)(s-x), \qquad x \in \mathcal{M}_{0,4} = \mathbb{P}^1 \setminus \{0, 1, \infty\}. \tag{3.50}$$

The second column gives the spectral curve of the Higgs bundle (E, ϕ). Since the Higgs fields have poles, the spectral curves are no longer contained in the cotangent bundle $T^*\mathbb{P}^1$. We need the compactified cotangent bundle

$$\overline{T^*\mathbb{P}^1} = \mathbb{P}(K_{\mathbb{P}^1} \oplus \mathcal{O}_{\mathbb{P}^1}) = \mathbb{F}_2,$$

which is a Hirzebruch surface. The parameter y is the fiber coordinate of the cotangent line $T^*_x\mathbb{P}^1$. The first line of the second column is the equation of the spectral curve in the (x, y) affine coordinate of $\mathbb{F}_2$. All but the last example produce a singular spectral curve. Let (u, w) be a coordinate system on another affine chart of $\mathbb{F}_2$ defined by

$$\begin{cases} x = 1/u \\ ydx = vdu, \qquad w = 1/v. \end{cases}$$

The singularity of Σ in the (u, w)-plane is given by the second line of the second column. The third line of the second column gives $\Sigma \in \mathrm{NS}(\mathbb{F}_2)$ as an element of the Néron-Severy group of $\mathbb{F}_2$. Here, C_0 is the class of the zero-section of $T^*\mathbb{P}^1$, and F represents the fiber class of $\pi : \mathbb{F}_2 \longrightarrow \mathbb{P}^1$. We also give the arithmetic and geometric genera of the spectral curve.

A solution $\Psi(x, \hbar)$ to the first example is given by the **Airy function**

$$Ai(x, \hbar) = \frac{1}{2\pi}\hbar^{-\frac{1}{6}} \int_{-\infty}^{\infty} \exp\left(\frac{ipx}{\hbar^{2/3}} + i\frac{p^3}{3}\right) dp, \tag{3.51}$$

which is an entire function in x for $\hbar \neq 0$, as discussed earlier in these lectures. The expansion coordinate $x^{\frac{3}{2}}$ of (1.38) indicates the class of the irregular singularity of the Airy differential equation.

The solutions to the second example are given by confluent hypergeometric functions, such as ${}_1F_1\left(\frac{1}{2\hbar}; \frac{1}{2}; -\frac{x^2}{2\hbar}\right)$, where

$$ {}_1F_1(a;c;z) := \sum_{n=0}^{\infty} \frac{(a)_n}{(c)_n} \frac{z^n}{n!} \tag{3.52}$$

is the **Kummer confluent hypergeomtric function**, and the **Pochhammer symbol** $(a)_n$ is defined by

$$(a)_n := a(a+1)(a+2)\cdots(a+n-1). \tag{3.53}$$

For $\hbar > 0$, the topological recursion determines the asymptotic expansion of a particular entire solution known as a **Tricomi confluent hypergeomtric function**

$$\Psi^{\mathrm{Catalan}}(x,\hbar) = \left(-\frac{1}{2\hbar}\right)^{\frac{1}{2\hbar}} \left(\frac{\Gamma[\frac{1}{2}]}{\Gamma[\frac{1}{2\hbar}+\frac{1}{2}]}\, {}_1F_1\left(\frac{1}{2\hbar}; \frac{1}{2}; -\frac{x^2}{2\hbar}\right) + \frac{\Gamma[-\frac{1}{2}]}{\Gamma[\frac{1}{2\hbar}]}\sqrt{-\frac{x^2}{2\hbar}}\, {}_1F_1\left(\frac{1}{2\hbar}+\frac{1}{2}; \frac{3}{2}; -\frac{x^2}{2\hbar}\right)\right).$$

The expansion is given in the form

$$\begin{aligned}\Psi^{\mathrm{Catalan}}(x,\hbar) &= \left(\frac{1}{x}\right)^{\frac{1}{\hbar}} \sum_{n=0}^{\infty} \frac{\hbar^n \left(\frac{1}{\hbar}\right)_{2n}}{(2n)!!} \cdot \frac{1}{x^{2n}} \\ &= \exp\left(\sum_{g=0}^{\infty}\sum_{n=1}^{\infty} \frac{1}{n!} \hbar^{2g-2+n} F^C_{g,n}(x,\ldots,x)\right),\end{aligned} \tag{3.54}$$

where $F^C_{g,n}$ is defined by (2.20), in terms of generalized Catalan numbers. The expansion variable x^2 in (3.54) indicates the class of irregularity of the Hermite differential equation at $x = \infty$. The cases for $(g,n) = (0,1)$ and $(0,2)$ require again a special treatment, as we discussed earlier.

The Hermite differential equation becomes simple for $\hbar = 1$, and we have the asymptotic expansion

$$\begin{aligned} i\sqrt{\frac{\pi}{2}} e^{-\frac{1}{2}x^2}\left[1 - \mathrm{erf}\left(\frac{ix}{\sqrt{2}}\right)\right] &= \sum_{n=0}^{\infty} \frac{(2n-1)!!}{x^{2n+1}} \\ = \exp\Bigg(\sum_{2g-2+n\geq -1} \frac{1}{n!} \sum_{\mu_1,\ldots,\mu_n>0} &\frac{C_{g,n}(\mu_1,\ldots,\mu_n)}{\mu_1\cdots\mu_n} \prod_{i=1}^{n} x^{-(\mu_1+\cdots+\mu_n)}\Bigg).\end{aligned} \tag{3.55}$$

Here, $\mathrm{erf}(x) := \frac{2}{\sqrt{\pi}}\int_0^x e^{-z^2}dz$ is the Gauß error function.

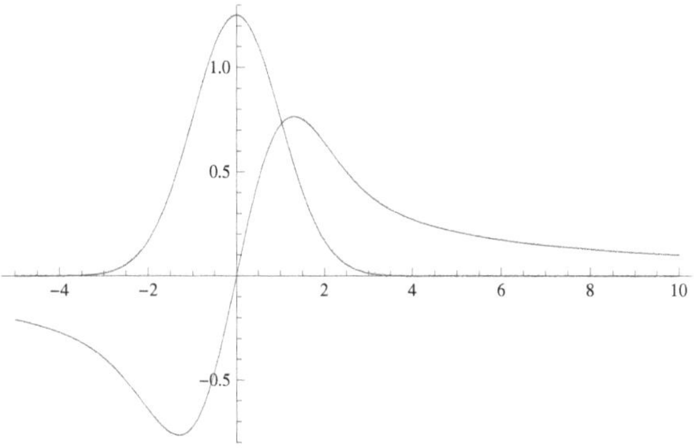

Fig. 3.2. The imaginary part and the real part of $\Psi^{\text{Catalan}}(x, 1)$. For $x >> 0$, the imaginary part dies down, and only the real part has a non-trivial asymptotic expansion. Thus (3.55) is a series with real coefficients.

One of the two independent solutions to the third example, the Gauß hypergeometric equation, that is holomorphic around $x = 0$, is given by

$$\Psi^{\text{Gauß}}(x, \hbar) = {}_2F_1\left(-\frac{\sqrt{(h-1)(h-3)}}{2h} + \frac{1}{h} - \frac{1}{2}, \frac{\sqrt{(h-1)(h-3)}}{2h} + \frac{1}{h} - \frac{1}{2}; \frac{1}{h}; x\right), \tag{3.56}$$

where

$${}_2F_1(a, b; c; x) := \sum_{n=0}^{\infty} \frac{(a)_n (b)_n}{(c)_n} \frac{x^n}{n!} \tag{3.57}$$

is the **Gauß hypergeometric function**. The topological recursion calculates the B-model genus expansion of the periods of the Legendre family of elliptic curves (3.50) at the point where the elliptic curve degenerates to a nodal rational curve. For example, the procedure applied to the spectral curve

$$y^2 + \frac{2x-1}{x(x-1)} y + \frac{1}{4x(x-1)} = 0$$

with a choice of

$$\eta = \frac{-(2x-1) - \sqrt{3x^2 - 3x + 1}}{2x(x-1)} dx,$$

which is an eigenvalue $\alpha_1(x)$ of the Higgs field ϕ, gives a genus expansion at $x = 0$:

$$\Psi^{\text{Gauß}}(x, \hbar) = \exp\left(\sum_{g=0}^{\infty} \sum_{n=1}^{\infty} \frac{1}{n!} \hbar^{2g-2+n} F_{g,n}^{\text{Gauß}}(x)\right). \tag{3.58}$$

At $\hbar = 1$, we have a topological recursion expansion of the period $\omega_1(x)$ defined in (3.46):

$$\frac{\omega_1(x)}{\pi} = \Psi^{\text{Gauß}}(x, 1) = \exp\left(\sum_{g=0}^{\infty}\sum_{n=1}^{\infty}\frac{1}{n!}F_{g,n}^{\text{Gauß}}(x)\right). \tag{3.59}$$

A subtle point we notice here is that while the Gauß hypergeometric equation has regular singular points at $x = 0, 1, \infty$, the Hermite equation has an irregular singular point of class 2 at ∞. The spectral curve of each case has an ordinary double point at $x = \infty$. But the crucial difference lies in the intersection of the spectral curve Σ with the divisor C_∞. For the Hermite case we have $\Sigma \cdot C_\infty = 4$ and the intersection occurs all at once at $x = \infty$. For the Gauß hypergeometric case, the intersection $\Sigma \cdot C_\infty = 4$ occurs once each at $x = 0, 1$, and twice at $x = \infty$. This *confluence* of regular singular points is the source of the irregular singularity in the Hermite differential equation.

The fourth row indicates an example of a quantum curve that has one regular singular point at $x = -1$ and one irregular singular point of class 1 at $x = \infty$. The spectral curve has an ordinary double point at $x = \infty$, the same as the Hermite case. As Figure 3.3 shows, the class of the irregular singularity at $x = \infty$ is determined by how the spectral curve intersects with C_∞.

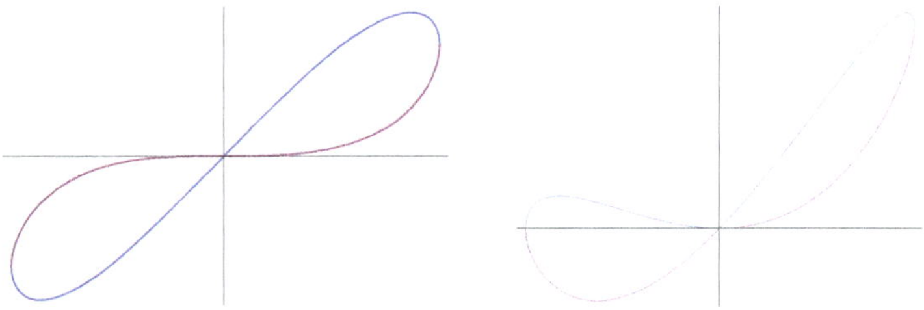

Fig. 3.3. The spectral curves of the second and the fourth examples. The horizontal line is the divisor C_∞, and the vertical line is the fiber class F at $x = \infty$. The spectral curve intersects with C_∞ a total of four times. The curve on the right has a triple intersection at $x = \infty$, while the one on the left intersects all at once.

The existence of the irregular singularity in the quantum curve associated with a spectral curve has nothing to do with the singularity of the spectral curve. The fifth example shows a non-singular spectral curve of genus 1 (Figure 3.4), for which the quantum curve has a class 1 irregular singularity at $x = \infty$.

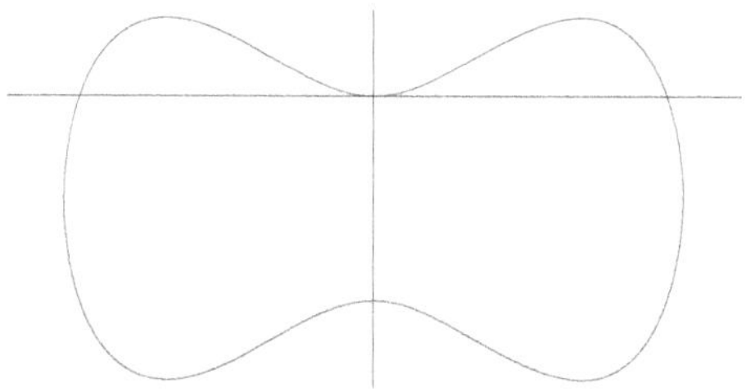

Fig. 3.4. The spectral curve of the fifth example, which is non-singular. The corresponding quantum curve has two regular singular points at $x = \pm 1$, and a class 1 irregular singular point at $x = \infty$.

4. Difference operators as quantum curves

Quantum curves often appear as infinite-order differential operators, or *difference* operators. In this section we present three typical examples: simple Hurwitz numbers, special double Hurwitz numbers, and the Gromov-Witten invariants of $\mathbb{P}^1$. These examples do not come from the usual Higgs bundle framework, because the rank of the Higgs bundle corresponds to the order of the quantum curves as a differential operator. Therefore, we ask:

Question 4.1: What is the geometric structure generalizing the Hitchin spectral curves that correspond to difference operators as their quantization?

In these lectures, we do not address this question, leaving it for a future investigation. We are content with giving examples here. We refer to [53] for a new and different perspective for the notion of quantum curves for difference operators.

4.1. *Simple and orbifold Hurwitz numbers*

The *simple Hurwitz number* $H_{g,n}(\vec{\mu})$ counts the automorphism weighted number of the topological types of simple Hurwitz covers of $\mathbb{P}^1$ of type $(g, \vec{\mu})$. A holomorphic map $\varphi : C \longrightarrow \mathbb{P}^1$ is a *simple Hurwitz cover* of type $(g, \vec{\mu})$ if C is a complete nonsingular algebraic curve defined over $\mathbb{C}$ of genus g, φ has n labeled poles of orders $\vec{\mu} = (\mu_1, \ldots, \mu_n)$, and all other critical points of φ are unlabeled simple ramification points.

In a similar way, we consider the *orbifold Hurwitz number* $H^{(r)}_{g,n}(\vec{\mu})$ for every positive integer $r > 0$ to be the automorphism weighted count of the topological types of smooth orbifold morphisms $\varphi : C \longrightarrow \mathbb{P}^1[r]$ with the same pole structure as the simple Hurwitz number case. Here, C is a connected 1-dimensional orbifold (a *twisted* curve) modeled on a non-singular curve of genus g with $(\mu_1 + \cdots + \mu_n)/r$ stacky points of the type $\left[p/(\mathbb{Z}/r\mathbb{Z})\right]$. We impose that the inverse image of the morphism φ of the unique stacky point $\left[0/(\mathbb{Z}/r\mathbb{Z})\right] \in \mathbb{P}^1[r]$ coincides with the set of stacky points of C. When $r = 1$ we recover the simple Hurwitz number: $H^{(1)}_{g,n}(\vec{\mu}) = H_{g,n}(\vec{\mu})$.

Theorem 4.2: *Cut-and-join equation, [10]. The orbifold Hurwitz numbers* $H^{(r)}_{g,n}(\mu_1, \ldots, \mu_n)$ *satisfy the following equation.*

$$\begin{aligned} sH^{(r)}_{g,n}(\mu_1, \ldots, \mu_n) &= \frac{1}{2} \sum_{i \neq j} (\mu_i + \mu_j) H^{(r)}_{g,n-1}\left(\mu_i + \mu_j, \mu_{[\hat{i},\hat{j}]}\right) \\ &+ \frac{1}{2} \sum_{i=1}^{n} \sum_{\alpha+\beta=\mu_i} \alpha\beta \Bigg[H^{(r)}_{g-1,n+1}\left(\alpha, \beta, \mu_{[\hat{i}]}\right) \\ &\qquad + \sum_{\substack{g_1+g_2=g \\ I \sqcup J = [\hat{i}]}} H^{(r)}_{g_1,|I|+1}(\alpha, \mu_I) H^{(r)}_{g_2,|J|+1}(\beta, \mu_J) \Bigg]. \end{aligned} \tag{4.1}$$

Here

$$s = s(g, \vec{\mu}) = 2g - 2 + n + \frac{\mu_1 + \cdots + \mu_n}{r} \tag{4.2}$$

is the number of simple ramification point given by the Riemann-Hurwitz formula. As before, we use the convention that for any subset $I \subset [n] = \{1, 2, \ldots, n\}$, $\mu_I = (\mu_i)_{i \in I}$. *The hat notation* $\hat{i}$ *indicates that the index* i *is removed. The last summation is over all partitions of* g *and set partitions of* $[\hat{i}] = \{1, \ldots, i-1, i+1, \ldots, n\}$.

Remark 4.3: There is a combinatorial description, in the same manner we have done for the Catalan numbers in Section 2, for simple and orbifold Hurwitz numbers. The cut-and-join equation is derived as the edge-contraction formula, exactly in the same way for the Catalan recursion (2.8). See [26] for more detail.

We regard $H^{(r)}_{g,n}(\vec{\mu})$ as a function in n integer variables $\vec{\mu} \in \mathbb{Z}^n_+$. Following the recipe of [27, 33, 68] that is explained in the earlier sections, we define the *free energies* as the Laplace transform

$$F^{(r)}_{g,n}(z_1, \ldots, z_n) = \sum_{\vec{\mu} \in \mathbb{Z}^n_+} H^{(r)}_{g,n}(\vec{\mu})\, e^{-\langle \vec{w}, \vec{\mu} \rangle}. \tag{4.3}$$

Here, $\vec{w} = (w_1, \ldots, w_n)$ is the vector of the Laplace dual coordinates of $\vec{\mu}$, $\langle \vec{w}, \vec{\mu} \rangle = w_1\mu_1 + \cdots + w_n\mu_n$, and variables z_i and w_i for each i are related by the *r-Lambert* function

$$e^{-w} = ze^{-z^r}. \tag{4.4}$$

It is often convenient to use a different variable $x = ze^{-z^r}$, with which the plane analytic curve called the r-**Lambert curve** is given by

$$\begin{cases} x = ze^{-z^r} \\ y = z^r. \end{cases} \tag{4.5}$$

Then the free energies $F_{g,n}^{(r)}$ of (4.3) are generating functions of the orbifold Hurwitz numbers. By abuse of notation, we also write

$$F_{g,n}^{(r)}(x_1, \ldots, x_n) = \sum_{\vec{\mu} \in \mathbb{Z}_+^n} H_{g,n}^{(r)}(\vec{\mu}) \prod_{i=1}^n x_i^{\mu_i} = \sum_{\vec{\mu} \in \mathbb{Z}_+^n} H_{g,n}^{(r)}(\vec{\mu}) \prod_{i=1}^n \left(z_i e^{-z_i^r} \right)^{\mu_i}. \tag{4.6}$$

For every (g, n), the power series (4.6) in $(x_1, \ldots, x_n)$ is convergent and defines an analytic function.

Theorem 4.4: *Differential recursion for Hurwitz numbers, [10]. In terms of the z-variables, the free energies are calculated as follows.*

$$F_{0,1}^{(r)}(z) = \frac{1}{r}z^r - \frac{1}{2}z^{2r}, \tag{4.7}$$

$$F_{0,2}^{(r)}(z_1, z_2) = \log \frac{z_1 - z_2}{x_1 - x_2} - (z_1^r + z_2^r), \tag{4.8}$$

where $x_i = z_i e^{-z_i^r}$. *For* (g, n) *in the stable range, i.e., when* $2g - 2 + n > 0$, *the free energies satisfy the differential recursion equation*

$$\begin{aligned}
&\left(2g - 2 + n + \frac{1}{r} \sum_{i=1}^n z_i \frac{\partial}{\partial z_i} \right) F_{g,n}^{(r)}(z_1, \ldots, z_n) \\
&= \frac{1}{2} \sum_{i \neq j} \frac{z_i z_j}{z_i - z_j} \left[\frac{1}{(1 - rz_i^r)^2} \frac{\partial}{\partial z_i} F_{g,n-1}^{(r)}\left(z_{[\hat{j}]}\right) - \frac{1}{(1 - rz_j^r)^2} \frac{\partial}{\partial z_j} F_{g,n-1}^{(r)}\left(z_{[\hat{i}]}\right) \right] \\
&\quad + \frac{1}{2} \sum_{i=1}^n \frac{z_i^2}{(1 - rz_i^r)^2} \left. \frac{\partial^2}{\partial u_1 \partial u_2} F_{g-1,n+1}^{(r)}\left(u_1, u_2, z_{[\hat{i}]}\right) \right|_{u_1 = u_2 = z_i} \\
&+ \frac{1}{2} \sum_{i=1}^n \frac{z_i^2}{(1 - rz_i^r)^2} \sum_{\substack{g_1 + g_2 = g \\ I \sqcup J = [\hat{i}]}}^{\text{stable}} \left(\frac{\partial}{\partial z_i} F_{g_1,|I|+1}^{(r)}(z_i, z_I) \right) \left(\frac{\partial}{\partial z_i} F_{g_2,|J|+1}^{(r)}(z_i, z_J) \right).
\end{aligned} \tag{4.9}$$

Remark 4.5: Since $F^{(r)}_{g,n}(z_1,\dots,z_n)\big|_{z_i=0}=0$ for every i, the differential recursion (4.9), which is a linear first order partial differential equation, uniquely determines $F^{(r)}_{g,n}$ inductively for all (g,n) subject to $2g-2+n>0$. This is a generalization of the result of [69] to the orbifold case.

Remark 4.6: The differential recursion of Theorem 4.4 is obtained by taking the Laplace transform of the cut-and-join equation for $H^{(r)}_{g,n}(\vec{\mu})$. The r-Lambert curve itself, (4.5), is obtained by computing the Laplace transform of $H^{(r)}_{0,1}(\mu)$, and solving the differential equation that arises from the cut-and-join equation. See also [26] for a different formulation of the r-Hurwitz numbers using the graph enumeration formulation and a universal mechanism to obtain the spectral curve.

The differential recursion produces two results, as we have seen for the case of the Catalan numbers. One is the quantum curve by taking the principal specialization, and the other the topological recursion of [34].

Theorem 4.7: *Quantum curves for r-Hurwitz numbers, [10]. We introduce the partition function, or the wave function, of the orbifold Hurwitz numbers as*

$$\Psi^{(r)}(z,\hbar)=\exp\left(\sum_{g=0}^{\infty}\sum_{n=1}^{\infty}\frac{1}{n!}\hbar^{2g-2+n}F^{(r)}_{g,n}(z,z,\dots,z)\right). \tag{4.10}$$

It satisfies the following system of (an infinite-order) linear differential equations.

$$\left(\hbar D-e^{r\left(-w+\frac{r-1}{2}\hbar\right)}e^{r\hbar D}\right)\Psi^{(r)}(z,\hbar)=0, \tag{4.11}$$

$$\left(\frac{\hbar}{2}D^2-\left(\frac{1}{r}+\frac{\hbar}{2}\right)D-\hbar\frac{\partial}{\partial\hbar}\right)\Psi^{(r)}(z,\hbar)=0, \tag{4.12}$$

where

$$D=\frac{z}{1-rz^r}\frac{\partial}{\partial z}=x\frac{\partial}{\partial x}=-\frac{\partial}{\partial w}.$$

Let the differential operator of (4.11) *(resp.* (4.12)*) be denoted by P (resp. Q). Then we have the commutator relation*

$$[P,Q]=P. \tag{4.13}$$

The semi-classical limit of each of the equations (4.11) *or* (4.12) *recovers the r-Lambert curve* (4.5).

Remark 4.8: The Schrödinger equation (4.11) is first established in [67].

Remark 4.9: The above theorem is a generalization of [68, Theorem 1.3] for an arbitrary $r > 0$. The restriction $r = 1$ reduces to the simple Hurwitz case.

Remark 4.10: Unlike the situation of Hitchin spectral curves, the results of the quantization of the analytic spectral curves are a **difference-differential equation**, and a PDE containing the differentiation with respect to the deformation parameter $\hbar$.

Now let us define

$$W_{g,n}^{(r)}(z_1, \ldots, z_n) := d_1 d_2 \cdots d_n F_{g,n}^{(r)}(z_1, \ldots, z_n). \tag{4.14}$$

Then we have

Theorem 4.11: *Topological recursion for orbifold Hurwitz numbers, [10]. For the stable range $2g-2+n > 0$, the symmetric differential forms* (4.14) *satisfy the following integral recursion formula.*

$$\begin{aligned} W_{g,n}^{(r)}(z_1, \ldots, z_n) &= \frac{1}{2\pi i} \sum_{j=1}^{r} \oint_{\gamma_j} K_j(z, z_1) \Bigg[W_{g-1,n+1}^{(r)}\big(z, s_j(z), z_2, \ldots, z_n\big) \\ &+ \sum_{i=2}^{n} \Big(W_{0,2}^{(r)}(z, z_i) \otimes W_{g,n-1}^{(r)}\big(s_j(z), z_{[\hat{1},\hat{i}]}\big) + W_{0,2}^{(r)}\big(s_j(z), z_i\big) \otimes W_{g,n-1}^{(r)}\big(z, z_{[\hat{1},\hat{i}]}\big) \Big) \\ &+ \sum_{\substack{g_1+g_2=g \\ I \sqcup J = \{2,\ldots,n\}}}^{\text{stable}} W_{g_1,|I|+1}^{(r)}\big(z, z_I\big) \otimes W_{g_2,|J|+1}^{(r)}\big(s_j(z), z_J\big) \Bigg]. \end{aligned} \tag{4.15}$$

Here, the integration is taken with respect to z along a small simple closed loop γ_j around p_j, and $\{p_1, \ldots, p_r\}$ are the critical points of the r-Lambert function $x(z) = ze^{-z^r}$ at $1 - rz^r = 0$. Since $dx = 0$ has a simple zero at each p_j, the map $x(z)$ is locally a double-sheeted covering around $z = p_j$. We denote by s_j the deck transformation on a small neighborhood of p_j. Finally, the integration kernel is defined by

$$K_j(z, z_1) = \frac{1}{2} \frac{1}{W_{0,1}^{(r)}\big(s_j(z_1)\big) - W_{0,1}^{(r)}(z_1)} \otimes \int_z^{s_j(z)} W_{0,2}^{(r)}(\,\cdot\,, z_1). \tag{4.16}$$

Remark 4.12: As mentioned earlier, the significance of the integral formalism is its universality. The differential equation (4.9) takes a different form depending on the counting problem, whereas the integral formula (4.15) depends only on the choice of the spectral curve.

Remark 4.13: The proof is based on the idea of [33]. The notion of the *principal part* of meromorphic differentials plays a key role in converting the Laplace transform of the cut-and-join equation into a residue formula.

4.2. *Gromov-Witten invariants of the projective line*

Hurwitz numbers and Gromov-Witten invariants of $\mathbb{P}^1$ are closely related [75]. However, their relations to the topological recursion is rather different. For example, the topological recursion for stationary Gromov-Witten invariants of $\mathbb{P}^1$ was conjectured by Norbury and Scott [74] as a concrete formula, but its proof [30] is done in a very different way than that of [33,69]. This is based on the fact that we do not have a counterpart of the cut-and-join equation for the case of the Gromov-Witten invariants of $\mathbb{P}^1$. Nonetheless, the quantum curve exists.

Let $\overline{\mathcal{M}}_{g,n}(\mathbb{P}^1, d)$ denote the moduli space of stable maps of degree d from an n-pointed genus g curve to $\mathbb{P}^1$. This is an algebraic stack of dimension $2g - 2 + n + 2d$. The dimension reflects the fact that a generic map from an algebraic curve to $\mathbb{P}^1$ has only simple ramifications, and the number of such ramification points, which we can derive from the Riemann-Hurwitz formula, gives the dimension of this stack. The descendant Gromov-Witten invariants of $\mathbb{P}^1$ are defined by

$$\left\langle \prod_{i=1}^{n} \tau_{b_i}(\alpha_i) \right\rangle_{g,n}^{d} := \int_{[\overline{\mathcal{M}}_{g,n}(\mathbb{P}^1,d)]^{vir}} \prod_{i=1}^{n} \psi_i^{b_i} ev_i^*(\alpha_i), \tag{4.17}$$

where $[\overline{\mathcal{M}}_{g,n}(\mathbb{P}^1, d)]^{vir}$ is the virtual fundamental class of the moduli space,

$$ev_i : \overline{\mathcal{M}}_{g,n}(\mathbb{P}^1, d) \longrightarrow \mathbb{P}^1$$

is a natural morphism defined by evaluating a stable map at the i-th marked point of the source curve, $\alpha_i \in H^*(\mathbb{P}^1, \mathbb{Q})$ is a cohomology class of the target $\mathbb{P}^1$, and ψ_i is the tautological cotangent class in $H^2(\overline{\mathcal{M}}_{g,n}(\mathbb{P}^1, d), \mathbb{Q})$. We denote by 1 the generator of $H^0(\mathbb{P}^1, \mathbb{Q})$, and by $\omega \in H^2(\mathbb{P}^1, \mathbb{Q})$ the Poincaré dual to the point class. We assemble the Gromov-Witten invariants into particular generating functions as follows. For every (g, n) in the stable sector $2g - 2 + n > 0$, we define the *free energy* of type (g, n) by

$$F_{g,n}(x_1, \ldots, x_n) := \left\langle \prod_{i=1}^{n} \left(-\frac{\tau_0(1)}{2} - \sum_{b=0}^{\infty} \frac{b!\tau_b(\omega)}{x_i^{b+1}} \right) \right\rangle_{g,n}. \tag{4.18}$$

Here the degree d is determined by the dimension condition of the cohomology classes to be integrated over the virtual fundamental class. We note that (4.18) contains the class $\tau_0(1)$. For unstable geometries, we introduce two functions

$$S_0(x) := x - x\log x + \sum_{d=1}^{\infty} \left\langle -\frac{(2d-2)!\tau_{2d-2}(\omega)}{x^{2d-1}} \right\rangle_{0,1}^{d}, \tag{4.19}$$

$$S_1(x) := -\frac{1}{2}\log x + \frac{1}{2}\sum_{d=0}^{\infty} \left\langle \left(-\frac{\tau_0(1)}{2} - \sum_{b=0}^{\infty} \frac{b!\tau_b(\omega)}{x^{b+1}} \right)^2 \right\rangle_{0,2}^{d}, \tag{4.20}$$

utilizing an earlier work of [27]. Then we have

Theorem 4.14: *The quantum curve for the Gromov-Witten invariants of $\mathbb{P}^1$, [29]. The wave function*

$$\Psi(x,\hbar) := \exp\left(\frac{1}{\hbar} S_0(x) + S_1(x) + \sum_{2g-2+n>0} \frac{\hbar^{2g-2+n}}{n!} F_{g,n}(x,\ldots,x) \right) \tag{4.21}$$

satisfies the quantum curve equation of an infinite order

$$\left[\exp\left(\hbar \frac{d}{dx} \right) + \exp\left(-\hbar \frac{d}{dx} \right) - x \right] \Psi(x,\hbar) = 0. \tag{4.22}$$

Moreover, the free energies $F_{g,n}(x_1,\ldots,x_n)$ as functions in n-variables, and hence all the Gromov-Witten invariants (4.17), *can be recovered from the equation* (4.22) *alone, using the mechanism of the* ***topological recursion*** *of [34].*

Remark 4.15: The appearance of the extra terms in S_0 and S_1, in particular, the $\log x$ terms, is trickier than the cases studied in these lectures. We refer to [29, Section 3].

Remark 4.16: Put

$$S_m(x) := \sum_{2g-2+n=m-1} \frac{1}{n!} F_{g,n}(x,\ldots,x). \tag{4.23}$$

Then our wave function is of the form

$$\Psi(x,\hbar) = \exp\left(\sum_{m=0}^{\infty} \hbar^{m-1} S_m(x) \right), \tag{4.24}$$

which provides the WKB approximation of the quantum curve equation (4.22). Thus the significance of (4.18) is that the exponential generating function (4.21) of the descendant Gromov-Witten invariants of $\mathbb{P}^1$ gives the solution to the WKB analysis in a closed formula for the difference equation (4.22).

Remark 4.17: For the case of Hitchin spectral curves [24, 25], the Schrödinger-like equation (4.22) is a direct consequence of the generalized topological recursion. In the $GW(\mathbb{P}^1)$ context, the topological recursion does not play any role in establishing (4.22).

We can recover the classical mechanics corresponding to (4.22) by taking its semi-classical limit, which is the singular perturbation limit

$$\lim_{\hbar\to 0}\left(e^{-\frac{1}{\hbar}S_0(x)}\left[\exp\left(\hbar\frac{d}{dx}\right)+\exp\left(-\hbar\frac{d}{dx}\right)-x\right]e^{\frac{1}{\hbar}S_0(x)}e^{\sum_{m=1}^{\infty}\hbar^{m-1}S_m(x)}\right)$$
$$=\left(e^{S_0'(x)}+e^{-S_0'(x)}-x\right)e^{S_1(x)}=0. \quad (4.25)$$

In terms of new variables $y(x)=S_0'(x)$ and $z(x)=e^{y(x)}$, the semi-classical limit gives us an equation for the spectral curve

$$z\in\Sigma=\mathbb{C}^*\subset\mathbb{C}\times\mathbb{C}^*\overset{\exp}{\longleftarrow}T^*\mathbb{C}=\mathbb{C}^2\ni(x,y)$$

by

$$\begin{cases} x=z+\frac{1}{z} \\ y=\log z \end{cases}. \quad (4.26)$$

This is the reason we consider (4.22) as the quantization of the Laudau-Ginzburg model

$$x=z+\frac{1}{z}.$$

It was conjectured in [74] that the stationary Gromov-Witten theory of $\mathbb{P}^1$ should satisfy the topological recursion with respect to the spectral curve (4.26). The conjecture is solved in [30] as a corollary to its main theorem, which establishes the correspondence between the topological recursion and the Givental formalism.

The key discovery of [29] is that the quantum curve equation (4.22) is equivalent to a recursion equation

$$\frac{x}{\hbar}\left(e^{-\hbar\frac{d}{dx}}-1\right)X_d(x,\hbar)+\frac{1}{1+\frac{x}{\hbar}}e^{\hbar\frac{d}{dx}}X_{d-1}(x,\hbar)=0 \quad (4.27)$$

for a rational function

$$X_d(x,\hbar)=\sum_{\lambda\vdash d}\left(\frac{\dim\lambda}{d!}\right)^2\prod_{i=1}^{\ell(\lambda)}\frac{x+(i-\lambda_i)\hbar}{x+i\hbar}. \quad (4.28)$$

Here λ is a partition of $d\geq 0$ with parts λ_i and $\dim\lambda$ denotes the dimension of the irreducible representation of the symmetric group S_d characterized by λ.

Acknowledgments

This chapter is based on the series of lectures that the authors have given in Singapore, Kobe, Hannover, Hong Kong, and Leiden in 2014–2015. They are indebted to Richard Wentworth and Graeme Wilkin for their kind invitation to the Institute for Mathematical Sciences at the National University of Singapore, where these lectures were first delivered at the IMS Summer Research Institute, *The Geometry, Topology and Physics of Moduli Spaces of Higgs Bundles*, in July 2014. The authors are also grateful to Masa-Hiko Saito for his kind invitation to run the Kobe Summer School consisting of an undergraduate course and a graduate course on the related topics at Kobe University, Japan, in July–August, 2014. A part of these lectures was also delivered at the *Advanced Summer School: Modern Trends in Gromov-Witten Theory*, Leibniz Universität Hannover, in September 2014.

The authors' special thanks are due to Laura P. Schaposnik, whose constant interest in the subject made these lecture notes possible.

The authors are grateful to the American Institute of Mathematics in California, the Banff International Research Station, Max-Planck-Institut für Mathematik in Bonn, and the Lorentz Center for Mathematical Sciences, Leiden, for their hospitality and financial support for the collaboration of the authors. They thank Jørgen Andersen, Philip Boalch, Gaëtan Borot, Vincent Bouchard, Andrea Brini, Leonid Chekhov, Bertrand Eynard, Laura Fredrickson, Tamás Hausel, Kohei Iwaki, Maxim Kontsevich, Andrew Neitzke, Paul Norbury, Alexei Oblomkov, Brad Safnuk, Albert Schwarz, Sergey Shadrin, Yan Soibelman, and Piotr Sułkowski for useful discussions. They also thank the referees for numerous suggestions to improve these lecture notes.

O.D. thanks the Perimeter Institute for Theoretical Physics, Canada, and M.M. thanks the University of Amsterdam, l'Institut des Hautes Études Scientifiques, Hong Kong University of Science and Technology, and the Simons Center for Geometry and Physics, for financial support and hospitality. The research of O.D. has been supported by GRK 1463 of Leibniz Universität Hannover and MPIM in Bonn. The research of M.M. has been supported by MPIM in Bonn, NSF grants DMS-1104734 and DMS-1309298, and NSF-RNMS: Geometric Structures And Representation Varieties (GEAR Network, DMS-1107452, 1107263, 1107367).

References

1. M. Aganagic, R. Dijkgraaf, A. Klemm, M. Mariño, and C. Vafa, *Topological Strings and Integrable Hierarchies*, [arXiv:hep-th/0312085], Commun. Math. Phys. **261**, 451–516 (2006).
2. G.B. Airy, *On the intensity of light in the neighbourhood of a caustic*, Transactions of the Cambridge Philosophical Society **6**, 379–402 (1838).
3. J. Andersen, G. Borot, and N. Orantin, *Modular functors, cohomological field theories and topological recursion*, arXiv:1509.01387 (2015).
4. D. Arinkin, *On λ-connections on a curve where λ is a formal parameter*, Mathematical Research Letters **12**, 551–565 (2005).
5. A. Beauville, M. S. Narasimhan, and S. Ramanan, *Spectral curves and the generalized theta divisor*, J. reine angew. Math. **398**, 169–179 (1989).
6. O. Biquard and P. Boalch, *Wild non-Abelian Hodge theory on curves*, Compos. Math. **140**, 179–204 (2004).
7. P. Boalch, *Symplectic manifolds and isomonodromic deformations*, Adv. in Math. **163**, 137–205 (2001).
8. P. Boalch, *Hyperkähler manifolds and nonabelian Hodge theory of (irregular) curves*, arXiv:1203.6607v1 (2012).
9. G. Borot and B. Eynard, *All-order asymptotics of hyperbolic knot invariants from non-perturbative topological recursion of A-polynomials*, arXiv:1205.2261v1 [math-ph] (2012).
10. V. Bouchard, D. Hernández Serrano, X. Liu, and M. Mulase, *Mirror symmetry for orbifold Hurwitz numbers*, arXiv:1301.4871 [math.AG] (2013).
11. V. Bouchard, A. Klemm, M. Mariño, and S. Pasquetti, *Remodeling the B-model*, Commun. Math. Phys. **287**, 117–178 (2009).
12. V. Bouchard, A. Klemm, M. Mariño, and S. Pasquetti, *Topological open strings on orbifolds*, Commun. Math. Phys. **296**, 589–623 (2010).
13. V. Bouchard and M. Mariño, *Hurwitz numbers, matrix models and enumerative geometry*, Proc. Symposia Pure Math. **78**, 263–283 (2008).
14. A. Brini, B. Eynard, and M. Mariño, *Torus knots and mirror symmetry* arXiv:1105.2012 (2011).
15. K. Chapman, M. Mulase, and B. Safnuk, *The Kontsevich constants for the volume of the moduli of curves and topological recursion*, Communications in Number Theory and Physics **5**, 643–698 (2011).
16. L. Chekhov, B. Eynard, and N. Orantin, *Free energy topological expansion for the 2-matrix model*, Journal of High Energy Physics **0612:053** (2006).
17. R. Dijkgraaf, H. Fuji, and M. Manabe, *The Volume Conjecture, Perturbative Knot Invariants, and Recursion Relations for Topological Strings*, Nucl. Phys. **B849**, 166–211 (2011).
18. R. Dijkgraaf, L. Hollands, and P. Sułkowski, *Quantum curves and $\mathcal{D}$-modules*, Journal of High Energy Physics **0810.4157**, 1–58 (2009).
19. R. Dijkgraaf, L. Hollands P. Sułkowski, and C. Vafa, *Supersymmetric gauge theories, intersecting branes and free Fermions*, Journal of High Energy Physics **0802.106**, (2008).

20. R. Dijkgraaf, E. Verlinde, and H. Verlinde, *Loop equations and Virasoro constraints in non-perturbative two-dimensional quantum gravity*, Nucl. Phys. **B348**, 435–456 (1991).
21. N. Do, O. Leigh, and P. Norbury, *Orbifold Hurwitz numbers and Eynard-Orantin invariants*, arXiv:1212.6850 (2012).
22. N. Do and D. Manescu, *Quantum curves for the enumeration of ribbon graphs and hypermaps*, arXiv:1312.6869 [math.GT] (2013).
23. O. Dumitrescu, L. Fredrickson, G. Kydonakis, R. Mazzeo, M. Mulase and A. Neitzke, *From the Hitchin section to opers through Nonabelian Hodge*, accepted in Journal of Differential Geometry, 23 pages (2016).
24. O. Dumitrescu and M. Mulase, *Quantum curves for Hitchin fibrations and the Eynard-Orantin theory*, Lett. Math. Phys. **104**, 635–671 (2014).
25. O. Dumitrescu and M. Mulase, *Quantization of spectral curves for meromorphic Higgs bundles through topological recursion*, to appear in Proceedings of the 2016 Von Neumann Symposium, American Mathematical Society (2014).
26. O. Dumitrescu and M. Mulase, *Edge contraction on dual ribbon graphs and 2D TQFT*, Journal of Algebra **494** (January 2018).
27. O. Dumitrescu, M. Mulase, A. Sorkin and B. Safnuk, *The spectral curve of the Eynard-Orantin recursion via the Laplace transform*, in "Algebraic and Geometric Aspects of Integrable Systems and Random Matrices," Dzhamay, Maruno and Pierce, Eds. Contemporary Mathematics **593**, 263–315 (2013).
28. P. Dunin-Barkowski, M. Kazarian, N. Orantin, S. Shadrin, and L. Spitz, *Polynomiality of Hurwitz numbers, Bouchard-Mariño conjecture, and a new proof of the ELSV formula*, arXiv:1307.4729 [math.AG (physics.hep-th physics.math-ph)] (2013).
29. P. Dunin-Barkowski, M. Mulase, P. Norbury, A. Popolitov, and S. Shadrin, *Quantum spectral curve for the Gromov-Witten theory of the complex projective line*, arXiv:1312.5336 [math-ph] (2013).
30. P. Dunin-Barkowski, N. Orantin, S. Shadrin, and L. Spitz, *Identification of the Givental formula with the spectral curve topological recursion procedure*, arXiv:1211.4021 [physics.math-ph (math.AG, physics.hep-th)] (2012).
31. T. Ekedahl, S. Lando, M. Shapiro, A. Vainshtein, *Hurwitz numbers and intersections on moduli spaces of curves*, Invent. Math. **146**, 297–327 (2001) [arXiv:math/0004096].
32. B. Eynard, *A short overview of the "Topological recursion"*, arXiv:1412.3286 (2014).
33. B. Eynard, M. Mulase and B. Safnuk, *The Laplace transform of the cut-and-join equation and the Bouchard-Mariño conjecture on Hurwitz numbers*, Publications of the Research Institute for Mathematical Sciences **47**, 629–670 (2011).
34. B. Eynard and N. Orantin, *Invariants of algebraic curves and topological expansion*, Communications in Number Theory and Physics **1**, 347–452 (2007).
35. B. Eynard and N. Orantin, *Computation of open Gromov-Witten invariants for toric Calabi-Yau 3-folds by topological recursion, a proof of the BKMP conjecture*, arXiv:1205.1103v2 [math-ph] (2013).

36. C. Faber, *A conjectural description of the tautological ring of the moduli space of curves*, in "Moduli of curves and abelian varieties," 109–129, Aspects Math., **E33**, Vieweg, Braunschweig, 1999.
37. B. Fang, C.-C. M. Liu, and Z. Zong, *All genus open-closed mirror symmetry for affine toric Calabi-Yau 3-orbifolds*, arXiv:1310.4818 [math.AG] (2013).
38. D. Gaiotto, G.W. Moore, and A. Neitzke, *Wall-crossing, Hitchin systems, and the WKB approximation*, arXiv:0907.3987 [hep-th] (2009).
39. S. Garoufalidis, *On the characteristic and deformation varieties of a knot*, Geometry & Topology Monographs **7**, 291–309 (2004).
40. S. Garoufalidis and T.T.Q. Lê, *The colored Jones function is qholonomic*, Geometry and Topology **9**, 1253–1293 (2005).
41. E. Getzler and M.M. Kapranov, *Modular operad*, Compositio Mathematica **110**, 65–126 (1998).
42. I.P. Goulden, *A differential operator for symmetric functions and the combinatorics of multiplying transpositions*, Trans. A.M.S., **344**, 421–440 (1994).
43. I.P. Goulden and D.M. Jackson, *Transitive factorisations into transpositions and holomorphic mappings on the sphere*, Proc. A.M.S., **125**, 51–60 (1997).
44. S. Gukov and P. Sułkowski, *A-polynomial, B-model, and quantization*, arXiv:1108.0002v1 [hep-th] (2011).
45. J. Harer and D. Zagier, *The Euler characteristic of the moduli space of curves*, Inventiones Mathematicae **85**, 457–485 (1986).
46. N.J. Hitchin, *The self-duality equations on a Riemann surface*, Proc. London Math. Soc. (Ser. 3) **55**, 59–126 (1987).
47. N.J. Hitchin, *Stable bundles and integrable systems*, Duke Math. J. **54**, 91–114 (1987).
48. L. Hollands, *Topological strings and quantum curves*, Ph.D. Thesis, University of Amsterdam arXiv:0911.3413 [hep-th] (2009).
49. E. Ionel, *Relations in the tautological ring of* $\mathcal{M}_g$, Duke Math. J. **129**, 157–186 (2005).
50. K. Iwaki and A. Saens-Rodriguez, *Quantum Curve and the First Painlevé Equation*, arXiv:1507.06557 (2015).
51. P. Johnson, R. Pandharipande, and H.H. Tseng, *Abelian Hurwitz-Hodge integrals*, Michigan Math. J. **60**, 171–198 (2011) [arXiv:0803.0499].
52. A. Kapustin and E. Witten, *Electric-magnetic duality and the geometric Langlands program*, Communications in Number Theory and Physics **1**, 1–236 (2007).
53. R. Kashaev and M. Mariño, *Operators from mirror curves and the quantum dilogarithm*, arXiv:1501.01014 [hep-th], (2015).
54. Y. Kodama and V.U. Pierce, *Combinatorics of dispersionless integrable systems and universality in random matrix theory*, Communications in Mathematical Physics **292**, 529–568 (2009).
55. M. Kontsevich, *Intersection theory on the moduli space of curves and the matrix Airy function*, Communications in Mathematical Physics **147**, 1–23 (1992).
56. M. Kontsevich and Yuri Manin, *Gromov-Witten classes, quantum cohomology, and enumerative geometry*, Communications in Mathematical Physics **164**, 525–562 (1994).

57. M. Kontsevich and Y. Soibelman, *Wall-crossing structures in Donaldson-Thomas invariants, integrable systems and Mirror Symmetry*, arXiv:1303.3253 [math.AG] (2013).
58. M. Kontsevich and D. Zagier, *Periods*, in " Mathematics unlimited–2001 and beyond", 771–808, Springer-verlag, Berlin, (2001).
59. M. Mariño, *Open string amplitudes and large order behavior in topological string theory*, J. High Energy Physics **0803-060**, 1–33 (2008).
60. M. Mirzakhani, *Simple geodesics and Weil-Petersson volumes of moduli spaces of bordered Riemann surfaces*, Invent. Math. **167**, 179–222 (2007).
61. M. Mirzakhani, *Weil-Petersson volumes and intersection theory on the moduli space of curves*, J. Amer. Math. Soc. **20**, 1–23 (2007).
62. M. Mulase, *Algebraic theory of the KP equations*, in "Perspectives in Mathematical Physics," R. Penner and S.-T. Yau, Editors, International Press Company, 157–223 (1994).
63. M. Mulase, *Lectures on the Asymptotic Expansion of a Hermitian Matrix Integral*, in Springer Lecture Notes in Physics **502**, H. Aratyn *et al.*, Editors, 91–134 (1998).
64. M. Mulase and M. Penkava, *Ribbon graphs, quadratic differentials on Riemann surfaces, and algebraic curves defined over* $\overline{\mathbb{Q}}$, The Asian Journal of Mathematics **2** (4), 875–920 (1998).
65. M. Mulase and M. Penkava, *Topological recursion for the Poincaré polynomial of the combinatorial moduli space of curves*, Advances in Mathematics **230**, 1322–1339 (2012).
66. M. Mulase and B. Safnuk, *Mirzakhani's Recursion Relations, Virasoro Constraints and the KdV Hierarchy*, Indian Journal of Mathematics **50**, 189–228 (2008).
67. M. Mulase, S. Shadrin, and L. Spitz, *The spectral curve and the Schrödinger equation of double Hurwitz numbers and higher spin structures*, Communications in Number Theory and Physics **7**, 1–19 (2013).
68. M. Mulase and P. Sułkowski, *Spectral curves and the Schrödinger equations for the Eynard-Orantin recursion*, arXiv:1210.3006 (2012).
69. M. Mulase and N. Zhang, *Polynomial recursion formula for linear Hodge integrals*, Communications in Number Theory and Physics **4**, 267–294 (2010).
70. D. Mumford, *Towards an enumerative geometry of the moduli space of curves*, in "Arithmetic and Geometry," M. Artin and J. Tate, eds., Part II, 271–328, Birkhäuser, 1983.
71. A. Neitzke, *Hitchin systems in* $\mathcal{N} = 2$ *field theory*, Preprint, November 2013.
72. P. Norbury, *Counting lattice points in the moduli space of curves*, arXiv:0801.4590 (2008).
73. P. Norbury, *String and dilaton equations for counting lattice points in the moduli space of curves*, arXiv:0905.4141 (2009).
74. P. Norbury and N. Scott, *Gromov-Witten invariants of* $\mathbb{P}^1$ *and Eynard-Orantin invariants*, arXiv:1106.1337 (2011).
75. A. Okounkov and R. Pandharipande, *Gromov-Witten theory, Hurwitz numbers, and matrix models, I*, Proc. Symposia Pure Math. **80**, 325–414 (2009).

76. H. Ooguri, P. Sulkowski, M. Yamazaki, *Wall Crossing As Seen By Matrix Models*, arXiv:1005.1293 (2010).
77. R. Pandharipande and A. Pixton, *Relations in the tautological ring of the moduli space of curves*, arXiv:1301.4561 [math.AG] (2013).
78. F. Petit, *Quantization of spectral curves and DQ-modules*, arXiv:1507.04315 (2015).
79. L. Schneps and P. Lochak, *Geometric Galois actions*, London Mathematical Society Lecture Notes Series **242**, 1997.
80. C.T. Simpson, *Higgs bundles and local systems*, Publications Mathématiques de l'I.H.E.S. **75**, 5–95 (1992).
81. D. D. Sleator, R. E. Tarjan, and W. P. Thurston, *Rotation distance, triangulations, and hyperbolic geometry*, Journal of the American Mathematical Society **1**, 647–681 (1988).
82. K. Strebel, *Quadratic differentials*, Springer-Verlag, 1984.
83. J. Teschner, *Supersymmetric gauge theories, quantisation of moduli spaces of flat connections, and Liouville theory*, arXiv:1412.7140 (2014).
84. R. Vakil, Harvard Thesis 1997.
85. T.R.S. Walsh and A.B. Lehman, *Counting rooted maps by genus. I*, Journal of Combinatorial Theory **B-13**, 192–218 (1972).
86. E. Witten, *Two dimensional gravity and intersection theory on moduli space*, Surveys in Differential Geometry **1**, 243–310 (1991).
87. E. Witten, *Gauge theory and wild ramification*, Analysis and Applications **6**, 429–501 (2008).
88. E. Witten, *Mirror symmetry, Hitchin's equations, and Langlands duality*, arXiv:0802.0999 (2008).
89. E. Witten, *Analytic continuation of Chern-Simons theory*, arXiv:1001.2933v4 [hep-th] (2010).
90. J. Zhou, *Local Mirror Symmetry for the Topological Vertex* arXiv:0911.2343 (2009).
91. J. Zhou, *Intersection numbers on Deligne-Mumford moduli spaces and quantum Airy curve*, arXiv:1206.5896 [math.AG] (2012).
92. J. Zhou, *Quantum Mirror Curves for* $\mathbb{C}^3$ *and the Resolved Confiold*, arXiv:1207.0598v1 [math.AG] (2012).
93. M. Zhou, *The differential recursion equation for the generalized Catalan numbers*, Senior Thesis, University of California, Davis (2013).

An Introduction to Moduli Stacks, with a View towards Higgs Bundles on Algebraic Curves

Sebastian Casalaina-Martin* and Jonathan Wise†

Department of Mathematics
University of Colorado, Boulder
Campus Box 395
Boulder, CO 80305-0395, USA
* *casa@math.colorado.edu*
† *jonathan.wise@math.colorado.edu*

This chapter is based in part on lecture notes prepared for the summer school "The Geometry, Topology and Physics of Moduli Spaces of Higgs Bundles" at the Institute for Mathematical Sciences at the National University of Singapore in July of 2014. The aim is to provide a brief introduction to algebraic stacks, and then to give several constructions of the moduli stack of Higgs bundles on algebraic curves. The first construction is via a "bootstrap" method from the algebraic stack of vector bundles on an algebraic curve. This construction is motivated in part by Nitsure's GIT construction of a projective moduli space of semi-stable Higgs bundles, and we describe the relationship between Nitsure's moduli space and the algebraic stacks constructed here. The third approach is via deformation theory, where we directly construct the stack of Higgs bundles using Artin's criteria.

Contents

Introduction

Stacks have been used widely by algebraic geometers since the 1960s for studying parameter spaces for algebraic and geometric objects [20]. Their popularity is growing in other areas as well (e.g., [59]). Nevertheless, despite their utility, and their having been around for many years, stacks still do not seem to be as popular as might be expected. Of course, echoing the introduction of [23], this may have something to do with the technical nature of the topic, which may dissuade the uninitiated, particularly when there may be less technical methods available that can be used instead. To capture a common sentiment, it is hard to improve on the following excerpt from Harris–Morrison [46, 3.D], which serves as an introduction to their section on stacks:

> *'Of course, here I'm working with the moduli stack rather than with the moduli space. For those of you who aren't familiar with stacks, don't worry: basically, all it means is*

that I'm allowed to pretend that the moduli space is smooth and there's a universal family over it.'

Who hasn't heard these words, or their equivalent spoken at a talk? And who hasn't fantasized about grabbing the speaker by the lapels and shaking him until he says what – exactly – he means by them?

At the same time, the reason stacks are used as widely as they are is that they *really* are a natural language for talking about parameterization. While there is no doubt a great deal of technical background that goes into the set-up, the pay-off is that once this foundation has been laid, stacks provide a clean, unified language for discussing what otherwise may require many caveats and special cases.

Our goal here is to give a brief introduction to algebraic stacks, with a view towards defining the moduli stack of Higgs bundles. We have tried to provide enough motivation for stacks that the reader is inclined to proceed to the definitions, and a sufficiently streamlined presentation of the definitions that the reader does not immediately stop at that point! In the end, we hope the reader has a good sense of what the moduli stack of Higgs bundles is, why it is an algebraic stack, and how the stack relates to the quasi-projective variety of Higgs bundles constructed by Nitsure [64].

Due to the authors' backgrounds, for precise statements, the topic will be treated in the language algebraic geometry, i.e., schemes. However, the aim is to have a presentation that is accessible to those in other fields, as well, particularly complex geometers, and most of the presentation can be made replacing the word "scheme" with the words "complex analytic space" (or even "manifold") and "étale cover" with "open cover". This is certainly the case up though, and including, the definition of a stack in §3. The one possible exception to this rule is the topic of algebraic stacks §6, for which definitions in the literature are really geared towards the category of schemes. In order to make our presentation as accessible as possible, in §5 and §6 we provide definitions of algebraic stacks that make sense for any presite; in particular, the definition gives a notion of an "algebraic" stack in the category of complex analytic spaces. There are other notions of analytic stacks in the literature, but for concision, we do not pursue the connection between the definitions.

The final sections of this survey (§10 and §11) study algebraic stacks infinitesimally, with the double purpose of giving modular meaning to infinitesimal motion in an algebraic stack and introducing Artin's criterion as a means to prove stacks are algebraic. Higgs bundles are treated as an extended example, and along they way we give cohomological interpretations of the tangent bundles of the moduli stacks of curves, vector bundles, and morphisms between them. In the end we obtain a direct proof of the algebraicity of the stack of Higgs bundles, complementing the one based on established general algebraicity results given earlier.

While these notes are meant to be somewhat self contained, in the sense that essentially all relevant definitions are included, and all results are referenced to the literature, these notes are by no means comprehensive. There are now a number of introductions to the topic of algebraic stacks that have appeared recently (and even more are in preparation), and these notes are inspired by various parts of those treatments. While the following list is not exhaustive, it provides a brief summary of some of the other resources available. To begin, we direct the reader to the now classic, concise introduction by Fantechi [23]. The reader may very well want to read that ten-page introduction before this one. A more detailed introduction to Deligne–Mumford stacks is given in the book [12]. The book [24] provides another detailed introduction, with less emphasis on the algebraic property of algebraic stacks. The book [54] provides a concise, detailed, but perhaps less widely accessible introduction to the topic. There is also the comprehensive treatise [77], which has a complete treatment of all of the details. The reader who would like to follow the early development of the subject might want to consult [29], [62], [20], [2, Exp. II], [4, Exp. XVII, XVIII], [53], and [9].

Notation

We denote by S the category of schemes over a fixed base, which the reader may feel free to assume is $\operatorname{Spec}\mathbb{C}$, where $\mathbb{C}$ is the field of complex numbers. All schemes will be members of this category. Until we discuss algebraic stacks, one could even take S to be the category of complex analytic spaces, or any other reasonable category of "spaces" with which one would like to work.

Our convention is generally to use roman letters for schemes (e.g., M), script letters for functors to sets (e.g., $\mathscr{M}$), and calligraphic letters for categories fibered in groupoids; (e.g., $\mathcal{M}$). We will typically use sans serif letters for various standard categories (e.g., M).

1. Moduli problems as functors

When one wants to parameterize some kind of algebraic or geometric objects, one says one has a moduli problem. The goal is to find another geometric object, called a moduli space, that parameterizes the objects of interest. One of the most natural ways to phrase a moduli problem is in terms of the corresponding moduli functor. From this perspective, the hope is that the moduli functor will be representable by a geometric object, which will be the moduli space. In fact, it is possible to say a great deal about the moduli space purely in terms of the moduli functor, without even knowing the moduli space exists! For example, the tangent space can be computed by evaluating the moduli functor on the spectrum of the ring of dual numbers (§1.3).

We make this precise in what follows. The best way to get a feel for this is through examples. For this reason, in this section we start by considering the familiar problems of parameterizing linear subspaces of a fixed vector space, and of parameterizing Riemann surfaces of a fixed genus.

1.1. *Grassmannians*

We expect this example is familiar to most of the readers, and the aim will be to motivate an approach to these types of problems, which we will continue to use throughout. We direct the reader to [30, Ch.1,§5] for details.

In this section we explicitly take S to be the category of $\mathbb{C}$-schemes, or the category of analytic spaces, for simplicity; a "scheme" refers to a member of this category. For an object S of S, we adopt the notation $\mathbb{C}^n_S := \mathbb{C}^n \times S$ for the trivial vector bundle of rank n over S.

Recall that Grassmannians arise from the moduli problem that consists of parameterizing r-dimensional complex subspaces of $\mathbb{C}^n$.

There is clearly a set of such spaces

$$G(r,n) := \{W \subseteq_{\text{lin}} \mathbb{C}^n : \dim W = r\}.$$

In fact, one can easily put a "natural" complex structure on $G(r,n)$ that makes this set into a smooth complex projective variety of dimension $r(n-r)$, which we call the Grassmannian. Of key importance from our perspective is the fact that there is a rank r vector bundle $\mathbb{U}$ over $G(r,n)$, and an inclusion of vector bundles:

$$0 \to \mathbb{U} \to \mathbb{C}^n_{G(r,n)}$$

such that for any $\mathbb{C}$-scheme S and any inclusion of vector bundles $\mathbb{F} \hookrightarrow \mathbb{C}^n_S$

over S with rank $\mathbb{F} = r$, there is a *unique* $\mathbb{C}$-morphism

$$f : S \to G(r, n)$$

such that the inclusion $\mathbb{F} \hookrightarrow \mathbb{C}^n_S$ is isomorphic to the pullback of the inclusion $\mathbb{U} \hookrightarrow \mathbb{C}^n_{G(r,n)}$ by f. Diagramatically:

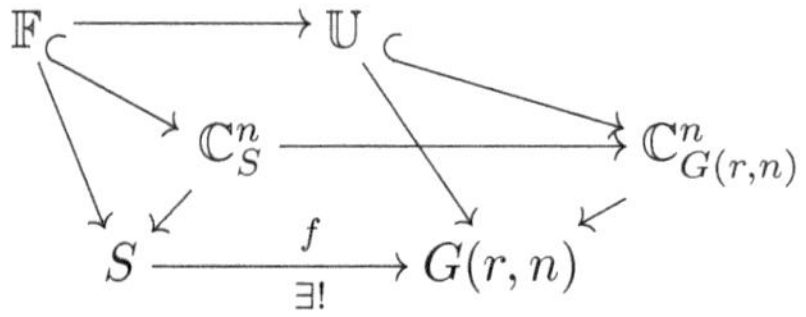

Set theoretically, we define $\mathbb{U}$ by setting the fiber of $\mathbb{U}$ over a point $[W] \in G(r, n)$ corresponding to a subspace $W \subseteq \mathbb{C}^n$ to be the subspace W itself; i.e., we have $\mathbb{U}_{[W]} = W \hookrightarrow \mathbb{C}^n = (\mathbb{C}^n_{G(r,n)})_{[W]}$. The bundle $\mathbb{U}$ is called the universal subbundle.

In fact, the topology and geometry of $G(r, n)$, as well as the vector bundle structure on $\mathbb{U}$, are all encoded in the universal property. To see the topology, consider a linear projection $p : \mathbb{C}^n \to \mathbb{C}^r$. If S is any scheme and $\mathbb{F} \hookrightarrow \mathbb{C}^n_S$ is a rank r vector subbundle, then for a general projection p, denoting by p_S the induced morphism $p_S : \mathbb{C}^n_S \to \mathbb{C}^r_S$, then the composition $p_{\mathbb{F}} : \mathbb{F} \to \mathbb{C}^n_S \stackrel{p_S}{\to} \mathbb{C}^r_S$ is an isomorphism over a nonempty open subset $S_p \subseteq S$. Choosing appropriate projections, these open subsets cover S. This applies in particular to the Grassmannian itself, so that we obtain an open cover of $G(r, n)$ by subsets U_p where $p_{\mathbb{U}}\big|_{U_p} : \mathbb{U}\big|_{U_p} \to \mathbb{C}^r_{U_p}$ is an isomorphism. Moreover, we can easily see that locally, after trivializing $\mathbb{U}$, a point of U_p may be identified with a section of the projection $p : \mathbb{C}^n \to \mathbb{C}^r$; or more precisely, a map $S \to U_p$ may be identified with a section of $p_S : \mathbb{C}^n_S \to \mathbb{C}^r_S$. Consequently, one can check from the universal property that U_p must be isomorphic to the space of sections of the projection, which is isomorphic to $\mathrm{Hom}_{\mathbb{C}}(\mathbb{C}^r, \mathbb{C}^{n-r}) = \mathbb{C}^{(n-r)\times r}$, the space of $(n-r) \times r$ complex matrices. This of course agrees with the standard method of constructing charts for the Grassmanian by choosing bases for the subvector space, and performing row reduction. We also have the identification $p_{\mathbb{U}}\big|_{U_p} : \mathbb{U}_{U_p} \to \mathbb{C}^r_{U_p}$, which gives the vector bundle structure on $\mathbb{U}$.

The take-away from this discussion is that there is an identification of families (up to isomorphism) of r-dimensional subspaces of $\mathbb{C}^n$ parameterized by S, with morphisms of $\mathbb{C}$-schemes S into $G(r, n)$ and, moreover, that *we can understand everything about $G(r, n)$ in terms of this identification.* As perverse as it might seem, we could throw away the topological space and sheaf of rings and consider instead the Grassmannian *functor*:

$$\mathscr{G}(r,n) : \mathsf{S}^{\mathrm{op}} \to (\mathsf{Set})$$

$$\mathscr{G}(r,n)(S) := \left\{ (\mathbb{F}, i) \,\middle|\, \begin{array}{l} \mathbb{F} \text{ is a vector bundle on } S \\ i : \mathbb{F} \hookrightarrow \mathbb{C}^n_S \text{ is a linear inclusion} \end{array} \right\} \Big/ \sim$$

where two inclusions of vector bundles $\mathbb{F} \hookrightarrow S \times \mathbb{C}^n$ and $\mathbb{F}' \hookrightarrow S \times \mathbb{C}^n$ over S are equivalent if there is a commutative diagram of the form

$$\begin{array}{ccc} \mathbb{F} & \hookrightarrow & \mathbb{C}^n_S \\ \wr\downarrow & & \| \\ \mathbb{F}' & \hookrightarrow & \mathbb{C}^n_S. \end{array}$$

The functor $\mathscr{G}(r,n)$ acts on morphisms in the following way. Associated to a morphism $f : S' \to S$, we have

$$f^* : \mathscr{G}(r,n)(S) \to \mathscr{G}(r,n)(S')$$

taking the class of the inclusion $\mathbb{F} \hookrightarrow S \times \mathbb{C}^n$ to the class of the inclusion $\mathbb{F}' \hookrightarrow S' \times \mathbb{C}^n$ defined via pullback diagrams:

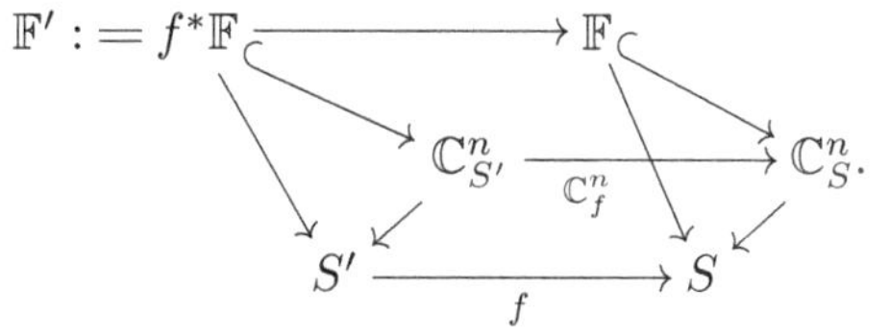

Here $f^*\mathbb{F}$ is the pullback of $\mathbb{F}$ to S' via f and $\mathbb{C}^n_f$ is the canonical morphism $\mathbb{C}^n_{S'} = \mathbb{C}^n \times S' \xrightarrow{\mathrm{id} \times f} \mathbb{C}^n \times S = \mathbb{C}^n_S$.

The assertion that there is an identification of families (up to isomorphism) of r-dimensional subspaces of $\mathbb{C}^n$ parameterized by S, with morphisms of schemes S into $G(r,n)$, can be formulated precisely as an isomorphism of functors

$$\begin{aligned} \mathrm{Hom}_{\mathsf{S}}(-, G(r,n)) &\longrightarrow \mathscr{G}(r,n) \\ (f : S \to G(r,n)) &\longmapsto f^*(\mathbb{U} \hookrightarrow \mathbb{C}^n_{G(r,n)}). \end{aligned} \tag{1.1}$$

In other words, the Grassmannian functor is representable by the Grassmannian scheme. We also say that the Grassmannian is a fine moduli space for the Grassmannian functor.

Our previous extraction of the geometric structure of $G(r,n)$ from the universal property, in other words, from the functor $\mathscr{G}(r,n)$, is a manifestation of the general principle encoded in Yoneda's lemma (stated below).

Indeed, if there is another space $G(r,n)'$ such that

$$\mathrm{Hom}_{\mathsf{S}}\big(-, G(r,n)'\big) \cong \mathscr{G}(r,n),$$

then

$$\mathrm{Hom}_{\mathsf{S}}\big(-, G(r,n)'\big) \cong \mathrm{Hom}_{\mathsf{S}}\big(-, G(r,n)\big),$$

and one may use Yoneda's lemma to conclude that $G(r,n) \cong G(r,n)'$.

Lemma 1.1 (Yoneda). *Let* S *be a category. There is a fully faithful functor*

$$\begin{aligned} h : \mathsf{S} &\to \mathsf{Fun}\big(\mathsf{S}^{\mathsf{op}}, (\mathsf{Set})\big) \\ S &\mapsto h_S := \mathrm{Hom}_{\mathsf{S}}(-, S) \\ (S \xrightarrow{\phi} S') &\mapsto (h_S \xrightarrow{\phi_*} h_{S'}), \end{aligned}$$

which identifies S *with a full subcategory of the category of functors* $\mathsf{Fun}\big(\mathsf{S}^{\mathsf{op}}, (\mathsf{Set})\big)$. *For each* $S \in \mathrm{Ob}(\mathsf{S})$, *the map*

$$\begin{aligned} \mathrm{Hom}_{\mathsf{Fun}(\mathsf{S}^{\mathsf{op}},(\mathsf{Set}))}(h_S, \mathscr{F}) &\to \mathscr{F}(S) \\ (\eta : h_S \to \mathscr{F}) &\mapsto \eta(S)(\mathrm{id}_S) \end{aligned}$$

is a bijection.

This is best worked through on one's own, preferably in private. Nevertheless, in the spirit of completeness that inspires these notes, we refer the reader to [24, Ch.2, §2.1] or [38, Ch. 0, §1.1] for more details.

Definition 1.2 (Representable functor). A functor isomorphic to one of the form $\mathrm{Hom}_{\mathsf{S}}(-, S)$ for some S in S is called a *representable functor.*

Remark 1.3. Due to the Yoneda Lemma, we will often write S in place of h_S or $\mathrm{Hom}_{\mathsf{S}}(-, S)$.

Remark 1.4. Note that under the identification of (1.1), the identity map $\mathrm{id}_{G(r,n)} \in \mathrm{Hom}_{\mathsf{S}}\big(G(r,n), G(r,n)\big)$ is sent to the class of the inclusion of the universal subbundle $\mathbb{U} \hookrightarrow \mathbb{C}^n_{G(r,n)}$ in $\mathscr{G}(r,n)\big(G(r,n)\big)$.

Summarizing the discussion above, the fact that the Grassmannian functor is representable is equivalent to the fact that there is a universal family, $\mathbb{U}$, of dimension r-subspaces of $\mathbb{C}^n$, parameterized by the Grassmannian $G(r,n)$, such that any other family of dimension r-subspaces of $\mathbb{C}^n$ is obtained from this universal family via pullback along a (unique) map to the Grassmannian.

Remark 1.5. This whole construction could have been dualized, considering instead r-dimensional quotients of the dual vector space $(\mathbb{C}^n)^\vee$. The problem is clearly equivalent, but this dual formulation is more obviously related to the Quot scheme we will discuss later. We leave it to the reader to make this comparison when the Quot scheme is introduced.

1.2. *Moduli of Riemann surfaces*

The example of the Grassmannian gives a moduli space that is representable by a projective variety. Another accessible example of a moduli problem that on the other hand captures the complications leading to the development of stacks is that of parameterizing Riemann surfaces.

We will work in this section with the algebraic analogues of compact Riemann surfaces of genus g; i.e., smooth proper complex algebraic curves of genus g. Our moduli problem is that of parameterizing all such curves up to isomorphism. Motivated by the discussion of the Grassmannian, we make this more precise by defining a moduli functor. To do this, we define a *relative curve of genus* g to be a surjective morphism of schemes $\pi : X \to S$ such that

- π is smooth,
- π is proper,
- every geometric fiber is a connected curve of genus g.

Recall that if X and S are smooth, then $\pi : X \to S$ is smooth if and only the differential is everywhere surjective. In other words, this includes the analogues of surjective morphisms of smooth complex manifolds, with surjective differential, where every (set theoretic) fiber is a compact Riemann surface of genus g.

We now define the moduli functor of genus g curves as:

$$\mathscr{M}_g : \mathsf{S}^{\mathsf{op}} \to (\mathsf{Set})$$
$$\mathscr{M}_g(S) := \{\pi : X \to S, \text{ a relative curve of genus } g\}/\sim$$

where two relative curves $\pi : X \to S$ and $\pi' : X' \to S$ are equivalent if there is a commutative diagram (1.2), in which the upper horizontal arrow is an isomorphism:

$$\begin{array}{ccc} X' & \xrightarrow{\sim} & X \\ {\scriptstyle \pi'}\downarrow & & \downarrow{\scriptstyle \pi} \\ S & = & S. \end{array} \tag{1.2}$$

The functor $\mathscr{M}_g$ acts on morphisms in the following way. Associated to a morphism $f : S' \to S$, we have

$$f^* : \mathscr{M}_g(S) \to \mathscr{M}_g(S')$$

defined via pullback diagrams; in other words, given $[\pi : X \to S] \in \mathscr{M}_g(S)$, we define $f^*[\pi : X \to S]$ as the class of the curve $\pi' : X' \to S'$, defined by the fiber product diagram:

$$\begin{array}{ccc} X' & \longrightarrow & X \\ {\scriptstyle \pi'}\downarrow & & \downarrow{\scriptstyle \pi} \\ S' & \xrightarrow{\ f\ } & S. \end{array}$$

Example 1.6 (Isotrivial family). Consider the relative curve $\pi : X \to S = \mathbb{C}^*$ of genus 1 given by the equation

$$y^2 = x^3 + t,\ \ t \neq 0.$$

Here X is the projective completion of $\{(x, y, t) : y^2 - x^3 - t = 0\}$ in $\mathbb{P}^2_{\mathbb{C}}$, and the map $\pi : X \to S$ is induced by $(x, y, t) \mapsto t$. It turns out this family is *isotrivial* (i.e., all of the fibers are abstractly isomorphic); this can be seen quickly by confirming that the j-invariant of each fiber is 0, or simply writing down the isomorphism $(x, y, t) \mapsto (\lambda^{-2}x, \lambda^{-3}y, 1)$, where $\lambda^6 = t$, between X_t and X_1. One can also show that this family is not equivalent to a trivial family; i.e., not isomorphic to $S \times X_1$. One way to see this is to check that the monodromy action of $\pi_1(\mathbb{C}^*, 1)$ on $H^1(X_1, \mathbb{C})$ is nontrivial (see e.g., [17, §4.1.2]), and therefore, that X/S is not a trivial family. Another way to see this is to observe that the explicit isomorphism given above trivializes X over the pullback via $\mathbb{C}^* \to \mathbb{C}^* : \lambda \mapsto \lambda^6$. We can therefore characterize X as the quotient of $X_1 \times \mathbb{C}^*$ by the action $\zeta.(x, y, \lambda) = (\zeta^{-2}x, \zeta^{-3}y, \zeta\lambda)$ of a 6-th root of unity $\zeta \in \mu_6$. The étale sheaf on S of isomorphisms between X and the $X_1 \times S$ is a torsor under the automorphism group of X_1, which is μ_6. It is therefore classified by an element of $H^1(S, \mu_6) = \mathrm{Hom}(\mu_6, \mu_6) = \mathbb{Z}/6\mathbb{Z}$, which one can check is 1. Since this corresponds to a nontrivial torsor, X/S is not a trivial family. One can generalize this example to show that for every g, there exist isotrivial, but nontrivial families of curves (see Example A.4).

Proposition 1.7. *The functor $\mathscr{M}_g$ is not representable.*

Proof. See e.g. [46] for more details. The main point is that if $\mathscr{M}_g$ were representable, then every isotrivial family of curves would be equivalent to a trivial family, which we have just seen is not the case.

To this end, suppose that $\mathscr{M}_g = \mathrm{Hom}_{\mathsf{S}}(-, M_g)$ for some scheme M_g. Let $C_g \to M_g$ be the family of curves corresponding to the identity morphism id_{M_g}. Now let $\pi : X \to S$ be an isotrivial family of curves over a connected base S, that is not equivalent to $S \times X_s$ for any $s \in S$, with $s = \operatorname{Spec} \mathbb{C}$ (here X_s is the fiber of X/S; see e.g., Examples 1.6 and A.4). The curve $X_s \to s$ corresponds to an element of $\mathscr{M}_g(\operatorname{Spec} \mathbb{C}) = \mathrm{Hom}_{\mathsf{S}}(\operatorname{Spec} \mathbb{C}, M_g)$; i.e., the isomorphism class of the curve corresponds to a point, say $[X_s] \in M_g$. Similarly, the family $\pi : X \to S$ corresponds to a morphism $f : S \to M_g$. Since every fiber of $\pi : X \to S$ is isomorphic to $[X_s]$, the image of S under f must be the point $[X_s]$. By definition of the representability of the functor, we must then have that the pullback of the universal family $C_g \to M_g$ along f, i.e., $X_s \times S$, is equivalent to $\pi : X \to S$, which we assumed was not the case. □

Remark 1.8. While the functor $\mathscr{M}_g$ is not representable, it does admit a coarse moduli space. In other words, there is a quasi-projective variety M_g and a morphism $\Phi : \mathscr{M}_g \to M_g$ (here we are using the convention mentioned in Remark 1.3 of denoting a scheme and its associated functor with the same letter) such that:

(1) for any scheme S and any morphism $\Psi : \mathscr{M}_g \to S$, there is a unique morphism $\eta : M_g \to S$ such that the following diagram commutes:

$$\begin{array}{ccc} \mathscr{M}_g & \xrightarrow{\Phi} & M_g \\ {\scriptstyle\Psi}\big\downarrow & \swarrow_{\eta,\ \exists!} & \\ S & & \end{array}$$

(2) Φ is a bijection when evaluated on any algebraically closed field.

A scheme representing a functor, i.e., a fine moduli space, is clearly a coarse moduli space. For many moduli functors that we consider, we will be able to find a scheme satisfying the first condition above; however, when the automorphisms of our objects are positive dimensional, it will not in general be possible to find a scheme satisfying the second condition, as well.

For brevity (or at least for lesser verbosity), we have not included a discussion of coarse moduli spaces. The interested reader may consult [46, Def. 1.3], or [6, Def. 4.1].

There was not anything special about curves in Proposition 1.7: all we needed was an isotrivial family that was nontrivial. In fact, nontrivial automorphism groups almost always give rise to nontrivial isotrivial families (see Corollary A.7) so we conclude that moduli problems involving objects with nontrivial automorphisms will *almost never* be representable by genuine spaces. Since most moduli problems of interest involve objects with nontrivial automorphisms, this means *fine moduli spaces almost never exist*; i.e., moduli problems are almost never representable by schemes.

There are a few ways one might try to go around this problem: one could *rigidify* the problem by imposing additional structure in the hopes of eliminating nontrivial automorphisms; one could give up on the idea of a fine moduli space and settle for a *coarse moduli space* that at least gets the points right, even if it botches the universal property (see Remark 1.8).

Stacks do not go around the problem of isomorphisms so much as they go through it. By remembering how objects are equivalent, as opposed to merely that they are equivalent, we can eliminate the issue that is responsible for Proposition 1.7. The price we must pay is to enlarge the class of objects we are willing to call spaces and sacrifice some of our geometric intuition. What we hope to illustrate in these notes is that the cost in geometric intuition is less than one might first expect, and that the payoff in newly available moduli spaces more than compensates for it. Indeed, stacks will allow us to bring our intuition about geometric families to bear on the geometry of their moduli spaces, giving us a new—and, we would argue, more powerful—sort of intuition to replace what we have sacrificed.

1.3. *The tangent space of a moduli functor*

Because of Yoneda's lemma, a fine moduli space is uniquely characterized by the functor it represents. In particular, the tangent space to a moduli space at a point can be determined directly from the moduli functor, without even knowing that the moduli space is representable! This is not merely a formal convenience: the tangent space is an essential tool in *proving* that moduli functors are representable (see §10 and §11).

As an example of the definition, we compute the tangent space of the Grassmannian. The same calculation actually computes the tangent space of the Quot scheme, as well (see also [24, §6.4]). Notice that we do not make use anywhere of the representability of the Grassmannian in this calculation. Further examples will appear in §10.1.

We have already seen on p.206 how the cover of $G(r,n)$ by open subsets U_p associated to projections $p : \mathbb{C}^n \to \mathbb{C}^r$ is visible from the functor $\mathscr{G}(r,n)$. From this, near a point of the Grassmannian $W \hookrightarrow \mathbb{C}^n$, we get a local identification of $\mathscr{G}(r,n)$ with the functor represented by $\mathrm{Hom}_{\mathbb{C}}(\mathbb{C}^r, \mathbb{C}^{n-r}) =$

$\mathrm{Hom}(W, \mathbb{C}^n/W)$. Since this is a vector space, this gives a local identification of the tangent space $T_{\mathscr{G}(r,n),W}$ with $\mathrm{Hom}(W, \mathbb{C}^n/W)$. As is well known, this identification can be made globally. Rather than verify the compatibility of this identification with gluing, which would require us to use the fact that $\mathscr{G}(r,n)$ is a sheaf, we will arrive at it via a global construction working directly from the definition of $\mathscr{G}(r,n)$.

Definition 1.9 (Ring of dual numbers). The *ring of dual numbers* (over $\mathbb{C}$) is $\mathbb{D} = \mathbb{C}[\epsilon]/(\epsilon^2)$. For any scheme S over $\mathbb{C}$, we write $S[\epsilon]$ for $S \times \operatorname{Spec} \mathbb{D}$.

Definition 1.10 (Tangent space to a moduli functor). If $\mathscr{F} : \mathsf{S}^{\mathsf{op}} \to (\mathsf{Set})$ is a functor, the *tangent bundle* to $\mathscr{F}$ is the following functor:

$$T_{\mathscr{F}} : \mathsf{S}^{\mathsf{op}} \to (\mathsf{Set})$$
$$T_{\mathscr{F}}(S) = \mathscr{F}\big(S[\epsilon]\big)$$

Reduction modulo ϵ gives a map $\mathbb{D} \to \mathbb{C}$ and therefore a closed embedding $S \to S[\epsilon]$ for any scheme S. This induces natural functions

$$T_{\mathscr{F}}(S) = \mathscr{F}\big(S[\epsilon]\big) \to \mathscr{F}(S)$$

and therefore a natural transformation

$$T_{\mathscr{F}} \to \mathscr{F}.$$

We allow ourselves the following notational shortcut, which some may feel is abusive: When $\xi \in \mathscr{F}(S)$, in other words, when (S, ξ) is an S-point of $\mathscr{F}$, we write $T_{\mathscr{F}}(S, \xi)$ or $T_{\mathscr{F}}(\xi)$ for the fiber of $T_{\mathscr{F}}(S)$ over $\xi \in \mathscr{F}(S)$.

For instance, if we have $\mathscr{F} = \mathrm{Hom}_{\mathsf{S}}(-, M)$ for some complex manifold M, and $\xi : \operatorname{Spec} \mathbb{C} \to M$ is a point of M, then $T_{\mathscr{F}}(\xi) = T_M(\xi)$, the holomorphic tangent space to M at x.

Warning 1.11. The tangent space to an arbitrary functor $\mathscr{F}$ may not be at all well-behaved! For example $T_{\mathscr{F}}(\xi)$ might not even be a vector space. When trying to characterize the functors that are representable by schemes (or algebraic spaces or algebraic stacks) one of the first axioms we impose is that $\mathscr{F}$ should behave infinitesimally like a scheme, and in particular that its tangent spaces should be vector spaces (see §10).

1.3.1. *Tangent space to the Grassmannian*

Suppose that S is a scheme and $\mathbb{F} \hookrightarrow \mathbb{C}^n_S$ is an S-point of $\mathscr{G}(r,n)$. Denote by F the sheaf of regular sections of $\mathbb{F}$. Note that F is a quasicoherent locally free sheaf of rank r, and recall that $\mathbb{F}$ can be recovered from F, either from transition functions, or directly as

$$\mathbb{F} = \underline{\mathrm{Spec}}_S(\mathrm{Sym}^\bullet F^\vee),$$

where $F^\vee = \underline{\mathrm{Hom}}_{\mathscr{O}_S}(F, \mathscr{O}_S)$ is the sheaf of sections of the dual vector bundle $\mathbb{F}^\vee = \mathrm{Hom}_S(\mathbb{F}, \mathbb{C}_S)$. By definition, $T_{\mathscr{G}(r,n)}(S, \mathbb{F})$ is the set of isomorphism classes of extensions of $\mathbb{F}$ to a vector subbundle $\mathbb{F}_1 \subseteq \mathbb{C}^n_{S[\epsilon]}$. Because vector bundles are determined by their sheaves of sections, deforming $\mathbb{F}$ is the same as finding a locally free deformation F to $S[\epsilon]$. One can easily check by dualizing that this is equivalent to finding a locally free deformation of $F^\vee$ to $S[\epsilon]$.

Since $\mathbb{F} \hookrightarrow \mathbb{C}^n_S$, we have a quotient $\mathscr{O}^n_S \to F^\vee$ by duality. We write $E^\vee$ for the kernel of this quotient. To find a locally free deformation of $F^\vee$ to $S[\epsilon]$ is the same as to complete the diagram below with a locally free $\mathscr{O}_{S[\epsilon]}$-module $F_1^\vee$:

$$\begin{array}{ccccccccc} & & & & \mathscr{O}^n_{S[\epsilon]} & \dashrightarrow & F_1^\vee & \dashrightarrow & 0 \\ & & & & \downarrow & & \vdots\!\downarrow & & \\ 0 & \longrightarrow & E^\vee & \longrightarrow & \mathscr{O}^n_S & \longrightarrow & F^\vee & \longrightarrow & 0. \end{array} \tag{1.3}$$

Since $F_1^\vee$ is locally free, tensoring the exact sequence

$$0 \to \epsilon\mathscr{O}_S \to \mathscr{O}_S[\epsilon] \to \mathscr{O}_S \to 0 \tag{1.4}$$

with $F_1^\vee$ we see that the kernel of $F_1^\vee \to F^\vee$ is $\epsilon F_1^\vee \simeq F^\vee$. Note, moreover, that a short computation shows that any quasicoherent sheaf $F_1^\vee$ fitting into the diagram (1.3) above with $\epsilon F_1^\vee \simeq F^\vee$ will be locally free, because a local basis for $F^\vee$ can be lifted via the projection $F_1^\vee \to F^\vee$ to a basis for $F_1^\vee$. This can also be seen using the infinitesimal criterion for flatness [22, Ex. 6.5].

Therefore our problem is, equivalently, to complete the diagram below with $\mathscr{O}_{S[\epsilon]}$-modules $E_1^\vee$ and $F_1^\vee$, so that the middle row is a short exact sequence, and the vertical arrows are those induced from (1.4):

$$\begin{array}{ccccccccc} & & 0 & & 0 & & 0 & & \\ & & \downarrow & & \downarrow & & \downarrow & & \\ 0 & \longrightarrow & \epsilon E^\vee & \longrightarrow & \epsilon \mathcal{O}_S^n & \longrightarrow & \epsilon F^\vee & \longrightarrow & 0 \\ & & \downarrow & & \downarrow & & \downarrow & & \\ 0 & \dashrightarrow & E_1^\vee & \dashrightarrow & \mathcal{O}_{S[\epsilon]}^n & \dashrightarrow & F_1^\vee & \dashrightarrow & 0 \\ & & \downarrow & & \downarrow & & \downarrow & & \\ 0 & \longrightarrow & E^\vee & \longrightarrow & \mathcal{O}_S^n & \longrightarrow & F^\vee & \longrightarrow & 0. \\ & & \downarrow & & \downarrow & & \downarrow & & \\ & & 0 & & 0 & & 0 & & \end{array} \tag{1.5}$$

In particular, all rows and columns above are short exact sequences.

Since $\epsilon E^\vee$ necessarily maps to zero in $F_1^\vee$, to produce a quotient $F_1^\vee$ of $\mathcal{O}_{S[\epsilon]}^n$ lifting the quotient $F^\vee$ of $\mathcal{O}_S^n$, it is equivalent to produce a quotient of $\mathcal{O}_{S[\epsilon]}^n/\epsilon E^\vee$. We are therefore trying to complete the diagram below:

$$\begin{array}{ccccccccc} 0 & \longrightarrow & E_1^\vee/\epsilon E^\vee & \dashrightarrow & \mathcal{O}_{S[\epsilon]}^n/\epsilon E^\vee & \dashrightarrow & F_1^\vee & \dashrightarrow & 0 \\ & & \| & & \downarrow & & \downarrow & & \\ 0 & \longrightarrow & E^\vee & \longrightarrow & \mathcal{O}_S^n & \longrightarrow & F^\vee & \longrightarrow & 0. \end{array}$$

But note that $E_1^\vee/\epsilon E^\vee$ projects isomorphically to $E^\vee$. Therefore $T_{\mathscr{G}(r,n)}(S, F)$ is isomorphic to the set of lifts of the following diagram:

$$\begin{array}{ccc} & & \mathcal{O}_{S[\epsilon]}^n/\epsilon E^\vee \\ & \nearrow & \downarrow \\ E^\vee & \longrightarrow & \mathcal{O}_S^n. \end{array}$$

We have one canonical lift by composing the inclusion $E^\vee \hookrightarrow \mathcal{O}_S^n$ with the section $\mathcal{O}_S^n \to \mathcal{O}_{S[\epsilon]}^n$. (Note that this is not an $\mathcal{O}_{S[\epsilon]}$ homomorphism until it is restricted to $E^\vee$.) An easy diagram chase in (1.5) shows that the difference between any two lifts is a homomorphism $E^\vee \to F^\vee$, so we get a canonical bijection:

$$T_{\mathscr{G}(r,n)}(S, \mathbb{F}) = \operatorname{Hom}_{\mathcal{O}_S}(E^\vee, F^\vee) = \operatorname{Hom}_{\mathcal{O}_S}(F, E) = \operatorname{Hom}_S(\mathbb{F}, \mathbb{E})$$

where $\mathbb{E} = \underline{\operatorname{Spec}}_S(\operatorname{Sym}^\bullet E^\vee)$ is the quotient of the trivial vector bundle $\mathbb{C}_S^n$ by the subbundle $\mathbb{F}$.

2. Moduli functors as categories fibered in groupoids

As nontrivial automorphisms tend to preclude the representability of a moduli problem (by a scheme), one plausible way to proceed is to account for these automorphisms by retaining them in the definition of the moduli functor. We are led to consider functors valued in groupoids (categories in which all morphisms are isomorphisms), rather than in sets, and immediately find ourselves in a morass of technicalities (see §2.1). In our judgment, a more elegant solution can be found in the notion of a category fibered in groupoids (see Remark 2.4 for the reasoning that motivates our point of view). In §2.1 we briefly discuss functors valued in groupoids, with the primary objective of convincing the reader that another approach would be preferable. We then introduce categories fibered in groupoids in §2.2.

2.1. *(Lax 2-)functors to groupoids*

Instead of studying a moduli problem by defining a functor whose value on a scheme is the *set* of isomorphism classes of families over that scheme, we try to define one whose value is the *category* of families parameterized by that scheme, with morphisms being isomorphisms of families. Contrary to the set of isomorphism classes, the category explicitly allows two families to be isomorphic in more than one way.

Definition 2.1 (Groupoid). A *groupoid* is a category in which all morphisms are isomorphisms.

Remark 2.2. In the literature, in the definition of a groupoid, it is common to require the additional condition that the category be small (the class of objects is a set). This is not required for our treatment, and so we drop the condition since almost every category we will consider will not be small. For instance, the category of sets with one element is obviously not small (for every set E there is a one element set $\{E\}$), and one can immediately generalize this to examples we consider here. However, the groupoids we study will usually be *essentially small*, meaning they are equivalent to small categories, so none of the pathologies that compel one to include a hypothesis of smallness will trouble us.

One way to avoid worrying about small categories (or at least, to transfer the worry to somewhere else) is to introduce Grothendieck universes. Essentially, one assumes axiomatically that there are very large sets, called universes, that are large enough to 'do set theory' within. All objects of interest occur within the universe, but one can still use set-theoretic language to speak about the universe itself. For example, the collection of all 1-element sets *within the universe* does form a set, namely one in bijection with the universe itself.

Ultimately, these considerations are technical from our perspective, and we will remain silent about them in the sequel. One may consult [77, Tag 0007] for yet another way around these technical issues.

Continuing on, at first glance, it seems that we want a "functor" to the category of groupoids:

$$\mathcal{M} : \mathsf{S}^{\mathsf{op}} \to (\mathsf{Groupoid}).$$

Making this precise leads to the morass of technicalities alluded to above. The issue here is the assignment for morphisms. In our examples, we pulled back families along morphisms. The fact that pullbacks are only defined up to isomorphism, albeit a canonical one, means that one does not have an equality of g^*f^* with $(fg)^*$ but only an isomorphism between them. One then has not one but two ways of identifying $h^*g^*f^* \simeq (fgh)^*$ and one must require these be the same.

Example 2.3. Here is how this plays out for the moduli of curves: Let

$$\mathscr{M}_g^{L2} : \mathsf{S}^{\mathsf{op}} \to (\mathsf{Groupoid})$$

be the moduli functor in groupoids for curves of genus g; by definition $\mathscr{M}_g^{L2}(S)$ is the *category* of families of curves over S, with S-isomorphisms as the morphisms. If $T \to S$ is a morphism of schemes we obtain a functor $\mathscr{M}_g^{L2}(S) \to \mathscr{M}_g^{L2}(T)$ sending C to $C \times_S T$. If we have a pair of morphisms $U \to T \to S$ then we obtain two maps $\mathscr{M}_g^{L2}(S) \to \mathscr{M}_g^{L2}(U)$, one sending C to $(C \times_S T) \times_T U$ and the other sending C to $C \times_S U$. These are canonically isomorphic, but they are not equal! Do we have to keep track of this canonical isomorphism, as well as the compatibilities it must satisfy when we encounter a sequence $V \to U \to T \to S$?

Pursuing this line of reasoning, one ultimately arrives at the definition of a *pseudo-functor* or *lax 2-functor*. However, just to define lax 2-functors is an unpleasant task, with little to do with the geometry that ultimately motivates us. (The reader who desires one may find a definition in [24, §3.1.2].) Fortunately, we will not have to think too hard about lax 2-functors because Grothendieck has supplied a more elegant solution: categories fibered in groupoids. The fundamental observation is that the pullbacks we need are canonically isomorphic because they satisfy universal properties that are literally the same. If we keep track of universal properties rather than the objects possessing them, we arrive at a more efficient definition.

Remark 2.4. According to the philosphy behind categories fibered in groupoids, the mistake in Example 2.3 was to *choose* a fiber product $C \times_S T$. We were then forced to carry it around and keep track of the compatibilities that it obviously satisfies. The category fibered in groupoids posits only that an object satisfying the universal property of the fiber product *exists*, i.e., that there is *some* cartesian diagram (2.1),

$$\begin{array}{ccc} D & \longrightarrow & C \\ \downarrow & & \downarrow \\ T & \longrightarrow & S \end{array} \tag{2.1}$$

without actually specifying a construction of one.

2.2. *Categories fibered in groupoids*

From our perspective, the motivation for a category fibered in groupoids is to avoid the technical complications of the definition of a lax 2-functor by essentially clumping all of the groupiods of interest into one large category over the category of schemes; the issues we ran into defining the lax 2-functor on morphisms are avoided by using the universal properties of fibered products in our category. We now make this precise. Temporarily, we let S denote any category.

Definition 2.5. A *category over* S is a pair $(\mathcal{M}, \pi)$ consisting of a category $\mathcal{M}$ together with a covariant functor $\pi : \mathcal{M} \to \mathsf{S}$. If S is an object of S, the *fiber of* $\mathcal{M}$ *over* S, denoted $\mathcal{M}_S$ or $\mathcal{M}(S)$ is defined to be the subcategory consisting of all objects over S, and all morphisms over the identity of S. An object X *in* $\mathcal{M}$ *is said to lie over* S if it is in $\mathcal{M}_S$. A *morphism* $\tilde{f} : X' \to X$ *in* $\mathcal{M}$ *is said to lie over a morphism* $f : S' \to S$ if $\pi(\tilde{f}) = f$. A morphism between categories $(\mathcal{M}, \pi)$ and $(\mathcal{M}', \pi')$ over S is a functor $F : \mathcal{M} \to \mathcal{M}'$ such that $\pi' \circ F = \pi$.

We indicate objects and morphisms lying above other objects and morphisms in diagrams like this:

$$\begin{array}{ccccc} \mathcal{M} & \quad & X' & \xrightarrow{\tilde{f}} & X \\ \downarrow{\scriptstyle \pi} & & \downarrow & & \downarrow \\ \mathsf{S} & & S' & \xrightarrow{f} & S. \end{array} \tag{2.2}$$

A fibered category is essentially one for which "pullback diagrams" exist. Keeping in mind the definition of a fibered product, a quick glance at (2.3) should make clear the meaning of a pullback diagram in this setting.

Definition 2.6 (Cartesian morphism). Let $(\mathcal{M}, \pi)$ be a category over S. A morphism $\tilde{f} : X' \to X$ in $\mathcal{M}$ is *cartesian* if the following condition holds. Denote by $f : S' \to S$ the morphism $\pi(\tilde{f})$ in S (as in (2.2)). Given any morphism $g : S'' \to S'$ in S, and any morphism $\widetilde{f \circ g} : X'' \to X$ in $\mathcal{M}$ lying over $f \circ g$, there is a unique morphism $\tilde{g} : X'' \to X'$ in $\mathcal{M}$ lying over g such that $\widetilde{f \circ g} = \tilde{f} \circ \tilde{g}$. Pictorially, every diagram (2.3) has a unique completion:

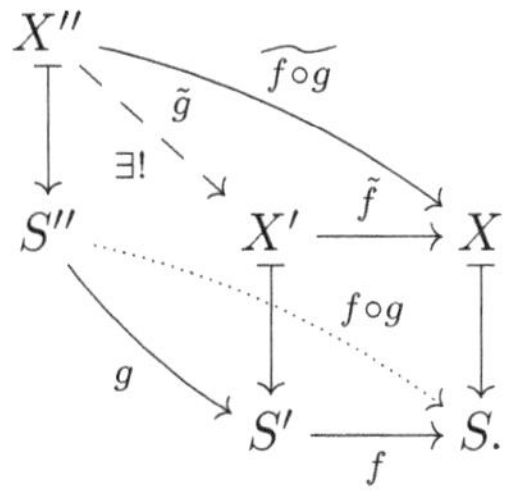

(2.3)

Definition 2.7 (Fibered category). A *category* $(\mathcal{M}, \pi)$ over S is said to be *fibered over* S if for any morphism $f : S' \to S$ in S and any object X of $\mathcal{M}$ lying over S, there exists a cartesian morphism $\tilde{f} : X' \to X$ in $\mathcal{M}$ lying over f.

Definition 2.8 (Category fibered in groupoids (CFG)). A category $(\mathcal{M}, \pi)$ over S is said to be *fibered in groupoids* if it is fibered over S, and for every S in S, we have that $\mathcal{M}(S)$ is a groupoid.

Remark 2.9. Categories fibered in groupoids are typically introduced as fibered categories in which all morphisms are *cartesian* [5, Exp. VI, §6, Remarques], [24, Def. 3.21], or equivalently, a fibered category where the fibers are all groupoids. In examples, as we have seen here, where we define all morphisms via fibered product diagrams, and pullbacks, one is led naturally to this definition. The added generality of fibered categories (not neccesarily fibered in groupoids) is important in order to formulate faithfully flat descent efficiently [42, §B.3], [5, Exp. VIII], [24, §4.2]. However, it is not particularly relevant to the study of moduli problems that is our focus here.

Fibered categories were originally defined in [42, §A.1.a, Def. 1.1] to be what we would call lax 2-functors (albeit valued in categories rather than in groupoids), and what others call *pseudo-functors* [5, Exp. VI, §8], [24, Def. 3.10]. The definition of a fibered category given in [5, Exp. VI, Def. 6.1], is equivalent to the slightly different formulation in [24, Def. 3.5]. Lax 2-functors are equivalent (we do not attempt to say precisely in what sense) to fibered categories with *cleavage* [5, Exp. VI, §7–8], [24, Def. 3.9, §3.1.3].

Remark 2.10. The notation $\mathcal{M}(S)$ for $\pi^{-1}(S)$ is meant to be suggestive of the relationship between categories fibered in groupoids and functors valued in groupoids. Indeed, one may construct an equivalence between the notions such that the groupoid-valued functor associated to $\mathcal{M}$ has value $\mathcal{M}(S)$ on $S \in \mathsf{S}$.

Example 2.11 (CFG associated to a presheaf). Let $\mathscr{F} : \mathsf{S}^{\mathsf{op}} \to (\mathsf{Set})$ be a functor in sets. One obtains a CFG $\pi : \mathcal{F} \to \mathsf{S}$ in the following way. For each $S \in \mathrm{Ob}(\mathsf{S})$, let $\mathcal{F}_S = \mathscr{F}(S)$. Let $S, S' \in \mathrm{Ob}(\mathsf{S})$ and suppose that $X_S \in \mathcal{F}_S$ and $X_{S'} \in \mathcal{F}_{S'}$. Then we assign a morphism $X_{S'} \to X_S$ in $\mathcal{F}$ if there is a morphism $f : S' \to S$ in S, and $\mathscr{F}(f)(X_S) = X'_S$. The functor $\pi : \mathcal{F} \to \mathsf{S}$ is given by sending X_S to S, and similarly for morphisms.

It is not difficult to generalize this construction to yield a CFG associated to a lax 2-functor [24, §3.1.3], provided that one has first given a precise definition of the latter.

Example 2.12 (CFG associated to an object of S). Let S be in S. The *slice category* S/S is the CFG defined to have objects that are pairs (S', f) where $S' \in \mathsf{S}$ and $f : S' \to S$ is a morphism. A morphism $(S', f) \to (S'', g)$ is a morphism $h : S' \to S''$ such that $gh = f$:

$$\begin{array}{ccc} S' & \xrightarrow{\ h\ } & S'' \\ & {\scriptstyle f}\searrow \quad \swarrow{\scriptstyle g} & \\ & S & \end{array}$$

The projection $\pi : \mathsf{S}/S \to \mathsf{S}$ is $\pi(S', f) = S'$. A CFG equivalent to one of the form S/S for some S in S is said to be *representable by* S.

Remark 2.13 (Agreement of CFGs for S and h_S). For any $S \in \mathrm{Ob}(\mathsf{S})$, we can assign the functor h_S, and the category fibered in groupoids associated to h_S. This agrees with the CFG S/S.

Example 2.14 (The CFG (sieve) associated to a family of maps). Given a collection of morphisms $\mathcal{R} = \{S_i \to S\}$ in S, we define an associated full sub-CFG of S/S, which will also be denoted by $\mathcal{R}$. The objects of $\mathcal{R}$ are the objects $S' \to S$ of S/S that factor through one of the $S_i \to S$.

Sub-CFGs of representable CFGs (i.e., a sub-CFGs of S/S for some S in S) are known as *sieves.*

To give the full statement of the Yoneda lemma for CFGs we need another definition. The correct language for discussing this is that of the 2-category of CFGs over S. We postpone this more technical discussion until later (§4). Here we give a working definition that suffices for our purposes.

Definition 2.15. Let $p_{\mathcal{M}} : \mathcal{M} \to \mathsf{S}$ and $p_{\mathcal{N}} : \mathcal{N} \to \mathsf{S}$ be CFGs over S. There is a category $\mathrm{Hom}_{\mathsf{CFG/S}}(\mathcal{M}, \mathcal{N})$ with objects being morphisms $\mathcal{M} \to \mathcal{N}$ of fibered categories over S and morphisms being natural isomorphisms.

Lemma 2.16 (2-Yoneda [24, §3.6.2]). *Let S be a category, and let $S \in \mathrm{Ob}(\mathsf{S})$. For any fibered category $\pi : \mathcal{M} \to \mathsf{S}$ the natural transformation*

$$\mathrm{Hom}_{\mathsf{CFG/S}}(\mathsf{S}/S, \mathcal{M}) \to \mathcal{M}(S)$$
$$F \mapsto F(S \xrightarrow{\mathrm{id}_S} S)$$

(defined similarly for morphisms) is an equivalence of categories.

Convention 2.17. In view of the 2-Yoneda lemma, we may introduce the following notation: Suppose that $S \in \mathsf{S}$ and that $\mathcal{M}$ is a CFG over S. We write $X : S \to \mathcal{M}$ to mean $X \in \mathrm{Hom}_{\mathsf{CFG/S}}(\mathsf{S}/S, \mathcal{M})$. By the Yoneda lemma, this is the same as specifying an object of $\mathcal{M}(S)$, and we frequently do not distinguish notationally between $X : S \to \mathcal{M}$ and $X \in \mathcal{M}(S)$.

Corollary 2.18 ([24, §3.6]). *Let S be a small category, and let $S, S' \in \mathrm{Ob}(\mathsf{S})$. The map*

$$\mathrm{Hom}_{\mathsf{S}}(S', S) \to \mathrm{Ob}\left(\mathrm{Hom}_{\mathsf{CFG/S}}(\mathsf{S}/S', \mathsf{S}/S)\right)$$

obtained by post-composition of arrows (e.g., a morphism $f : S' \to S$ is sent to the functor that assigns to an arrow $(g : S'' \to S') \in \mathrm{Ob}(\mathsf{S}/S')$, the composition $(f \circ g : S'' \to S) \in \mathrm{Ob}(\mathsf{S}/S)$) is a bijection.

Remark 2.19. Let $\mathscr{M}, \mathscr{N} : \mathsf{S}^{\mathsf{op}} \to (\mathsf{Set})$ be two functors. Let $\mathcal{M}, \mathcal{N}$ be the associated categories fibered in groupoids over S. In a similar way, there is a bijection

$$\mathrm{Hom}_{\mathsf{Fun}(\mathsf{S}^{\mathsf{op}}, (\mathsf{Set}))}(\mathscr{M}, \mathscr{N}) \to \mathrm{Ob}\left(\mathrm{Hom}_{\mathsf{CFG/S}}(\mathcal{M}, \mathcal{N})\right).$$

We delay introducing the notion of 2-categories until later (§4), but the consequence of the 2-Yoneda Lemma, and this observation, is that the category S, and the category of functors $\mathsf{Fun}(\mathsf{S}^{\mathsf{op}}, (\mathsf{Set}))$ can be viewed as full 2-subcategories of the 2-category of CFGs over S. Consequently, we will frequently identify objects S of S, and functors to sets $\mathscr{M} : \mathsf{S}^{\mathsf{op}} \to (\mathsf{Set})$ with their associated CFGs, S/S, and $\mathcal{M}$, respectively.

2.2.1. *The Grassmannian as a category fibered in groupoids*

For this section, let S be the category of schemes over $\mathbb{C}$. The Grassmannian CFG

$$\pi : \mathcal{G}(r, n) \to \mathsf{S}$$

is defined as follows. Objects of $\mathcal{G}(r, n)$ are pairs $(S, \mathbb{F})$ where S is a scheme and $\mathbb{F} \hookrightarrow \mathbb{C}^n_S$ is a vector subbundle of rank r. Morphisms in $\mathcal{G}(r, n)$ are defined via pullback. More precisely, a morphism $(S', \mathbb{F}') \to (S, \mathbb{F})$ is a diagram

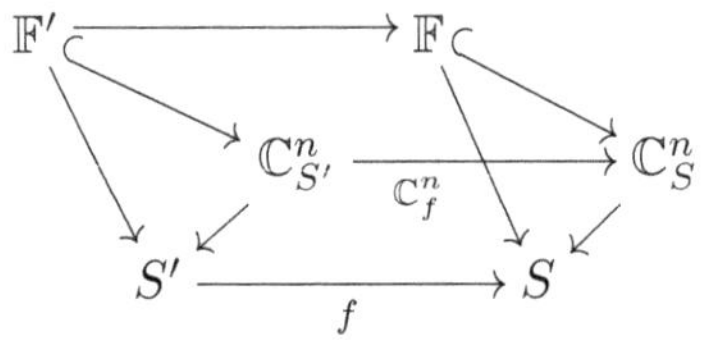

where $f : S' \to S$ is a morphism in S and the rectangles are all cartesian. The map $\pi : \mathcal{G}(r, n) \to \mathsf{S}$ is the forgetful functor $\pi(S, \mathbb{F}) = S$.

Remark 2.20. To dispel any confusion that may arise from the notation in Example 2.11, we emphasize that the CFG associated to the functor $\mathscr{G}(r, n)$ is equivalent to $\mathcal{G}(r, n)$. The former is fibered in sets whereas the latter is fibered in groupoids that are equivalent, but not isomorphic, to sets. We use the two notations to emphasize that one is a CFG and one is a functor.

2.2.2. *The CFG of curves of genus g*

Again we take S to be the category of schemes over $\mathbb{C}$. We define the CFG of genus g curves

$$\pi : \mathcal{M}_g \to \mathsf{S}$$

in the following way. Objects of $\mathcal{M}_g$ are pairs (S, X) where $X \to S$ is a relative curve of genus g (see §1.2). Morphisms in $\mathcal{M}_g$ are defined via pullback. More precisely, a morphism from $(S', X') \to (S, X)$ is a cartesian diagram

$$\begin{array}{ccc} X' & \longrightarrow & X \\ \downarrow & & \downarrow \\ S' & \xrightarrow{f} & S \end{array}$$

where $f : S' \to S$ is a morphism in S. The map $\pi : \mathcal{M}_g \to \mathsf{S}$ is $\pi(S, X) = S$.

Remark 2.21. We emphasize that the CFG associated to the functor $\mathscr{M}_g$ is not equivalent to $\mathcal{M}_g$. The former is fibered in sets whereas the latter is fibered in groupoids that are not equivalent to sets.

3. Stacks

Stacks are the categories fibered in groupoids that respect topology, in the sense that compatible locally defined morphisms into a stack can be glued together into global morphisms. This is the most basic requirement a category fibered in groupoids must satisfy in order to be studied geometrically.

As usual, algebraic geometry introduces a troublesome technicality here: the Zariski topology is much too coarse to do much interesting gluing. Indeed, suppose that $\mathcal{F}$ is a CFG that 'deserves to be studied geometrically' and consider a scheme S with a free action of $G = \mathbb{Z}/2\mathbb{Z}$ and a $\xi \in \mathcal{F}(S)$ that is equivariant with respect to the G-action. If $\mathcal{F}$ were a stack in the complex analytic topology, we could descend ξ to an element of $\mathcal{F}(S/G)$ because $S \to S/G$ is a covering space, an in particular a local homeomorphism, hence a cover in the analytic topology. There is no such luck in the Zariski topology, where $S \to S/G$ may fail to be a local homeomorphism and may therefore also fail to be a cover in the Zariski topology.

It is important to be able to do this kind of descent, so one must introduce an abstract replacement for the concept of a topology, called a Grothendieck topology. The essence of the definition is to isolate exactly the aspects of topology that are necessary to speak about gluing. It is possible to express the sheaf conditions without ever making reference to points, or even to open sets. All that is needed is the concept of a cover. Grothendieck's definition actually goes further, replacing even the concept of a cover with the abstract notion of a sieve. Even though topologies afford a few pleasant properties that pretopologies do not (see §A.3 for more details), we will primarily limit our discussion to pretopologies in this introduction.

Remark 3.1. Lest it appear genuinely pointless to remove the points from topology, consider the vast expansion of settings that can be considered topologically by way of Grothendieck topologies. To take just one, Quillen [69] and Rim [1, Exp. VI] were able to understand extensions of commutative rings—that is, deformations of affine schemes—by putting a Grothendieck topology on the category of *commutative rings* (see [28,83,84] for further developments of this idea).

We begin this section by reviewing the definition of a Grothendieck pretopology, using sheaves on a topological space as our motivation. We then define the isomorphism presheaf, and descent data, and finally, we give the definition of a stack.

3.1. *Sheaves and pretopologies*

We take as motivation for Grothendieck pretopologies the definition of a sheaf on a topological space.

3.1.1. *Sheaves on a topological space*

Let X be a topological space. For any open subsets $U' \subseteq U$ of X, let $\iota_{U',U}$ denote the inclusion of U' inside of U. Define the category of open sets on X, O_X, with the following objects and morphisms:

$$\operatorname{Obj} \mathsf{O}_X = \{U \subseteq X : U \text{ open}\}$$

$$\operatorname{Hom}_{\mathsf{O}_X}(U', U) = \begin{cases} \{\iota_{U',U}\} & U' \subseteq U \\ \varnothing & \text{else.} \end{cases}$$

A *presheaf (of sets)* is a functor

$$\mathscr{F} : \mathsf{O}_X^{\mathsf{op}} \to (\mathsf{Set}).$$

Given $a \in \mathscr{F}(U)$, and a subset $\iota_{U',U} : U' \subseteq U$, we denote by $a|_{U'}$, or $\iota^*_{U',U}(a)$, the image of a under the map $\mathscr{F}(\iota_{U',U}) : \mathscr{F}(U) \to \mathscr{F}(U')$.

A presheaf is:

(1) *separated* if, given an open cover $\{U_i \to U\}$ and two sections a and b in $\mathscr{F}(U)$ such that $a|_{U_i} = b|_{U_i}$ in $\mathscr{F}(U_i)$ for all i, one has $a = b$.
(2) a *sheaf* if, given an open cover $\{U_i \to U\}$ with intersections $U_{ij} = U_i \cap U_j$, and elements $a_i \in \mathscr{F}(U_i)$ for all i, satisfying $a_i\big|_{U_{ij}} = a_j\big|_{U_{ij}}$ for all i and j, there is a unique $a \in \mathscr{F}(U)$ such that $a\big|_{U_i} = a_i$ for all i.

A morphism of presheaves, separated presheaves, or sheaves is a natural transformation of functors.

Now, to motivate the definition of a pretopology, a presite, and a sheaf on a presite, we rephrase this definition in the language of equalizers. Recall that if U' and U'' are open subsets of U, then $U' \cap U'' = U' \times_U U''$. Given any open cover $\mathcal{U} = \{U_i \to U\}$, denote by $\mathscr{F}(\mathcal{U})$ the equalizer of diagram (3.1):

$$\prod_i \mathscr{F}(U_i) \underset{\mathrm{pr}_2^*}{\overset{\mathrm{pr}_1^*}{\rightrightarrows}} \prod_{i,j} \mathscr{F}(U_i \underset{U}{\times} U_j). \tag{3.1}$$

Recall that the equalizer is the categorical limit for morphisms into the diagram; in other words, we obtain a diagram

$$\mathscr{F}(\mathcal{U}) \longrightarrow \prod_i \mathscr{F}(U_i) \underset{\mathrm{pr}_2^*}{\overset{\mathrm{pr}_1^*}{\rightrightarrows}} \prod_{ij} \mathscr{F}(U_i \times_U U_j),$$

and the arrow on the left is universal (terminal) for morphisms into (3.1). The natural map on the left in the diagram below is induced by the restriction maps:

$$\mathscr{F}(U) \longrightarrow \prod_i \mathscr{F}(U_i) \underset{\mathrm{pr}_2^*}{\overset{\mathrm{pr}_1^*}{\rightrightarrows}} \prod_{ij} \mathscr{F}(U_i \times_U U_j).$$

By the universal property of the equalizer, this induces a map:

$$\mathscr{F}(U) \to \mathscr{F}(\mathcal{U})$$

The sheaf conditions have the following translations into the language of equalizers.

Lemma 3.2 ([24, Cor. 2.40]). *Let* $\mathscr{F} : \mathsf{O}_X^{\mathsf{op}} \to (\mathsf{Set})$ *be a presheaf.*

(1) $\mathscr{F}$ *is separated if and only if* $\mathscr{F}(U) \to \mathscr{F}(\mathcal{U})$ *is injective for all open* U *in* X *and all open covers* $\mathcal{U}$ *of* U.

(2) $\mathscr{F}$ *is a sheaf if and only if* $\mathscr{F}(U) \to \mathscr{F}(\mathcal{U})$ *is a bijection for all open* U *in* X *and all open covers* $\mathcal{U}$ *of* U.

The main takeaway from this discussion is that we can repackage the sheaf condition in terms of fibered products and equalizers. This provides the motivation for a Grothendieck pretopology, and a sheaf on a presite.

3.1.2. *Pretopologies*

Temporarily denote by S any category. A *Grothendieck pretopology* $\mathscr{T}$ on S consists of the following data: for each object S in S, a collection of families of maps $\{S_\alpha \to S\}$, called *covers* of S in $\mathscr{T}$ (or *covering families* in $\mathscr{T}$), such that:

(PT 0) For all objects S in S, and for all morphisms $S_\alpha \to S$ which appear in some covering family of S, and for all morphisms $S' \to S$, the fibered product $S_\alpha \times_S S'$ exists.
(PT 1) For all objects S in S, all morphisms $S' \to S$, and all covering families $\{S_\alpha \to S\}$, the family $\{S_\alpha \times_S S' \to S'\}$ is a covering family.
(PT 2) If $\{S_\alpha \to S\}$ is a covering family, and if for all α, $\{S_{\beta\alpha} \to S_\alpha\}$ is a covering family, then the family of composites $\{S_{\beta\alpha} \to S_\alpha \to S\}$ is a covering family.
(PT 3) If $S' \to S$ is an isomorphism, then it is a covering family.

Example 3.3. Let X be a topological space. The category O_X together with open covers is a Grothendieck pretopology.

Example 3.4. Let X be a topological space, define a category S to have as objects $\mathscr{P}(X)$, the set of all subsets of X, and as morphisms the inclusions. We give every subset $S \subseteq X$ the induced topology. Then the collection of all open covers of subsets of X gives a Grothendieck pretopology on S. Indeed, (PT0) is satisfied, since the fibered product is given by intersection. (PT1) holds since we are giving every subset the induced topology, so an open cover of a superset gives an open cover of a subset. (PT2) holds since refinements of open covers are open covers. (PT3) holds since isomorphisms are equalities.

Example 3.5. Let $\pi : \mathcal{X} \to \mathsf{S}$ be a CFG over a category S equipped with a Grothendieck pretopology. Call a family of maps $\{X_i \to X\}$ in $\mathcal{X}$ covering if $\{\pi(X_i) \to \pi(X)\}$ is covering in S. Then this determines a Grothendieck pretopology on $\mathcal{X}$. To verify this, it may be helpful to observe that the induced morphism of CFGs $\mathcal{X}/X \to \mathsf{S}/\pi(X)$ is an equivalence of categories; indeed, by the definition, $\mathcal{X}$ being a CFG implies the morphism is essentially surjective, and fully faithful.

Definition 3.6 (Presite). A pair $(\mathsf{S}, \mathscr{T})$ consisting of a category S together with a Grothendieck pretopology $\mathscr{T}$ is called a *presite*. Often $\mathscr{T}$ is left tacit and one uses S to stand for both the presite and its underlying category of objects.

Example 3.7 (Covers in the étale pretopology). The primary example of a Grothendieck pretopology that we will use is the étale pretopology on the category of schemes. We denote the associated presite by S_{et}. Covers in S_{et} are collections of jointly surjective étale morphisms. Recall that étale morphisms are the algebro-geometric analogue of local isomorphisms in the complex analytic category.

Given a pretopology on S and a scheme S in S, we obtain the category S/S, and an induced pretopology defined in the obvious way. For instance, $(\mathsf{S}/S)_{\mathrm{et}}$ has covers given by jointly surjective étale morphisms in that category.

Remark 3.8 (Analytic category). Readers who prefer working in the analytic category, should feel free to take S to be the category of complex analytic spaces, and to work with the pretopology $\mathscr{T}$ generated by the usual open covers of complex analytic spaces, in the analytic topology. In fact, one could as easily take S to be the category of complex manifolds, with smooth morphisms (so that fibered products remain in the category), and work with the pretopology $\mathscr{T}$ generated by the usual open covers of complex manifolds.

Example 3.9 (Standard pretopologies on schemes). The most commonly used Grothendieck pretopologies on the category of schemes are the:

- *Zariski pretopology,*
- *étale pretopology,*
- *faithfully flat finite presentation (fppf) pretopology,*
- *faithfully flat quasicompact (fpqc) pretopology.*

Each of these pretopologies is a refinement of the one preceding it. We will write $\mathsf{S}_{\mathrm{Zar}}$, S_{et}, etc., for the respective presites. The Zariski pretopology (covers by Zariski open sets) is too coarse for most of the applications we have in mind. For simplicity, we will work almost exclusively with the étale pretopology.

3.1.3. *Sheaves on a presite*

The definition of a pretopology is exactly set up to allow us to make a definition of a sheaf following our discussion of sheaves on topological spaces. A *presheaf (of sets)* on S is just a functor

$$\mathscr{F} : \mathsf{S}^{\mathsf{op}} \to (\mathsf{Set}).$$

Given any cover $\mathcal{R} = \{S_i \to S\}$, denote by $\mathscr{F}(\mathcal{R}) = \mathscr{F}(\{S_i \to S\})$ the equalizer of the diagram

$$\textstyle\prod_i \mathscr{F}(S_i) \underset{\mathrm{pr}_2^*}{\overset{\mathrm{pr}_1^*}{\rightrightarrows}} \prod_{ij} \mathscr{F}(S_i \times_S S_j).$$

As before, there is a natural map

$$\mathscr{F}(S) \to \mathscr{F}(\mathcal{R}).$$

Following Lemma 3.2, we make the following definition:

Definition 3.10 (Sheaf on a site). Let $\mathscr{F}$ be a presheaf on a presite $(\mathsf{S}, \mathscr{T})$.

(1) $\mathscr{F}$ is *separated* if $\mathscr{F}(S) \to \mathscr{F}(\mathcal{R})$ is an injection for every covering family $\mathcal{R}$ of every object S of S.
(2) $\mathscr{F}$ is a *sheaf* if $\mathscr{F}(S) \to \mathscr{F}(\mathcal{R})$ is a bijection for every covering family $\mathcal{R}$ of every object S of S.

Remark 3.11. On occasion we will be more precise about how fine a pretopology can be used to obtain a given statement. For instance, we may specify that a presheaf is a sheaf with respect to the fpqc pretopology, which implies it is also a sheaf with respect to all of the other pretopologies mentioned above.

Definition 3.12 (Subcanonical presite). A pretopology $\mathscr{T}$ on a category S is called *subcanonical* if every representable functor on S is a sheaf with respect to $\mathscr{T}$. A presite $(\mathsf{S}, \mathscr{T})$ is called subcanonical if $\mathscr{T}$ is subcanonical.

Theorem 3.13 (Grothendieck [24, Thm. 2.55]). *Let S be a scheme. The presite $(\mathsf{S}/S)_{\mathrm{fpqc}}$ is subcanonical; in particular $(\mathsf{S}/S)_{\mathrm{et}}$ is subcanonical.*

3.2. *The isomorphism presheaf*

Let $\pi : \mathcal{M} \to \mathsf{S}$ be a CFG. If we view $\mathcal{M}$ as a space, as is our intention, and view an object $X \in \mathcal{M}(S)$ as a map $S \to \mathcal{M}$ then we expect that morphisms into $\mathcal{M}$ that agree locally should agree globally. The only proviso is that because two maps into $\mathcal{M}$ can agree with one another in more than one way, we must interpret local agreement of X and Y to include not only choices of local isomorphisms between X and Y over a cover, but also compatibility of these choices on the overlaps in the cover.

In order to state this condition precisely, we introduce the presheaf of witnesses to the agreement of objects of $\mathcal{M}$. The condition we want to impose is that this presheaf be a sheaf.

Definition 3.14 (Isomorphism presheaf). Let $\pi : \mathcal{M} \to \mathsf{S}$ be a CFG. Given S in S, and $X, Y \in \mathcal{M}(S)$, we obtain a presheaf

$$\mathscr{I}\!som_{\mathcal{M}}(X, Y) : (\mathsf{S}/S)^{\mathsf{op}} \to (\mathsf{Set})$$

in the following way. For every object $(S' \to S)$ in S/S, we set

$$\mathscr{I}\!som_{\mathcal{M}}(X,Y)(S' \to S) := \mathrm{Isom}_{\mathcal{M}(S')}(X\big|_{S'}, Y\big|_{S'})$$

Thus $\mathscr{I}\!som_{\mathcal{M}}(X,Y)(S' \to S)$ consists of all isomorphisms $\alpha : X\big|_{S'} \to Y\big|_{S'}$ in $\mathcal{M}$ that lie over the identity $\mathrm{id}_{S'}$ in S. The assignment for morphisms is left to the reader (see [24, p.62]).

Remark 3.15. An observant reader will note that the restrictions in the definition of $\mathscr{I}\!som_{\mathcal{M}}(X,Y)$ depend, albeit only up to a canonical isomorphism, on a choice of inverse to the functor $\mathrm{Hom}(S,\mathcal{M}) \to \mathcal{M}(S)$, as guaranteed by the Yoneda lemma.

Remark 3.16. Concretely, to say that isomorphisms form a sheaf means the following. Given a cover $\{S_i \to S\}$ in the pretopology on S, and any collection of isomorphisms $\alpha_i : X|_{S_i} \to Y|_{S_i}$ over the identity on S_i such that $\alpha_i|_{S_{ij}} = \alpha_j|_{S_{ij}}$, there is a unique isomorphism $\alpha : X \to Y$ such that $\alpha|_{S_i} = \alpha_i$. Here we are using the shorthand $S_{ij} := S_i \times_S S_j$.

Definition 3.17 (Prestack). A CFG $\mathcal{M} \to \mathsf{S}$ such that for every S in S, and every X, Y in $\mathcal{M}(S)$, the presheaf $\mathscr{I}\!som_{\mathcal{M}}(X,Y)$ is a sheaf, is called a *prestack*. We will also say that isomorphisms are a sheaf.

Remark 3.18. The notation would be more consistent with sheaf notation (see Proposition 3.32) if we called categories fibered in groupoids (or fibered categories) 'prestacks' and called prestacks 'separated prestacks'. But we will keep to tradition.

It is typically very easy to prove that a category fibered in groupoids arising from a moduli problem is a prestack in a subcanonical topology. This is because an object $Y \in \mathcal{M}(S)$ is usually representable by a scheme, perhaps with some extra structure or properties, and descending local isomorphisms between $X \in \mathcal{M}(S)$ and Y amounts to descending locally defined morphisms from X to Y, which is automatic because Y represents a sheaf in any subcanonical topology!

Example 3.19. Let us see how this works concretely in the example of $\mathcal{M}_g$. Let S be a scheme in S, and let $X \to S$ and $Y \to S$ be relative curves of genus g. Suppose that there exists an étale cover $\{S_i \to S\}$ so that for each S_i there are S_i-isomorphisms $\alpha_i : X_{S_i} \to Y_{S_i}$ such that $\alpha_i|_{S_{ij}} = \alpha_j|_{S_{ij}}$ (using the shorthand $S_{ij} = S_i \times_S S_j$). These correspond by the universal property of fiber product to morphisms $X_{S_i} \to Y$ and $X_{S_{ij}} \to Y$ satisfying the same compatibility condition. As Y represents a sheaf and $\{X_{S_i} \to X\}$

is a cover of X in the pretopology, these glue to a morphism $\alpha : X \to Y$ such that $\alpha\big|_{S_i} = \alpha_{S_i}$. To check that this is in fact a morphism over S, note that the commutativity of the diagrams (3.2)

$$\begin{array}{ccc} X_{S_i} & \longrightarrow & Y \\ \downarrow & & \downarrow \\ S_i & \longrightarrow & S \end{array} \tag{3.2}$$

for all i implies the commutativity of diagram (3.3),

$$\begin{array}{ccc} X & \xrightarrow{\alpha} & Y \\ & \searrow \quad \swarrow & \\ & S & \end{array} \tag{3.3}$$

this time because morphisms into S are a sheaf.

In order to make a precise general statement along these lines, we make a very general definition:

Definition 3.20 (Stable class of arrows [24, Def. 3.16, p.48]). A class of arrows $\mathbf{P}$ in a category S is *stable (under base change)* if morphisms in $\mathbf{P}$ can be pulled back via arbitrary morphisms in S, and the result of any such pullback is also in $\mathbf{P}$.

Example 3.21 (CFG associated to a stable class of arrows). Given any stable class of arrows $\mathbf{P}$ in a category S, one may make $\mathbf{P}$ into a category $\mathcal{P}$ by setting objects to be arrows in $\mathbf{P}$ and setting morphisms to be cartesian squares. There is a morphism $\mathcal{P} \to \mathsf{S}$ given by sending an object $X \to S$ to the target S (and similarly for morphisms). Then one can check that $\mathcal{P} \to \mathsf{S}$ is a CFG if and only if $\mathbf{P}$ is a stable class of arrows.

Theorem 3.22 ([24, Prop. 4.31, p.88]). *Let $(\mathsf{S}, \mathscr{T})$ be a subcanonical presite, and $\mathbf{P}$ a stable class of arrows. Let $\pi : \mathcal{P} \to \mathsf{S}$ be the associated CFG (Example 3.21). Then $(\mathcal{P}, \pi)$ is a prestack.*

Remark 3.23. The discussion in Example 3.19 adapts in a straightforward way to a proof of Theorem 3.22.

Corollary 3.24. *The CFGs $\mathcal{G}(r, n)$ and $\mathcal{M}_g$ are prestacks in the étale topology on schemes.*

Proof. The case of $\mathcal{M}_g$ follows directly from the theorem with $(\mathsf{S}, \mathscr{T}) = \mathsf{S}_{\text{et}}$ and $\mathcal{P}$ the class of relative curves of genus g. The case of $\mathcal{G}(r, n)$ requires a slight modification (since there is more than one arrow in the definition of the objects), but is essentially the same. □

3.3. *Descent for categories fibered in groupoids*

Recall that we can view vector bundles either as global geometric objects admitting local trivializations, or alternatively, as collections of locally trivial objects together with transition functions, which satisfy the cocycle condition. We now use this motivation to define the notion of descent datum in the setting of CFGs. This is the last definition we will need before defining a stack.

3.3.1. *Vector bundles on open subsets of a complex manifold*

Let X be a complex manifold. Let O_X be the category of open sets. Give O_X the structure of a presite in the usual way, by taking covering families to be open covers. Define a CFG of vector bundles

$$\pi : \mathcal{V}_X^r \to \mathsf{O}_X$$

in the following way. For each $U \subseteq X$ open, the objects in the fiber $\mathcal{V}_U^r$ are the rank r, holomorphic vector bundles on U. Morphisms in $\mathcal{V}^r$ are given by pullback diagrams. Consider an open covering $\{U_i \subseteq U\}$. We use the notation $U_{ij} = U_i \cap U_j$ and $U_{ijk} = U_i \cap U_j \cap U_k$ for the double and triple overlaps. Suppose we are given a vector bundle E_i over U_i for every i, and an isomorphism $\alpha_{ij} : E_i|_{U_{ij}} \to E_j|_{U_{ij}}$ for every i, j, all of which satisfy the cocycle condition $\alpha_{ik} = \alpha_{jk} \circ \alpha_{ij}$ over U_{ijk}. Then there exists a vector bundle E lying over U, together with isomorphisms $\alpha_i : E|_{U_i} \to E_i$ such that $\alpha_{ij} = \alpha_j|_{U_{ij}} \circ (\alpha_i|_{U_{ij}})^{-1}$.

This means precisely that the category fibered in groupoids of vector bundles on a topological space satisfies descent and therefore is a stack in the usual topology. The formulation of descent and the definition of a stack axiomatize this familiar gluing process.

3.3.2. *Intuitive definition of descent*

In this section we will give a direct translation of the gluing condition for vector bundles encountered in §3.3.1 in the context of categories fibered in groupoids. While intuitive, this formulation has both technical and practical definiencies. We correct these in §A.2.1 and §A.2.2, but a reader looking to develop intuition about stacks may safely ignore these matters, especially in a first reading.

Definition 3.25 (Descent using gluing data). Let $\mathcal{M}$ be a category fibered in groupoids over a presite $(\mathsf{S}, \mathscr{T})$ with a cleavage (Definition 2.8). A *descent datum for* $\mathcal{M}$ *over a space* S is the following: a covering $\{S_i \to S\}$; for every i, an object X_i over S_i; for every i, j an isomorphism $\alpha_{ij} : X_i|_{S_{ij}} \to X_j|_{S_{ij}}$ in the fiber $\mathcal{M}(S_{ij})$, which satisfies the cocycle condition $\alpha_{ik} = \alpha_{jk} \circ \alpha_{ij}$ over S_{ijk}. The descent datum is said to be *effective* if there exists an X lying over S, together with isomorphisms $\alpha_i : X|_{S_i} \to X_i$ in the fiber such that $\alpha_{ij} = \alpha_j|_{S_{ij}} \circ (\alpha_i|_{S_{ij}})^{-1}$.

Remark 3.26. From the example in §3.3.1, we see that every descent datum for the CFG $\mathcal{V}_X^r \to \mathsf{O}_X$ is effective.

We now discuss the category of descent data, and the meaning of effective descent data in this context.

Definition 3.27 (The category of descent data). Let $(\mathsf{S}, \mathscr{T})$ be a presite, and let $\pi : \mathcal{M} \to \mathsf{S}$ be a CFG. Let $\mathcal{R} = \{S_i \to S\}$ be a covering in S. An *object with descent data on* $\mathcal{R}$, or *descent datum on* $\mathcal{R}$, is a collection $(\{X_i\}, \{\alpha_{ij}\})$ of objects $X_i \in \mathcal{M}(S_i)$, together with isomorphisms $\alpha_{ij} : \mathrm{pr}_2^* X_j \cong \mathrm{pr}_1^* X_i$ in $\mathcal{M}(S_i \times_S S_j)$, such that the following cocycle condition is satisfied: For any triple of indices i, j, k, we have the equality

$$\mathrm{pr}_{13}^* \alpha_{ik} = \mathrm{pr}_{12}^* \alpha_{ij} \circ \mathrm{pr}_{23}^* \alpha_{jk} : \mathrm{pr}_3^* X_k \to \mathrm{pr}_1^* X_i. \tag{3.4}$$

An *arrow between objects with descent data*

$$\{\phi_i\} : (\{X_i\}, \{\alpha_{ij}\}) \to (\{X_i'\}, \{\alpha_{ij}'\})$$

is a collection of arrows $\phi_i : X_i \to X_i'$ with the property that for each pair of indices, i, j, the diagram

$$\begin{array}{ccc} \mathrm{pr}_2^* X_j & \xrightarrow{\mathrm{pr}_2^* \phi_j} & \mathrm{pr}_2^* X_j' \\ {\scriptstyle \alpha_{ij}} \downarrow & & {\scriptstyle \alpha_{ij}'} \downarrow \\ \mathrm{pr}_1^* X_i & \xrightarrow{\mathrm{pr}_1^* \phi_i} & \mathrm{pr}_1^* X_i' \end{array}$$

commutes. The objects and morphisms above determine a *category of descent data* $\mathcal{M}(\{S_i \to S\})$.

Remark 3.28. We have deliberately omitted a number of canonical isomorphism from the notation here. This obscures some technical issues (which can, of course, be resolved: see §A.2.1). For example, different choices of fiber products S_{ij} and S_{ijk} will lead to different but equivalent categories $\mathcal{M}(\{S_i \to S\})$.

Remark 3.29. If $\mathcal{R}$ is a covering, as in Definition 3.27, it is technically convenient to indicate the category of descent data with respect to $\mathcal{R}$ as a map $\mathcal{R} \to \mathcal{M}$. This notation is consistent with the Yoneda identification between $\mathcal{M}(S)$ and the category of morphisms $S \to \mathcal{M}$, which is the special case where $\mathcal{R}$ is the trivial covering $\{S \to S\}$. In general, the notation is justified by replacing the covering with the associated sieve or simplicial object (see Section A.2).

Note that whenever a cover $\mathcal{R}'$ refines a cover $\mathcal{R}$ there is an induced morphisms $\mathcal{M}(\mathcal{R}) \to \mathcal{M}(\mathcal{R}')$, which we notate as a composition:

$$\mathcal{R}' \to \mathcal{R} \to \mathcal{M}$$

We identify the object S with the trivial cover of itself, so that it makes sense to write $\mathcal{R} \to S$ for any cover $\mathcal{R}$ of S.

Given $X \in \mathcal{M}(S)$, we can construct an object with descent data on $\{\sigma_i : S_i \to S\}$ as follows. The objects are the pullbacks $\sigma_i^* X$; the isomorphisms $\alpha_{ij} : \mathrm{pr}_2^* \sigma_j^* X \cong \mathrm{pr}_1^* \sigma_i^* X$ are the isomorphisms that come from the fact that both $\mathrm{pr}_2^* \sigma_j^* X$ and $\mathrm{pr}_1^* \sigma_i^* X$ are pullbacks of X to $S_i \times_S S_j$ (they are both equipped with canonical isomorphisms with $(\sigma_i \circ \mathrm{pr}_1)^* X = (\sigma_j \circ \mathrm{pr}_2)^* X$). (If we identify $\mathrm{pr}_2^* \sigma_j^* = (\sigma_j \circ \mathrm{pr}_2)^*$, etc., as is common, then the α_{ij} are identity morphisms.) If $\phi : X \to X'$ is an arrow in $\mathcal{M}(S)$, then we get arrows $\sigma_i^* X \xrightarrow{\sigma_i^*(\phi)} \sigma_i^* X'$ yielding an arrow from the object with descent data associated to X with the object with descent data associated to X'. In short, we have defined a functor $\mathcal{M}(S) \to \mathcal{M}(\{S_i \to S\})$.

Definition 3.30 (Effectivity of descent data). A descent datum $(\{X_i\}, \{\alpha_{ij}\}) \in \mathcal{M}(\{S_i \to S\})$ is said to be *effective* if it is isomorphic to the image of an object of $\mathcal{M}(S)$.

3.4. *The long-awaited definition of a stack*

We are finally ready for the definition of a stack:

Definition 3.31 (Stack). A *stack* is a category fibered in groupoids (Definition 2.8) over a presite $(\mathsf{S}, \mathscr{T})$ (Definition 3.6) such that isomorphisms are a sheaf (Definition 3.17) and every descent datum is effective (Definition 3.25). A *prestack* is a category fibered in groupoids over $(\mathsf{S}, \mathscr{T})$ such that isomorphisms are a sheaf. A *morphism of (pre)stacks* over $(\mathsf{S}, \mathscr{T})$ is a morphism of the underlying CFGs over S.

The definition can be rephrased in the following way, emphasizing the connection with sheaves.

Proposition 3.32 ([24, Prop. 4.7, p.73]). *Let $\mathcal{M} \to \mathsf{S}$ be a category fibered in groupoids over a presite $(\mathsf{S}, \mathscr{T})$.*

(1) *$\mathcal{M}$ is a prestack over S if and only if for each covering $\mathcal{R} = \{S_i \to S\}$, the functor $\mathcal{M}(S) \to \mathcal{M}(\mathcal{R})$ is fully faithful.*
(2) *$\mathcal{M}$ is a stack over S if and only if for each covering $\mathcal{R} = \{S_i \to S\}$, the functor $\mathcal{M}(S) \to \mathcal{M}(\mathcal{R})$ is an equivalence of categories.*

This essentially leads driectly to the following proposition.

Proposition 3.33 ([24, Prop. 4.9, p.73]). *Let $(\mathsf{S}, \mathscr{T})$ be a presite, and let $\mathscr{M} : \mathsf{S}^{\mathsf{op}} \to (\mathsf{Sets})$ be a presheaf. Denote by $\mathcal{M} \to \mathsf{S}$ the associated category fibered in groupoids.*

(1) *$\mathcal{M}$ is a prestack if and only if $\mathscr{M}$ is a separated presheaf.*
(2) *$\mathcal{M}$ is a stack if and only if $\mathscr{M}$ is a sheaf.*

Corollary 3.34. *If $(\mathsf{S}, \mathscr{T})$ is subcanonical, then every object in S represents a stack. More precisely, given an object S, we associate to it the category fibered in groupoids S/S, which is a stack over S.*

Remark 3.35. In fact, using the language of 2-categories, for a subcanonical presite $(\mathsf{S}, \mathscr{T})$, the category S, and the category of sheaves on $(\mathsf{S}, \mathscr{T})$ can be viewed as full 2-subcategories of the 2-category of stacks over S.

Definition 3.36 (Local class of arrows). A class $\mathbf{P}$ of arrows in a presite $(\mathsf{S}, \mathscr{T})$ is *local (on the target)* if it is stable and has the "converse" property that for any cover $\{S_i \to S\}$ and any arrow $X \to S$, if the pullbacks $S_i \times_S X \to S_i$ are in $\mathbf{P}$ for all i, then $X \to S$ is also in $\mathbf{P}$.

Remark 3.37. Fix a morphism $X \to Y$ in S and consider the collection of all $S \to Y$ such that $X \times_Y S \to S$ has property $\mathbf{P}$. This is a category fibered in groupoids, in fact a sieve of Y, that we denote $\mathcal{P}(X \to Y)$. Then $\mathbf{P}$ is a local property if and only if $\mathcal{P}(X \to Y)$ is a stack for all morphisms $X \to Y$.

This observation can be used to prove the following theorem:

Theorem 3.38 ([24, Thm. 4.38, p.93]). *Let S be a scheme. Let $\mathbf{P}$ be a class of flat projective canonically polarized morphisms of finite presentation in $(\mathsf{S}/S)_{\mathrm{fpqc}}$, which is local. Then the associated CFG $\mathcal{P} \to (\mathsf{S}/S)_{\mathrm{fpqc}}$ (from Definition 3.21) is a stack.*

Remark 3.39. Canonically polarized morphisms include smooth morphisms such that the determinant of the relative cotangent bundle is relatively ample; for instance, families of smooth curves of genus $g \geq 2$. The theorem in fact holds in more generality for polarized morphisms, but then one must add a compatibility condition for the polarizations, which lengthens the statement (see [24, Thm. 4.38, p.93]). The above will be sufficient for many of our applications.

Corollary 3.40. *The CFG $\mathcal{M}_g$ is a stack in the fpqc topology (and therefore in the étale topology) for $g \geq 2$.*

Example 3.41. Let us make the descent condition concrete in the case of $\mathcal{M}_g$. Suppose we are given an étale cover $\{S_i \to S\}$ and for each S_i a relative curve $X_i \to S_i$. Suppose moreover that for each i, j, we are given an S_{ij}-isomorphism $\alpha_{ij} : X_i|_{S_{ij}} \to X_j|_{S_{ij}}$, which satisfies the cocycle condition $\alpha_{ik} = \alpha_{jk} \circ \alpha_{ij}$ over S_{ijk}. Then there exists a relative curve $X \to S$, together with S_i-isomorphisms $\alpha_i : X|_{S_i} \to X_i$ such that $\alpha_{ij} = \alpha_j|_{S_{ij}} \circ (\alpha_i|_{S_{ij}})^{-1}$.

Remark 3.42. A similar argument can be used to show that $\mathcal{G}(r, n)$ is a stack. For brevity, we omit the details as we will also give references for Quot stacks, of which $\mathcal{G}(r, n)$ is a special case.

Remark 3.43. One can also show that $\mathcal{M}_g$ is a stack for $g = 0$; it does not follow immediately from the theorem above since the canonical bundle of such curves is not ample. However, for $g = 0$, the anti-canonical bundle is ample, and this gives a polarization that can be used in a more general formulation of the theorem [24, Thm. 4.38]. However, $\mathcal{M}_1$ is not a stack! See Section A.4 for a demonstration and a discussion of the fix.

4. Fibered products of stacks

Because of the ubiquity of base change in algebraic geometry, it is essential to know that one can take fiber products of stacks. In this section we present a construction of the fiber product of categories fibered in groupoids, which yields a fiber product of stacks when applied to CFGs that are stacks.

A reader who wants to get to algebraic stacks as quickly as possible may prefer to look briefly at §4.1 and then skip the remainder of §4, referring back as necessary. A more detailed discussion of 2-categories and universal properties within them can be found in the Stacks Project [77, Tags 003G, 02X8, 003O].

4.1. A working definition of 2-fibered products

Here we give a working definition of a 2-fibered product of CFGs. This should suffice for understanding the definitions of an algebraic stack. Again, on a first pass, the reader is encouraged to look at this section, and then skip the remainder of §4. Our presentation here is taken largely from [77].

Recall that a fiber product of morphisms of sheaves $f : \mathscr{X} \to \mathscr{Z}$ and $g : \mathscr{Y} \to \mathscr{Z}$ is defined by setting $(\mathscr{X} \times_{\mathscr{Z}} \mathscr{Y})(S) = \mathscr{X}(S) \times_{\mathscr{Z}(S)} \mathscr{Y}(S)$ for all S. The same definition could be used for lax 2-functors valued in groupoids, except one must first define a fiber product of groupoids. Here there is a subtle, but crucial point, since objects of $\mathscr{Z}(S)$ can be 'equal' to each other in more than one way. The fiber product must therefore keep track of all of the different ways $f(X)$ and $g(Y)$ are equal to each other. One defines $\mathscr{X}(S) \times_{\mathscr{Z}(S)} \mathscr{Y}(S)$ to be the category of triples (X, Y, α) where $X \in \mathscr{X}(S)$, $Y \in \mathscr{Y}(S)$, and $\alpha : f(X) \simeq g(Y)$ is an isomorphism. Morphisms $(X, Y, \alpha) \to (X', Y', \alpha')$ in this groupoid are pairs (u, v) with $u : X \to X'$ and $v : Y \to Y'$ are morphisms such that $\alpha' f(u) = f(v)\alpha$ as morphisms $f(X) \to g(Y')$.

Remark 4.1. This construction is analogous to one construction of the homotopy fiber product in algebraic topology, with isomorphism playing the role of homotopy. This is not an accident, as homotopy fiber products are intended to be invariant under replacement of the spaces involved with homotopy equivalent spaces; the fiber product of groupoids is intended to be invariant under equivalence of categories. Even more directly, groupoids can be realized as topological spaces by way of the geometric representation of the simplicial nerve, under which the fiber product of groupoids is transformed into the homotopy fiber product.

Since we are not defining stacks in terms of lax 2-functors here, we make the straightforward translation of the above idea to categories fibered in groupoids:

Definition 4.2 ([77, Tag 0040]). Let $f : \mathcal{X} \to \mathcal{Z}$ and $g : \mathcal{Y} \to \mathcal{Z}$ be morphisms of CFGs over S. Then the *fiber product* of $\mathcal{X}$ and $\mathcal{Y}$ over $\mathcal{Z}$, denoted $\mathcal{X} \times_{\mathcal{Z}} \mathcal{Y}$, is the category over S whose objects are quadruples (S, X, Y, α) where $S \in \mathsf{S}$, $X \in \mathcal{X}(S)$, $Y \in \mathcal{Y}(S)$, and $\alpha : f(X) \simeq g(Y)$ is an isomorphism in $\mathcal{Z}(S)$. A morphism $(S, X, Y, \alpha) \to (S', X', Y', \alpha')$ is triple (φ, u, v) where $\varphi : S \to S'$ is a morphism in S, $u : X \to X'$ is a morphism in $\mathcal{X}$ lying above φ, and $v : Y \to Y'$ is a morphism in $\mathcal{Y}$ lying above φ, and $\alpha' f(u) = g(v)\alpha$.

Lemma 4.3 ([77, Tag 0040]). *The fiber product of CFGs is a CFG.*

Lemma 4.4 ([54, (3.3), p.16]). *The fibered product of stacks is a stack.*

The 2-fibered product of CFGs $\mathcal{X} \times_{\mathcal{Z}} \mathcal{Y}$ has a universal property similar to that satisfied by a fiber product of sheaves. Indeed, there are forgetful morphisms $p : \mathcal{X} \times_{\mathcal{Z}} \mathcal{Y} \to \mathcal{X}$ and $q : \mathcal{X} \times_{\mathcal{Z}} \mathcal{Y} \to \mathcal{Y}$, respectively sending (S, X, Y, α) to X and to Y. Then α gives an isomorphism $fp(S, X, Y, \alpha) = f(X) \simeq g(Y) = gq(S, X, Y, \alpha)$. By the definition of morphisms in $\mathcal{X} \times_{\mathcal{Z}} \mathcal{Y}$, this isomorphism is natural in (S, X, Y, α). We denote this natural isomorphism $\psi : fp \simeq gq$. In standard terminology, the following diagram is 2-*commutative*:

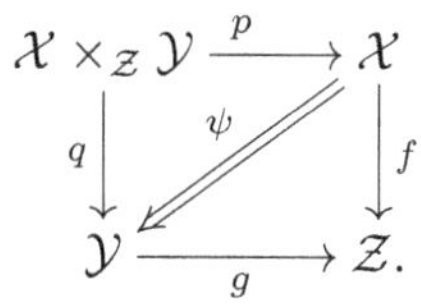

The universal property of the 2-fibered product is that $(\mathcal{X} \times_{\mathcal{Z}} \mathcal{Y}, p, q, \psi)$ is the universal completion of the diagram of solid arrows below to a 2-commutative diagram

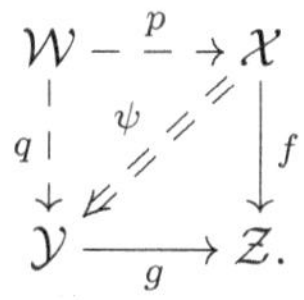

(4.1)

In other words, given a 2-commutative diagram (4.1), there is a 2-commutative diagram:

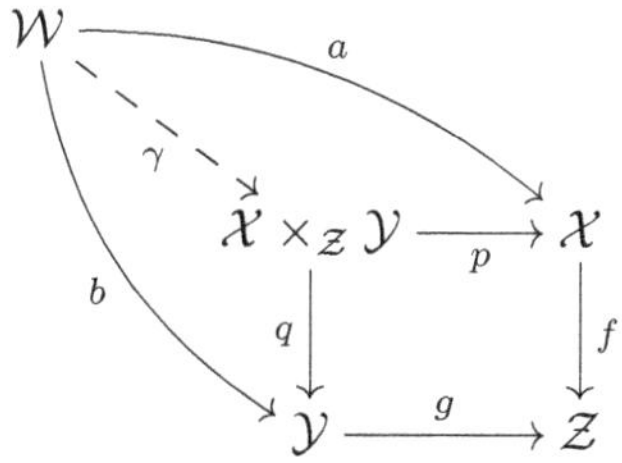

Note that the 2-commutativity of this diagram includes the tacit specification of natural isomorphisms $p\gamma \simeq a$ and $q\gamma \simeq b$. The functor γ is determined by these data uniquely up to a unique natural transformation. Using this one can show that if $\mathcal{W}$ also satisfies the universal property of $\mathcal{X} \times_{\mathcal{Z}} \mathcal{Y}$ then $\gamma : \mathcal{W} \to \mathcal{X} \times_{\mathcal{Z}} \mathcal{Y}$ is an equivalence.

Example 4.5. Suppose that $X, Y, Z \in \mathsf{S}$ and S has fibered products. Then it follows from the 2-Yoneda Lemma that $\mathsf{S}/X \times_{\mathsf{S}/Z} \mathsf{S}/Y$ is equivalent to $\mathsf{S}/(X \times_Z Y)$. Similarly, suppose that $\mathscr{X}, \mathscr{Y}, \mathscr{Z}$ are pre-sheaves on S with associated CFGs $\mathcal{X}, \mathcal{Y}, \mathcal{Z}$. Then $\mathcal{X} \times_{\mathcal{Z}} \mathcal{Y}$ is equivalent to the CFG associated to $\mathscr{X} \times_{\mathscr{Z}} \mathscr{Y}$.

4.2. *The diagonal*

The following example is used repeatedly.

Example 4.6 (The diagonal and the sheaf of isomorphisms). Let $\mathcal{M}$ be an S-stack over S, let S be in S, and let X, Y in $\mathcal{M}(S)$ be two objects corresponding under the 2-Yoneda Lemma to S-morphisms $X, Y : S \to \mathcal{M}$. Then, using the notation from §3.2, there is 2-cartesian diagram

$$\begin{array}{ccc} \mathscr{I}som_{\mathcal{M}}(X,Y) & \longrightarrow & S \\ \downarrow & & \downarrow {\scriptstyle (X,Y)} \\ \mathcal{M} & \xrightarrow{\ \Delta\ } & \mathcal{M} \times \mathcal{M}. \end{array}$$

We will verify this by way of the universal property (4.1) (see also [12, Prop. 5.12]). Suppose that we have a map $f : T \to S$, an object $Z \in \mathcal{M}(T)$, and an isomorphism

$$(\varphi, \psi) : (Z, Z) = \Delta(Z) \simeq f^*(X, Y) = (f^*X, f^*Y), \quad \text{in } (M \times \mathcal{M})(Z).$$

Then the composition $\psi \circ \varphi^{-1}$ is an isomorphism $f^*X \simeq f^*Y$, hence yields a section of $\mathscr{I}som_{\mathcal{M}}(X, Y)$ over T. Conversely, given such an isomorphism $\alpha : f^*X \simeq f^*Y$, we obtain 2-commutative diagram by taking $Z = (f^*X, f^*X)$ and $(\varphi, \psi) = (\mathrm{id}_{f^*X}, \alpha)$.

Intuitively, $\mathscr{I}som_{\mathcal{M}}(X, Y)$ is the sheaf of witnesses to the equality of X and Y, in the same way that the diagonal is the moduli space of pairs of objects of $\mathcal{M}$ that are equal to one another. Notably, $\mathscr{I}som_{\mathcal{M}}(X, Y)$ is not a subobject of S, reflecting the fact that Δ is not an embedding. This is because a pair of objects of a groupoid can be equal—that is, isomorphic—to each other in more than one way.

From the perspective of moduli, the isomorphisms of the objects of study were of central importance. On the other hand, the diagonal map is central in the definition of many properties of schemes. The diagram above relates the two notions.

Definition 4.7 (Injective morphism of stacks). A morphism $f : \mathcal{X} \to \mathcal{Y}$ of stacks over S is called *injective* (resp. an *isomorphism*) if for each $S \in \mathsf{S}$, the functor $f(S) : \mathcal{X}(S) \to \mathcal{Y}(S)$ is fully faithful (resp. an equivalence of categories). A *substack* is an injective morphism of stacks.

Lemma 4.8. *A stack $\mathcal{X}$ has injective diagonal if and only if $\mathcal{X}$ is representable by a sheaf (i.e., equivalent to the stack associated to a sheaf).*

Proof. Injective diagonal means that $\mathscr{I}som_{\mathcal{X}}(x, y)$ is a subobject of S, or, equivalently, that for any $x, y \in \mathcal{X}(S)$ there is at most one isomorphism between x and y. Thus $\mathcal{X}(S)$ is equivalent to a set. □

4.3. *Fibered products and the stack condition*

As an application of the formalism of fiber products introduced above, we give a simple but often useful criterion for a CFG over a presite to be stack, which is simply a translation of Proposition 3.32 (see also Definition A.15).

Lemma 4.9. *Let $\pi : \mathcal{M} \to \mathsf{S}$ be a CFG over a presite $(\mathsf{S}, \mathscr{T})$.*

(1) $\mathcal{M}$ is a prestack if and only for every covering family $\mathcal{R} = \{S_i \to S\}$, every morphism $f : \mathcal{R} \to \mathcal{M}$ (see Remark 3.29), and every pair of morphisms $f_1, f_2 : S \to \mathcal{M}$ making diagram (4.2) 2-commutative, there is a unique 2-isomorphism $f_1 \Rightarrow f_2$.

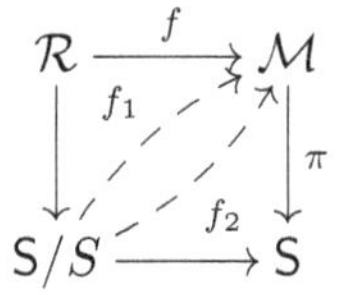

(4.2)

(2) $\mathcal{M}$ is a stack if and only for every sieve $\mathcal{R}$ associated to a covering family $\{S_i \to S\}$, with natural map $\mathcal{R} \to S$, any morphism $f : \mathcal{R} \to \mathcal{M}$, there exists a morphism $f_1 : S \to \mathcal{M}$ making the diagram below 2-commutative, which is unique up to 2-isomorphism (as explained in (1)).

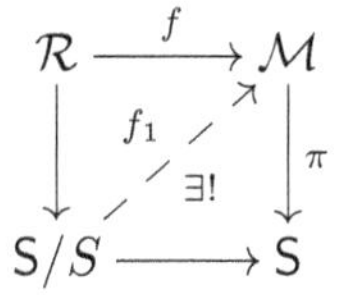

(4.3)

The lemma can be summarized with the following efficient characterization of stacks:

Corollary 4.10. *Let $\mathcal{M}$ be a category fibered in groupoids over* S. *Then $\mathcal{M}$ is a stack over* S *if and only if for every object S of* S *and every cover $\mathcal{R}$ of S, the functor*

$$\mathrm{Hom}(S, \mathcal{M}) \to \mathrm{Hom}(\mathcal{R}, \mathcal{M})$$

is an equivalence of categories.

Proof. The first part of Lemma 4.9 is the full faithfulness, and the second part is the essential surjectivity. □

Remark 4.11. Recall that the notation in the corollary is based on the abuse explained in Remark 3.29. The statement requires no such abuse when formulated with sieves (see A.2.2), the efficiency of which is one reason we like using sieves to think about Grothendieck topologies. For example, Corollary 4.10 generalizes immediately to give a definition of higher stacks, while the other formulations of descent from §3.3 become even more combinatorially complicated.

The following proposition is quite useful for making boot-strap arguments. It allows us to show that a CFG is a stack by showing it is a stack *relative* to a CFG already known to be a stack.

Proposition 4.12. *Suppose that $p : \mathcal{X} \to \mathcal{Y}$ is a morphism of CFGs over* S *and that $\mathcal{Y}$ is a stack over* S. *Then $\mathcal{X}$ is a stack over* S *if and only if it is a stack over $\mathcal{Y}$, where $\mathcal{Y}$ is given the pretopology of Example 3.5.*

Proof. We will verify the second part of the criterion in Lemma 4.9; the first part is very similar, so we omit it. Assume first that $\mathcal{X}$ is a stack over $\mathcal{Y}$. Consider the following diagram:

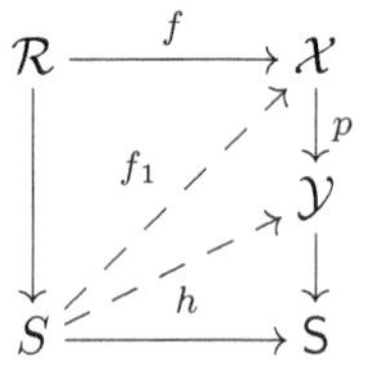

We would like to find a morphism f_1 rendering the outer square 2-commutative. Since $\mathcal{Y}$ is a stack over S, we can find an appropriate lift h as in the diagram. But then the assumption that $\mathcal{X}$ is a stack over $\mathcal{Y}$ guarantees the existence of the desired lifting f_1 of h. Thus $\mathcal{X}$ is a stack over S. The converse is similar. □

Corollary 4.13. *Suppose that $p : \mathcal{X} \to \mathcal{Y}$ is a morphism of CFGs over a presite* S. *Assume that $\mathcal{Y}$ is a stack over* S. *Then $\mathcal{X}$ is a stack over* S *if and only if for $S \in \mathsf{S}$ and every $y : S \to \mathcal{Y}$, the fiber product $\mathcal{X}_S = \mathcal{X} \times_{\mathcal{Y}} S$ is a stack on* S/S.

Proof. We show the harder direction, that $\mathcal{X}$ is a stack over S. From Proposition 4.12, it is enough to show that $\mathcal{X}$ is a stack over $\mathcal{Y}$. From Lemma 4.9 and the definition of the fibered product, one can easily deduce that $\mathcal{X}$ is a stack over $\mathcal{Y}$ if and only if $\mathcal{X} \times_{\mathcal{Y}} (\mathcal{Y}/Y)$ is a stack over $\mathcal{Y}/Y$ for all $Y \in \mathcal{Y}$. Now using the equivalence of categories $\mathcal{Y}/Y \cong \mathsf{S}/S$ (Example 3.5), we are done. □

Example 4.14. As an example of the utility of Proposition 4.12 and Corollary 4.13, consider the CFG $\mathcal{X}$ whose S-points are triples (C, C', f) where C and C' are families of smooth curves over S of genera g and h, both ≥ 2, and $f : C \to C'$ an S-morphism. Then there is a projection $p : \mathcal{X} \to \mathcal{M}_g \times \mathcal{M}_h$, where $\mathcal{M}_g$ denotes the CFG of families of smooth curves of genus g. We know that $\mathcal{M}_g$ and $\mathcal{M}_h$ are stacks in the étale topology by Corollary 3.40, and therefore the product $\mathcal{M}_g \times \mathcal{M}_h$ is a stack (Lemma 4.4). Applying Corollary 4.13 to p, to verify that $\mathcal{X}$ is a stack it is therefore sufficient to show that, for any *fixed* pair of smooth curves C and C' over a scheme S, the functor $\mathscr{M}or(C, C')(-) : (\mathsf{S}/S)^{\mathsf{op}} \to (\mathsf{Set})$ that assigns to an S-scheme T the set of T-morphisms $C_T \to C'_T$, is a sheaf. This is easily verified using the fact that the étale topology is subcanonical (see Example 3.19).

5. Stacks adapted to a presite

In this section, we take the most naive approach to defining an algebraic stack on a presite, namely, we define what we call a stack adapted to a presite (Definition 5.2). This is simply a stack with a representable cover by an object in the presite (in the sense of Definition 6.17). While there are technical reasons (see Example B.24) that in total generality this is not really the 'right' definition, it nevertheless provides a quick definition, that immediately suffices for many of the standard examples one sees (e.g., algebraic spaces, Fantechi Deligne–Mumford stacks, and Laument–Moret-Bailly Deligne–Mumford stacks). In fact, by iteratively enlarging the presite, and then reducing back down to the original presite, one can obtain all of the definitions of algebraic stacks we discuss in this chapter (see §B.9).

We hope that in this generality, and brevity, the salient points of an algebraic stack will be apparent to readers who prefer to work in categories other than schemes. The reader interested in moving quickly to the definition of the algebraic stack of smooth curves, or the algebraic stack of Higgs bundles, may prefer to read this section, and then skip directly to §7. Technically, the stack of Higgs bundles is adapted to the smooth presite of algebraic spaces, rather than to the étale presite of schemes; nevertheless, the main aspects of the formalism should already be apparent to the reader after this section.

5.1. *Definiton of stacks adapted to a presite*

In order to define a stack adapted to a presite, we want the definition of a representable morphism:

Definition 5.1 (S-representable morphism). Let S be a subcanonical presite that admits fibered products. We say a morphism $\mathcal{X} \to \mathcal{Y}$ of CFGs is S-representable if for every S in S, the fiber product $\mathcal{X} \times_{\mathcal{Y}} S$ is in S (i.e., equivalent to a stack S/S' for some S' in S).

$$\begin{array}{ccc} \mathcal{X} \times_{\mathcal{Y}} S & \longrightarrow & S \\ \downarrow & & \downarrow \\ \mathcal{X} & \longrightarrow & \mathcal{Y}. \end{array}$$

With this definition, we define an S-adapted stack by requiring that the stack admit an 'S-representable cover' in S. More precisely:

Definition 5.2 (S-adapted stack). Let S be a subcanonical presite that admits fibered products. Then an S-*adapted stack* (or a stack adapted to the presite S) is a stack $\mathcal{M}$ over S admitting an S-representable cover in S of the following form: there exists a U in S and an S-representable morphism

$$U \xrightarrow{\ p\ } \mathcal{M}$$

such that for every S in S and every $S \to \mathcal{M}$, the morphism $U \times_{\mathcal{M}} S \to S$ in S obtained from base change is a cover in the pretopology on S. Such a morphism p is called a *presentation* of $\mathcal{M}$. A *morphism of* S-*adapted stacks* is a morphism of stacks.

Remark 5.3. We will define more generally a covering morphism of CFGs in Definition 6.2. The cover defined in Definition 5.2 above is a cover in that sense as well (Remark 6.4).

Example 5.4 (Algebraic spaces and adapted Deligne–Mumford stacks). If S is a scheme and S is the étale presite on schemes over S, then the S-adapted stacks are called Fantechi Deligne–Mumford (F DM) stacks (over S). The S-adapted stacks that are representable by sheaves are called algebraic spaces.

Example 5.5. If S is the presite of complex analytic spaces with the pretopology induced from usual open covers, then one obtains a notion of an adapted complex analytic stack. Similarly, if S is the presite of topological spaces with the presite induced from usual open covers, then one obtains a notion of an adapted topological stack. There are other definitions of complex analytic and topological stacks appearing in the literature; we do not pursue the relationship among the definitions.

Remark 5.6 (Conditions on the diagonal). Other notions of algebraic stacks are determined by requiring the diagonal $\Delta : \mathcal{M} \to \mathcal{M} \times_{\mathsf{S}} \mathcal{M}$ to be S-representable, and putting further geometric conditions on the diagonal. For instance, an Laument–Moret-Bailly Deligne–Mumford (LMB DM) stack over the étale presite S of schemes over a fixed scheme S is an S-adapted stack, with S-representable, separated, and quasicompact diagonal.

To see how all the other algebraic stacks discussed in this chapter can be defined using stacks adapted to a presite, see §B.9.

5.2. *Bootstrapping stacks adapted to a presite*

It can often be useful to show that a stack is S-adapted using bootstrap methods; in other words, it will often be the case that one can exhibit a morphism from a stack of interest to a well-known S-adapted stack, so that it is easy to check the S-adapted condition on the fibers. The following proposition states that in this situation the stack is an S-adapted stack.

Proposition 5.7. *Let $f : \mathcal{X} \to \mathcal{Y}$ be a morphism of stacks over a subcanonical presite S that admits fibered products. Assume that $\mathcal{Y}$ is an S-adapted stack and that for all S in S and all morphisms $S \to \mathcal{Y}$ we have that $\mathcal{X} \times_{\mathcal{Y}} S$ is S-adapted. Then $\mathcal{X}$ is an S-adapted stack.*

Proof. Later we will prove Proposition 6.31, whose proof can be readily modified to fit this situation. We sketch the argument here. Choose a presentation $Y \to \mathcal{Y}$. Then set $\mathcal{Z} = \mathcal{X} \times_{\mathcal{Y}} Y$. One can check that S-representable morphisms are stable under base change (see e.g., Lemma 6.8; in that notation, take $\mathsf{C} = \mathsf{S}$). Thus the projection $\mathcal{Z} \to \mathcal{X}$ is S-representable. One can then check directly from the definition that for every S in S and every morphism $S \to \mathcal{X}$, the morphism $\mathcal{Z} \times_{\mathcal{X}} S \to S$ is a cover in S (use $Y \times_{\mathcal{Y}} S = Y \times_{\mathcal{Y}} \mathcal{X} \times_{\mathcal{X}} S = \mathcal{Z} \times_{\mathcal{X}} S$). Furthermore, $\mathcal{Z}$ is S-adapted, by assumption, so there is a presentation $Z \to \mathcal{Z}$. One can check that S-representable morphisms are stable under composition (see e.g., Lemma 6.8; in that notation, take $\mathsf{C} = \mathsf{S}$). Thus the composition $Z \to \mathcal{Z} \to \mathcal{X}$ is S-representable. One can then check directly from the definition that this morphism is a cover in the sense of Definition 5.2, and thus provides the desired presentation for $\mathcal{X}$. □

Remark 5.8. A result similar to Proposition 5.7 holds when additional conditions are placed on the diagonal of the stacks (see Corollary B.26).

6. Algebraic stacks

In the previous section we introduced the notion of a stack adapted to a presite. This provided a quick definition, that suffices in many cases. However, in general, that approach is a little too naive, particularly if one does not enlarge the presite iteratively. In this section, we take a slightly more lengthy approach, which considers stacks that are adapted to larger classes of morphisms in the presite. After iterating this process, we arrive at the definition of an algebraic stack (adapted to a class of morphisms in the presite); stacks adapted to the presite are algebraic stacks.

We now also provide a lengthier motivation to the study of algebraic stacks than we provided in the previous section. Suppose that S is a subcanonical presite, meaning that every $S \in \mathsf{S}$ represents a sheaf, or equivalently that S/S is a stack on S. A stack on S is fundamentally a topological object, and the category of stacks on S is therefore too inclusive a milieu for our geometric purposes. Even when S is a category of geometric objects, the stacks (and even the sheaves) on S need not behave at all geometrically. To take just one example, a stack on the category of schemes always has a tangent space at a point, but this tangent space may not have the structure of a vector space (see §10.1). Nevertheless, some sheaves and stacks that are not representable *do* behave geometrically, and our goal will be to identify those that do: algebraic spaces and algebraic stacks.

Remark 6.1. Granting our post hoc reasoning for the definition of algebraic stacks, it might have been more sensible to call them 'geometric stacks'. This terminology has caught on in some places [79], but we stick to the traditional nomenclature here.

The essential idea in the definition of algebraic stacks is that a stack that resembles geometric objects locally can itself be studied geometrically, provided that the meaning of 'local' is sufficiently geometric. In the algebraic category, 'locally' is interpreted 'over a smooth cover' and the 'geometric objects' are taken to be schemes.

Here a technicality arises: there are étale *sheaves*, known as algebraic spaces, that resemble schemes étale-locally, but are not themselves schemes. More worryingly, the class of algebraic stacks modelled locally by algebraic spaces is strictly larger than the class of those modelled locally by schemes. In order to make the whole theory satisfyingly formal, one takes the smallest class of stacks that includes affine schemes and is stable under disjoint unions and groupoid quotients. Thankfully, this turns out to be the same as the class of algebraic stacks with smooth covers by algebraic spaces.

This section will be quite formal, and applies to essentially any presite. The main idea is that of a **P**-*adapted stack* in Section 6.4 over a presite S, which is precisely a stack modelled locally, according to a suitable property **P**, on morphisms in S. In Section 6.5, we iterate the construction to arrive at the class of algebraic stacks.

In fact, most stacks we encounter in practice (e.g., quasiseparated Deligne–Mumford stacks) are adapted to the class of morphisms of schemes (or even to affine schemes), and the iteration in Section 6.5 is useful only to have a category with good formal properties. A reader interested in seeing stacks in geometric action may prefer to skip Section 6.5.

The key point is an elementary notion of a cover, and to this end, we must first discuss representable morphisms (§6.2). We also present Proposition 6.31, which is a useful tool for bootstrapping from one adapted stack to another, analogous to Corollary 4.13 for stacks.

6.1. *Covers*

We explain what it means for a morphism of stacks over a presite S to be a cover:

Definition 6.2 (Covering morphism of CFGs). A morphism of CFGs $\mathcal{X} \to \mathcal{Y}$ on a presite S is said to be *covering* if, for any morphism $S \to \mathcal{Y}$, there is a cover $\{S_i \to S\}$ such that the induced maps $S_i \to \mathcal{Y}$ lift to $\mathcal{X}$:

$$\begin{array}{ccc} & & S_i \\ & \swarrow & \downarrow \\ \mathcal{X} \times_{\mathcal{Y}} S & \longrightarrow & S \\ \downarrow & & \downarrow \\ \mathcal{X} & \longrightarrow & \mathcal{Y}. \end{array}$$

Remark 6.3. Note that a morphism of objects of S is covering according to Definition 6.2 if and only if it is covering in the topology associated to the pretopology of S (Definition A.16).

Remark 6.4. The presentation $P : U \to \mathcal{M}$ in Definition 5.2 is covering the sense of Definition 6.2. Indeed, choose an S in S and a morphism $S \to \mathcal{M}$, and consider the fibered product

$$\begin{array}{ccc} U \times_{\mathcal{M}} S & \longrightarrow & S \\ \downarrow & & \downarrow \\ U & \xrightarrow{P} & \mathcal{M}. \end{array}$$

In Definition 6.2 we require that there is a cover $\{S_i \to S\}$ so that for each i, the map $S_i \to S$ factors through $U \times_{\mathcal{M}} S$. Of course, since $U \times_{\mathcal{M}} S \to S$ is a cover by Definition 5.2, this condition is automatically satisfied taking $\{S_i \to S\} = \{U \times_{\mathcal{M}} S \to S\}$.

Lemma 6.5. *Coverings are stable under composition, base change, and fibered products of morphisms.*

Proof. We prove stability under composition: Suppose that $\mathcal{X} \to \mathcal{Y} \to \mathcal{Z}$ is a composition of covering morphism and W is a scheme over $\mathcal{Z}$. Since $\mathcal{Y}$ covers $\mathcal{Z}$, there is a cover $\{W_i \to W\}$ such that the maps $W_i \to \mathcal{Z}$ lift to $\mathcal{Y}$. Now, since $\mathcal{X}$ covers $\mathcal{Y}$ there are covers $\{W_{ij} \to W_i\}$ such that $W_{ij} \to \mathcal{Y}$ lifts to $\mathcal{X}$. The family $\{W_{ij} \to W\}$ is a cover of a cover, hence is covering.

We prove stability under base change: Suppose that $\mathcal{X} \to \mathcal{Y}$ is covering and $\mathcal{Z} \to \mathcal{Y}$ is an arbitrary morphism. Let W be a scheme over Z. Then there is a cover $\{W_i \to W\}$ such that the maps $W_i \to \mathcal{Y}$ lift to $\mathcal{X}$, as $\mathcal{X}$ covers $\mathcal{Y}$. But then by definition of the fiber product, the maps $W_i \to \mathcal{Z}$ lift to $\mathcal{Z} \times_{\mathcal{Y}} \mathcal{X}$.

Stability under fiber product now follows formally from the parts already demonstrated (e.g., the proof of Lemma 6.8 can easily be adapted to this purpose). □

6.2. *Representable morphisms*

A representable morphism of stacks on S is roughly one whose fibers are representable by some prescribed class of stacks:

Definition 6.6 (Representable morphism). Let C be a class of CFGs over a subcanonical presite S such that C is stable under isomorphism and fiber product and every CFG over S can be covered (Defintion 6.2) by objects of C. We say that a morphism $\mathcal{X} \to \mathcal{Y}$ of CFGs is *representable* *by objects of* C or C*-representable* (or *representable*, when C is clear from context) if, for every $\mathcal{Z} \in \mathsf{C}$, the fiber product $\mathcal{X} \times_{\mathcal{Y}} \mathcal{Z}$ is in C.

$$\begin{array}{ccc} \mathcal{X} \times_{\mathcal{Y}} \mathcal{Z} & \longrightarrow & \mathcal{Z} \\ \downarrow & & \downarrow \\ \mathcal{X} & \longrightarrow & \mathcal{Y}. \end{array}$$

When S is the category of schemes and C is also the category of schemes, i.e., the collection of stacks on the étale site of schemes representable by schemes, we call a C-representable morphism *schematic* (i.e., we recover Definition 5.1).

Remark 6.7. For obvious reasons, schematic morphisms are sometimes also called representable morphisms, without qualification. However, it is also common to use the term representable morphism for morphisms representable by algebraic spaces, so for clarity we use 'schematic' to refer to morphisms representable by schemes.

Lemma 6.8. *For* C *a class of CFGs over a category* S *that is stable under isomorphism and fiber product and such that every CFG over* S *can be covered (Definition 6.2) by objects of* C*:*

(1) The composition of C*-representable morphisms is* C*-representable.*
(2) The base change of a C*-representable morphism is* C*-representable.*
(3) The fibered product of C*-representable morphisms is* C*-representable.*
(4) If $f : \mathcal{X} \to \mathcal{Y}$ *and* $g : \mathcal{Y} \to \mathcal{Z}$ *are morphisms such that* gf *is* C*-representable and the diagonal of* g *is* C*-representable then* f *is* C*-representable.*

Proof. The verification is formal, so the proof in [12, Prop. 5.8, p.87] for schematic morphisms applies here. Variants appear in [54, Lem. 3.11 and 3.12], [77, Tags 0300, 0301, 0302]. For completeness, we give a proof.

(1) Suppose $\mathcal{X} \to \mathcal{Y}$ and $\mathcal{Y} \to \mathcal{Z}$ are C-representable. To see that the composition is C-representable, consider a $\mathcal{W} \in \mathsf{C}$ and a morphism $\mathcal{W} \to \mathcal{Z}$. Then $\mathcal{W} \times_{\mathcal{Z}} \mathcal{X} = (\mathcal{W} \times_{\mathcal{Z}} \mathcal{Y}) \times_{\mathcal{Y}} \mathcal{X}$ is C-representable using first the C-representability of $\mathcal{Y} \to \mathcal{Z}$, and then the C-representability of $\mathcal{X} \to \mathcal{Y}$.

(2) If $\mathcal{X} \to \mathcal{Y}$ is a C-representable morphism, $\mathcal{W} \to \mathcal{Z} \to \mathcal{Y}$ are morphisms of CFGs with $\mathcal{W} \in \mathsf{C}$, then $\mathcal{W} \times_{\mathcal{Z}} (\mathcal{Z} \times_{\mathcal{Y}} \mathcal{X}) = \mathcal{W} \times_{\mathcal{Y}} \mathcal{X}$, hence is in C.

(3) is essentially [39, Rem. (1.3.9) p.33], which observes that the conclusion follows from (1) and (2), together with the fact that given

morphisms $f : \mathcal{X} \to \mathcal{X}'$ and $g : \mathcal{Y} \to \mathcal{Y}'$ over a stack $\mathcal{Z}$, the fibered product $\mathcal{X} \times_{\mathcal{Z}} \mathcal{Y} \xrightarrow{f \times_{\mathrm{id}_{\mathcal{Z}}} g} \mathcal{X}' \times_{\mathcal{Z}} \mathcal{Y}'$ is given by the composition of morphisms obtained from fibered product diagrams:

$$\mathcal{X} \times_{\mathcal{Z}} \mathcal{Y} \xrightarrow{f \times_{\mathrm{id}_{\mathcal{Z}}} \mathrm{id}_{\mathcal{Y}}} \mathcal{X}' \times_{\mathcal{Z}} \mathcal{Y} \xrightarrow{\mathrm{id}_{\mathcal{X}'} \times_{\mathrm{id}_{\mathcal{Z}}} g} \mathcal{X}' \times_{\mathcal{Z}} \mathcal{Y}'.$$

(4) For any $\mathcal{W} \to \mathcal{Y}$ with $\mathcal{W}$ in C, we have a cartesian diagram

$$\begin{array}{ccc} \mathcal{W} \times_{\mathcal{Y}} \mathcal{X} & \longrightarrow & \mathcal{W} \times_{\mathcal{Z}} \mathcal{X} \\ \downarrow & & \downarrow \\ \mathcal{Y} & \xrightarrow{\Delta_g} & \mathcal{Y} \times_{\mathcal{Z}} \mathcal{Y}. \end{array}$$

We know $\mathcal{W} \times_{\mathcal{Z}} \mathcal{X}$ is in C since gf is representable, so $\mathcal{W} \times_{\mathcal{Y}} \mathcal{X}$ is in C by representability of the diagonal of g. □

Definition 6.9 (Locality to the target for representable morphisms). Let **P** be a property of morphisms between CFGs in C that is stable under base change (Definition 3.20). We say **P** is local to the target if, for any morphism $\mathcal{X} \to Y$ and any cover $\mathcal{Z} \to \mathcal{Y}$, the morphism $\mathcal{X} \times_{\mathcal{Z}} \mathcal{Y} \to \mathcal{Z}$ has property **P** if and only if $\mathcal{X} \to \mathcal{Y}$ does.

Now that we have these definitions, we can translate properties of morphisms in S into properties of representable morphisms of stacks:

Definition 6.10 (P-representable morphism). Let C be a class of CFGs over a subcanonical presite S such that C is stable under isomorphism and fiber product and every CFG over S can be covered by objects of C. Let **P** be a property of morphisms between CFGs in C that is stable under base change (Definition 3.20), stable under composition, and local on the target (Definition 6.9). A C-representable morphism $\mathcal{X} \to \mathcal{Y}$ of stacks over S (Definition 6.6) is said to have property **P** if for every $\mathcal{Z}$ in C and every $\mathcal{Z} \to \mathcal{Y}$, the morphism $\mathcal{X} \times_{\mathcal{Y}} \mathcal{Z} \to \mathcal{Z}$ of CFGs between objects of C has property **P**. We also call this a **P**-representable morphism.

Example 6.11. Here are some examples of classes of morphisms of schemes that are stable under base change and local to the target for the étale topology on schemes:

- quasiseparated [35, Prop. (2.7.1) (ii)];
- quasicompact [35, Cor. (2.6.4) (v)];
- flat [37, Prop. (17.7.4) (iii)];
- smooth [37, Prop. (17.7.4) (v)];
- étale [37, Prop. (17.7.4) (vi)];

- unramified [37, Prop. (17.7.4) (iv)];
- separated [35, Prop. (2.7.1) (i)];
- proper [35, Prop. (2.7.1) (vii)];
- finite type [35, Prop. (2.7.1) (v)];
- locally of finite type [35, Prop. (2.7.1) (iii)];
- finite presentation [35, Prop. (2.7.1) (vi)];
- locally of finite presentation [35, Prop. (2.7.1) (iv)];
- locally of finite type and pure relative dimension d [35, Cor. (4.1.4)];
- surjective [38, Props. (3.6.1) and (3.6.2) (i)] or [35, Prop. (2.6.1)].

For more references, see [54, (3.10)] or [12, Prop. 5.5].

6.3. *Locality to the source*

Definition 6.12 (Local to the source). Let **P** be a property of morphisms in S that is local (on the target) and stable under base change and composition. We call **P** *local to the source* if a morphism $X \to Y$ in S has the property **P** if and only if, for any covering family $U_i \to X$ in the pretopology S, all of the composed morphisms $U_i \to Y$ have property **P**.

Example 6.13. Let **P** be the class of *étale morphisms* of schemes. Then **P** is local to the source in the étale topology.

Example 6.14. Let **P** be the class of *local isomorphisms* of complex analytic spaces. Then **P** is local to the source in the analytic topology.

Example 6.15. Recall that smooth surjections are covering (in the sense of Definition 6.2) in the étale topology (Example A.22). The class of smooth morphisms is local to the source in the étale topology [77, Tag 06F2]. We will prove a more general version of this statement in Lemma 9.20.

Example 6.16. The same proof shows that smooth morphisms in the category of complex analytic spaces, and submersions in the category of C^∞-manifolds, are local to the source in the usual topologies.

6.4. *Adapted stacks*

Let C be a class of CFGs over a subcanonical presite S such that C is stable under isomorphism and fiber product and every CFG over S can be covered by objects of C. Let **P** be a property of morphisms between CFGs in C that is stable under base change (Definition 3.20), stable under composition, local to the source (Definition 6.12), and local to the target (Definition 6.9).

In this section we will introduce the class of stacks that admit **P**-representable covers by objects of C, calling these (Definition 6.17) stacks **P**-*adapted (to* C*)*, or just *adapted stacks*, if the context is clear (in particular, if it will not be confused with a stack adapted to the presite). In the next section, we will show that once one has the class of **P**-adapted stacks, the property **P** can always be defined for morphisms between **P**-adapted stacks, which will allow us to iterate this procedure in the next section and arrive at the definition of an algebraic stack (with respect to a class of morphisms **P** in S).

Perhaps remarkably, many of the algebraic stacks we consider in this chapter are smooth-adapted to schemes (the class C is taken to be S and the class of morphisms **P** is taken to be the class of smooth morphisms), so the reader who so desires may safely ignore the question of iteration and proceed after this section to Section 9. In fact, for the purposes of this chapter, we will only need to iterate once, to obtain the class of algebraic spaces, and smooth morphisms between algebraic spaces, as all of the stacks we will work with are smooth-adapted to algebraic spaces.

Definition 6.17 (Stacks P-adapted to C). Let $\mathsf{C} \supseteq \mathsf{S}$ be a class of CFGs over a subcanonical presite S such that C is stable under isomorphism and fiber product and every CFG over S can be covered by objects of C. Let **P** be a property of morphisms between CFGs in C that is stable under base change (Definition 3.20), stable under composition, local to the source (Definition 6.12), and local to the target (Definition 6.9). Then a stack on S is **P**-*adapted to* C if it admits a **P**-representable (Definition 6.10) cover (Definition 6.2) by an object of C.

In other words, $\mathcal{M}$ is **P**-adapted to C if there exists a U in C and a morphism

$$p : U \longrightarrow \mathcal{M}$$

such that for every $\mathcal{Z}$ in C and every $\mathcal{Z} \to \mathcal{M}$, we have that $U \times_{\mathcal{M}} \mathcal{Z}$ is in C, and the morphism $U \times_{\mathcal{M}} \mathcal{Z} \to \mathcal{Z}$ has property **P**, and is covering. Such a morphism p is called a *presentation* of $\mathcal{M}$.

When **P**, S, and C are clear, we abbreviate the terminology to *adapted stacks*. A *morphism* of adapted stacks is a morphism of stacks.

Remark 6.18. The presence of both S and C in the definition is for technical reasons. We will later wish to iterate this construction, and we do not wish to undertake a definition of Grothendieck topologies on 2-categories. Had we given such a definition, we would have simply taken C to be a 2-site and defined $\mathbf{P}$-adapted stacks to C without any reference to S. This would be similar to the approach taken in §5; note that in that approach one can avoid 2-cagetories by simply working with presites on the category of algebraic spaces (rather than with all the stacks that arise from the adaption process to the presite).

Example 6.19. If $\mathsf{C} = \mathsf{S}$ (as categories) and $\mathbf{P}$ is the class of coverings in the presite $(S, \mathscr{T})$, then $\mathbf{P}$-adapted stacks are the same as the S-adapted stacks of Definition 5.2. In particular, if S is a scheme and S is the étale presite on schemes over S (i.e., $(\mathsf{S}/S)_{\text{et}}$), we take $\mathsf{C} = \mathsf{S}$ and $\mathbf{P}$ to be the class of étale covers, then the $\mathbf{P}$-adapted stacks are called adapted Deligne–Mumford (F DM) stacks. The $\mathbf{P}$-adapted stacks that are representable by sheaves are called algebraic spaces (see §B).

Example 6.20. If $\mathsf{S} = \mathsf{C}$ is the presite of complex analytic spaces, and $\mathbf{P}$ is the class of smooth morphisms, with the pretopology induced from usual open covers, then one obtains a notion of an adapted complex analytic stack. Similarly, if $\mathsf{S} = \mathsf{C}$ is the presite of topological spaces with the presite induced from usual open covers, then one obtains a notion of an adapted topological stack. There are other definitions of complex analytic and topological stacks appearing in the literature; we do not pursue the relationship among the definitions.

Lemma 6.21. *Suppose that $\mathcal{X} \to \mathcal{Z}$ and $\mathcal{Y} \to \mathcal{Z}$ are morphisms of $\mathbf{P}$-adapted stacks. Then $\mathcal{X} \times_{\mathcal{Z}} \mathcal{Y}$ is a $\mathbf{P}$-adapted stack.*

Proof. Choose $\mathbf{P}$ covers $X_0 \to \mathcal{X}$, $Y_0 \to \mathcal{Y}$, and $Z_0 \to \mathcal{Z}$. Then the map

$$X_0 \times_{\mathcal{Z}} Y_0 \times_{\mathcal{Z}} Z_0 \to \mathcal{X} \times_{\mathcal{Z}} \mathcal{Y} \times_{\mathcal{Z}} \mathcal{Z} = \mathcal{X} \times_{\mathcal{Z}} \mathcal{Y}$$

is a composition of changes of base of representable $\mathbf{P}$ covers, hence is a representable $\mathbf{P}$ cover. □

6.5. *Iterated adaptation*

In order to iterate the definition of $\mathbf{P}$-adapted stacks, we must extend the property $\mathbf{P}$ to morphisms representable by adapted stacks. The main content of this section is that there is a unique way to do this so that $\mathbf{P}$ remains stable under base change and composition, and local to the source and target.

Remark 6.22. Note that Lemma 9.20 gives a canonical way of extending the definition of smooth morphisms to all morphisms of CFGs over the category of schemes in a way that is still stable under composition and base change and local to source and target. This definition necessarily agrees with Definition 6.23, below, which is valid for an arbitrary class of morphisms that is stable under composition to base change and local to source and target. The reader should feel free to skip to Definition 6.27 if he or she is only interested in algebraic stacks over schemes and is willing to rely on Lemma 9.20.

Definition 6.23 (Bootstrapping property P). Let $\mathsf{C} \supseteq \mathsf{S}$ be a class of CFGs over a subcanonical presite S such that C is stable under isomorphism and fiber product and every CFG over S can be covered by objects of C. Let $\mathbf{P}$ be a property of morphisms in C that is stable under base change (Definition 3.20), stable under composition, local to the source (Definition 6.12), and local to the target (Definition 6.9). Suppose that $\mathcal{X}$ and $\mathcal{Y}$ are $\mathbf{P}$-adapted stacks, so that there are presentations $X_0 \to \mathcal{X}$ and $Y_0 \to \mathcal{Y}$; i.e., $\mathbf{P}$-representable covers from objects in C. We say that $\mathcal{X} \to \mathcal{Y}$ *is in class* $\mathbf{P}$ if the map $X_0 \times_{\mathcal{Y}} Y_0 \to Y_0$ is in class $\mathbf{P}$.

For brevity in what follows, we will sometimes write $\mathcal{X} \to \mathcal{Y}$ is $\mathbf{P}$ if it is in class $\mathbf{P}$.

Lemma 6.24. *The definition of the class* $\mathbf{P}$ *of morphisms between* $\mathbf{P}$-*adapted stacks given above does not depend on the choices of presentations* $X_0 \to \mathcal{X}$ *and* $Y_0 \to \mathcal{Y}$.

Proof. Suppose that $X_0' \to \mathcal{X}$ is a different presentation. Then let $X_0'' = X_0' \times_{\mathcal{X}} X_0$. The projection $X_0'' \to X_0'$ is a $\mathbf{P}$ cover, since $\mathbf{P}$ covers are stable under base change, and $X_0 \to \mathcal{X}$ is a $\mathbf{P}$ cover. Pulling back to Y_0 over $\mathcal{Y}$, we have a diagram:

$$\begin{array}{ccc} X_0'' \times_{\mathcal{Y}} Y_0 & \xrightarrow{f} & X_0' \times_{\mathcal{Y}} Y_0 \\ {\scriptstyle g}\downarrow & & \downarrow{\scriptstyle p} \\ X_0 \times_{\mathcal{Y}} Y_0 & \xrightarrow{q} & Y_0 \end{array}$$

By assumption q is $\mathbf{P}$, and g is the base change of a $\mathbf{P}$ morphism, so it is $\mathbf{P}$. Therefore $qg = pf$ is $\mathbf{P}$. But $\mathbf{P}$ is local to the source, and f is the base change of the $\mathbf{P}$ cover, $X_0'' \to X_0'$, so p is $\mathbf{P}$, as required. This proves the independence of the choice of X_0.

Now suppose that $Y_0' \to \mathcal{Y}$ is another presentation. Again we form $Y_0'' = Y_0' \times_{\mathcal{Y}} Y_0$ and note that the two projections are **P** covers, as is the map $Y_0'' \to \mathcal{Y}$. Now consider the diagram with cartesian squares:

$$\begin{array}{ccc} X_0 \times_{\mathcal{Y}} Y_0 & \xrightarrow{f} & Y_0 \\ \uparrow & & \uparrow \\ X_0 \times_{\mathcal{Y}} Y_0'' & \xrightarrow{h} & Y_0'' \\ {\scriptstyle p}\downarrow & & \downarrow{\scriptstyle q} \\ X_0 \times_{\mathcal{Y}} Y_0' & \xrightarrow{g} & Y_0' \end{array}$$

The map f is **P**, by assumption, so h is **P** as well. As q is **P**, so is $qh = gp$. But p is a **P** cover (being the base change of q), so it follows that g is **P**, as required. □

Lemma 6.25. *The property* **P** *of morphisms of stacks admitting* **P***-representable covers is stable under composition, stable under base change, local to the target, and local to the source.*

Proof. Consider a composition $\mathcal{X} \to \mathcal{Y} \to \mathcal{Z}$ where $X_0 \to X$, $Y_0 \to Y$, and $Z_0 \to Z$ are all **P**-representable covering maps, where $\mathcal{X} \to \mathcal{Y}$ is **P**. Form the following diagram:

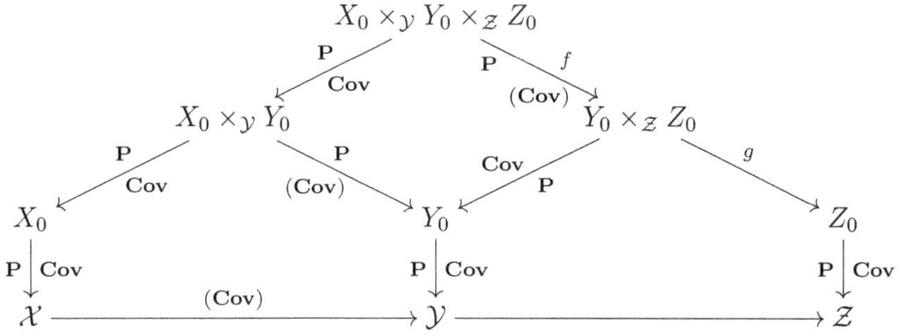

Morphisms that are **P** or covering without any further assumptions are labelled with a **P** or **Cov**. All of these morphisms are C-representable. If $\mathcal{X} \to \mathcal{Y}$ is covering then the other arrows labelled (**Cov**) are covering as well.

To prove stability under composition, suppose that $\mathcal{Y} \to \mathcal{Z}$ is **P**. Then g is **P**, so gf is **P**.

On the other hand, we have a cartesian diagram:

$$\begin{array}{ccccc} X_0 \times_{\mathcal{Y}} Y_0 \times_{\mathcal{Z}} Z_0 & \xrightarrow[\mathbf{P}\quad \mathbf{Cov}]{p} & X_0 \times_{\mathcal{Z}} Z_0 & \xrightarrow[\mathbf{Cov}]{q} & Z_0 \\ \downarrow & & \downarrow & & \\ Y_0 & \xrightarrow{\mathbf{P}\quad \mathbf{Cov}} & \mathcal{Y} & & \end{array}$$

Since Y_0 is $\mathbf{P}$ over $\mathcal{Y}$, we deduce that p is $\mathbf{P}$ and therefore that $qp = gf$ is $\mathbf{P}$. But p is covering, so by locality to the source, q is $\mathbf{P}$. This means that $\mathcal{X} \to \mathcal{Z}$ is $\mathbf{P}$, by definition. This proves stability under composition.

To prove locality to the source, we use the same diagrams. Suppose that $\mathcal{X} \to \mathcal{Z}$ is $\mathbf{P}$. Then q is $\mathbf{P}$, so $qp = gf$ is $\mathbf{P}$. But then f is covering and $\mathbf{P}$, so we deduce that g is $\mathbf{P}$, as required.

Now we prove stability under base change. Consider a cartesian diagram

$$\begin{array}{ccc} \mathcal{X}' & \longrightarrow & \mathcal{X} \\ \downarrow & & \downarrow \\ \mathcal{Y}' & \longrightarrow & \mathcal{Y} \end{array} \tag{6.1}$$

where $\mathcal{X}$ is $\mathbf{P}$ over $\mathcal{Y}$. Let $\mathcal{X}' \to \mathcal{X}$ be the base change. Suppose that $X_0 \to \mathcal{X}$ and $Y_0 \to \mathcal{Y}$ are $\mathbf{P}$ covers. Then $p : X_0 \times_{\mathcal{Y}} Y_0 \to Y_0$ is $\mathbf{P}$, by assumption. Changing base to $\mathcal{Y}'$, we get $\mathbf{P}$ covers $X_0' \to \mathcal{X}$ and $Y_0' \to \mathcal{Y}$. The map $q : X_0' \times_{\mathcal{Y}'} Y_0' \to Y_0'$ is the base change of p. But both $X_0 \times_{\mathcal{Y}} Y_0$ and Y_0 are in C, by assumption, and $\mathbf{P}$ is stable under base change for morphisms in C. Therefore q is $\mathbf{P}$, from which it follows that $\mathcal{X}' \to \mathcal{Y}'$ is $\mathbf{P}$, by definition.

Finally, we prove locality on the target. Suppose that we have a cartesian diagram (6.1). Assume that $\mathcal{Y}'$ is a cover of $\mathcal{Y}$ and that $\mathcal{X}'$ is $\mathbf{P}$ over $\mathcal{Y}$. Let $X_0 \to \mathcal{X}$ and $Y_0 \to \mathcal{Y}$ be $\mathbf{P}$ covers. Assume that $\mathcal{X}'$ is $\mathbf{P}$ over $\mathcal{Y}'$. Set $Y_0' = Y_0 \times_{\mathcal{Y}} \mathcal{Y}'$ and $X_0' = X_0 \times_{\mathcal{X}} \mathcal{X}'$. Then $X_0' \times_{\mathcal{Y}'} Y_0' \to Y_0'$ is the base change of $X_0 \times_{\mathcal{Y}} Y_0$. By assumption $Y_0' \to Y_0$ is a cover, so by locality to the target for morphisms between objects of C, we deduce that $X_0 \times_{\mathcal{Y}} Y_0$ is $\mathbf{P}$ over Y_0. This means that $\mathcal{X}$ is $\mathbf{P}$ over $\mathcal{Y}$, by definition. □

Lemma 6.26. *Given a class of morphisms $\mathbf{P}'$ of $\mathbf{P}$-adapted stacks that extends the class of morphisms $\mathbf{P}$ to the class of $\mathbf{P}$-adapted stacks, and which is stable under base change and composition, and local to the source and target, $\mathbf{P}' = \mathbf{P}$.*

Proof. We leave the proof to the reader. □

6.6. *Algebraic stacks*

The lemmas in the previous section imply that the property **P** makes sense for stacks that are **P**-adapted to C. Replacing C with the category of stacks on S that are **P**-adapted to C, we may iterate the procedure, enlarging the class of stacks under consideration and ensuring that the property **P** makes sense for them at each step. Taking the union of all of these classes, we arrive at the class of algebraic stacks:

Definition 6.27 (Algebraic stack). Let **P** be a property of morphisms of S that is stable under base change, stable under composition, local to the target, and local to the source. An *algebraic stack* (with respect to the property **P**) is a member of the smallest class C of stacks on S such that $\mathsf{C} \supseteq \mathsf{S}$, C has property **P** defined on it, **P** is stable under base change and composition, and local to the source and target, and the class of stacks on S that are **P**-adapted to C (Definition 6.17) is C.

Definition 6.28 (Deligne–Mumford stacks and algebraic spaces). A stack on the category of schemes is called an *algebraic stack* if it is algebraic with respect to the étale topology and the class of smooth maps. It is called a *Deligne–Mumford stack* if it is algebraic with respect to the étale topology and the class of étale maps. It is called a *algebraic space* if it is an algebraic stack and it is representable by a sheaf.

Remark 6.29. One can also use the usual topology on complex analytic spaces and take **P** again to be smooth morphisms. Much of what we do here in the algebraic category works analytically as well, but we will not address the analytic category directly in what follows. Likewise, there is a class of 'algebraic stacks' on the category of manifolds, obtained by taking the usual topology on the category of manifolds and taking **P** to be the class of submersions.

6.7. *Fiber products and bootstrapping*

Corollary 6.30. *Fiber products of algebraic stacks are algebraic.*

Proof. This is immediate by iteration of Lemma 6.21. □

It can often be useful to show that a stack is **P**-adapted to C using bootstrap methods; in other words, it will often be the case that one can exhibit a morphism from a stack of interest to a well-known adapted stack, so that it is easy to check the **P**-adapted condition on the fibers. The following proposition states that in this situation the stack is a **P**-adapted stack. We make this precise in the following remark and proposition.

The class of **P**-adapted stacks is stable under isomorphism and fibered products (Lemma 6.21). Therefore, from Definition 6.6, we have the notion of a morphism of stacks representable by adapted stacks.

Proposition 6.31. *Let $f : \mathcal{X} \to \mathcal{Y}$ be a morphism of stacks over a subcanonical presite* S. *Assume that $\mathcal{Y}$ is a* **P**-*adapted stack and that f is representable by* **P**-*adapted stacks. Then $\mathcal{X}$ is a* **P**-*adapted stack.*

Proof. Choose a **P**-representable cover $Y_0 \to \mathcal{Y}$, with $Y_0 \in \mathsf{C}$. Then set $\mathcal{Z} = Y_0 \times_{\mathcal{Y}} \mathcal{X}$. The projection $\mathcal{Z} \to \mathcal{X}$ is a **P**-representable cover. Furthermore, $\mathcal{Z}$ is **P**-adapted, by assumption, so there is a **P**-representable cover $X_0 \to \mathcal{Z}$, with $X_0 \in \mathsf{C}$. Then $X_0 \to \mathcal{X}$ is a composition of **P**-representable covers, hence is a **P**-representable cover, as required. □

Corollary 6.32. *Suppose that $\mathcal{Y}$ is an algebraic stack and $f : \mathcal{X} \to \mathcal{Y}$ is a morphism of stacks that is representable by algebraic stacks. Then $\mathcal{X}$ is an algebraic stack.*

Proof. This is immediate by iteration of Proposition 6.31. □

7. Moduli stacks of Higgs bundles

In this section we construct the moduli stack of Higgs bundles on a smooth complex projective curve. We also construct several related moduli stacks of interest. In what follows, the reader should always feel free to assume that the morphism $\pi : X \to S$ is a projective morphism between schemes, and that all sheaves are coherent (or even vector bundles). For instance, one case of special interest that will always satisfy the given hypotheses will be the case where $S = \operatorname{Spec} \mathbb{C}$, X is a smooth complex projective curve (or compact Riemann surface), and the sheaves are taken to be the sheaves of sections of holomorphic vector bundles on X.

In order not to avoid repetition below when defining various categories fibered in groupoids, we fix the following notation. Suppose that $f : T \to S$ is an S-scheme. If ξ is an object of some CFG on S/S, we denote by ξ_T a pullback of ξ to T. For example, if X is an S-scheme, we denote by X_T a

T-scheme making the following diagram cartesian:

$$\begin{array}{ccc} X_T & \longrightarrow & X \\ \downarrow & & \downarrow \\ T & \xrightarrow{\ f\ } & S. \end{array}$$

If L is a line bundle on X, we denote by $L_T = f^*L$ a pullback of that line bundle to X_T. Note that we have somewhat abusively used f^*L to denote the pullback of L via the morphism $X_T \to X$ induced from f. We shall make this abuse repeatedly in what follows.

When T is denoted by decorating S, we abbreviate our notation further and denote X_T by decorating X the same way as S; i.e., if $T = S'$ then we allow ourselves to write X' instead of $X_{S'}$.

7.1. *The moduli space of curves*

We will show that the stack of smooth curves is algebraic using the representability of the Hilbert scheme. We give another proof in Section 10, using Artin's criteria (see Theorem 11.25).

The following is a useful criterion to check properties of the diagonal of an algebraic moduli stack of schemes.

Theorem 7.1 ([24, Thm. 5.23, p.133]). *Let S be a noetherian scheme, let $X \to S$ be a flat, projective scheme over S, and let $Y \to S$ be a quasi-projective scheme over S. Then the functor $\mathscr{I}som_{\mathsf{S}/S}(X, Y)$ on schemes over S is representable by a scheme.*

Remark 7.2. In fact $\mathscr{I}som_{\mathsf{S}/S}(X, Y)$ is representable by an open subscheme of the Hilbert scheme $\mathrm{Hilb}_{X\times_S Y/S}$.

Remark 7.3. A special case of a theorem of Olsson ([66, Thm. 1.1]) allows one to drop the noetherian and projective hypotheses, in exchange for requiring finite presentation and settling for an algebraic space rather than scheme. *Let S be a scheme, let X be a flat, proper scheme of finite presentation over S, and let Y be a separated scheme of finite presentation over S. Then the functor $\mathscr{I}som_{\mathsf{S}/S}(X, Y)$ is representable by an algebraic space over S.*

Theorem 7.4 ([20, Prop. (5.1)]). *For $g \geq 2$, the stack $\mathcal{M}_g \to (\mathsf{S}/\mathbb{C})_{\mathsf{et}}$ is a Deligne–Mumford stack.*

Proof. The previous theorem asserts that $\Delta : \mathcal{M}_g \to \mathcal{M}_g \times_{\mathbb{C}} \mathcal{M}_g$ is schematic. More detailed analysis of the isomorphisms of algebraic curves establishes that the diagonal is unramified. Since we are in characterstic 0, this essentially follows from the fact that the automorphism group of a smooth curve of genus $g \geq 2$ is finite, and group schemes in characteristic 0 are smooth.

Now let H_g be the open subset of the Hilbert scheme parameterizing smooth, ν-canonically embedded curves for $\nu \gg 0$. The universal family $C_g \to H_g$ determines a morphism $P : H_g \to \mathcal{M}_g$, that is schematic by virtue of the fact that the diagonal is schematic (see Lemma B.12). One can check that P is smooth, and therefore provides a presentation of the stack. Therefore $\mathcal{M}_g$ is DM, since the diagonal is unramified (see Lemma B.23). □

7.2. *The* Quot *stack and the* Quot *scheme*

Let $\pi : X \to S$ be a proper morphism of finite presentation between schemes. Let E be a quasicoherent sheaf of finite presentation on X. We define a category fibered in groupoids

$$\mathsf{Quot}_{E/X/S} \to \mathsf{S}/S$$

as follows. For each S-scheme $f : S' \to S$, we set the objects of $\mathsf{Quot}_{E/X/S}(S')$ to be surjections

$$E' \xrightarrow{q'} F' \longrightarrow 0$$

where, as was introduced at the beginning of Section 7, E' denotes the pullback of E to $X' = X \times_S S'$. The sheaf F' is required to be an S'-flat quasicoherent sheaf of finite presentation on X'. We will denote such an object by the pair (F', q').

Given an S-morphism $g : S'' \to S'$ and object $(F'', q'') \in \mathsf{Quot}_{E/X/S}(S'')$, a morphism over g from (F'', q'') to (F', q') is a commutative diagram of quasicoherent sheaves on X'':

$$\begin{array}{ccccc} E'' & \xrightarrow{q''} & F'' & \longrightarrow & 0 \\ \| & & \downarrow \wr & & \\ g^*E' & \xrightarrow{g^*q'} & g^*F' & \longrightarrow & 0 \end{array}$$

(The equality on the left indicates the canonical isomorphism induced by pullback.)

As quotients of a coherent sheaf cannot have any nontrivial automorphisms respecting the quotient map, the $\mathsf{Quot}_{E/X/S}$ CFG is equivalent to the CFG associated to the presheaf

$$\mathscr{Q}uot_{E/X/S} : (\mathsf{S}/S)^{\mathsf{op}} \to (\mathsf{Set})$$

wherein $\mathscr{Q}uot_{E/X/S}(T)$ is defined to be the set of isomorphism classes of objects of $\mathsf{Quot}_{E/X/S}(T)$. Of course, $\mathsf{Quot}_{E/X/S}$ is not literally equal to $\mathscr{Q}uot_{E/X/S}$, since one is a CFG and the other is a presheaf.

Remark 7.5. By dualizing, one can check that the Grassmannian $\mathcal{G}(r,n)$ is a substack of $\mathsf{Quot}_{\mathscr{O}^{\oplus n}_{\operatorname{Spec}\mathbb{C}}/\operatorname{Spec}\mathbb{C}/\operatorname{Spec}\mathbb{C}}$.

The following is a special case of a result of Lieblich:

Theorem 7.6 ([56, Prop. 2.7], [77, Tag 08KA]). *Let $\pi : X \to S$ be a proper morphism of finite presentation between schemes. The CFG $\mathsf{Quot}_{E/X/S}$ is an algebraic space that is locally of finite presentation over S.*

More can be said for projective morphisms $\pi : X \to S$. Let L be a relatively very ample line bundle on X/S. If $S' = \operatorname{Spec} k$ for a field k, and F' is a coherent sheaf on $X_{S'}$, then the Hilbert function of F' with respect to L' is defined to be

$$\Phi(m) := \chi(X', F'(m)) = \sum_{i=0}^{\dim X} (-1)^i \dim_k H^i(X', F' \otimes (L')^{\otimes m}).$$

This is in fact a polynomial in $\mathbb{Q}[m]$ (see [24, §5.1.4] for more details). There is a decomposition

$$\mathsf{Quot}_{E/X/S} = \coprod_{\Phi \in \mathbb{Q}[m]} \mathsf{Quot}^{\Phi,L}_{E/X/S}$$

where $\mathsf{Quot}^{\Phi,L}_{E/X/S}(S')$ is the set of equivalence classes $\langle F', q' \rangle$ over S' such that for each $s' \in S'$, the Hilbert polynomial of F'_t with respect to L'_t is equal to Φ. Crucially, cohomology and base change implies that the inclusions of the subfunctors $\mathsf{Quot}^{\Phi,L}_{E/X/S} \subseteq \mathsf{Quot}_{E/X/S}$ are representable by *open and closed* subfunctors of $\mathsf{Quot}_{E/X/S}$. This reduces the study of $\mathsf{Quot}_{E/X/S}$ to that of the subfunctors $\mathsf{Quot}^{\Phi,L}_{E/X/S}$.

The main theorem is due to Grothendieck [40] (see also [7], and [24, Thm. 5.14, p.127]).

Theorem 7.7 ([40]). *Let S be a noetherian scheme, $\pi : X \to S$ a projective morphism, and L a relatively very ample line bundle on X/S. Then for any coherent sheaf E and any polynomial $\Phi \in \mathbb{Q}[m]$, the CFG $\mathsf{Quot}^{\Phi,L}_{E/X/S}$ is representable by a projective S-scheme $\mathrm{Quot}^{\Phi,L}_{E/X/S}$.*

Remark 7.8. By dualizing one can check that the Grassman CFG $\mathcal{G}(r,n)$ is isomorphic to the CFG $\mathsf{Quot}^{r,\mathcal{O}_{\operatorname{Spec}\mathbb{C}}}_{\mathcal{O}^{\oplus n}_{\operatorname{Spec}\mathbb{C}}/\operatorname{Spec}\mathbb{C}/\operatorname{Spec}\mathbb{C}}$, and the Grassmannian $G(r,n)$ is isomorphic to the Quot scheme $\mathrm{Quot}^{r,\mathcal{O}_{\operatorname{Spec}\mathbb{C}}}_{\mathcal{O}^{\oplus n}_{\operatorname{Spec}\mathbb{C}}/\operatorname{Spec}\mathbb{C}/\operatorname{Spec}\mathbb{C}}$. Of course, one cannot use this observation to construct the Grassmannian, as Grothendieck's proof of Theorem 7.7 relies on the representability of the Grassmannian by a projective scheme.

7.3. *Stacks of quasicoherent sheaves*

Let S be a scheme. Let $\pi : X \to S$ be a proper morphism of finite presentation between schemes. We define a category fibered in groupoids

$$\mathsf{QCoh}_{X/S} \to \mathsf{S}/S$$

in the following way. For an S-scheme $f : S' \to S$ in S/S, we take $\mathsf{QCoh}_{X/S}(S')$ to consist (in the notation introduced at the beginning of Section 7) of the S'-flat quasicoherent sheaves [77, Tag 01BE] on $X_{S'}$. Morphisms in $\mathsf{QCoh}_{X/S}$ are defined by pullback; i.e., if $g : S'' \to S'$ is a morphism, F'' is an object of $\mathsf{QCoh}_{X/S}(S'')$ and F' is an object of $\mathsf{QCoh}_{X/S}(S')$, we define a morhpism $F'' \to F'$ to be an isomorphism $F'' \to g^*F'$. Following [56], we use $\mathsf{Coh}_{X/S}$ to denote the substack of $\mathsf{QCoh}_{X/S}$ consisting of quasicoherent sheaves *of finite presentation* [77, Tag 01BN] (when the structure sheaf is coherent, finitely presented sheaves coincide with coherent sheaves [77, Tag 01BZ]). We similarly define $\mathsf{Fib}_{X/S}$ and $\mathsf{Fib}(r)_{X/S}$ by restricting to locally free sheaves of finite rank and locally free sheaves of rank r, respectively.

The first statement we want is descent for quasicoherent sheaves. This is [5, VIII, Thm.1.1, Cor.1.2, p.196]. See also [54, (3.4.4)], and [24, Thm. 4.23, p.82]. The same argument shows that the statements hold for coherent sheaves, and for vector bundles.

Theorem 7.9 ([5, Exp. VIII, Thm. 1.1, p.196]). *Let S be a scheme, and let X be a scheme over S. The CFG $\mathsf{QCoh}_{X/S} \to \mathsf{S}/S$ is a stack with respect to the fpqc topology. The same is true for $\mathsf{Coh}_{X/S}$, $\mathsf{Fib}_{X/S}$ and $\mathsf{Fib}(r)_{X/S}$.*

Moreover, these are algebraic stacks. The following is a special case of a result of Lieblich:

Theorem 7.10 ([56, Thm. 2.1]). *Let $\pi : X \to S$ be a proper morphism of finite presentation between schemes, with S an excellent scheme (see e.g., [77, Tag 07QS]; for instance S can be a scheme of finite type over a field). The stack $\mathsf{Coh}_{X/S}$ is an algebraic stack, locally of finite presentation over S. The same is true for $\mathsf{Fib}_{X/S}$ and $\mathsf{Fib}(r)_{X/S}$.*

For a projective morphism of noetherian schemes, the Quot scheme can be used to establish the presentation of the Artin stack (see Remark 7.13). For our purposes, the benefit of this perspective will be in comparing the moduli stack of Higgs bundles to the moduli scheme of semistable Higgs bundles constructed via GIT.

Theorem 7.11 ([54, Thm. 4.6.2.1, p.29]). *Let S be a noetherian scheme, and let $\pi : X \to S$ be a projective morphism. Assume that $\pi_*\mathscr{O}_X = \mathscr{O}_S$ universally (i.e., $\pi'_*\mathscr{O}_{X'} = \mathscr{O}'_S$ for all $S' \to S$). Then the S-stack $\mathsf{Coh}_{X/S}$ is an algebraic S-stack locally of finite type. The same is true for $\mathsf{Fib}_{X/S}$ and $\mathsf{Fib}(r)_{X/S}$.*

Remark 7.12. The hypothesis that $\pi_*\mathcal{O}_X = \mathcal{O}_S$ universally is satisfied whenever X is projective and flat over S with reduced, connected geometric fibers [24, Ex. 9.3.11, p.303].

Remark 7.13. For each coherent sheaf E on X, a universal quotient over the scheme $\mathrm{Quot}_{E/X/S}$ induces a morphism $\mathrm{Quot}_{E/X/S} \to \mathsf{Coh}_{X/S}$. Taking appropriate open subsets of these Quot schemes gives a presentation of the algebraic stack (see [54, p.30] for more details).

Remark 7.14. Suppose that $S = \mathrm{Spec}\, k$ for an algebraically closed field k, and that X is a smooth projective curve. We can define the CFG $\mathsf{Fib}_{X/S}(r,d)$ by restricting to vector bundles of degree d. All of the statements above hold in this setting; i.e., the CFG $\mathsf{Fib}_{X/S}(r,d)$ of vector bundles of rank r and degree d on X is an algebraic stack, of finite type over k, that admits a presentation from open subsets of Quot schemes.

7.4. *The stack of Higgs bundles over a smooth projective curve*

Let X be a smooth, projective curve over $\mathbb{C}$, of genus g. Fix $r \geq 1$ and $d \in \mathbb{Z}$. A *Higgs bundle* on X of rank r and degree d consists of a pair

$$(E, \phi)$$

where E is a vector bundle (locally free sheaf of finite rank) on X with $\operatorname{rank} E = r$ and $\deg E = d$, and

$$\phi \in \operatorname{Hom}_{\mathscr{O}_X}(E, E \otimes K_X),$$

where $K_X = \Omega^1_X = \omega_X$ is the canonical bundle on X. The aim of this section is to construct an *algebraic stack* $\mathcal{H}_{X/\operatorname{Spec}\mathbb{C}}(r,d)$ *of Higgs bundles on* X *of rank* r *and degree* d*, over the étale site* $\mathsf{S}/\operatorname{Spec}\mathbb{C}$.

7.4.1. *Higgs bundles as a category fibered in groupoids*

We begin by defining the CFG underlying the stack $\mathcal{H}_{X/\operatorname{Spec}\mathbb{C}}(r,d)$. Given a $\mathbb{C}$-scheme $f : S' \to \operatorname{Spec}\mathbb{C}$, we start by defining $\mathcal{H}_{X/\operatorname{Spec}\mathbb{C}}(r,d)(S')$ to have objects consisting of pairs

$$(E', \phi')$$

where $E' \in \mathsf{Fib}_{X/\operatorname{Spec}\mathbb{C}}(r,d)(S')$ is a relative vector bundle of rank r and degree d on X'/S', and

$$\phi' \in \operatorname{Hom}_{\mathcal{O}_{X'}}(E', E' \otimes f^*K_X).$$

Given a $\mathbb{C}$-morphism $g : S'' \to S'$, and (E', ϕ') in $\mathcal{H}_{X/\operatorname{Spec}\mathbb{C}}(r,d)(S')$, we obtain a pulled-back family

$$(E'', \phi'') := g^*(E', \phi')$$

in $\mathcal{H}_{X/\operatorname{Spec}\mathbb{C}}(r,d)(S'')$, where (in the notation from the beginning of Section 7) $E'' = g^*E'$ and $\phi'' = g^*\phi'$.

We now define the morphisms of $\mathcal{H}_{X/\operatorname{Spec}\mathbb{C}}(r,d)$ in the following way. Given (E'', ϕ'') in $\mathcal{H}_{X/\operatorname{Spec}\mathbb{C}}(F)(S'')$, and (E', ϕ') in $\mathcal{H}_{X/\operatorname{Spec}\mathbb{C}}(F)(S')$, then a morphism $(E'', \phi'') \to (E', \phi')$ over $g : S'' \to S'$ consists of an isomorphism

$$\alpha : E'' \to g^*E'$$

such that the following diagram commutes:

$$\begin{array}{ccc} E'' & \xrightarrow{\phi''} & E'' \otimes (fg)^*K_X \\ & & \big\downarrow \alpha\otimes\mathrm{id} \\ \alpha \big\downarrow & & g^*E' \otimes (fg)^*K_X \\ & & \big\| \\ g^*E' & \xrightarrow{g^*\phi'} & g^*(E' \otimes f^*K_X). \end{array}$$

We now have a category fibered in groupoids:

$$\mathcal{H}_{X/\operatorname{Spec}\mathbb{C}}(r,d) \to \mathsf{S}/\operatorname{Spec}\mathbb{C}.$$

7.4.2. *Higgs bundles as an algebraic stack*

The key point is the following well-known lemma:

Lemma 7.15. *Let $\pi : X \to S$ be a flat, proper morphism of finite presentation between schemes. Let G be an S-flat quasicoherent sheaf of $\mathscr{O}_X$-modules that is of finite presentation. There exists a quasicoherent $\mathscr{O}_S$-module M of finite presentation, such that the linear scheme $V = \underline{\operatorname{Spec}}_S\left(\operatorname{Sym}^\bullet_{\mathscr{O}_S} M\right)$ defined by M represents the functor*

$$(S' \xrightarrow{f} S) \mapsto \Gamma(X', f^*G),$$

(see the beginning of Section 7 for notation). The formation of V commutes with all base changes $S' \to S$. This functor will be denoted by $\mathscr{H}om_{X/S}(\mathscr{O}_X, G)$.

Lemma 7.15 is essentially contained in [34, Cor. 7.7.8, Rem. 7.7.9, p.202–3]. This can also be found in [71, p.28], [64, Lem. 3.5], and detailed proofs are given in [16, p.206–207] and [24, Thm. 5.8, p.120].

Corollary 7.16. *The forgetful morphism of CFGs*

$$\mathcal{H}_{X/\operatorname{Spec}\mathbb{C}}(r,d) \to \mathsf{Fib}_{X/\operatorname{Spec}\mathbb{C}}(r,d)$$

is schematic.

Proof. An S-morphism $S' \to \mathsf{Fib}_{X/\operatorname{Spec}\mathbb{C}}(r,d)$ corresponds to a vector bundle E' over X'. Using the construction of the fibered product $\mathcal{H}_{X/\operatorname{Spec}\mathbb{C}}(r,d) \times_{\mathsf{Fib}_{X/S}(r,d)} S'$ in Definition 4.2, one can establish that this is equivalent to the functor on S/S'

$$\mathscr{H}om_{X'/S'}(\mathscr{O}_{X'}, E'^{\vee} \otimes E' \otimes f^*K_X).$$

This is representable by a scheme, by virtue of the lemma above. □

Corollary 7.17. *The CFG* $\mathcal{H}_{X/\operatorname{Spec}\mathbb{C}}(r,d) \to \mathsf{S}/\operatorname{Spec}\mathbb{C}$ *is a stack.*

Proof. We have seen in Theorem 7.9 that $\mathsf{Fib}_{X/\operatorname{Spec}\mathbb{C}}(r,d)$ is a stack. Therefore, the corollary follows from Corollary 7.16 and Corollary 4.13. □

Theorem 7.18. *The CFG of Higgs bundles* $\mathcal{H}_{X/\operatorname{Spec}\mathbb{C}}(r,d) \to \mathsf{S}/\operatorname{Spec}\mathbb{C}$ *is an algebraic stack, locally of finite type over* $\operatorname{Spec}\mathbb{C}$.

Proof. We have seen in Corollary 7.17, that $\mathcal{H}_{X/\operatorname{Spec}\mathbb{C}}(r,d)$ is a stack. Under the hypotheses here, it follows from Theorem 7.11 (see also Remark 7.12) that $\mathsf{Fib}_{X/\operatorname{Spec}\mathbb{C}}(r,d)$ is an algebraic stack locally of finite type over $\mathbb{C}$. Thus from Corollary 7.16 and Corollary 6.32 we have that $\mathcal{H}_{X/\operatorname{Spec}\mathbb{C}}(r,d)$ is an algebraic stack, locally of finite type over $\mathbb{C}$. □

7.5. *Meromorphic Higgs bundles, and the stack of sheaves and endomorphisms*

There is also interest in considering so-called meromorphic Higgs bundles on a smooth projective curve X; that is pairs (E,ϕ) where E is a vector bundle and $\phi : E \to E \otimes K_X(D)$ is a morphism of sheaves, for some fixed divisor D on X. The construction of such a stack can be made in essentially the same way as for Higgs bundles. Since it is not much more work, we provide here a more general construction.

In this setting, the input data are the following:

(1) $\pi : X \to S$ a proper morphism of finite presentation between schemes, such that either
   (a) S is excellent, or,
   (b) S is noetherian, π is projective, and $\pi_*\mathscr{O}_X = \mathscr{O}_S$ universally.

(2) F an S-flat quasicoherent sheaf of $\mathscr{O}_X$-modules of finite presentation.

The output from this data will be an *algebraic stack* $\mathcal{E}_{X/S}(F)$ *of endomorphisms of coherent sheaves with values in* F*, over the site* $(\mathsf{S}/S)_{\text{et}}$. Taking X/S to be a smooth complex projective curve over $S = \operatorname{Spec}\mathbb{C}$, $F = K_X(D)$, and restricting to vector bundles, one obtains the stack of meromorphic Higgs bundles associated to the divisor D.

Remark 7.19. While the construction above generalizes the construction of Higgs bundles for curves, it seems that for many applications these stacks are not the correct generalizations. First in higher dimension (even in the smooth case over $S = \operatorname{Spec}\mathbb{C}$ and with $F = \Omega^1_X$), one should at least include the integrability condition $\phi \wedge \phi = 0$ (in $\underline{\mathrm{End}}(E) \otimes \Omega^2_X$). Second, in any dimension, it is often better to put derived structures on X and S so that the relative cotangent complex becomes perfect, and then consider the moduli problem on the derived scheme X. The result will be a derived stack (a notion which we will not introduce here). One case where the generalized construction we take here does suffice is for families of stable curves, where one can consider endomorphisms with values in the relative dualizing sheaf $\omega_{X/S}$. This will be discussed further below.

7.5.1. *The CFG* $\mathcal{E}_{X/S}(F)$

We begin by defining the category fibered in groupoids underlying the stack $\mathcal{E}_{X/S}(F)$. Given a morphism $f : S' \to S$, we start by defining $\mathcal{E}_{X/S}(F)(S')$ to have objects those pairs

$$(E', \phi')$$

where E' is an S'-flat coherent sheaf on X' and

$$\phi' \in \mathrm{Hom}_{\mathcal{O}_{X'}}(E', E' \otimes F'),$$

where $F' = f^*F$ and $X' = X \times_S S'$ (as always, we use the notation introduced at the beginning of Section 7).

Given an S-morphism $g : S'' \to S'$, we again denote the composition $f \circ g = f'$, and will use the notation introduced at the beginning of Section 7. Given (E', ϕ') in $\mathcal{E}_{X/S}(F)(S')$, we obtain a pulled-back family

$$g^*(E', \phi') = (g^*E', g^*\phi')$$

in $\mathcal{E}_{X/S}(F)(S'')$,

We now define morphisms in the following way. Given (E'', ϕ'') in $\mathcal{E}_{X/S}(F)(S'')$, and (E', ϕ') in $\mathcal{E}_{X/S}(F)(S')$, a morphism $(E'', \phi'') \to (E', \phi')$ over $g : S'' \to S'$ consists of an isomorphism

$$\alpha : E'' \to g^*E'$$

such that the following diagram commutes:

$$\begin{array}{ccc}
E'' & \xrightarrow{\phi''} & E'' \otimes F'' \\
 & & \big\downarrow{\scriptstyle \alpha \otimes \mathrm{id}} \\
\big\downarrow{\scriptstyle \alpha} & & g^*E' \otimes F'' \\
 & & \Big\| \\
g^*E' & \xrightarrow{g^*\phi'} & g^*(E' \otimes f^*F')
\end{array}$$

Now we have the category fibered in groupoids

$$\mathcal{E}_{X/S}(F) \to \mathsf{S}/S.$$

7.5.2. *$\mathcal{E}_{X/S}(F)$ is a stack*

In this setting, we replace Lemma 7.15 with a special case of a result due to Lieblich. Let $X \to S$ be a proper morphism of finite presentation between schemes. Let E and G be finitely presented, quasicoherent sheaves on X such that G is S-flat. We define a CFG

$$\mathscr{H}om_{X/S}(E, G) \to \mathsf{S}/S$$

by assigning to each morphism $f : S' \to S$ the set of homomorphisms

$$\mathrm{Hom}_{\mathscr{O}_{X'}}(E_{S'}, G_{S'}).$$

Morphisms are defined by pullback.

Proposition 7.20 (Leiblich [56, Prop. 2.3]). *Let $X \to S$ be a proper morphism of finite presentation between schemes. Let E and G be finitely presented, quasicoherent sheaves on X such that G is S-flat. The CFG $\mathscr{H}om_{X/S}(E, G) \to \mathsf{S}/S$ is representable by an algebraic space over S, locally of finite type.*

Corollary 7.21. *The forgetful morphism of CFGs*

$$\mathcal{E}_{X/S}(F) \to \mathsf{Coh}_{X/S}$$

is representable by algebraic spaces, locally of finite type over S.

Proof. An S-morphism $S' \to \mathsf{Coh}_{X/S}$ corresponds to E' in $\mathsf{Coh}_{X/S}(S')$. Therefore the fibered product $\mathcal{E}_{X/S}(F) \times_{\mathsf{Coh}_{X/S}} S'$ is the CFG $\mathscr{H}om_{X/S}(E', E' \otimes F')$, which is representable by virtue of Lieblich's result above. □

Corollary 7.22. *The CFG $\mathcal{E}_{X/S}(F)$ is a stack.*

Proof. We have seen in Theorem 7.9 that $\mathsf{Coh}_{X/S}$ is a stack. Therefore, the corollary follows from Corollary 7.21 and Corollary 4.13. □

7.5.3. *$\mathcal{E}_{X/S}(F)$ is an algebraic stack*

Theorem 7.23. *$\mathcal{E}_{X/S}(F)$ is an algebraic stack, locally of finite presentation over S.*

Proof. We have seen in Corollary 7.22, that $\mathcal{E}_{X/S}(F)$ is a stack. Under the hypotheses here, it follows from Theorem 7.10 that $\mathsf{Coh}_{X/S}$ is an algebraic stack, locally of finite presentation over S. Thus from Corollary 7.21 and Corollary 6.32 we have that $\mathcal{E}_{X/S}(F)$ is an algebraic stack, locally of finite presentation over S. □

7.6. *The stack of Higgs bundles over the moduli of stable curves*

We now construct a moduli stack of Higgs bundles over the moduli stack of stable curves:

$$\mathcal{HS}h_{\overline{\mathcal{M}}_g} \to \overline{\mathcal{M}}_g.$$

We start by fixing a base S in S (for instance $S = \operatorname{Spec}\mathbb{C}$, or even $S = \operatorname{Spec}\mathbb{Z}$).

Definition 7.24 (Stable curve). A *stable curve* of genus $g \geq 2$ over an algebraically closed field k is a connected curve of arithmetic genus g, with no singularities apart from nodes, and with finite automorphism group (every component isomorphic to $\mathbb{P}^1_k$ meets the rest of the curve in at least 3 points).

For an S-scheme $S' \to S$, a *relative stable curve of genus g* over S' is a surjective morphism of schemes $\pi : X' \to S'$ that is flat, proper and whose every geometric fiber is a stable curve of genus g.

We define the CFG $\overline{\mathcal{M}}_g$ over S/S. For every S-scheme $S' \to S$ we take the objects of $\overline{\mathcal{M}}_g(S')$ to be the relative stable curves over S'. The morphisms in $\overline{\mathcal{M}}_g$ are given by pullback diagrams, exactly as in the definition of $\mathcal{M}_g$. Using the fact that for every relative stable curve of genus $g \geq 2$ the relative dualizing sheaf $\omega_{X/S}$ is relatively ample, one can show that $\overline{\mathcal{M}}_g$ is an algebraic stack (in fact DM) exactly as was done for $\mathcal{M}_g$. In fact $\mathcal{M}_g$ is an open substack of $\overline{\mathcal{M}}_g$, since smoothness is an open condition.

Now we define the CFG $\mathsf{Coh}_{\overline{\mathcal{M}}_g/S}$. Over an S-scheme $S' \to S$, the objects of $\mathsf{Coh}_{\overline{\mathcal{M}}_g/S}(S')$ are pairs $(X'/S', E')$ where $X' \to S'$ is a relative stable curve, and E' is an S'-flat finitely, presented quasicoherent sheaf on X'. Morphism are defined by pullback in both entries. There is a natural morphism of CFGs

$$\mathsf{Coh}_{\overline{\mathcal{M}}_g/S} \to \overline{\mathcal{M}}_g$$

given by forgetting the sheaf. For every S-scheme $S' \to S$, and every morphism $S' \to \overline{\mathcal{M}}_g$ induced by a relative stable curve $X' \to S'$, the fibered product $\mathsf{Coh}_{\overline{\mathcal{M}}_g/S} \times_{\overline{\mathcal{M}}_g} S'$ is equivalent to $\mathsf{Coh}_{X'/S'}$. Since this is a algebraic stack, we have immediately from Corollary 4.13 and Corollary 6.32 that $\mathsf{Coh}_{\overline{\mathcal{M}}_g/S}$ is a algebraic stack (this is also a special case of [56, Thm. 2.1]). Similarly, we have stacks $\mathsf{Coh}_{\overline{\mathcal{M}}_g/S}(r,d)$ (resp. $\mathsf{Fib}_{\overline{\mathcal{M}}_g/S}(r,d)$) where we restrict to sheaves (resp. bundles) of relative rank r and degree d. Recall that the rank and degree for sheaves on a reducible curve are defined via the Hilbert polynomial with respect to the relative dualizing sheaf.

Finally we define the CFG $\mathcal{HS}h_{\overline{\mathcal{M}}_g/S}$. Over an S-scheme $S' \to S$, let the objects of $\mathcal{HS}h_{\overline{\mathcal{M}}_g/S}(S')$ be triples $(X'/S', E', \phi')$ where $X' \to S'$ is a relative stable curve, and E' is an S'-flat finitely, presented quasicoherent sheaf on X', and $\phi' \in \mathrm{Hom}_{\mathscr{O}_{X'}}(E', E' \otimes \omega_{X/S})$. Morphism are defined by pullback in the first two entries, and in the same way as in the definition of Higgs bundles in the last entry. There is a natural morphism of CFGs

$$\mathcal{HS}h_{\overline{\mathcal{M}}_g/S} \to \mathsf{Coh}_{\overline{\mathcal{M}}_g/S}$$

given by forgetting ϕ'. For every S-scheme $S' \to S$, and every morphism $S' \to \mathsf{Coh}_{\overline{\mathcal{M}}_g/S}$ induced by a pair $(X'/S', E')$, the fibered product $\mathcal{HS}h_{\overline{\mathcal{M}}_g/S} \times_{\mathsf{Coh}_{\overline{\mathcal{M}}_g/S}} S'$ is equivalent to $\mathcal{E}_{X'/S'}(\omega_{X'/S'})$. Since this is an algebraic stack, we have immediately from Corollary 4.13 and Corollary B.26 that $\mathcal{HS}h_{\overline{\mathcal{M}}_g/S}$ is an algebraic stack. Similarly, we have stacks $\mathcal{HS}h_{\overline{\mathcal{M}}_g/S}(r,d)$ (resp. $\mathcal{H}_{\overline{\mathcal{M}}_g/S}(r,d)$) where we restrict to sheaves (resp. bundles) of relative rank r and degree d.

Remark 7.25. In the discussion above we omitted some noetherian hypotheses on the test S-schemes S'. This is possible via the standard noetherian reduction arguments of [36, §8]. Similar arguments are made in §11.2, and we direct the reader there for more details on these types of arguments.

7.7. *Semi-stable Higgs bundles and the quotient stack*

In [64] Nitsure constructs a moduli scheme of semi-stable Higgs bundles on a smooth projective curve taking values in a line bundle L. The construction of the moduli stack of Higgs bundles on an algebraic curve above essentially follows Nitsure's construction in the setting of stacks. Here we introduce Nitsure's space, and compare it to the moduli stack. Nitsure's construction uses geometric invariant theory, so our stack-oriented perspective will rely on quotient stacks (Section C.1).

7.7.1. *Nitsure's construction*

Recall that the slope of a bundle E on a smooth curve X is defined to be $\mu(E) := \deg(E)/\operatorname{rank}(E)$. We will say that (E', ϕ') is a sub-Higgs bundle of (E, ϕ) if $E' \overset{\iota}{\hookrightarrow} E$ and there is a commutative diagram

$$\begin{array}{ccc} E & \xrightarrow{\phi} & E \otimes L \\ \uparrow{\scriptstyle\iota} & & \uparrow{\scriptstyle\iota\otimes \mathrm{id}_L} \\ E' & \xrightarrow{\phi'} & E' \otimes L. \end{array}$$

A Higgs bundle (E, ϕ) is said to be slope stable (resp. semi-stable) if for every sub-Higgs bundle $(E', \phi') \subseteq (E, \phi)$, one has $\mu(E') < \mu(E)$ (resp. $\mu(E') \leq \mu(E)$).

Fix positive integers r and d. By [48, Prop. 2.3.1], semistability is an open condition on a flat family of coherent sheaves. Therefore there is an *open* substack $\mathcal{H}^{ss}_{X/\operatorname{Spec}\mathbb{C}}(L)(r, d) \subseteq \mathcal{H}_{X/\operatorname{Spec} C}(r, d)$ of semi-stable Higgs bundles over X of rank r and degree d, taking values in L. Here we aim to relate this stack to a quasi-projective variety constructed by Nitsure. We review Nitsure's construction briefly.

Let $\mathscr{O}_X(1)$ be an ample line bundle on X. Let $N \in \mathbb{Z}$ be the minimal positive integer such that

$$\frac{d}{r} + N \deg \mathscr{O}_X(1) > \max\left\{2g - 1,\ 2g - 1 + \frac{(r-1)^2}{r} \deg L\right\}.$$

Set $p = (d + rN \deg \mathscr{O}_X(1)) + r(g - 1)$. Let Q be the component of the Quot scheme containing quotients $\mathscr{O}_X(-N)^{\oplus p} \to E \to 0$ where E is a rank r, degree d vector bundle on X. Let Q° be the locus in Q where E is locally free, the quotient $\mathscr{O}^p_X \to E(N) \to 0$ obtained by twisting by $\mathscr{O}_X(N)$ induces an isomorphism on the space of global sections, and $H^1(X, E(N)) = 0$. It is known that Q is projective, and that Q° is reduced and open in Q (see [64, p. 281]). Let $\mathscr{E}$ be the universal bundle on $X \times Q^\circ$ obtained from the universal quotient. Lemma 7.15 implies there is a linear scheme $F \to Q^\circ$ parameterizing pairs $(\mathscr{O}_X(-N)^{\oplus p} \to E \to 0, \phi)$, where $\phi : E \to E \otimes L$. [64, Cor. 3.4] implies that every slope semi-stable Higgs bundle of rank r and degree d (together with a presentation as a quotient) shows up in this family.

Let F^{ss} be the locus of pairs where the bundle and the endomorphism form a slope semi-stable Higgs bundle. The universal family over F induces a morphism

$$F^{ss} \to \mathcal{H}^{ss}_{X/\operatorname{Spec}\mathbb{C}}(L)(r, d).$$

It is essentially the content of [64, Prop. 3.6] that this morphism is smooth. Moreover, the obvious action of $\mathbb{P}\,\mathrm{GL}_p$ on Q given by changing coordinates for the choice of generators of the bundle lifts to an action of $\mathbb{P}\,\mathrm{GL}_p$ on F [64, p. 281]. The closed orbits in F^{ss} are given by S-equivalence classes of slope semi-stable Higgs bundles [64, §4].

Nitsure constructs a quotient of F in the following way. There is a quasi-projective variety H equipped with a $\mathbb{P}\,\mathrm{GL}_p$-linearized ample line bundle $\mathcal{L}$, and a $\mathbb{P}\,\mathrm{GL}_p$-equivariant morphism $\hat{\tau} : F \to H$ such that slope (semi-) stability on F corresponds to GIT (semi-)stability on H [64, §5]. Via $\hat{\tau}$, the quasiprojective GIT quotient $H/\!\!/_{\mathcal{L}}\mathbb{P}\,\mathrm{GL}_p$ induces a quasiprojective scheme structure on the set $F^{ss}/\mathbb{P}\,\mathrm{GL}_p$ of S-equivalence classes of slope semi-stable Higgs bundles [64, p. 290], which we will denote by $H^{ss}_{X/\operatorname{Spec}\mathbb{C}}(L)(r,d)$. (We expect one can also obtain the quotient directly via GIT on F with respect to the semiample line bundle obtained from $\mathcal{L}$ by pullback via $\hat{\tau}$.) This is the moduli scheme of Higgs bundles constructed by Nitsure. It is shown in [64, p.290, Thm. 5.10] that $H^{ss}_{X/\operatorname{Spec}\mathbb{C}}(L)(r,d)$ is a good quotient for F^{ss} by the action of $\mathbb{P}\,\mathrm{GL}_p$, and is a categorical moduli scheme for the associated moduli functor $\mathscr{H}^{ss}_{X/\operatorname{Spec}\mathbb{C}}(L)(r,d)$ of Higgs bundles (defined in the obvious way). Consequently there is a diagram

$$\begin{array}{ccccc} & & \mathcal{H}^{ss}_{X/\operatorname{Spec}\mathbb{C}}(L)(r,d) & \longrightarrow & \mathscr{H}^{ss}_{X/\operatorname{Spec}\mathbb{C}}(L)(r,d) \\ & \nearrow & & & \downarrow \\ F^{ss} & & & & \\ & \searrow & & & \\ & & [F^{ss}/\mathbb{P}\,\mathrm{GL}_p] & \longrightarrow & H^{ss}_{X/\operatorname{Spec}\mathbb{C}}(L)(r,d). \end{array}$$

The composition $\mathcal{H}^{ss}_{X/\operatorname{Spec}\mathbb{C}}(L)(r,d) \to H^{ss}_{X/\operatorname{Spec}\mathbb{C}}(L)(r,d)$ is a categorical moduli scheme for the stack $\mathcal{H}^{ss}_{X/\operatorname{Spec}\mathbb{C}}(L)(r,d)$, as well, in the sense that it is initial among all morphisms to schemes.

7.8. *The stack of principal G-Higgs bundles*

The focus of our presentation has been on Higgs vector bundles. For completeness we include a brief section of principal Higgs bundles on smooth complex projective curves. We direct the reader to [21] for more on the topic. We will give a deformation-theoretic perspective on G-Higgs bundles in Section 10.2.7.

Let X be a smooth complex projective curve, and let G be a complex semisimple Lie group. Let $\mathfrak{g}$ be the complex Lie algebra associated to G, and let

$$\mathrm{Ad} : G \to \mathrm{Aut}(\mathfrak{g})$$

be the adjoint representation of G. Given an algebraic principal G-bundle P on X, the adjoint bundle of P is the associated algebraic vector bundle

$$\mathrm{ad}P := P \times_G \mathfrak{g} := (P \times \mathfrak{g})/G$$

where the action of G is the product action via the natural action of G on P and the adjoint action of G on $\mathfrak{g}$.

Definition 7.26 (G-Higgs bundle). Let X be a smooth complex projective curve, and let G be a complex semisimple Lie group. A *G-Higgs bundle* on X is a pair (P, Φ) where P is a principal G-bundle over X and $\Phi \in H^0(X, \mathrm{ad}\, P \otimes K_X)$.

Remark 7.27. If $G \hookrightarrow \mathrm{GL}_n$, then the data of a G-Higgs bundle (P, Φ) gives rise to a Higgs bundle (E, ϕ), where $E := P \times_G \mathbb{C}^n$, and $\phi : E \to E \otimes K_C$ is described as follows. First observe that $\mathrm{End}(E) = P \times_G \mathrm{End}(\mathbb{C}^n)$. The embedding $G \hookrightarrow \mathrm{GL}_n$ induces on tangent spaces a G-equivariant map $\mathfrak{g} \to \mathrm{End}(\mathbb{C}^n)$, where by definition G acts by the adjoint representation on $\mathfrak{g}$, and by conjugation on $\mathrm{End}(\mathbb{C}^n)$ via the embedding $G \hookrightarrow \mathrm{GL}_n$. Thus we obtain a morphism

$$\mathrm{ad}\, P = P \times_G \mathfrak{g} \longrightarrow P \times_G \mathrm{End}(\mathbb{C}^n) = \mathrm{End}(E).$$

Tensoring by K_X and taking global sections, one obtains the morphism ϕ induced by Φ.

Following the construction in §7.4, one can define the CFG $\mathcal{H}^G_{X/\operatorname{Spec}\mathbb{C}}(r, d)$ over the étale site $\mathsf{S}/\operatorname{Spec}\mathbb{C}$ consisting of principal G-Higgs bundles whose associated vector bundle is of rank r and degree d.

Theorem 7.28. *The category fibered in groupoids of principal G-Higgs bundles, $\mathcal{H}^G_{X/\operatorname{Spec}\mathbb{C}}(r, d) \to \mathsf{S}/\operatorname{Spec}\mathbb{C}$, is an algebraic stack locally of finite type over* $\operatorname{Spec}\mathbb{C}$.

Proof. The CFG $\mathsf{Prin}^G_{X/\operatorname{Spec}\mathbb{C}}(r, d)$, of principal G-bundles over X with associated vector bundle of rank r and degree d, is an algebraic stack locally of finite type over $\mathbb{C}$ (see e.g., [54, Exa. (4.6)]). The forgetful functor $\mathcal{H}^G_{X/\operatorname{Spec}\mathbb{C}}(r, d) \to \mathsf{Prin}^G_{X/\operatorname{Spec}\mathbb{C}}(r, d)$ is schematic (this is similar to Corollary 7.16). It follows from Corollary 6.32 that $\mathcal{H}^G_{X/\operatorname{Spec}\mathbb{C}}(r, d)$ is an algebraic stack locally of finite type over $\mathbb{C}$. □

Remark 7.29. As in §7.5, one can easily generalize this construction to the case of a proper morphism $\pi : X \to S$ of finite presentation between schemes over $\mathbb{C}$, and pairs (P, Φ), where P is again a principal G-bundle, but

Φ is a global section of $\operatorname{ad} P \otimes F$, where F is some fixed S-flat quasicoherent sheaf of $\mathscr{O}_X$-modules of finite presentation. As in the case of Higgs bundles, for many applications this is not the correct generalization. In the smooth case, the following is standard: Let X be a smooth projective manifold. A *G-Higgs bundle* is a pair (P, Φ) where P is a principal G-bundle over X and $\Phi \in H^0(X, \operatorname{ad} P \otimes \Omega^1_X)$ is such that $[\Phi, \Phi] = 0 \in H^0(X, \operatorname{ad} P \otimes \bigwedge^2 \Omega^1_X)$.

8. The Hitchin fibration

In this section we describe the Hitchin fibration at the level of stacks.

8.1. *Characteristic polynomials and the Hitchin morphism*

In this section we introduce characteristic polynomials and the Hitchin morphism.

8.1.1. *Characteristic polynomials*

Let X be a scheme, let E be a locally free sheaf of rank n on X, let L be a locally free sheaf of rank 1 on X, and let

$$\phi : E \to E \otimes L$$

be a morphism of sheaves. Let $\mathbb{L} = \underline{\operatorname{Spec}}_X (\operatorname{Sym}^\bullet L^\vee)$ be the geometric line bundle associated to L, and let

$$p : \mathbb{L} \to X$$

be the structure map. Let

$$\mathscr{O}_{\mathbb{L}} \xrightarrow{T} p^* L \tag{8.1}$$

be the tautological section (this is the section corresponding to the tautological map of geometric line bundles $\mathbb{L} \times_X \mathbb{A}^1_X \to \mathbb{L} \times_X \mathbb{L}$, given heuristically by "$(v_x, \lambda_x) \mapsto (v_x, \lambda_x v_x)$ for $v_x \in \mathbb{L}_x$ and $\lambda_x \in \mathbb{A}^1_{X,x}$"; see §8.1.3 below).

Tensoring by $p^* E$, we obtain

$$p^* E \xrightarrow{T} p^* E \otimes p^* L.$$

Consequently we obtain the endomorphism

$$p^* E \xrightarrow{T - p^* \phi} p^* E \otimes p^* L.$$

Taking the determinant we obtain

$$p^* \det E \xrightarrow{\det(T - p^* \phi)} p^* \det E \otimes p^* L^{\otimes n}.$$

We can view this as a global section

$$\mathcal{O}_{\mathbb{L}} \xrightarrow{\det(T-p^*\phi)} p^*L^{\otimes n}.$$

Therefore we have

$$\det(T-p^*\phi) \in \Gamma(\mathbb{L}, p^*L^{\otimes n}) = \Gamma(X, p_*\mathcal{O}_{\mathbb{L}} \otimes L^{\otimes n}) = \bigoplus_{m\geq 0} T^m\Gamma(X, L^{\otimes(n-m)}). \tag{8.2}$$

On the right, T is a formal variable, which we introduce to make the book-keeping and some local computations easier to follow.

We call $\det(T-p^*\phi)$ the *characteristic polynomial of* ϕ (it is an element of the graded ring of global sections of tensor powers of L, which we formally view as a polynomial by introducing the formal variable T).

The component of $\det(T - p^*\phi)$ in $\Gamma(X, L^{\otimes i})$ can be obtained in the following way. The morphism

$$\phi : E \to E \otimes L$$

determines a global section of L via the composition

$$\mathcal{O}_X \to E^\vee \otimes E \to L,$$

where the first map takes 1 to id_E and the second is induced by ϕ. We define the trace $\mathrm{Tr}(\phi) \in H^0(X, L)$ to be this global section. The component of $\det(T - p^*\phi)$ in $H^0(X, L^{\otimes i})$ is given by $(-1)^i\,\mathrm{Tr}(\wedge^i\phi)$. In particular, $\det(T - p^*\phi) \in \bigoplus_{i=0}^n T^{n-i}\Gamma(X, L^{\otimes i})$. Moreover, the component in $\Gamma(X, \mathcal{O}_X)$ is always equal to 1, and so we will drop this term in what follows.

8.1.2. *The Hitchin morphism*

From the discussion above we can define the Hitchin morphism

$$\begin{aligned} h : \mathcal{H}_{X/S}(L)(n) &\to \bigoplus_{i=1}^{n} T^{n-i}\Gamma(X, L^{\otimes i}) \\ (E, \phi) &\mapsto \det(T - p^*\phi). \end{aligned}$$

Here we are denoting by $\mathcal{H}_{X/S}(L)(n)$ the sub-algebraic stack of $\mathcal{E}_{X/S}(L)(n)$ consisting of endomorphisms of locally free coherent sheaves of rank n with values in L, over the site $(\mathsf{S}/S)_{\mathrm{et}}$. On the right, we suggestively indicate the functor that on an S-scheme $S' \to S$ takes values $\bigoplus_{i=1}^n T^{n-i}\Gamma(X', L'^{\otimes i})$, where $X' = S' \times_S X$ and L' is the pullback to X'. In other words it is the functor $\mathscr{H}om_{X/S}(\mathcal{O}_X, \bigoplus_{i=1}^n T^{n-i} L^{\otimes i})$ of Lemma 7.15. By virtue of Lemma 7.15, this functor is representable by a scheme $\mathbb{A}_{X/S}(L)(n)$ over S. We call $\mathbb{A}_{X/S}(L)(n)$ the Hitchin base for $\mathcal{H}_{X/S}(L)(n)$. We obtain the Hitchin morphism

$$h : \mathcal{H}_{X/S}(L)(n) \to \mathbb{A}_{X/S}(L)(n).$$

Remark 8.1. When X/S is a smooth projective curve over $S = \operatorname{Spec} \mathbb{C}$ (and we consider $\mathbb{C}$-points of the moduli problem) the Hitchin map takes a rank n Higgs bundle (E, ϕ) on X with values in a line bundle L, and sends it to the corresponding n-tuple of "coefficients" of the characteristic polynomial of ϕ in the complex vector space $\bigoplus_{i=1}^n \Gamma(X, L^{\otimes i})$.

8.1.3. *The tautological section of p^*L*

The main goal of this subsection is to define the tautological section (8.1), and give a local description of the map. There are several equivalent ways to define it. One can use adjunction to identify the groups $\operatorname{Hom}_{\mathcal{O}_{\mathbb{L}}}(\mathcal{O}_{\mathbb{L}}, p^*L) = \operatorname{Hom}_{\mathcal{O}_X}(L^\vee, \operatorname{Sym}^\bullet L^\vee)$ and then use the tautological morphism of sheaves on the right (see Remark 8.2). Alternatively, one could consider the global section of the geometric line bundle $p^*\mathbb{L}$ on $\mathbb{L}$ given pointwise by assigning to $v \in \mathbb{L}$ the point v in the fiber of $p^*\mathbb{L}$ over v (see Remark 8.3). Finally, one can describe it from a morphism of geometric line bundles $T : \mathbb{A}^1_X \times_X \mathbb{L} \to \mathbb{L} \times_X \mathbb{L}$ over $\mathbb{L}$; since this is the how we will use the tautological section, we consider this approach in detail.

The $\mathcal{O}_X$-module structure map for the rank 1, locally free sheaf L

$$\mathcal{O}_X \times L \to L$$

induces a multiplication map

$$\mathbb{A}^1_X \times_X \mathbb{L} \xrightarrow{\ \mu\ } \mathbb{L},$$

as is seen easily from the following diagram, where U is an open set, and the dashed arrow indicates a section of $\mathcal{O}_X \times L$ over U:

$$\begin{array}{ccccc} & & \mathbb{A}^1_X \times_X \mathbb{L} & \xrightarrow{\mu} & \mathbb{L} \\ & \nearrow & \downarrow & & \downarrow \\ U & \hookrightarrow & X & = & X. \end{array}$$

This in turn induces a diagram

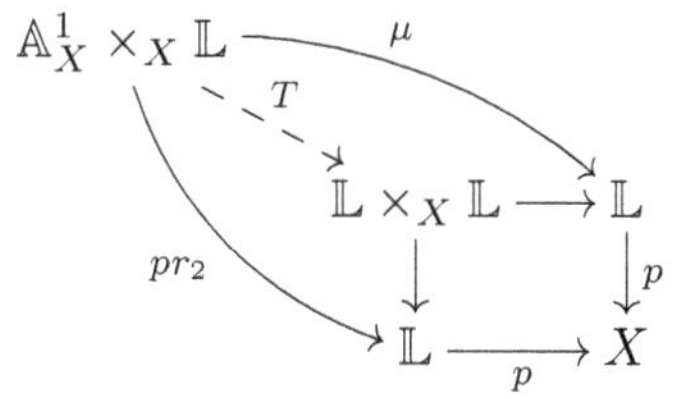

(8.3)

The map T is the geometric version of the tautological global section. More precisely, $\mathbb{A}^1_X \times_X \mathbb{L}$ is the pullback of the trivial geometric line bundle on X (i.e., $p^*\mathbb{A}^1_X$), and is hence the trivial geometric line bundle on $\mathbb{L}$ (i.e., $\mathbb{A}^1_{\mathbb{L}}$), and $\mathbb{L} \times_X \mathbb{L}$ is $p^*\mathbb{L}$. The associated morphism of sheaves is a morphism

$$\mathcal{O}_{\mathbb{L}} \xrightarrow{T} p^*L,$$

which corresponds to a global section $T \in H^0(\mathbb{L}, p^*L)$.

It can be useful to describe the tautological section locally. Let $U = \operatorname{Spec} R \subseteq X$ be an affine open subset. Assume that L is trivialized over U, corresponding to the trivial R-module R. We can identify the R-algebra $\operatorname{Sym}^\bullet L^\vee(U)$ with $R[T]$, so that $p : \mathbb{L} \to X$ is identified over U as $p : L|_U = \mathbb{A}^1_U = \operatorname{Spec} R[T] \to U = \operatorname{Spec} R$.

The multiplication map $\mu : \mathbb{A}^1_U \times_U \mathbb{L}|_U \to \mathbb{L}|_U$ is then identified with the R-algebra map

$$R[T, T'] = R[T] \otimes_R R[T'] \xleftarrow{\mu} R[T']$$

given by $T' \mapsto TT'$. The diagram (8.3) defining the geometric tautological section, i.e., the map $T : \mathbb{A}^1_U \times_U \mathbb{L}|_U \to \mathbb{L}|_U \times_U \mathbb{L}|_U$, is given at the level of R-algebras by the diagram:

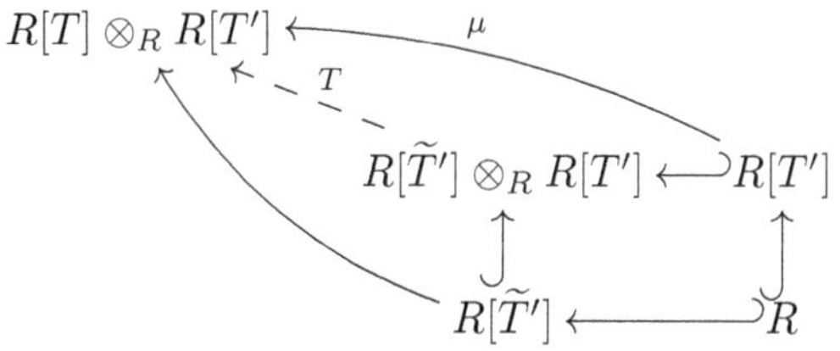

The tautological map T is given here by $T' \mapsto TT'$ and $\widetilde{T}' \mapsto T'$.

Now $p^{-1}(U) = \mathbb{L}|_U = \operatorname{Spec} R[T]$. We have that $\mathbb{A}^1_U \times_U \mathbb{L}|_U$ and $\mathbb{L}|_U \times_U \mathbb{L}|_U$ are geometric line bundles, which on $\mathbb{L}|_U$, where they are trivialized, correspond to the trivial rank 1, locally free module $R[T]$. The tautological section $T : \mathcal{O}_{\mathbb{L}} \to p^*L$ is given under these identifications by the multiplication map

$$R[T] \xrightarrow{\cdot T} R[T]. \tag{8.4}$$

Remark 8.2. Alternatively, one can describe the tautological section as follows. We have identifications

$$\mathrm{Hom}_{\mathcal{O}_{\mathbb{L}}}(\mathcal{O}_{\mathbb{L}}, p^*L) = \mathrm{Hom}_{\mathcal{O}_{\mathbb{L}}}(p^*\mathcal{O}_X, p^*L) = \mathrm{Hom}_{\mathcal{O}_X}(\mathcal{O}_X, p_*(\mathcal{O}_{\mathbb{L}} \otimes p^*L))$$

$$= \mathrm{Hom}_{\mathcal{O}_X}(\mathcal{O}_X, (\mathrm{Sym}^\bullet L^\vee) \otimes L) = \mathrm{Hom}_{\mathcal{O}_X}(L^\vee, \mathrm{Sym}^\bullet L^\vee).$$

There is a natural inclusion $L^\vee \hookrightarrow \mathcal{O}_X \oplus L^\vee \oplus (L^\vee)^{\otimes 2} \oplus \cdots$ onto the second factor, and this corresponds to the tautological section. Locally, if we set $U = \mathrm{Spec}\, R \subseteq X$ to be an open affine over which $\mathbb{L}$ is trivial, then $\mathrm{Sym}^\bullet L^\vee$ is identified with $R[T]$, and the natural map $L^\vee \to \mathrm{Sym}^\bullet L^\vee$ is associated to the map $R \to R[T]$ by multiplication by T.

Remark 8.3. One can also describe the tautological section geometrically as follows. The sheaf p^*L is the sheaf of sections of the line bundle $p^*\mathbb{L} := \mathbb{L} \times_X \mathbb{L} \to \mathbb{L}$, where the structure morphism to $\mathbb{L}$ is the first projection. There is a global section of the structure morphism given by the diagonal map $\mathbb{L} \to \mathbb{L} \times_X \mathbb{L}$. This is the tautological global section.

8.1.4. *The characteristic polynomial locally*

Let (E, ϕ) be a Higgs bundle on X with values in L. Let $U = \mathrm{Spec}\, R \subseteq X$ be an affine open subset. Assume that E is trivial over U, corresponding to R^n, and that L is also trivialized, corresponding to R. Then $\phi : E \to E \otimes L$ can be identified as a map $R^n \to R^n$, and thus with an $n \times n$ matrix (ϕ_{ij}) over R. We can also identify $\mathrm{Sym}^\bullet L^\vee$ with $R[T]$, so that $p : \mathbb{L} \to X$ is identified over U as $p : \mathbb{A}^1_U \to U$, and the tautological section $T : \mathcal{O}_{\mathbb{L}} \to p^*L$ is given under these identifications by the multiplication map

$$R[T] \xrightarrow{\cdot T} R[T].$$

The map $(T - p^*\phi) : p^*E \to p^*E \otimes L$ can then be identified over $\mathbb{L}|_U$ with the map $R[T]^n \to R[T]^n$ given by the matrix

$$T - p^*\phi = \begin{pmatrix} T - \phi_{11} & \cdots & -\phi_{1n} \\ \vdots & \ddots & \vdots \\ -\phi_{n1} & \cdots & T - \phi_{nn} \end{pmatrix}.$$

We then have

$$\det(T - p^*\phi)|_U = T^n - \mathrm{Tr}(\phi)T^{n-1} + \cdots + (-1)^n \det \phi \in R[T]. \tag{8.5}$$

Recall that globally we had $\det(T - p^*\phi) \in \Gamma(\mathbb{L}, p^*L^{\otimes n}) = \Gamma(X, p_*\mathcal{O}_{\mathbb{L}} \otimes L^{\otimes n}) = \bigoplus_{m \geq 0} T^m \Gamma(X, L^{\otimes(n-m)})$; the coefficients of the powers of T in this description and in (8.5) agree.

8.2. *Spectral covers and fibers of the Hitchin morphism*

Here we describe spectral covers, and the connection with the fibers of the Hitchin morphism. The main point for Higgs bundles on smooth curves are the results of [10, 73] reviewed in Remark 8.9. We give a weaker statement that holds in more generality in Proposition 8.4 and Lemma 8.6.

8.2.1. *Spectral covers*

Every $\sigma : \mathcal{O}_X \to \bigoplus_{i=1}^n L^{\otimes i}$ (corresponding to a map $\sigma : S \to \mathbb{A}_{X/S}(L)(n)$) determines via (8.2) a global section $\sigma : \mathcal{O}_{\mathbb{L}} \to p^*L^{\otimes n}$ of the line bundle $p^*L^{\otimes n}$ on $\mathbb{L}$, and consequently a zero set of the section:

$$\widetilde{X}(\sigma) := V(\sigma) \subseteq \mathbb{L}.$$

The map $\widetilde{X}(\sigma) \to X$ (obtained by composition from the map $p : \mathbb{L} \to X$) is called the *spectral cover associated to* σ. One can check (see the local computation below) that the spectral cover is a finite morphism of degree n. Associated to the scheme $\widetilde{X}(\sigma) = V(\sigma)$ is an ideal sheaf $\mathcal{I}_\sigma$, defined via the short exact sequence:

$$0 \to \mathcal{I}_\sigma \to \mathrm{Sym}^\bullet L^\vee \to \mathcal{O}_{V(\sigma)} \to 0. \tag{8.6}$$

More generally, the computation shows that for any $\sigma' : S' \to \mathbb{A}_{X/S}(L)(n)$, corresponding to $\sigma' : \mathcal{O}_{X'} \to \bigoplus_{i=1}^n L^{\otimes i}|_{X'}$, where $X' = X \times_S S'$, there is a corresponding spectral cover $p' : \widetilde{X}'(\sigma') \to X'$. In particular, there is a universal spectral cover

$$\widetilde{X}(\sigma_{\mathrm{id}_{\mathbb{A}_{X/S}(L)(n)}}) \to X \times_S \mathbb{A}_{X/S}(L)(n)$$

determined by the identity morphism on $\mathbb{A}_{X/S}(L)(n)$, from which all the spectral covers are obtained by pullback.

8.2.2. *Local description of the spectral cover*

To describe $\widetilde{X}(\sigma)$, it can be useful to consider the construction locally on X. Let $U = \operatorname{Spec} R \subseteq X$ be an affine open subset. Assume that L is trivialized over U, corresponding to the R-module R. Given $\sigma \in \bigoplus_{i=1}^{n} T^{n-i}\Gamma(X, L^{\otimes i})$, with components $\sigma_i \in \Gamma(X, L^{\otimes i})$, then we can write

$$\sigma|_U = T^n - \sigma_1|_U T^{n-1} + \cdots + (-1)^n \sigma_n|_U.$$

Since L, and hence $L^{\otimes i}$ is trivalized over U, we may view the $\sigma_i|_U$ as elements of R. The short exact sequence (8.6) is then written locally as

$$0 \longrightarrow R[T] \xrightarrow{\sigma|_U} R[T] \longrightarrow R[T]/\mathcal{I}_\sigma|_U \longrightarrow 0.$$

In other words, $\mathcal{I}_\sigma|_U$ is given by $(\sigma|_U) \subseteq R[T]$, and $\widetilde{X}(\sigma)|_U = \operatorname{Spec} R[T]/(\sigma|_U)$.

8.2.3. *Minimal ideals for endomorphisms*

Let X be a scheme, L a rank 1 locally free sheaf on X, and E a quasicoherent sheaf on X. A morphism of $\mathcal{O}_X$-modules

$$\phi : L^\vee \to \operatorname{End}(E)$$

is equivalent to a morphism of $\mathcal{O}_X$-algebras

$$\phi^\bullet : \operatorname{Sym}^\bullet L^\vee \to \operatorname{End}(E).$$

The *minimal ideal of* ϕ, denoted $\mathcal{I}_\phi$, is defined to be the kernel of $\phi^\bullet$. The *minimal cover* $\widetilde{X}(\phi) \to X$ associated to ϕ is defined to be the subscheme of $\mathbb{L}$ defined by $\mathcal{I}_\phi$. Note that E induces a quasicoherent sheaf M on $\widetilde{X}(\phi)$ such that the pushforward of M to X is equal to E. If E is locally finitely generated, the support of M is exactly $\widetilde{X}(\phi)$ (i.e., locally, the support is the set of primes containing the annihilator of the finitely generated module).

More generally, for any ideal sheaf $\mathcal{I} \subseteq \operatorname{Sym}^\bullet L^\vee$, a quasicoherent sheaf M on the scheme $V(\mathcal{I}) \subseteq \mathbb{L}$ is equivalent to a quasicoherent sheaf E on X together with a morphism $\phi : L^\vee \to \operatorname{End}(E)$ such that $\mathcal{I} \subseteq \mathcal{I}_\phi$; the identification is made via pushforward; i.e., E is the pushforward of M.

Locally on X, for locally free sheaves this is described as follows. Suppose that E is locally free of rank n, and is trivial over $U = \operatorname{Spec} R \subseteq X$. Then $\phi : L^\vee \to \operatorname{End}(E)$ induces an evaluation morphism $\phi^\bullet|_U : R[T] \to \operatorname{End}_R(R^n)$ sending T to $\phi|_U$. The kernel $\ker \phi^\bullet|_U$ is the restriction of the minimal ideal $\mathcal{I}_\phi|_U$. We have that R^n is an $R[T]/\mathcal{I}_\phi|_U$-module, with support $\operatorname{Spec} R[T]/\mathcal{I}_\phi|_U$. In addition, so long as $\sigma|_U \in \mathcal{I}_\phi|_U$, we can view R^n as an

$R[T]/(\sigma|_U)$-module, as well. In other words, E is obtained by push forward from a sheaf on the spectral cover associated to σ (although it may only be supported on the possibly smaller minimal cover associated to ϕ).

This discussion allows us to identify Higgs bundles with given minimal ideal with certain sheaves on the spectral cover.

Proposition 8.4 ([10, Prop. 3.6]). *Let $X \to S$ be a proper morphism of finite presentation between schemes with S excellent, and let L be a locally free sheaf of rank 1 on X. Given an S-morphism*

$$\sigma' : S' \to \mathbb{A}_{X/S}(L)(n),$$

to give a pair (E', ϕ') in $\mathcal{H}_{X/S}(L)(n)(S')$ such that ϕ' has minimal ideal $\mathcal{I}_{\phi'}$ equal to $\mathcal{I}_{\sigma'}$, it is equivalent to give a coherent sheaf M' on $\widetilde{X}'(\sigma')$ (where $X' = X \times_S S'$) such that the pushforward of M' to X' is a rank n locally free sheaf and the support of M' is $\widetilde{X}'(\sigma')$.

Proof. As we have seen above, $(E', \phi') \in \mathcal{H}_{X/S}(L)(n)(S')$ corresponds to a quasicoherent sheaf M' on $\mathbb{L}'$ (the total space of L') whose support is $\widetilde{X}'(\phi')$. But $\widetilde{X}'(\phi')$ coincides with $\widetilde{X}'(\sigma')$ if and only if $\mathcal{I}_{\sigma'} = \mathcal{I}_{\phi'}$. □

Remark 8.5. It is easy to see that every locally free sheaf M' of rank 1 on $\widetilde{X}'(\sigma)$ satisfies the conditions in Proposition 8.4. Note that one can easily construct examples (even with $\widetilde{X}'(\sigma')$ a union of smooth complex projective curves) where there are sheaves M' on $\widetilde{X}'(\sigma')$ that push forward to rank n locally free sheaves E' on X such that the induced endomorphism $\phi' : E' \to E' \otimes L'$ does not have minimal polynomial equal to σ'; this is the reason for the hypothesis on the support of M'.

8.2.4. *Fibers of the Hitchin map*

The following lemma asserts that the category of sheaves in Proposition 8.4 above is contained in the fiber of the Hitchin map.

Lemma 8.6. *A pair (E', ϕ') in $\mathcal{H}_{X/S}(L)(n)(S')$ such that the minimal ideal $\mathcal{I}_{\phi'}$ is equal to $\mathcal{I}_{\sigma'}$ has characteristic polynomial given by σ'. Moreover, if X' is integral and the fiber of $\widetilde{X}'(\sigma')$ over the generic point of X is geometrically reduced (e.g., X' is a variety over $\mathbb{C}$ and $\widetilde{X}'(\sigma')$ is reduced), then the converse holds.*

In other words, under these hypotheses, Proposition 8.4 describes the fiber of the Hitchin morphism in terms of sheaves on the spectral cover.

Proof. Let (E', ϕ') be a Higgs bundle as in the statement of the lemma, and let $p_{\phi'}(T)$ be the characteristic polynomial of ϕ'.

Assume first that $\mathcal{I}_{\phi'} = \mathcal{I}_{\sigma'}$. Viewing the characteristic polynomial as a global section $p_{\phi'}(T) : \mathcal{O}_{X'} \to \bigoplus_{i=1}^{n} L^{\otimes i}|_{X'}$ defines a locally principal ideal $\mathcal{I}_{p_{\phi'}(T)}$ of $\mathrm{Sym}^{\bullet} L^{\vee}$ (§8.2.1). The Cayley–Hamilton theorem says that $\mathcal{I}_{p_{\phi'}(T)} \subseteq \mathcal{I}_{\phi'}$. But we are assuming that $\mathcal{I}_{\phi'} = \mathcal{I}_{\sigma'}$. So we have $\mathcal{I}_{p_{\phi'}(T)} \subseteq \mathcal{I}_{\sigma'}$, with both being locally principal generated by monic polynomials of degree n. So we have reduced to the local statement: Given a ring R, and two principal ideals $(f(T))$ and $(g(T))$ in $R[T]$ with both $f(T)$ and $g(T)$ *monic* of degree n, if $(f(T)) \subseteq (g(T))$, then $f(T) = g(T)$. Thus the characteristic polynomial $p_{\phi'}(T)$ is given by σ.

Conversely, suppose X is integral, the fiber of $\widetilde{X}'(\sigma')$ over the generic point of X is geometrically reduced, and that the characteristic polynomial $p_{\phi'}(T)$ is given by σ'. We want to show that $\mathcal{I}_{\phi'} = \mathcal{I}_{\sigma'}$. It is enough to do this locally. In other words, we assume that R is an integral domain with field of fractions K contained in an algebraic closure $\overline{K}$, that $\sigma(T) \in R[T]$ is a monic polynomial of degree n, that $\overline{K}[T]/(\sigma(T))$ is reduced, and that $\phi : R^n \to R^n$ is an endomorphism with characteristic polynomial $\sigma(T)$. We want to show the containment $\mathcal{I}_{\phi} \supseteq (\sigma(T))$ is an equality. So consider $0 \neq f(T) \in \mathcal{I}_{\phi}$. The claim is that $f(T)$ is divisible by $\sigma(T)$ in $R[T]$. To see this, note that with $\phi_K : K^n \to K^n$ the induced morphism, we have that $f(\phi_K) = 0$. Thus $f(T)$ is divisible in $K[T]$ by the minimal polynomial for ϕ_K. Note also that from the local definition of the characteristic polynomial, it is clear that the characteristic polynomial of ϕ_K is the same as for ϕ, namely $\sigma(T)$. Our assumption that $\overline{K}[T]/(\sigma(T))$ is reduced, i.e., that the characteristic polynomial of ϕ_K has distinct roots, implies that the characteristic polynomial for ϕ_K agrees with the minimal polynomial. Thus $f(T)$ is divisible by $\sigma(T)$ in $K[T]$. But since $\sigma(T)$ is monic, one can then conclude (see the remark below) that $f(T)$ is divisible by $\sigma(T)$ in $R[T]$. □

Remark 8.7. *Let R be an integral domain and let K be its field of fractions. Suppose that $f(T), g(T) \in R[T]$ with $g(T)$* monic, *and there exists $h(T) \in K[T]$ such that $f(T) = g(T)h(T) \in K[T]$. Then $h(T) \in R[T]$.* Indeed, let us write

$$f(T) = \sum_{i=0}^{n+m} a_i T^i, \quad g(T) = \sum_{i=0}^{n} b_i T^i, \quad h(T) = \sum_{i=0}^{m} c_i T^i.$$

Then we have

$$\begin{aligned} a_{n+m} &= b_n c_m \\ a_{n+m-1} &= b_n c_{m-1} + b_{n-1} c_m \\ &\vdots \\ a_n &= b_n c_0 + b_{n-1} c_1 + \cdots + b_0 c_n. \end{aligned}$$

(Here c_j is taken to be 0 if $j > m$.) Since b_n is assumed to be 1, the first equality shows that $c_m \in R$. The second then shows that $c_{m-1} \in R$, and so on until we have established that $c_0 \in R$.

Example 8.8. It can be instructive to consider Proposition 8.4 and Lemma 8.6 the case where $X = S = \operatorname{Spec} \mathbb{C}$ and $L = \mathcal{O}_X$.

Remark 8.9. In the case where $\pi : X \to S$ is a smooth relative curve; i.e., a smooth proper morphism such that every geometric fiber is connected, and dimension 1, more can be said. If $\sigma' : S' = \operatorname{Spec}(k) \to \mathbb{A}_{X/S}(L)(n)$ is a geometric point inducing a reduced spectral curve $\widetilde{X}'(\sigma')$ over $\operatorname{Spec} k$, then $h^{-1}(\sigma')$ is the compactified Picard stack, parameterizing rank 1, torsion-free sheaves on $\widetilde{X}'(\sigma')$ [73, Prop. 5.1]. If $\widetilde{X}'(\sigma')$ is assumed further to be an integral curve, which is the case considered in [10, Prop. 3.6], then the key point is that in the notation of Proposition 8.4, for $\pi'_* M$ to be locally free, and thus torsion-free, it must be that M is torsion-free. Then, since X' is a smooth curve over k, any torsion-free sheaf is locally free, and thus for any torsion-free sheaf on X', the push-forward is locally free. The case of a general spectral curve $\widetilde{X}'(\sigma')$ is considered in [73]; the connection between slope stability of Higgs bundles on X' and slope stability of rank 1 torsion free sheaves on $\widetilde{X}'(\sigma')$ in the sense of Oda–Seshadri is also considered there. See also [64, Thm. 6.1].

9. Morphisms of stacks in algebraic geometry

As explained in Definition 6.10, many reasonable properties of morphisms in our presite S may be extended to S-representable morphisms of stacks. However, not every morphism of algebraic stacks that deserves to be called quasicompact or étale or smooth (to name just a few) is necessarily schematic.

In Section 6.5, we saw how to extend the notion of smoothness to morphisms between algebraic stacks, but this definition required the choice of a presentation of the stack in question. Since stacks representing moduli problems rarely come with an easily described presentation, it would be better to have an intrinsic definition of smoothness.

To give such a definition, as well as definitions of other geometric properties or morphisms between algebraic stacks, will be the purpose of this section.

Warning 9.1. We caution that although many of the definitions given here make sense for arbitrary morphisms of CFGs, or for arbitrary morphisms of stacks, they should not necessarily be regarded as reasonable generalizations from schemes without a further algebraicity assumption.

The characterization of smoothness we give in Section 9.7, for example, is only reasonable for morphisms of algebraic stacks (or at least morphisms representable by algebraic stacks) and not necessarily for all stacks.

9.1. *Injections, isomorphisms, and substacks*

In Definition 4.7 we gave the definition of an injection and of an isomorphism of stacks. We have not given a special name to morphisms of algebraic stacks that are only faithful (rather than being fully faithful or equivalences, respectively) when viewed as functors, because we already have one:

Lemma 9.2. *A morphism of algebraic stacks $f : \mathcal{X} \to \mathcal{Y}$ that is objectwise a faithful functor is representable by algebraic spaces.*

By saying $f : \mathcal{X} \to \mathcal{Y}$ is objectwise a faithful functor we mean that for each scheme S the induced functor of groupoids $f(X) : \mathcal{X}(S) \to \mathcal{Y}(S)$ is faithful.

Proof. Suppose that Z is an algebraic space and $Z \to \mathcal{Y}$ is a morphism. Let $\mathcal{X}_Z$ be the base change, which is algebraic by Corollary 6.32. The projection $\mathcal{X}_Z \to Z$ is also faithful. But if Z is viewed as a stack then, for any scheme S, the fiber $Z(S)$ is equivalent to a set. Since $\mathcal{X}_Z(S)$ is equivalent to a subcategory of $Z(S)$, this means that $\mathcal{X}_Z(S)$ is equivalent to a set, which means that $\mathcal{X}_Z$ is an algebraic space. □

Definition 9.3 (Open and closed substacks). Let $\mathcal{X}$ be a stack over S. A substack $\mathcal{U} \subseteq \mathcal{X}$ (resp. $\mathcal{Z} \subseteq \mathcal{X}$) is called an *open substack* (resp. *closed substack*) if for every scheme W, the fiber product $\mathcal{U} \times_{\mathcal{X}} W$ (resp. $\mathcal{Z} \times_{\mathcal{X}} W$) is representable by an open (resp. closed) subscheme of W.

9.2. *The underlying topological space*

Definition 9.4 (Topological space of a stack). Let $\mathcal{X}$ be a stack on the category S of schemes. For each scheme S, write $|S|$ for the underlying topological space of S. The *underlying topological space* of $\mathcal{X}$ is the universal topological space $|\mathcal{X}|$ that receives a continuous map $|S| \to |\mathcal{X}|$ for each map $S \to \mathcal{X}$.

In other words, $|\mathcal{X}| = \varinjlim_{S \to \mathcal{X}} |S|$ is the colimit of the spaces $|S|$, taken over all maps from schemes S to $\mathcal{X}$.

Remark 9.5. In [77, Tag 04XE], the underlying topological space was defined only for algebraic stacks, but [77, Tag 04XG] makes sense for arbitrary categories fibered in groupoids and agrees with the definition given here because the underlying set of a colimit of topological spaces is the colimit of the underlying sets.

The topology on the underlying set was only defined in [77, Tag 04XL] for algebraic stacks. However, the topology of *loc. cit.* agrees with Definition 9.4. Recall that a subset of $|\mathcal{X}|$ is called open if its preimage under a fixed flat, finite presentation, surjective map $U \to \mathcal{X}$ is open. This topology is clearly at least as fine than the topology we have defined, so we verify that every subset of $|\mathcal{X}|$ that is open in the sense of [77, Tag 04XL] is open in the sense of Definition 9.4.

Indeed, if $U \to \mathcal{X}$ is flat and locally of finite presentation then for any $V \to \mathcal{X}$, the map $U \times_{\mathcal{X}} V \to V$ is also flat and of finite presentation. The preimage in $|V|$ of the image of $|U|$ in $|\mathcal{X}|$ is the same as the image of $|U \times_{\mathcal{X}} V|$ in $|V|$. Since $U \times_{\mathcal{X}} V$ is flat of finite presentation over V, its image in $|V|$ is open, so the image of $|U|$ in $|\mathcal{X}|$ pulls back to an open subset of $|V|$. This holds for any $V \to \mathcal{X}$ so the image of $|U|$ in $|\mathcal{X}|$ is open, by definition of the colimit topology. As the topology in *loc. cit.* is uniquely characterized by this property, it must agree with $|\mathcal{X}|$, as defined in Definition 9.4.

9.3. *Quasicompact and quasiseparated morphisms*

Definition 9.6 (Quasicompact morphisms). We call a stack $\mathcal{X}$ over S *quasicompact* if every covering of $\mathcal{X}$ (Definition 6.2) by open substacks has a finite subcover. A morphism of stacks $\mathcal{X} \to \mathcal{Y}$ is *quasicompact* if $\mathcal{X} \times_{\mathcal{Y}} Z$ is quasicompact for all quasicompact schemes Z and all morphisms $Z \to \mathcal{Y}$. A morphism of stacks $\mathcal{X} \to \mathcal{Y}$ is *quasiseparated* if the diagonal morphism $\mathcal{X} \to \mathcal{X} \times_{\mathcal{Y}} \mathcal{X}$ is quasicompact.

9.4. *Separation and properness*

We will not actually discuss the separatedness or properness of morphisms of algebraic stacks in any examples in this survey, but we include the definitions for the sake of completeness.

Since smooth morphisms of schemes are always locally of finite type, Section 6.5 shows that there is a unique way to make sense of locally finite type morphisms of algebraic stacks that is stable under base change and composition and local to the source and target. Technically, this breaks our promise to give only *intrinsic* definitions in this section. However, in the noetherian situation, finite type coincides with finite presentation, which is characterized intrinsically in Section 9.6.

Definition 9.7 (Proper and separated morphisms of algebraic stacks). A morphism of *algebraic* stacks $\mathcal{X} \to \mathcal{Y}$ is said to be *proper* if it is an isomorphism or it is separated, of finite type, and universally closed. It is said to be separated if its diagonal $\mathcal{X} \to \mathcal{X} \times_{\mathcal{Y}} \mathcal{X}$ is proper.

Remark 9.8. Definition 9.7 is not as circular as it appears. The diagonal morphism of a morphism of algebraic stacks is representable by algebraic spaces (Lemma B.15, [77, Tag 04XS]), so the definition of separatedness for algebraic stacks depends only on the definition of properness for algebraic spaces. Iterating Definition 9.7, we see that definition of properness for algebraic spaces depends on separatedness for morphisms of algebraic spaces, and therefore, by iterating again, the definition depends on the definition of properness for diagonals of algebraic spaces. Continuing further, we see that we must define separatedness for diagonals of algebraic spaces. But the diagonal of a morphism of algebraic spaces is injective, so the definition ultimately depends on the definition of separatedness for injections of algebraic spaces. But the diagonal of an injection is an isomorphism, so is automatically proper.

9.5. *Formal infinitesimal properties*

Definition 9.9 (Infinitesimal extension [37, Def. (17.1.1)]). An *infinitesimal extension* or *nilpotent extension* of a scheme S is a closed embedding $S \subseteq S'$ such that the ideal of S in S' is nilpotent.

Lemma 9.10. *Suppose that $S \subseteq S'$ is an infinitesimal extension and that $T \to S$ is étale. Then there is an infinitesimal extension $T \subseteq T'$ and étale map $T' \to S'$ inducing T as the fiber product $T' \times_{S'} S$. Moreover, T' is unique up to unique isomorphism.*

Proof. Since T' will be unique when it is constructed, this is a local problem in the Zariski topology on S': if T' has been constructed over a suitable open cover, the uniquenes will imply that the various T's can be glued together. We can therefore work Zariski-locally in S', or equivalently in S, since S and S' have the same Zariski topology.

The same reasoning shows that we can work Zariski-locally in T as well. This permits us to assume that S, S', and T are all affine. By the Jacobian criterion (e.g., [77, Tag 00TA, 00T6]), we can assume that $S = \operatorname{Spec} A$ and that $T = \operatorname{Spec} B$ where $B = A[x_1, \dots, x_n]/(f_1, \dots, f_n)$ and the determinant $\det \frac{\partial f_i}{\partial x_j}$ is a unit of B. If $S' = \operatorname{Spec} A'$, we take $T' = \operatorname{Spec} B'$ where $B' = A'[x_1, \dots, x_n]/(g_1, \dots, g_n)$ where g_i is an arbitrary lift of f_i to a polynomial with coefficients in A'. Then $\det \frac{\partial g_i}{\partial x_j}$ reduces to $\det \frac{\partial f_i}{\partial x_j}$ in B. Since B' is an infinitesimal extension of B, this means that $\det \frac{\partial g_i}{\partial x_j}$ is a unit in B', so by the Jacobian criterion, B' is étale over A'.

This establishes that T' exists locally in the Zariski topology of S'. It remains to prove the uniqueness of T'. Suppose that T' and T'' are two such extensions. Then by the infinitesimal lifting criterion for étale maps, applied to the diagram

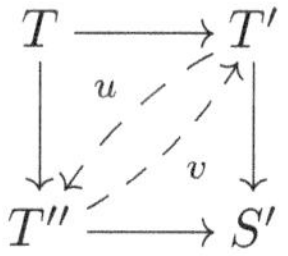

and there are unique lifts $u : T' \to T''$ and $v : T'' \to T'$. These must compose to the identity maps, again by the infinitesimal criterion for étale maps, this time applied to the diagrams below:

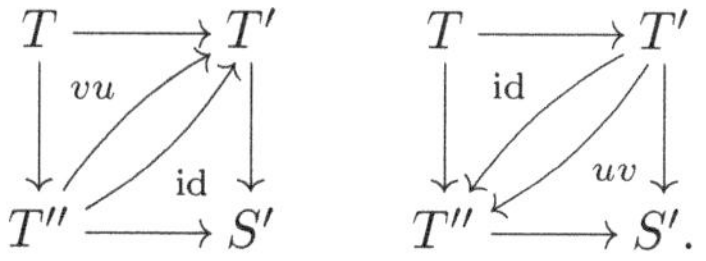

□

Corollary 9.11. *Suppose that $S \subseteq S'$ is an infinitesimal extension of schemes. Then the étale sites of S and S' are equivalent.*

Definition 9.12. A morphism of CFGs $f : \mathcal{X} \to \mathcal{Y}$ is said, respectively, to be *formally unramified*, *formally étale*, or *formally smooth* if every commutative diagram (9.1) below, in which $S \subseteq S'$ is an infinitesimal extension of schemes, admits at most one (up to unique isomorphism), exactly one (up to unique isomorphism), or at least one lift *étale-locally in* S.

$$\begin{array}{ccc} S & \longrightarrow & \mathcal{X} \\ \downarrow & \nearrow & \downarrow \\ S' & \longrightarrow & \mathcal{Y}. \end{array} \tag{9.1}$$

We explicate a bit the meaning of 'étale-locally' in the definition. To say that $\mathcal{X} \to \mathcal{Y}$ is formally smooth means that, given any lifting problem (9.1), there is an étale cover $\{S_\alpha \to S\}$ such that, denoting by S'_α the unique infinitesimal extension of S_α lifting S (by Lemma 9.10), the diagram

$$\begin{array}{ccccc} S_\alpha & \longrightarrow & S & \longrightarrow & \mathcal{X} \\ \downarrow & & \downarrow & \nearrow & \downarrow \\ S'_\alpha & \longrightarrow & S' & \longrightarrow & \mathcal{Y} \end{array} \tag{9.2}$$

admits a lift for every α.

To say that $\mathcal{X} \to \mathcal{Y}$ is formally unramified means, first, that given any two lifts of (9.1), there is a cover of S by S_α such that the induced lifts of (9.2) are isomorphic, and, second, that any two isomorphisms between lifts of (9.1) agree after passage to a suitable étale cover of S.

To be formally étale is the conjunction of these properties.

9.6. *Local finite presentation*

It was observed in [36, Prop. 8.14.2] that a scheme X is locally of finite presentation if and only if whenever $A = \varinjlim A_i$ is a filtered colimit of commutative rings, the natural map

$$\varinjlim X(A_i) \to X(A) \tag{9.3}$$

is a bijection. The same formula characterizes algebraic stacks that are locally of finite presentation, provided one interprets a filtered colimit of groupoids correctly. It is therefore reasonable to use (9.3) as the *definition* of local finite presentation for stacks that are not known to be algebraic.

To make sense of the filtered colimit of groupoids in equation (9.3), one can take

$$\mathsf{Obj} \varinjlim \mathcal{X}(A_i) = \bigcup_i \mathsf{Obj}\mathcal{X}(A_i)$$

and, for any objects $\xi \in \mathcal{X}(A_i)$ and $\eta \in \mathcal{X}(A_j)$,

$$\mathrm{Hom}(\xi, \eta) = \varinjlim_{k \geq i,j} \mathrm{Hom}_{\mathcal{X}(A_k)}(\xi_k, \eta_k)$$

where ξ_k and η_k denote pullbacks of ξ and η, respectively, to $\mathcal{X}(A_k)$.

In order to formulate local finite presentation for morphisms of stacks, it is useful to introduce the pro-object "$\varprojlim$" $\operatorname{Spec} A_i$ associated to a filtered system of commutative rings A_i. By definition, "$\varprojlim$" $\operatorname{Spec} A_i$ is the covariant functor on schemes (and stacks) obtained by taking the filtered colimit of the functors represented by the $\operatorname{Spec} A_i$. What this actually means is that one should interpret a morphism

$$\text{“}\varprojlim\text{”}\operatorname{Spec} A_i \to X$$

to a scheme (or stack) as an object of

$$\varinjlim \operatorname{Hom}(\operatorname{Spec} A_i, X) = \varinjlim X(A_i);$$

i.e., as compatible systems of morphisms $\operatorname{Spec} A_i \to X$.

Remark 9.13. In more technical terms, the quotation marks indicate that one is taking a colimit in the category of covariant functors valued in sets (or groupoids). We absolve ourselves of responsibility for the notation [2, p. 81].

Definition 9.14. A morphism $\mathcal{X} \to \mathcal{Y}$ of stacks in the étale topology on schemes is said to be *locally of finite presentation* if whenever $A = \varinjlim A_i$ is a filtered colimit of commutative rings, then every commutative diagram of solid lines (9.4) can be completed uniquely by a dashed arrow.

$$\begin{array}{ccc} \operatorname{Spec} A & \longrightarrow & \mathcal{X} \\ \downarrow & \nearrow & \downarrow \\ \text{“}\varprojlim\text{”}\operatorname{Spec} A_i & \longrightarrow & \mathcal{Y} \end{array} \tag{9.4}$$

Remark 9.15. Observe the resemblance between the lifting diagram (9.4) and the lifting diagram (9.1). This allows us to reason formally about étale maps and local finite presentation maps at the same time.

Under this analogy, Lemma 9.16 below is the analogue of Lemma 9.10.

Lemma 9.16. *Suppose that a ring A is the filtered colimit of rings A_ℓ. Set $S = \operatorname{Spec} A$ and $S_\ell = \operatorname{Spec} A_\ell$. Then for any affine*[a] *étale map $T \to S$ there is for some ℓ an affine scheme T_ℓ admitting an étale map $T_\ell \to S_\ell$, and this map is unique up to unique isomorphism and enlargement of ℓ.*

Proof. Suppose that $T = \operatorname{Spec} B$ is affine and étale over S. By the Jacobian criterion, there is an open cover of T by finitely many subsets $U = \operatorname{Spec} C$

[a]In fact, quasicompact and quasiseparated would suffice.

where $C = A[x_1, \ldots, x_n]/(f_1, \ldots, f_n)$ and $\det \frac{\partial f_i}{\partial x_j}$ is a unit of C. Since we are proving a uniqueness statement and only finitely many such subsets are involved, we may treat the subsets individually. We may therefore assume that $T = U$.

For k sufficiently large, the coefficients of the f_j all appear in the image of A_k in A, so that we may form the ring $B_k = A[x_1, \ldots, x_n]/(g_1, \ldots, g_n)$ for some lifts $g_1, \ldots, g_n$ of $f_1, \ldots, f_n$ to A_k. The determinant $\det \frac{\partial g_i}{\partial x_j}$ maps to the unit $\det \frac{\partial f_i}{\partial x_j}$ in B. Pick $t \in B$ inverse to $\det \frac{\partial f_i}{\partial x_j}$. As $B = \varinjlim_{\ell \geq k} B_\ell$, the element t must be in the image of some B_ℓ with $\ell \geq k$. Furthermore $t \frac{\partial g_i}{\partial x_j}$ must equal 1 in all sufficiently large B_ℓ. This implies that by taking ℓ sufficiently large, the image of $\det \frac{\partial g_i}{\partial x_j}$ is a unit. Thus B_ℓ is étale over A_ℓ for all sufficiently large ℓ, and B_ℓ induces B over A.

We must still prove that B_ℓ is unique up to enlargement of ℓ. Suppose that C_k were étale over A_k, also inducing B over A. Since B_ℓ is of finite presentation over A_ℓ and C_k is of finite presentation over A_k, the isomorphism $\varprojlim B_\ell \simeq B \simeq \varprojlim C_k$ must come from an isomorphism $B_m \simeq C_m$ defined over some m that is $\geq k$ and $\geq \ell$. This map is necessarily unique up to further enlargement of m (see [36, Thm. (8.8.2) (i)] for more details). $\square$

Definition 9.17. By an étale map "$\varprojlim$"$U_i \to$ "$\varprojlim$" $\operatorname{Spec} A_i$, we mean a family of maps $U_i \to \operatorname{Spec} A_i$ that are étale for all sufficiently large i.

In terms of this definition, Lemma 9.16 says any étale map $U \to$ "$\varprojlim$" $\operatorname{Spec} A_i$ is induced from an étale map "$\varprojlim$"$U_i \to$ "$\varprojlim$" $\operatorname{Spec} A_i$, and that this map is unique up to a unique, suitably defined, isomorphism.

Remark 9.18. Lemma 9.16 says, in a sense that we do not attempt to make precise, that the étale site of $\operatorname{Spec} A$ is the colimit of the étale sites of the $\operatorname{Spec} A_i$, whenever A is the filtered colimit of the A_i. See [3, Thm. 8.3.13].

9.7. *Smooth, étale, and unramified morphisms*

Definition 9.19 (Unramified, étale, smooth][37, Def. (17.3.1)]). A morphism of algebraic stacks $f : \mathcal{X} \to \mathcal{Y}$ is said, respectively, to be *unramified*, *étale*, or *smooth* if it is locally of finite presentation and formally unramified, formally étale, or formally smooth.

We have now defined smooth morphisms between algebraic stacks in two ways. On one hand, we have defined smooth morphisms in Definition 6.23 as the unique extension of smoothness from morphisms of schemes in a way that is stable under composition and base change and local to the source and target. On the other hand, we have defined smoothness intrinsically for morphisms of algebraic stacks in Definition 9.19.

We are obliged to verify that they are equivalent. It is sufficient to show that local finite presentation and smoothness are stable under composition and base change and local to the source and target, as these definitions clearly agree with the usual definitions in the category of schemes and the extension in Definition 6.23 was uniquely characterized by these properties.

Lemma 9.20. *Let* **P** *be one of the following properties of morphisms of stacks in the étale topology on schemes:*

(i) formal smoothness,
(ii) formal unramifiedness,
(iii) formal étaleness, or
(iv) local finite presentation.

Then **P** *is stable under composition and base change and is local to the source and target.*

Proof. We omit the verification for composition and base change, since these are formal and straightforward.

All of these properties may be phrased in terms of existence or uniqueness (or both) of lifts of a diagram

$$\begin{array}{ccc} S & \longrightarrow & \mathcal{X} \\ \downarrow & \nearrow & \downarrow \\ S' & \longrightarrow & \mathcal{Y} \end{array} \tag{9.5}$$

In the case of formal unramifiedness, formal étaleness, or formal smoothness, S' is an infinitesimal extension of S. In the case of local finite presentation, we have a ring A that is the filtered colimit of rings A_i, and $S = \operatorname{Spec} A$ and $S' = \text{“}\varprojlim\text{”} \operatorname{Spec} A_i$.

The only fact we will use here is that if $S \to S'$ is the left side of one of these diagrams and $T \to S$ is an étale map, then there is a unique (up to unique isomorphism) extension of T to an étale map $T' \to S'$. In the case where $S \to S'$ is an infinitesimal extension, this is Lemma 9.10; in the case where $S = \operatorname{Spec} A$ and $S' = \text{“}\varprojlim\text{”} \operatorname{Spec} A_i$ with the A_i filtered and $\varinjlim A_i = A$, this is Lemma 9.16.

We prove locality to the target. Consider a lifting problem (9.5), and assume that $\mathcal{Y}' \to \mathcal{Y}$ is a covering map such that the base change $\mathcal{X}' \to \mathcal{Y}'$ is **P**. We may freely replace S by an étale cover, so we may assume that the map $S' \to \mathcal{Y}$ factors through $\mathcal{Y}'$. This induces a factorization of $S \to \mathcal{X}$ through $\mathcal{X}'$. Property **P** for $\mathcal{X}'$ over $\mathcal{Y}'$ gives a lift f in diagram (9.6), which yields g by composition.

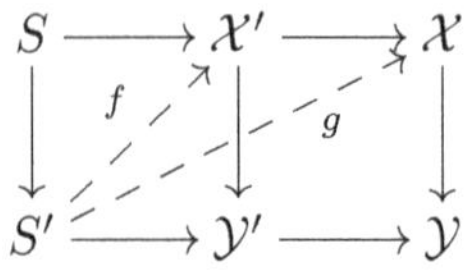

(9.6)

Now we prove locality to the source. Again, consider a lifting problem (9.5) and suppose that $X_0 \to X$ is a cover that is formally smooth over $\mathcal{Y}$. Since we can replace S by an étale cover, we can assume that $S \to \mathcal{X}$ factors through X_0. Formal smoothness of X_0 over $\mathcal{Y}$ gives a lift f in diagram (9.7), which gives us g by composition.

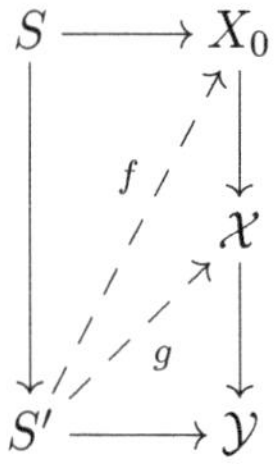

(9.7)

□

10. Infinitesimal deformation theory

In this section we introduce some deformation theory, with a view towards Artin's criterion for a stack to be algebraic. Roughly speaking, deformation theory is the study of families of objects over Artin rings. To get a concrete idea, consider the Kodaira–Spencer approach to deforming a complex manifold X over the unit disk by extending the given transition functions τ_{ij} of the manifold to transition functions $\widetilde{\tau}_{ij} = \tau_{ij} + \tau_{ij}^{(1)}t + \tau_{ij}^{(2)}t^2 + \cdots$, where t is the parameter on the disk (for each fixed t, one obtains a manifold with the given transition functions). Deformation theory would then be the problem of iteratively accomplishing this formally, by first extending up to t (e.g., $\tau_{ij} + \tau_{ij}^{(1)}t \mod t^2$), and then extending up to t^2, and so on. In the process, one would observe that the first order deformations are governed by $H^1(X, T_X)$ (called the tangent space), and that once one has extended to first order, the ability to extend to second order would be governed by $H^2(X, T_X)$ (called an obstruction space).

In general, deformation theory aims to abstract this to the setting of any stack over schemes (and to deforming over bases other than the unit disk). The heart of the matter turns out to be defining the tangent space to a stack, and an obstruction theory to a stack. We take an abstract point of view and identify precisely the condition, homogeneity, on a CFG that gives it a well-behaved tangent space. The advantage of this abstraction is that the existence of a well-behaved tangent space, and obstruction theory, can be used to prove that a stack is algebraic, as we will discuss in Section 11.

10.1. *Homogeneous categories fibered in groupoids*

After introducing homogeneity and demonstrating its basic properties, our goal will be to show that the stack of Higgs bundles is homogeneous, without relying on its algebraicity. In this first section we stick to the abstract setting, and also introduce the tangent space to a stack. We discuss the pertinent deformation theory in the following section, §10.2, where we establish that the stack of Higgs bundles in homogeneous.

10.1.1. *The tangent bundle of a stack*

The following definition reprises Definition 9.9:

Definition 10.1 (Square-zero extension). An *infinitesimal extension* of schemes is a closed embedding $S \subseteq S'$ such that the ideal $I_{S/S'}$ is nilpotent. A *square-zero extension* is an infinitesimal extension $S \subseteq S'$ such that $I^2_{S/S'} = 0$.

The utility of square-zero extensions is twofold: every infinitesimal extension can be factored as a sequence of square-zero extensions, and square-zero extensions behave 'linearly', in the sense that deformations and obstructions over square-zero extensions are classified by linear-algebraic data. These observations may be viewed as a functorial perspective on Taylor series.

The most important examples of infinitesimal extensions are the trivial ones:

Definition 10.2 (Dual numbers). The *ring of dual numbers* is $\mathbb{D} = \mathbb{Z}[\epsilon]/(\epsilon^2)$. For any scheme S, we write $S[\epsilon] = S \times_{\operatorname{Spec}\mathbb{Z}} \operatorname{Spec}\mathbb{D}$. More generally, if V is a quasicoherent sheaf of $\mathcal{O}_S$-modules, we write $\mathbb{D}(V)$ for the sheaf of $\mathcal{O}_S$-algebras,

$$\mathcal{O}_S + \epsilon V = \operatorname{Sym}^\bullet(\epsilon V)/(\epsilon^2 \operatorname{Sym}^2 V),$$

whose underlying sheaf of abelian groups consists locally of symbols $f + \epsilon v$ with $f \in \mathcal{O}_S$ and $v \in V$ and has multiplication law

$$(f + \epsilon v)(g + \epsilon w) = fg + \epsilon(fw + gv).$$

We write $S[\epsilon V]$ for the scheme whose underlying topological space is S and whose sheaf of rings is $\mathbb{D}(V)$; i.e., $S[\epsilon V] = \underline{\mathrm{Spec}}_S \mathbb{D}(V)$.

There is a canonical morphism $S[\epsilon V] \to S$ corresponding to the following homomorphism of sheaves of rings:

$$\begin{aligned} \mathcal{O}_S &\to \mathcal{O}_S + \epsilon V \\ f &\mapsto f + 0\epsilon. \end{aligned}$$

Unless otherwise specified, when it is necessary to equip $S[\epsilon V]$ with the structure of a scheme over S, we do so with this morphism.

Remark 10.3. The construction of $S[\epsilon V] \to S$ commutes with base change in S. It can therefore be extended to apply to any CFG, as follows. If $\mathcal{X}$ is a CFG and V is a quasicoherent sheaf on $\mathcal{X}$, then we define $\mathcal{X}[\epsilon V](S)$ to the category of pairs (ξ, δ) where $\xi \in \mathcal{X}(S)$ and δ is a section of $S[\epsilon\xi^* V]$ over S. When $V = \mathcal{O}_{\mathcal{X}}$, we write $\mathcal{X}[\epsilon]$ rather than $\mathcal{X}[\epsilon\mathcal{O}_{\mathcal{X}}]$; in this case

$$\mathcal{X}[\epsilon] = \mathcal{X} \times_{\mathrm{Spec}\,\mathbb{Z}} \mathrm{Spec}\,\mathbb{D}.$$

Definition 10.4 (Tangent bundle). Let $\mathcal{X}$ be a CFG over S/S. We give $R[\epsilon]$ the structure of an S-scheme via the canonical projection $R[\epsilon] \to R \to S$, and the *relative tangent bundle* of $\mathcal{X}$ over S is the category fibered in groupoids $T_{\mathcal{X}/S}$ over S/S with

$$T_{\mathcal{X}/S}(R) = \mathcal{X}\big(R[\epsilon]\big)$$

for each S-scheme R.

Remark 10.5. If $\mathcal{X}$ is a presheaf, then $T_{\mathcal{X}/S}$ is also a presheaf. In the case of $S = \mathrm{Spec}\,\mathbb{Z}$ we write $T_{\mathcal{X}} = T_{\mathcal{X}/\,\mathrm{Spec}\,\mathbb{Z}}$.

The idea to study the tangent space this way, and to think of the spectrum of the ring of dual numbers as a pair of infinitesimally nearby points, goes back at least to Weil [81, §2].

Proposition 10.6. *For any morphism of schemes $X \to S$, we have*

$$T_{X/S} = \underline{\mathrm{Spec}}_X(\mathrm{Sym}^\bullet \Omega_{X/S}).$$

Proof. This reduces to the affine situation, where it comes down to the following identities: given a ring k and algebras $k \to A \xrightarrow{\phi} B$, then

$$\mathrm{Hom}_{k\text{-alg},\phi}(A, B[\epsilon]) = \mathrm{Der}_k(A, B) = \mathrm{Hom}_{A\text{-mod}}(\Omega_{A/k}, B)$$

$$= \mathrm{Hom}_{A\text{-alg}}(\mathrm{Sym}^{\bullet}\, \Omega_{A/k}, B),$$

where the first group of homomorphisms consists of the k-algebra homomorphisms that reduce to ϕ modulo ϵ. □

Corollary 10.7. *When X is a smooth scheme over $\mathbb{C}$, then $T_{X/\operatorname{Spec}\mathbb{C}}$ is a vector bundle over X and coincides with any familiar definition of the tangent bundle.*

Corollary 10.8. *When $\mathcal{X}$ is an algebraic stack over S, so is $T_{\mathcal{X}/S}$.*

Proof. It is almost immediate that $T_{\mathcal{X}/S}$ is a stack in the étale topology: If we have an étale cover of R by U_i then the $U_i[\epsilon]$ form an étale cover of $R[\epsilon]$. Étale descent for the maps $U_i[\epsilon] \to \mathcal{X}$ to $R[\epsilon] \to \mathcal{X}$ yields étale descent for $U_i \to T_{\mathcal{X}/S}$ to $R \to T_{\mathcal{X}/S}$. To see that $T_{\mathcal{X}/S}$ is an *algebraic* stack, note that if $X_0 \to \mathcal{X}$ is a smooth cover of $\mathcal{X}$ by a scheme then $T_{X_0/S} \to T_{\mathcal{X}/S}$ is also a smooth cover. □

Warning 10.9. We will see in a moment (item (i) on p. 296) that there is a projection $T_{\mathcal{X}/S} \to \mathcal{X}$, as one expects, but that it is not necessarily representable by algebraic spaces! This is because the objects parameterized by $\mathcal{X}$ may possess *infinitesimal automorphisms.* Nevertheless we will see that the fibers of $T_{\mathcal{X}/S}$ over $\mathcal{X}$ behave like 'groupoids with vector space structure'. This is why we insist on calling $T_{\mathcal{X}/S}$ the tangent *bundle*: it has all the features of a vector bundle except for being a set!

10.1.2. *Homogeneity*

Theorem 10.10 ([82, Prop. 2.1]). *Let S be a scheme. Suppose that $Q \subseteq Q'$ is an infinitesimal extension of S-schemes and $f : Q \to R$ is an* affine *S-morphism. Then there is a universal (initial) S-scheme R' completing diagram* (10.1)*:*

$$\begin{array}{ccc} Q & \xrightarrow{\ \textit{inf.}\ } & Q' \\ {\scriptstyle \textit{affine}}\ \big\downarrow f & & \big\downarrow \\ R & \dashrightarrow & R' \end{array} \tag{10.1}$$

Furthermore, R' is also universal (initial) among algebraic S-stacks completing the diagram. The underlying topological space of R' is the same as that of R and viewing $\mathcal{O}_{Q'}$ and $\mathcal{O}_{R'}$ as sheaves on Q and R respectively,

$$\mathcal{O}_{R'} = \mathcal{O}_R \underset{f_*\mathcal{O}_Q}{\times} f_*\mathcal{O}_{Q'}.$$

In particular, if $R = \operatorname{Spec} A$ (hence Q and Q' are also affine, say $Q = \operatorname{Spec} B$ and $Q' = \operatorname{Spec} B'$) then

$$R' = \operatorname{Spec}(A \underset{B}{\times} B').$$

The universality here means that for any *algebraic stack $\mathcal{X}$* the natural map displayed in equation (10.2) is an equivalence of groupoids:

$$\mathcal{X}(R') \to \mathcal{X}(R) \underset{\mathcal{X}(Q)}{\times} \mathcal{X}(Q') \tag{10.2}$$

Proof. We will prove this in the case where R is also an infinitesimal extension of Q, which implies that the étale sites of Q, Q', R, and R' are all equivalent (Corollary 9.11).

First, note that schemes are homogeneous. Indeed, suppose we have a commutative diagram

$$\begin{array}{ccc} Q & \hookrightarrow & Q' \\ \downarrow & & \downarrow \\ R & \longrightarrow & X \end{array}$$

where X is a scheme. All of Q, Q', and R have the same underlying topological space, so we denote by f the continuous map from that space to the underlying topological space of X. The maps of schemes give homomorphisms of sheaves of rings:

$$\begin{array}{ccc} f^{-1}\mathcal{O}_X & \longrightarrow & \mathcal{O}_{Q'} \\ \downarrow & & \downarrow \\ \mathcal{O}_R & \longrightarrow & \mathcal{O}_Q \end{array}$$

By the universal property of $\mathcal{O}_{R'} = \mathcal{O}_R \times_{\mathcal{O}_Q} \mathcal{O}_{Q'}$, there is a unique factorization $f^{-1}\mathcal{O}_X \to \mathcal{O}_{R'}$, which gives the desired, uniquely determined, map $R' \to X$.

To extend this to algebraic stacks, it is sufficient by Definition 6.27 to show that *whenever $\mathcal{X}$ is a stack admitting a smooth cover by a homogeneous stack, $\mathcal{X}$ is homogeneous.*

Suppose that we have a commutative diagram

$$\begin{array}{ccc} Q & \hookrightarrow & Q' \\ \downarrow & & \downarrow \\ R & \longrightarrow & \mathcal{X} \end{array}$$

where $\mathcal{X}$ has a smooth cover $X_0 \to \mathcal{X}$ and X_0 is homogeneous. Since we are trying to prove a uniqueness assertion, it is sufficient to work étale-locally on Q. Replace Q by an étale cover such that the map $Q \to \mathcal{X}$ factors through X_0. Since X_0 is formally smooth over $\mathcal{X}$, we can assume, after refining the cover further, that the maps $Q' \to \mathcal{X}$ and $R \to \mathcal{X}$ also lift to X_0, extending the lift already constructed over Q. By the homogeneity of X_0, there is a unique map $R' \to X_0$ compatible with the maps from Q, Q', and R. Composing with the projection to $\mathcal{X}$ gives the desired map $R' \to \mathcal{X}$. □

Definition 10.11 (Pushout). Suppose that $Q \subseteq Q'$ is an infinitesimal extension of S-schemes and $f : Q \to R$ is an affine S-morphism. We refer to the scheme R' in Theorem 10.10 as the *pushout* of $Q \subseteq Q'$ along the map $f : Q \to R$.

A CFG is said to be homogeneous if it behaves like an algebraic stack with respect to (10.2):

Definition 10.12 (Homogeneous CFG). A CFG (not necessarily an algebraic stack) is said to be *homogeneous* if for every pushout as in Definition 10.11, we have (10.2) is an equivalence of categories.

Remark 10.13. This definition is a natural extension of [1, Def. VI.2.5], which was stated only in the context of artinian algebras. For much of what we have to say, we will only be interested in homogeneity with respect to morphisms $Q \to R$ that are isomorphisms on topological spaces. However, in practice, it is little more difficult to verify homogeneity in the generality we have formulated it.

As we have seen in Theorem 10.10, homogeneity is a necessary condition for a stack over schemes to be algebraic. It is not sufficient, but in the presence of a reasonable finiteness condition, it does guarantee *formal* representability by an algebraic stack (see Theorem 11.10).

The following theorem is very helpful for checking homogeneity. It was asserted by Schlessinger in [74, Lem. 3.4] but proved there only for free modules; Milnor gave a proof for projective modules [61, Thms. 2.1–2.3]; a proof in the general case can be found in [25, Thm. 2.2].

Theorem 10.14 (Schlessinger, Milnor, Ferrand). *Let $\mathcal{Q}$ be the CFG whose fiber over a scheme S is the category of flat, quasicoherent $\mathcal{O}_S$-modules. Then $\mathcal{Q}$ is homogeneous.*

In language closer to the original statements, this says that if one has a cartesian diagram of commutative rings

$$\begin{array}{ccc} B & \xleftarrow{\text{inf.}} & B' \\ \uparrow & & \uparrow \\ A & \longleftarrow & A' \end{array}$$

in which $B' \to B$ (and therefore also $A' \to A$) is a nilpotent extension then it is equivalent to specify either of the following data:

(i) a flat A'-module M', or
(ii) a flat A-module M, a flat B'-module N', a flat B-module N, and an A-module map $M \to N$ and a B'-module map $N' \to N$ inducing by adjunction isomorphisms $M \otimes_A B \to N$ and $N' \otimes_{B'} B \to N$ of B-modules.

One direction of the correspondence sends M' to $M = M' \otimes_{A'} A$, $N' = M' \otimes_{A'} B'$, $N = M' \otimes_{A'} B$ with the canonical maps $M' \otimes_{A'} A \to M' \otimes_{A'} B$ and $M' \otimes_{A'} B' \to M' \otimes_{A'} B$. The other direction has $M' = M \times_N N'$.

10.1.3. *Tangent bundle to a homogeneous stack*

To begin to appreciate the significance of homogeneity, consider that in general, the tangent bundle of a stack, and even the tangent bundle of a presheaf, need not be a very well behaved object: it might not even be a vector space. However, the tangent space of a *homogeneous* stack has all of the familiar structure one expects:

(i) (projection to base) The map $\mathbb{Z}[\epsilon]/(\epsilon^2) \to \mathbb{Z}$ sending ϵ to 0 induces closed embeddings $R \to R[\epsilon]$ for all schemes R. For any S-scheme R, we use this to obtain a morphism $T_{\mathcal{X}/S}(R) = \mathcal{X}(R[\epsilon]) \to \mathcal{X}(R)$, whence a projection $T_{\mathcal{X}/S} \to \mathcal{X}$. (This does not require homogeneity.)
(ii) (zero section) The map $\mathbb{Z} \to \mathbb{Z}[\epsilon]/(\epsilon^2)$ induces $R[\epsilon] \to R$. This gives a map $\mathcal{X}(R) \to \mathcal{X}(R[\epsilon]) = T_{\mathcal{X}/S}(R)$, whence a section $\mathcal{X} \to T_{\mathcal{X}/S}$ of the aforementioned projection. This is the zero section in the vector

bundle when $T_{\mathcal{X}/S}$ is a vector bundle. (This does not require homogeneity.)

(iii) (addition law) We have $\mathbb{Z}[\epsilon_1, \epsilon_2]/(\epsilon_1^2, \epsilon_1\epsilon_2, \epsilon_2^2) \xrightarrow{\sim} \mathbb{D} \times_{\mathbb{Z}} \mathbb{D}$ given by

$$x + y\epsilon_1 + z\epsilon_2 \mapsto (x + y\epsilon_1, x + z\epsilon_2).$$

In addition to the two projections $\mathbb{D} \times_{\mathbb{Z}} \mathbb{D} \to \mathbb{D}$, there is a third map sending both ϵ_1 and ϵ_2 to ϵ. We obtain a commutative diagram whose outer square is cartesian (in fact the upper right and lower left are cartesian as well):

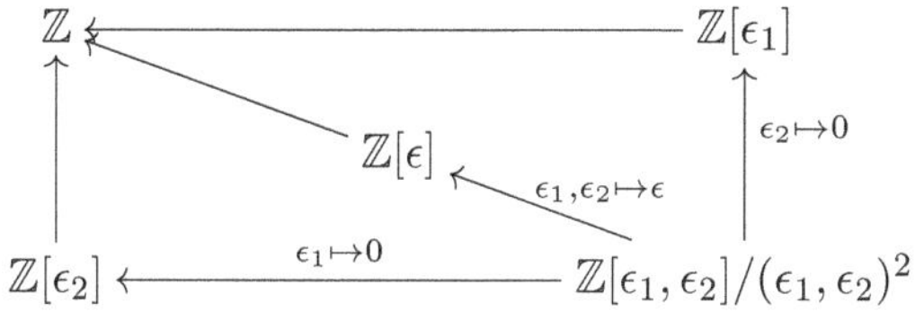

Applying this to any S-scheme R, we obtain a pushout diagram with an extra morphism $\sigma : R[\epsilon] \to R[\epsilon_1, \epsilon_2]$, where $R[\epsilon_1, \epsilon_2] = R \times \operatorname{Spec}(\mathbb{D} \times_{\mathbb{Z}} \mathbb{D})$ (it is also the vanishing locus of $\epsilon_1\epsilon_2$ in $R[\epsilon_1][\epsilon_2]$):

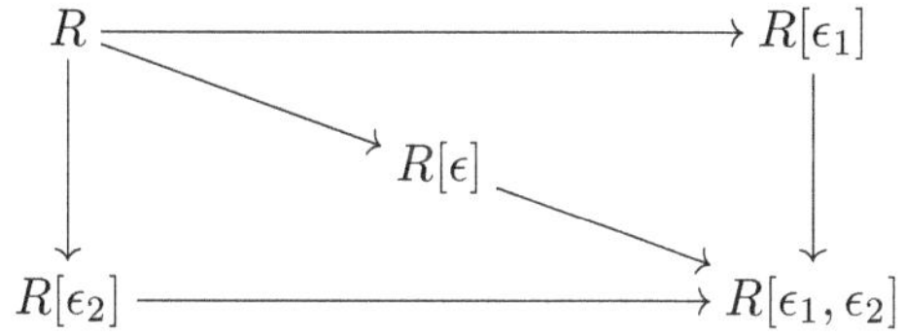

Now consider maps from the diagram above into our homogeneous stack $\mathcal{X}$; the map $\mathcal{X}(R[\epsilon_1, \epsilon_2]) \to \mathcal{X}(R[\epsilon_1]) \times_{\mathcal{X}(R)} \mathcal{X}(R[\epsilon_2])$ is an equivalence of categories.

Combining these facts, and abbreviating $T_{\mathcal{X}/S}$ to T, we have a diagram:

$$T(R) \underset{\mathcal{X}(R)}{\times} T(R) = \mathcal{X}(R[\epsilon_1]) \underset{\mathcal{X}(R)}{\times} \mathcal{X}(R[\epsilon_2]) \xleftarrow{\sim} \mathcal{X}(R[\epsilon_1, \epsilon_2]) \xrightarrow{\mathcal{X}(\sigma)} \mathcal{X}(R[\epsilon]) = T(R)$$

This gives the fibers of $T_{\mathcal{X}/S}(R) \to \mathcal{X}(R)$ an addition law; i.e., given an R-point $R \to \mathcal{X}$, the collection of all extensions $R[\epsilon] \to \mathcal{X}$ has an induced addition law. This coincides with the addition law of the vector space structure in the familiar situation where $\mathcal{X}$ is a smooth scheme and R is the spectrum of a field.

(iv) (scalar multiplication) If $\lambda \in \Gamma(R, \mathcal{O}_R)$, there is an induced map:

$$\begin{aligned} \mathcal{O}_R + \epsilon\mathcal{O}_R &\longrightarrow \mathcal{O}_R + \epsilon\mathcal{O}_R \\ f + \epsilon v &\longmapsto f + \epsilon\lambda v \end{aligned}$$

giving a map $R[\epsilon] \to R[\epsilon]$. Applying $\mathcal{X}$, this induces a map

$$T_{\mathcal{X}/S}(R) \to T_{\mathcal{X}/S}(R)$$

commuting with the projection to $\mathcal{X}(R)$. This is the action of scalars in the vector space structure when R is a point and $\mathcal{X}$ is smooth.

(v) (vector space structure) The axioms of a vector space can be verified in similar ways, occasionally making use of $\mathbb{D} \times_{\mathbb{Z}} \mathbb{D} \times_{\mathbb{Z}} \mathbb{D} = \mathbb{Z}[\epsilon_1, \epsilon_2, \epsilon_3]/(\epsilon_1, \epsilon_2, \epsilon_3)^2$. We leave these verifications to the reader.

(vi) (Lie bracket) The Lie bracket will not be used in the rest of the chapter, and may be skipped. There is a map

$$\mathbb{D} \underset{\mathbb{Z}}{\otimes} \mathbb{D} = \mathbb{Z}[\epsilon_1, \epsilon_2]/(\epsilon_1^2, \epsilon_2^2) \to \mathbb{Z}[\epsilon_1, \epsilon_2]/(\epsilon_1, \epsilon_2)^2 = \mathbb{D} \underset{\mathbb{Z}}{\times} \mathbb{D}$$

sending $\epsilon_1\epsilon_2$ to 0. There is also an isomorphism

$$(\mathbb{D} \underset{\mathbb{Z}}{\otimes} \mathbb{D}) \underset{\mathbb{Z}[\epsilon_1,\epsilon_2]/(\epsilon_1,\epsilon_2)^2}{\times} (\mathbb{D} \underset{\mathbb{Z}}{\otimes} \mathbb{D}) \cong (\mathbb{D} \underset{\mathbb{Z}}{\otimes} \mathbb{D}) \underset{\mathbb{Z}}{\times} \mathbb{Z}[\epsilon_1\epsilon_2]/(\epsilon_1\epsilon_2)^2$$

sending (a, b) on the left to $(a, (a \bmod (\epsilon_1, \epsilon_2)) + b - a)$ on the right (this choice of isomorphism will give rise to the choice of a sign for the Lie bracket). This all leads to a commutative diagram, in which both the small square and the outer rectangle are cartesian:

$$\begin{array}{ccccc} \mathbb{Z}[\epsilon_1, \epsilon_2] & \longleftarrow & \mathbb{Z}[\epsilon_1][\epsilon_2] & & \mathbb{Z}[\epsilon_1\epsilon_2] \\ \uparrow & & \uparrow & & \uparrow \\ \mathbb{Z}[\epsilon_1][\epsilon_2] & \longleftarrow & \mathbb{Z}[\epsilon_1][\epsilon_2] \times_{\mathbb{Z}[\epsilon_1,\epsilon_2]} \mathbb{Z}[\epsilon_1][\epsilon_2] & \overset{\sim}{\longrightarrow} & \mathbb{Z}[\epsilon_1][\epsilon_2] \times_{\mathbb{Z}} \mathbb{Z}[\epsilon_1\epsilon_2] \end{array} \tag{10.3}$$

(with a curved arrow $\mathbb{Z}[\epsilon_1\epsilon_2] \to \mathbb{Z}[\epsilon_1, \epsilon_2]$ across the top)

Suppose $v, w \in T_{\mathcal{X}}(\mathcal{X})$ are vector fields on $\mathcal{X}$. We view them as maps (see Definition 10.2 and Remark 10.3 for notation):

$$v : \mathcal{X}[\epsilon_1] \to \mathcal{X}$$
$$w : \mathcal{X}[\epsilon_2] \to \mathcal{X}.$$

We obtain two morphisms

$$\mathrm{id} \times v : \mathcal{X}[\epsilon_1] \times_{\mathcal{X}} \mathcal{X}[\epsilon_2] \to \mathcal{X}[\epsilon_2]$$
$$w \times \mathrm{id} : \mathcal{X}[\epsilon_1] \times_{\mathcal{X}} \mathcal{X}[\epsilon_2] \to \mathcal{X}[\epsilon_1];$$

note that the fibered product is over the standard projections on both factors. Note that $\mathcal{X}[\epsilon_1] \times_{\mathcal{X}} \mathcal{X}[\epsilon_2] = \mathcal{X}[\epsilon_1][\epsilon_2] = \mathcal{X} \times \operatorname{Spec}(\mathbb{D} \otimes \mathbb{D}) = \mathcal{X} \times \operatorname{Spec} \mathbb{Z}[\epsilon_1, \epsilon_2]$.

The two compositions $v(\mathrm{id} \times w)$, $w(v \times \mathrm{id}) : \mathcal{X}[\epsilon_1] \times_{\mathcal{X}} \mathcal{X}[\epsilon_2] \to \mathcal{X}$ are retractions (they agree with the identity when restricted to the canonical inclusion of $\mathcal{X}$ in $\mathcal{X}[\epsilon_1] \times_{\mathcal{X}} \mathcal{X}[\epsilon_2]$) that agree with one another when restricted to $\mathcal{X}[\epsilon_1, \epsilon_2] \subseteq \mathcal{X}[\epsilon_1] \times_{\mathcal{X}} \mathcal{X}[\epsilon_2]$. In other words, they give us a commutative diagram, dual to (10.3):

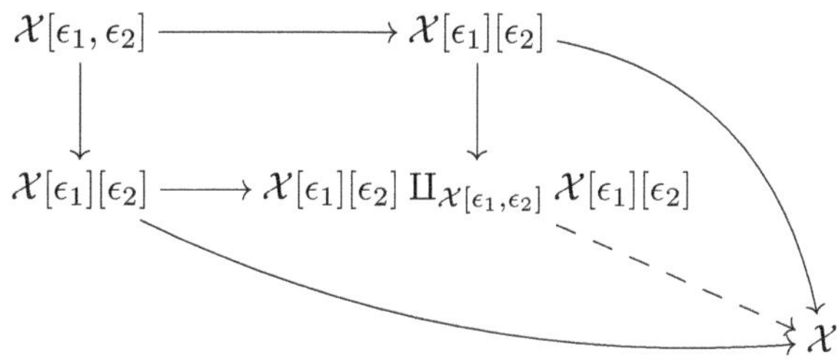

(10.4)

All of the morphisms restrict to the identity on $\mathcal{X}$. Now we restrict under the inclusion $\mathcal{X}[\epsilon_1\epsilon_2] \subset \mathcal{X}[\epsilon_1][\epsilon_2] \times_{\mathcal{X}[\epsilon_1,\epsilon_2]} \mathcal{X}[\epsilon_1][\epsilon_2]$ to get

$$[v, w] : \mathcal{X}[\epsilon_1\epsilon_2] \subset \mathcal{X}[\epsilon_1][\epsilon_2] \amalg_{\mathcal{X}} \mathcal{X}[\epsilon_1\epsilon_2] \xrightarrow{\sim} \mathcal{X}[\epsilon_1][\epsilon_2] \amalg_{\mathcal{X}[\epsilon_1,\epsilon_2]} \mathcal{X}[\epsilon_1][\epsilon_2] \to \mathcal{X}$$

The point is that the homogeneity of $\mathcal{X}$ ensures the existence of the dashed arrow above, and the composition of the dashed arrow with the canonical inclusion of $\mathcal{X}[\epsilon_1\epsilon_2]$.
For the reader's convenience, we include a verification that this does indeed compute the Lie bracket when $\mathcal{X}$ is representable by an affine scheme Spec A. In that case, v and w correspond to derivations δ_1 and δ_2 from A to itself. We consider these as ring homomorphisms:

$$\mathrm{id} + \epsilon_i \delta_i : A \to A + \epsilon_i A$$

The map $v(\mathrm{id} \times w)$ is dual to the composition

$$A \xrightarrow{\mathrm{id}+\epsilon_1\delta_1} A + \epsilon_1 A \xrightarrow{\mathrm{id}\otimes(\mathrm{id}+\epsilon_2\delta_2)} A + \epsilon_1 A + \epsilon_2 A + \epsilon_1\epsilon_2 A$$

sending $f \in A$ to $f + \epsilon_1\delta_1(f) + \epsilon_2\delta_2(f) + \epsilon_1\epsilon_2\delta_2 \circ \delta_1(f)$. To be clear, the second map is obtained from $\mathrm{id} + \epsilon_2\delta_2$ by application of $A[\epsilon_1] \otimes_A (-) = \mathbb{Z}[\epsilon_1] \otimes_{\mathbb{Z}} (-) = (-)[\epsilon_1]$ and carries $f_0 + \epsilon_1 f_1$ to $(f_0 + \epsilon_2\delta_2(f_0)) + \epsilon_1(f_1 + \epsilon_2\delta_2(f_1))$. The map $w(v \times \mathrm{id})$ sends f to $f + \epsilon_1\delta_1(f) + \epsilon_2\delta_2(f) + \epsilon_1\epsilon_2\delta_1 \circ \delta_2(f)$. Taking the difference of these recovers the derivation $\delta_2\delta_1 - \delta_1\delta_2$.

Suppose ξ is a k-point of $\mathcal{X}$. The discussion above shows that $T_{\mathcal{X}}(\xi)$ has all the trappings of a k-vector space structure, except that $T_{\mathcal{X}}(\xi)$ is a groupoid that may not be equivalent to a set. Nevertheless, it is useful to think of $T_{\mathcal{X}}(\xi)$ as a '2-vector space'. In fact, given a scheme R and a morphism $\xi : R \to \mathcal{X}$, we can extract two important invariants from this groupoid:

$$T_{\mathcal{X}}^{-1}(\xi) := \mathrm{Aut}_{T_{\mathcal{X}}(\xi)}(0)$$
$$T_{\mathcal{X}}^{0}(\xi) := T_{\mathcal{X}}(\xi)/\mathrm{isom}.$$

Here 0, the zero section $0 : R \to T_{\mathcal{X}}(\xi)$, corresponds to an object of the groupoid $T_{\mathcal{X}}(\xi)$, and $\mathrm{Aut}_{T_{\mathcal{X}}(\xi)}(0)$ is the automorphism group of that object. Likewise, $T_{\mathcal{X}}(\xi)/\mathrm{isom}$ is the set of isomorphism classes of objects of the groupoid. The following lemma is a formal consequence of the discussion above:

Lemma 10.15. *When k is a field and $\xi : \mathrm{Spec}\, k \to \mathcal{X}$, the sets $T_{\mathcal{X}}^{-1}(\xi)$ and $T_{\mathcal{X}}^{0}(\xi)$ are k-vector spaces.*

It will be important to have a relative variant of the tangent bundle:

Definition 10.16 (Relative tangent bundle). Let $f : \mathcal{X} \to \mathcal{Y}$ be a morphism of CFGs over S. We define the *relative tangent bundle* $T_f = T_{\mathcal{X}/\mathcal{Y}}$ to be the kernel of the morphism of stacks $T_{\mathcal{X}/S} \to T_{\mathcal{Y}/S}$. In other words, the relative tangent bundle is the fiber product of stacks

$$T_{\mathcal{X}/\mathcal{Y}} = T_{\mathcal{X}/S} \underset{T_{\mathcal{Y}/S}}{\times} \mathcal{Y} = T_{\mathcal{X}/S} \underset{f^{-1}T_{\mathcal{Y}/S}}{\times} \mathcal{X}$$

where the morphisms $\mathcal{Y} \to T_{\mathcal{Y}/S}$ and $\mathcal{X} \to f^{-1}T_{\mathcal{Y}/S}$ are the zero sections. In slightly more concrete terms, a section of $T_{\mathcal{X}/\mathcal{Y}}$ is a section of $T_{\mathcal{X}/S}$, together with an isomorphism between its image in $T_{\mathcal{Y}}$ and the zero section.

10.1.4. *Relative homogeneity*

Definition 10.17 (Relative homogeneity). Let $f : \mathcal{X} \to \mathcal{Y}$ be a morphism of CFGs over S. We say that f is *homogeneous*, or that $\mathcal{X}$ is *homogeneous* over $\mathcal{Y}$, if for any scheme S and any morphism $S \to \mathcal{Y}$, the CFG $\mathcal{X}_S = \mathcal{X} \times_{\mathcal{Y}} S$ is homogeneous.

Another way of formulating the definition is that for any cocartesian diagram (10.1) and any compatible objects (we leave it to the reader to formulate compatibility precisely)

$$\overline{\eta}' \in \mathcal{Y}(R') \qquad \eta \in \mathcal{X}(R) \qquad \xi' \in \mathcal{X}(Q') \qquad \xi \in \mathcal{X}(Q)$$

there is a $\eta \in \mathcal{X}(R')$ inducing all of them, and this η is unique up to unique isomorphism.

The following lemma is proved by a standard formal argument:

Lemma 10.18. *(i) Let $f : \mathcal{X} \to \mathcal{Y}$ and $g : \mathcal{Y} \to \mathcal{Z}$ be morphisms of CFGs over* S. *If g is homogeneous then f is homogeneous if and only if gf is.*
(ii) The base change of a homogeneous morphism of CFGs is homogeneous.

10.2. *Deformation theory*

When we speak of the deformation theory of a stack $\mathcal{X}$ we mean extending morphisms $R \to \mathcal{X}$ to morphisms $R' \to \mathcal{X}$, where R' is an infinitesimal extension of R. The definition of the tangent space of a stack in §10.1.1 connects the tangent space of a stack $\mathcal{X}$ to the deformation theory of the objects it parameterizes. Indeed, if $\xi \in \mathcal{X}(k)$ then $T_{\mathcal{X}}(\xi)$ over ξ is precisely the groupoid of extensions of ξ to $\xi' \in \mathcal{X}\big(k[\epsilon]/(\epsilon^2)\big)$.

We now work out several examples that will be used in the construction of the moduli of Higgs bundles. Our main focus is on using the deformation theory to show that various stacks are homogeneous.

10.2.1. *Deformations of morphisms of vector bundles*

Here we discuss the deformation theory for morphisms of vector bundles. From our perspective, the main point is Lemma 10.20 (see also Remark 10.21) on homogeneity.

Let $\pi : X \to S$ be a flat family of schemes over S and let E and F be two vector bundles on X. Let $\mathscr{H} = \mathscr{H}om_{X/S}(E, F)$ be the S-sheaf of morphisms from E to F (see §7.5.2).

We compute the tangent space $T_{\mathscr{H}/S}(\xi)$ of $\mathscr{H}$ over S at an R-point $\xi : E_R \to F_R$, where $R \to S$ is an S-scheme, and E_R and F_R are the pullbacks of E and F to R. An element of $T_{\mathscr{H}/S}(\xi)$ is an extension of the S-map $R \to \mathscr{H}$ to $R[\epsilon]$ such that the composition $R[\epsilon] \to \mathscr{H} \to S$ is the same as the composition $R[\epsilon] \to R \to S$ of the canonical retraction $R[\epsilon] \to R$ and the fixed map $R \to S$. In other words, it is a morphism of vector bundles $E_{R[\epsilon]} \to F_{R[\epsilon]}$ that reduces to u modulo ϵ.

The vector bundle $E_{R[\epsilon]}$ is identified with $E_R + \epsilon E_R$, and similarly for $F_{R[\epsilon]}$. Notice that $\xi(e_0 + \epsilon e_1) = \xi(e_0) + \epsilon\xi(e_1)$ gives one such morphism $E_{R[\epsilon]} \to F_{R[\epsilon]}$ that reduces to ξ modulo ϵ. If ξ' is any other then the difference $\xi' - \xi$ is a map $E_R \to \epsilon F_R \cong F_R$, so we get an identification $T_{\mathscr{H}/S}(\xi) \simeq \mathscr{H}(R)$. As this identification is natural, we get

$$T_{\mathscr{H}/S} = \mathscr{H} \underset{S}{\times} \mathscr{H}.$$

Of course, it is no surprise that the tangent space of $\mathscr{H}$ is $\mathscr{H}$ itself, since $\mathscr{H}$ already has a linear structure. We have

$$T^{-1}_{\mathscr{H}/S}(\xi) = 0$$
$$T^{0}_{\mathscr{H}/S}(\xi) = \mathrm{Hom}(E_R, F_R).$$

This gives the tangent space to $\mathscr{H}$ over S, but does not explain the higher order infinitesimal structure. For that we must pose a more general lifting problem: given $\xi \in \mathscr{H}(R)$ and an arbitrary square-zero extension R' of R as an S-scheme, can ξ be lifted to $\xi' \in \mathscr{H}(R')$? To answer this question, we consider it locally in X_R, that is, we cover X by open subsets U with corresponding extensions U' over R' and ask for extensions of $E_R\big|_U \to F_R\big|_U$ to $E_{R'}\big|_{U'} \to F_{R'}\big|_{U'}$. We make two observations:

(i) there is a cover of $X_{R'}$ by open sets U' such that $\xi\big|_{X_R \cap U'}$ extends to a morphism $E_{U'} \to F_{U'}$.
(ii) if ξ' and ξ'' are any two extensions, then $\xi' - \xi''$ may be viewed as a morphism $E_R \to F_R \otimes \pi^* J$ where J is the ideal of R in R' and, by abuse of notation, $\pi : X_R \to R$ is the morphism obtained from $\pi : X \to S$ by pullback.

These observations combine to imply that there is a $\underline{\mathrm{Hom}}(E_R, F_R \otimes \pi^* J)$-*torsor* P on X (in the Zariski topology) whose sections are in bijection with the lifts of ξ to $H(R')$ (those who would rather avoid the language of torsors may obtain the following lemma by a Čech cohomology calculation). This yields a deformation-obstruction theory:

Lemma 10.19 (cf. [24, Thm. 8.5.3 (a)]). *Let X be an S-scheme, let E and F be vector bundles on X, and let $R \subseteq R'$ be a square-zero extension of S-schemes with ideal J. Associated to any homomorphism $\xi : E_R \to F_R$ there is an obstruction $\omega \in H^1(X_R, \underline{\mathrm{Hom}}(E_R, F_R \otimes \pi^* J))$ whose vanishing is equivalent to the existence of an extension of ξ to some $\xi' : E_{R'} \to F_{R'}$. If there is at least one extension then the set of all extensions possesses a simply transitive action of $H^0(X_R, \underline{\mathrm{Hom}}(E_R, F_R \otimes \pi^* J)) = \mathrm{Hom}(E_R, F_R \otimes \pi^* J)$.*

Beyond its finite dimensionality, the particular deformation-obstruction theory is not actually necessary for the proof of algebraicity. What is important is homogeneity:

Lemma 10.20. *The functor $\mathscr{H}om_{X/S}(E, F)$ is homogeneous.*

Proof. If $f : Q \to R$ is an affine morphism of S-schemes and $Q \subseteq Q'$ is a square-zero extension of S-schemes, let R' be the pushout (as in Theorem 10.10), with $f' : Q' \to R'$ denoting the tautological morphism. (Since Q and Q' have the same underlying topological space, we do not bother to introduce notation for the inclusion $Q \subseteq Q'$.) Since Q and Q', as well as R and R' have the same underlying topological space, and we have a natural morphism $F_{Q'} \to F_Q$, we can push forward to obtain a morphism $f'_* F_{Q'} \to f_* F_Q$. Using adjunction we have have that the identity $f^* F_R = F_Q$ induces a morphism $F_R \to f_* F_Q$. Thus we obtain a fibered product $f'_* F_{Q'} \times_{f_* F_Q} F_R$.

Assume we have $v \in H(R)$ and $u' \in H(Q')$, both extending $u = f^* v$. Then u' gives us

$$f'^* E_{R'} = E_{Q'} \xrightarrow{u'} F_{Q'} = f'^* F_{R'},$$

whence $E_{R'} \to f'_* f'^* F_{R'} = f'_* F_{Q'}$ by adjunction. We also have a map $E_{R'} \to E_R \xrightarrow{v} F_R$. These induce the same map $E_{R'} \to f_* F_Q$ so we get a map

$$\eta' : E_{R'} \to f'_* F_{Q'} \underset{f_* F_Q}{\times} F_R$$

by the universal property of the fiber product.

On the other hand, the canonical map $\gamma : F_{R'} \to f'_* F_{Q'} \times_{f_* F_Q} F_R$ is an isomorphism, by Theorem 10.14, or directly: γ is an isomorphism modulo J, since it reduces to the identity $F_R = f_* F_Q \times_{f_* F_Q} F_R$, and the kernel of reduction is the fiber product of the kernels, namely $(f_* F_Q \otimes \pi^* J) \times_0 0 = F_R \otimes \pi^* J$. Therefore γ is an isomorphism by the 5-lemma. To conclude we can view η' as a map $E_{R'} \to F_{R'}$ by composition with γ^{-1}. □

Remark 10.21. As was pointed out above (see Proposition 7.20), under mild hypotheses, a result of Lieblich implies that $\mathscr{H}om_{X/S}(E, F)$ is an algebraic space over S, locally of finite type. In particular, this also shows that $\mathscr{H}om_{X/S}(E, F)$ is homogeneous. However, Lieblich's proof relies on the homogeneity, so this reasoning is actually circular.

10.2.2. *Deformations of vector bundles*

Here we discuss the deformation theory for vector bundles; see also [24, Prop. 6.5.1]. The main points are Corollary 10.23 and Lemma 10.25 (see also Remark 10.26).

Let $\pi : X \to S$ be a family of schemes over S. Suppose that R is an S-scheme and $E \in \mathsf{Fib}_{X/S}(R)$ is a vector bundle over X_R. Let R' be a square-zero S-scheme extension with ideal J. We ask whether E can be extended to $E' \in \mathsf{Fib}_{X/S}(R')$, and if so, in how many ways. Again we make several observations:

(i) there is a cover of $X_{R'}$ by open subsets U' such that $E\big|_{X_R \cap U'}$ can be extended in at least one way to a vector bundle on U';

(ii) if E' and E'' are two extensions of $E\big|_{X_R \cap U'}$ to U' then there is a cover of U' by open subsets V' such that $E'\big|_{V'} \simeq E''\big|_{V'}$ as extensions of $E\big|_{X_R \cap V'}$;

(iii) if $u, v : E'\big|_{V'} \to E''\big|_{V'}$ are two isomorphisms of extensions of $E\big|_{X_R \cap V'}$ then u and v differ by a homomorphism $E\big|_{X_R \cap V'} \to E\big|_{X_R \cap V'} \otimes \pi^* J$.

Choosing a suitable cover, we therefore obtain a Čech 2-cocycle for the sheaf of groups $\underline{\mathrm{Hom}}(E, E \otimes \pi^* J)$. This is a coboundary if and only if a deformation exists. A more careful analysis shows that the isomorphism classes of all deformations then correspond to 1-cocycles modulo coboundaries. Here is the statement in its usual form:

Lemma 10.22 (cf. [24, Thm. 8.5.3 (b)]). *Fix a flat family of schemes $\pi : X \to S$ and a square-zero S-extension $R \subseteq R'$ with ideal J. A vector bundle E on X_R induces an obstruction $\omega \in H^2\big(X_R, \underline{\mathrm{Hom}}(E, E \otimes \pi^* J)\big)$ whose vanishing is equivalent to the existence of an extension of E to $X_{R'}$. If $\omega = 0$, the set of isomorphism classes of extensions is a principal homogeneous set under a natural action of $H^1\big(X_R, \underline{\mathrm{Hom}}(E, E \otimes \pi^* J)\big)$. The automorphisms (as an extension) of any given extension are canonically $H^0\big(X_R, \underline{\mathrm{Hom}}(E, E \otimes \pi^* J)\big)$.*

In particular, this lemma gives the tangent space of $\mathsf{Fib}_{X/S}$:

Corollary 10.23. *Suppose that R is an S-scheme and E is a vector bundle over X_R. We have $T_{\mathsf{Fib}_{X/S}}(E) = \mathrm{B}\underline{\mathrm{Hom}}(E, E)(X_R)$ where E is the universal vector bundle on $\mathsf{Fib}_{X/S}$ and the prefix* B *denotes the classifying stack (Section C.1). In particular,*

$$T^{-1}_{\mathsf{Fib}_{X/S}}(E) = \underline{\mathrm{Hom}}(E, E)$$
$$T^{0}_{\mathsf{Fib}_{X/S}}(E) = \underline{\mathrm{Ext}}^1(E, E).$$

Proof. We have just seen that sections of $T_{\mathsf{Fib}_{X/S}}$ correspond to 1-cocycles for $\underline{\mathrm{Hom}}(E, E)$ modulo coboundaries. But 1-cocycles modulo coboundaries also classify torsors. □

The Čech calculations can be abstracted into the observation that properties (i) – (iii) above imply there is a gerbe $\mathscr{G}$ over X_R (see e.g., [54, (3.15), p.22] for the definition of a gerbe) whose sections correspond to extensions of E to $E' \in \mathsf{Fib}_{X/R}(R')$, and that this gerbe is banded by the sheaf of abelian groups $\underline{\mathrm{Hom}}(E_R, E_R \otimes \pi^* J)$. Giraud classifies banded gerbes cohomologically:

Theorem 10.24 ([29, Thm. IV.3.4.2]). *Let $\mathscr{G}$ be a gerbe on X, banded by an abelian group A. There is an obstruction $\omega \in H^2(X, A)$ to the existence of a global section of $\mathscr{G}$. Should this obstruction vanish, global sections up to isomorphism form a principal homogeneous set under the action of $H^1(X, A)$. Automorphisms of any given section are in canonical bijection with $H^0(X, A)$.*

We also record the homogeneity of $\mathsf{Fib}_{X/S}$:

Lemma 10.25. *For any scheme X, the CFG $\mathsf{Fib}_{X/S}$ is homogeneous.*

Proof. Since S is representable, it is homogeneous by Theorem 10.10. Therefore by Lemma 10.18, it is sufficient to show that $\mathsf{Fib}_{X/S}$ is homogeneous over S. We will content ourselves to sketch the construction of the inverse to the functor

$$\mathsf{Fib}_{X/S}(R') \to \mathsf{Fib}_{X/S}(Q') \underset{\mathsf{Fib}_{X/S}(Q)}{\times} \mathsf{Fib}_{X/S}(R) \tag{10.5}$$

associated to a cocartesian diagram

$$\begin{array}{ccc} Q & \hookrightarrow & Q' \\ \downarrow & & \downarrow \\ R & \longrightarrow & R' \end{array}$$

of S-schemes, where $Q \to R$ is affine and $Q \hookrightarrow Q'$ is an infinitesimal extension. An object of the right side of (10.5) consists of vector bundles E' on $X_{Q'}$, F on X_R, and E on X_Q, along with identifications $E'\big|_{X_Q} = E = F\big|_{X_Q}$. By applying [25, Thm. 2.2] to a cover of $X_{R'}$ by open affines and then to covers of the intersections by open affines, we obtain a vector bundle F' on $X_{R'}$ restricting to E' on $X_{Q'}$, to E on X_Q, and to F on X_R, as required. $\square$

Remark 10.26. This also follows from the fact that $\mathsf{Fib}_{X/S}$ is an algebraic stack, but, as in Remark 10.21, this reasoning is circular.

Corollary 10.27. *Let $\mathcal{F}$ be the stack of triples (E, F, σ) where E and F are vector bundles on X and $\sigma : E \to F$ is a morphism of vector bundles. Then $\mathcal{F}$ is homogeneous.*

Proof. We have a morphism $\mathcal{F} \to \mathsf{Fib}_{X/S} \times_S \mathsf{Fib}_{X/S}$. We have just seen that $\mathsf{Fib}_{X/S}$ is homogeneous, so by two applications of Lemma 10.18, it follows that $\mathsf{Fib}_{X/S} \times_S \mathsf{Fib}_{X/S}$ is homogeneous. Lemma 10.20 says that $\mathcal{F}$ is relatively homogeneous over $\mathsf{Fib}_{X/S} \times_S \mathsf{Fib}_{X/S}$, so we may conclude by another application of Lemma 10.18. □

10.2.3. *Deformations of nodal curves*

We now discuss the deformation theory of nodal curves.

Definition 10.28 (Nodal curve). Let S be a scheme. A *nodal curve* over S is an algebraic space C and a projection $\pi : C \to S$ that is étale-locally isomorphic to $\operatorname{Spec} \mathcal{O}_S[x, y]/(xy - t)$ for a local section t of $\mathcal{O}_S$. That is, there is an étale cover of S by affines $U = \operatorname{Spec} A$ and an étale cover of $\pi^{-1}U$ by schemes V, each of which admits an étale map to $\operatorname{Spec} A[x, y]/(xy - t)$ for some $t \in A$.

We shall write $\mathcal{N}$ for the stack over schemes whose S-points are the families of nodal curves over S.

Remark 10.29. A more conventional definition of a nodal curve over S is as a flat family $\pi : C \to S$ whose fibers are 1-dimensional, reduced schemes whose only singularities are ordinary double points. It follows from [27, Prop. III.2.8] that the two notions are equivalent.

We will show that nodal curves form a homogeneous stack (Lemma 10.31), compute the tangent space of this stack (Lemmma 10.30), and show that it is formally smooth (Corollary 10.32). This whole section could easily be adapted to curves with locally planar singularities (or to more general schemes of finite type with hypersurface singularities), but for concreteness, we stick to nodal curves.

The method used in the last section to study deformations of vector bundles works quite well for deformations of *smooth* curves, and even families of smooth schemes $\pi : X \to S$. If S' is a square-zero extension of S with ideal J, one discovers that extensions of X to S' always exist locally, that any two deformations are locally isomorphic, and that any two deformations are related by a section of $T_{X/S} \otimes \pi^* J$, from which it follows by a Čech calculation (as in the last section) that (see [24, Thm. 8.5.9 (b)]):

(i) there is obstruction to the existence of a flat extension X' over S extending X lying in $H^2(X, T_{X/S} \otimes \pi^* J)$;
(ii) should a deformation exist, the isomorphism classes of flat extensions form a torsor under $H^1(X, T_{X/S} \otimes \pi^* J)$;
(iii) automorphisms of a fixed flat extension are canonically $H^0(X, T_{X/S} \otimes \pi^* J)$.

The deformation theory of nodal curves introduces a new complication: It is still the case that deformations exist locally, and one can easily compute that automorphisms of deformations can be identified with sections of $T_{X/S} \otimes J$. However, it is no longer the case any two deformations are locally isomorphic, since the node $xy = 0$ has a 1-parameter family of first order deformations $xy = \lambda\epsilon$. A naive Čech calculation will therefore not suffice, but with a little more effort, we will see that the deformations and obstructions for nodal curves can be classified in much the same way as for smooth curves.

Suppose that C is an S-point of $\mathcal{N}$. An S-point of $T_{\mathcal{N}}$ lying above C is a cartesian diagram (10.6), in which C' is an $S[\epsilon]$-point (see Definition 10.2 for notation) of $\mathcal{N}$:

$$\begin{array}{ccc} C & \longrightarrow & C' \\ {\scriptstyle\pi}\downarrow & & \downarrow \\ S & \longrightarrow & S[\epsilon]. \end{array} \tag{10.6}$$

We consider the more general problem of extending an S-point C of $\mathcal{N}$ to a $S[\epsilon J]$-point:

$$\begin{array}{ccc} C & \longrightarrow & C' \\ {\scriptstyle\pi}\downarrow & & \downarrow \\ S & \underset{\dashleftarrow}{\longrightarrow} & S[\epsilon J]. \end{array}$$

Here J is a quasicoherent sheaf on S and C' is an $S[\epsilon J]$-point of $\mathcal{N}$. Since there is a canonical retraction $S[\epsilon J] \to S$, we can consider C' as an S-scheme. As C is the closed sub-scheme of C' determined by $\pi^* J$ (with $\pi^* J^2 = 0$), we may form the exact sequence (10.7):

$$0 \to \pi^* J \to \mathcal{O}_C \underset{\mathcal{O}_{C'}}{\otimes} \Omega_{C'/S} \to \Omega_{C/S} \to 0. \tag{10.7}$$

The right exactness of this sequence follows from general principles and does not depend on the fact that C is a nodal curve. To see the left exactness, we may work étale-locally and assume:

$$\begin{aligned} S &= \operatorname{Spec} A & S' &= \operatorname{Spec} A' \\ C &= \operatorname{Spec} B & C' &= \operatorname{Spec} B' \\ B &= A[x,y]/I & B' &= A'[x,y]/I' \\ I &= (xy - t) & I' &= (xy - t') \end{aligned}$$

In fact, the local structure of nodes implies we can arrange for C and C' to be *étale* over $A[x,y]/I$ and $A'[x,y]/I'$, respectively. Then the sequence (10.7) on C is pulled back via the étale (and in particular flat) map $C \to \operatorname{Spec} B$ from the corresponding sequence over $\operatorname{Spec} B$. For the sake of proving (10.7) is exact, we can therefore replace C with $\operatorname{Spec} B$ and C' with $\operatorname{Spec} B'$ without loss of generality.

Our diagram of rings is

$$\begin{array}{ccc} B = A[x,y]/(xy-t) & \longleftarrow & B' = A'[x,y]/(xy-t') \\ \uparrow & & \uparrow \\ A & \overset{\dashrightarrow}{\longleftarrow} & A' = A[\epsilon J]. \end{array}$$

We are trying to show the exactness of

$$0 \to B \otimes_A J \to B \underset{B'}{\otimes} \Omega_{B'/A} \to \Omega_{B/A} \to 0. \tag{10.8}$$

The left exactness of (10.8) is equivalent to showing that, for every injective[b] B-module M, the map $\operatorname{Hom}_B(B \otimes_{B'} \Omega_{B'/A}, M) \to \operatorname{Hom}_B(B \otimes_A J, M)$ is surjective; i.e., every B-module homomorphism $B \otimes_A J \to M$ extends to a morphism $B \otimes_{B'} \Omega_{B'/A} \to M$. Using tensor-hom adjunction, and the universal property of the module of Kähler differentials, this translates into the statement that, every A-module homomorphism $J \to M$ should extend to an A-derivation $B' \to M$:

$$\begin{array}{ccc} B' = \frac{A[\epsilon J][x,y]}{(xy-t')} & & \\ {\scriptstyle \cdot\epsilon}\uparrow & \overset{\partial}{\searrow} & \\ J & \xrightarrow{\ \varphi\ } & M. \end{array}$$

[b]Recall that a morphism $M' \to M$ of modules over a ring A is injective if and only if for every injective A-module I, $\operatorname{Hom}_A(M, I) \to \operatorname{Hom}_A(M', I)$ is surjective. To see this, choose an inclusion $M' \hookrightarrow I$ for some injective module I. Then we get a map $M \to I$ extending the inclusion $M' \subseteq I$, so $M' \to M$ must be injective.

First we will extend φ to a derivation

$$\delta' : A[\epsilon J][x,y] \to M.$$

To do this, we will invoke the identifications

$$\mathrm{Hom}_A(J,M) = \mathrm{Hom}_{A\text{-alg},\mathrm{Id}_A}(A[\epsilon J], A[\epsilon M]) = \mathrm{Der}_A(A[\epsilon J], M).$$

On the left, we take a homomorphism $\varphi : J \to M$ and send it to the ring homomorphism $\mathrm{Id}_A \oplus \varphi : A \oplus \epsilon J \to A \oplus \epsilon M$. The group in the middle consists of the A-algebra homomorphisms that are the identity on the first term. On the right, we just project to get a derivation. Similarly, we have indentifications

$$\mathrm{Hom}_{A\text{-alg},\mathrm{Id}_A}(A[\epsilon J][x,y], A[\epsilon M]) = \mathrm{Der}_A(A[\epsilon J][x,y], M).$$

In summary, the homomorphism $\varphi : J \to M$ induces a ring homomorphism $\tilde{\varphi} : A[\epsilon J] \to A[\epsilon M]$. We can extend this to $\hat{\varphi} : A[\epsilon J][x,y] \to A[\epsilon M]$ by sending x and y to any element of M. This then gives a derivation δ' as desired.

The choices of such extensions δ' are a torsor under $\mathrm{Der}_A(A[x,y], M)$, and we will adjust δ' by such a derivation so that it descends to a derivation $B' \to M$. The obstruction to descending δ' to B' is the homomorphism $\delta'\big|_{I'}$, so we want to find a derivation $\delta \in \mathrm{Der}(A[x,y], M)$ such that the restriction of δ to I' (via the composition $I' \subseteq A'[x,y] \to A[x,y]$) agrees with δ'.

To execute this, let us take a fixed extension δ'. Then $\delta'(I'J) \subseteq I'\delta(J) + J\delta(I') = 0$ because M is a B-module, and therefore $I'M = JM = 0$. This implies $\delta\big|_{I'}$ descends to $I'/JI' = I$. We also have $\delta'(I'^2) = I'\delta'(I') = 0$, for the same reason, so δ' gives us a homomorphism $u : I/I^2 \to M$.

Now, we have an exact sequence

$$0 \to I/I^2 \to B \underset{A[x,y]}{\otimes} \Omega_{A[x,y]/A} \to \Omega_{B/A} \to 0,$$

either by [75, Thm. D.2.7, p. 310] or a direct verification. Since M is injective, the homomorphism $u : I/I^2 \to M$ extends to a derivation $\delta : A[x,y] \to M$ (again we are using tensor-hom adjunction and the universal property of the sheaf of Kähler differentials). Trivially extending δ to $A'[x,y]$ we obtain a derivation $\delta' - \delta$ on $A'[x,y]$. We have that $\delta' - \delta = 0$ on I' essentially by construction. Moreover, $\delta' - \delta$ agrees with $\delta'\big|_J = \varphi$ on J because δ vanishes there. Thus $\delta' - \delta$ descends to a derivation in $\mathrm{Der}_A(B, M)$ that agrees with φ on J, as required. We conclude that (10.7) is exact.

Conversely, given any extension of $\mathcal{O}_C$-modules

$$0 \to \pi^* J \to \Omega' \to \Omega_{C/S} \to 0$$

we can define an extension of C/S to $S[\epsilon J]$. Namely, one may define

$$\mathcal{O}_{C'} = \mathcal{O}_C \times_{\Omega_{C/S}} \Omega',$$

where the map from $\mathcal{O}_C$ to $\Omega_{C/S}$ is the universal differential, d. This is clearly an $\mathcal{O}_C$-module, but it is not a priori obvious that $\mathcal{O}_{C'}$ is equipped with a ring structure. The ring structure can be constructed easily by factoring d as

$$\mathcal{O}_C \xrightarrow{\mathrm{id}+\epsilon d} \mathcal{O}_C + \epsilon\Omega_{C/S} \longrightarrow \Omega_{C/S}$$

and recognizing that

$$\mathcal{O}_{C'} \simeq \mathcal{O}_C \times_{\mathcal{O}_C + \epsilon\Omega_{C/S}} (\mathcal{O}_C + \epsilon\Omega').$$

Here the map on the right is

$$\mathcal{O}_C + \epsilon\Omega' \to \mathcal{O}_C + \epsilon\Omega_{C/S}$$

$$a + \omega' \mapsto a + \tilde{\omega}'$$

where we are using the given map $\Omega' \to \Omega_{C/S}$, $\omega' \mapsto \tilde{\omega}'$; this is the standard map taking

$$\mathrm{Hom}_{\mathcal{O}_C}(\Omega', \Omega_{C/S}) \to \mathrm{Hom}_{\mathcal{O}_C\text{-alg},\mathrm{Id}}(\mathcal{O}_C[\epsilon\Omega'], \mathcal{O}_C[\epsilon\Omega_{C/S}]).$$

This above description of $\mathcal{O}_{C'}$ is as a fiber product of rings, hence is naturally equipped with a ring structure.

The map

$$C' \to S[\epsilon J]$$

is given topologically by $C \to S$, and at the level of sheaves by a map

$$\mathcal{O}_{S[\epsilon]} = \mathcal{O}_S \oplus \epsilon J \to \mathcal{O}_{C'} = \mathcal{O}_C \times_{\mathcal{O}_C \oplus \epsilon\Omega_{C/S}} (\mathcal{O}_C \oplus \epsilon\Omega').$$

We are given maps $\mathcal{O}_S \to \mathcal{O}_C$, and $J \to \Omega'$. The map $\mathcal{O}_S \oplus \epsilon J \to \mathcal{O}_C$ is defined by the given map on the first term, plus the zero map on the second term. The map $\mathcal{O}_S[\epsilon J] \to \mathcal{O}_C[\epsilon\Omega']$ is given by the map $\mathcal{O}_S[\epsilon J] \to \mathcal{O}_C[\epsilon\pi^* J]$, and then the natural map $\mathcal{O}_C[\epsilon\pi^* J] \to \mathcal{O}_C[\epsilon\Omega']$. These maps agree on composition to $\mathcal{O}_C[\epsilon\Omega_{C/S}]$, and so define a morphism to the fibered product.

One can check that these processes are inverses of one another, and so this yields an equivalence of categories between the tangent space of $\mathcal{N}$ at C/S and the category of extensions of $\Omega_{C/S}$ by $\pi^* J$.

Lemma 10.30. *Let $\pi : C \to S$ be an S-point of the stack $\mathcal{N}$ of all nodal curves. Then*

$$T_{\mathcal{N}}(C/S) = \mathbf{Ext}(\Omega_{C/S}, \mathcal{O}_C),$$

where we write $\mathbf{Ext}(A, B)$ *for the* groupoid *of extensions of B by A. In particular,*

$$\begin{aligned} T_{\mathcal{N}}^{-1}(C/S) &= \mathrm{Hom}_{\mathcal{O}_C}(\Omega_{C/S}, \mathcal{O}_C) = \Gamma(C, T_{C/S}) \\ T_{\mathcal{N}}^{0}(C/S) &= \mathrm{Ext}^1_{\mathcal{O}_C}(\Omega_{C/S}, \mathcal{O}_C). \end{aligned}$$

Observe that the discussion above never made use of the assumption that C be proper over S. Thus $\underline{\mathrm{Ext}}^1(\Omega_{C/S}, \mathcal{O}_C)$ may be interpreted as the associated sheaf of the presheaf of isomorphism classes of local deformations of C. This allows us to interpret the 5-term exact sequence of the local-to-global spectral sequence for $\mathrm{Ext}(\Omega_{C/S}, \mathcal{O}_C)$ deformation theoretically:

$$0 \to H^1(C, T_{C/S}) \to \mathrm{Ext}^1(\Omega_{C/S}, \mathcal{O}_C) \to \Gamma(C, \underline{\mathrm{Ext}}^1(\Omega_{C/S}, \mathcal{O}_C)) \to H^2(C, T_{C/S}).$$

If S is affine then $H^2(C, T_{C/S}) = 0$ because C has relative dimension 1 over S. The first term of the sequence may be interpreted, via a Čech calculation, as the set of isomorphism classes of locally trivial deformations of C. Somewhat imprecisely, we have:

$$0 \to (\text{loc. triv. defs. of } C) \to (\text{defs. of } C) \to (\text{defs. of nodes of } C) \to 0.$$

When S is not affine, the final term $H^2(C, T_{C/S})$ may be interpreted as the obstruction to finding a deformation of the curve realizing specified deformations of the nodes.

Now we turn to the more general problem of extending a curve over a general square-zero extension $S \subseteq S'$ with ideal J:

$$\begin{array}{ccc} C & \dashrightarrow & C' \\ {\scriptstyle\pi}\downarrow & & \downarrow \\ S & \longrightarrow & S' \end{array} \qquad (10.9)$$

We will make use of homogeneity:

Lemma 10.31. *The CFG of all nodal curves (not necessarily proper!) is homogeneous, as are the substack of all proper nodal curves, the substack of curves of fixed genus, and the substack of canonically polarized curves.*

Proof. Let $Q \subseteq Q'$ be an infinitesimal extension of schemes and $Q \to R$ an affine morphism. Form the pushout R' of Q' and R under Q using Theorem 10.10. Letting $\mathcal{N}$ denote the stack of all nodal curves, we construct the inverse to the map (10.10).

$$\mathcal{N}(R') \to \mathcal{N}(Q') \underset{\mathcal{N}(Q)}{\times} \mathcal{N}(R). \tag{10.10}$$

Suppose that C is a nodal curve over Q, that C' is an extension of C to Q', and that $C = D \times_R Q$ for nodal curve D over R and observe that $C \subseteq C'$ is an infinitesimal extension and $C \to D$ is affine, so we can also form the pushout D' of C' and D under C (Theorem 10.10).

By Theorem 10.14, D' is flat over R' and $D' \times_{R'} R = D$, $D' \times_{R'} Q' = C'$, and $D' \times_{R'} Q = C$. We argue that D' is a nodal curve over R'. Since the fibers of D' over R' are the same as the fibers of D over R, the fibers of D' over R' are nodal curves. We have already seen that D' is flat over R', so we therefore only need to check that D' is locally of finite presentation over R'.

One can verify easily that local generators and relations for $\mathcal{O}_D$ as an algebra over $\mathcal{O}_R$ lift to local generators and relations for $\mathcal{O}_{D'}$ as an algebra over $\mathcal{O}_{R'}$, implying it is locally of finite presentation as well.

This proves the homogeneity of the stack of all nodal curves. For the remaining statements, note that they are stable under infinitesimal deformation. □

Now we return to the problem of completing (10.9):

(i) The problem of completing (10.9) can be solved locally in C. Indeed, one may find a cover of S by affine open subschemes $U = \operatorname{Spec} A$ and elements $t \in A$ such that $\pi^{-1}U$ has a cover by open affines V, each of which is étale over $\operatorname{Spec} B$, with $B = A[x, y]/(xy - t)$. If $U' = \operatorname{Spec} A'$ is the open subset of S' whose preimage in S is U then one may form $B' = A'[x, y]/(xy - t')$ where t' is any lift of t to A'. Then $\operatorname{Spec} B'$ is a deformation of $\operatorname{Spec} B$ to $\operatorname{Spec} A'$ and this lifts uniquely to a deformation of V by [37, Thm. (18.1.2)].
We may thus select local deformations V' for each V in an open cover of C and attempt to glue.

(ii) Observe that

$$\mathcal{O}_{S'} \underset{\mathcal{O}_S}{\times} \mathcal{O}_{S'} \simeq \mathcal{O}_{S'} + \epsilon J,$$

via the map $(f, g) \mapsto f + \epsilon(g - f)$, so that $S' \amalg_S S' \simeq S' \amalg_S S[\epsilon J]$ and therefore by homogeneity, we have

$$\mathcal{N}(S') \underset{\mathcal{N}(S)}{\times} \mathcal{N}(S') = \mathcal{N}(S' \underset{S}{\amalg} S') \simeq \mathcal{N}(S' \underset{S}{\amalg} S[\epsilon J]) = \mathcal{N}(S') \underset{\mathcal{N}(S)}{\times} \mathcal{N}(S[\epsilon J]).$$

In other words, any two extensions of an open subset of C to a nodal curve over S' differ by a uniquely determined element of $\mathcal{N}(S[\epsilon J])$.

(iii) We have already seen that $\mathcal{N}(S[\epsilon J])$ may be identified with the category of extensions $\mathbf{Ext}(\Omega_{C/S}, \pi^* J)$. Thus the first obstruction to gluing the deformations U' comes from making sure that the isomorphism classes of the deformations U' agree on overlaps. Since the isomorphism classes form a torsor under $\underline{\mathrm{Ext}}^1(\Omega_{C/S}, \pi^* J)$, this obstruction lies in $H^1\big(C, \underline{\mathrm{Ext}}^1(\Omega_{C/S}, \pi^* J)\big)$.

(iv) One can arrange very easily for this obstruction to vanish by working locally in S. Indeed, $\underline{\mathrm{Ext}}^1(\Omega_{C/S}, \pi^* J)$ is quasicoherent and is supported on the nodes of C, which is finite (and in particular affine) over S. Therefore when S is affine the first obstruction vanishes.

(v) Now assuming that the first obstruction vanishes we can fix a compatible system of isomorphism classes of local deformations. We must determine whether one can find genuine local deformations within those isomorphism classes in a compatible way. For each open $U \subseteq C$ in a suitable cover we select a deformation U' in the specified isomorphism class. If U_i and U_j are two such open sets then write U'_{ij} for the restriction of U'_i to $U_i \cap U_j$. Because we have chosen the isomorphism classes compatibly, $U'_{ij} \simeq U'_{ji}$, and we may select such an isomorphism φ_{ij}. Over triple overlaps the cocycle condition $\varphi_{ki} \circ \varphi_{jk} \circ \varphi_{ij} = \mathrm{id}$ obstructs the gluing. As the element $\varphi_{ki} \circ \varphi_{jk} \circ \varphi_{ij}$ lies in the automorphism group of U'_{ijk}, which is canonically identified with $\Gamma\big(U_{ijk}, \underline{\mathrm{Hom}}(\Omega_{C/S}, \pi^* J)\big)$ we obtain a Čech 2-cocycle in $H^2\big(C, \underline{\mathrm{Hom}}(\Omega_{C/S}, \pi^* J)\big)$. Once again, this obstruction vanishes when S is affine, this time because C has relative dimension 1 over S and the coefficients are taken in a quasicoherent sheaf.

Corollary 10.32. *The stack of all nodal curves satisfies the formal criterion for smoothness.*

Proof. We have just seen that all obstructions to infinitesimal deformation vanish over an affine base, which is precisely the formal criterion for smoothness. □

Corollary 10.33. *Any algebraic substack of the stack of all nodal curves* *that is locally of finite presentation and stable under infinitesimal extensions is smooth. In particular, the stack of all proper nodal curves is smooth and the open substack of all canonically polarized nodal curves is smooth.*

Remark 10.34. An observant reader will have noticed that the obstructions to (10.9) constructed above lie in the graded pieces of $\mathrm{Ext}^2(\Omega_{C/S}, \pi^* J)$ induced by the local-to-global spectral sequence. This is not an accident: There is a single obstruction in $\mathrm{Ext}^2(\Omega_{C/S}, \pi^* J)$ whose vanishing is equivalent to the existence of a solution to (10.9). We will discuss one way to obtain this obstruction in §10.3.

In fact, one can dispense with the assumption that C be a curve over S and obtain an obstruction to deformation in $\mathrm{Ext}^2(\mathbf{L}_{C/S}, \pi^* J)$ where $\mathbf{L}_{C/S}$ is the relative cotangent complex of C over S, constructed by Illusie [49, Cor. 2.1.3.3, Thm. 2.1.7]. For a nodal curve, one has $\mathbf{L}_{C/S} = \Omega_{C/S}$. In general, $\mathbf{L}_{C/S}$ is concentrated in nonpositive degrees and $\Omega_{C/S}$ is the 0-th homology group.

10.2.4. *Simultaneous deformation of curves and vector bundles*

Let S be a scheme, C a curve over S, and E a vector bundle on C. We consider the problem of extending C and E to the scheme $S[\epsilon J]$ where J is a quasicoherent sheaf of $\mathcal{O}_S$-modules. We begin by defining a sheaf of modules $\Upsilon_C(E)$ on C to play a role analogous to the one played by the module of differentials when we studied deformations of curves.

For each $\mathcal{O}_C$-module F, define $\Phi(F)$ to be the set of pairs (δ, φ) where $\delta : \mathcal{O}_C \to F$ is an $\mathcal{O}_S$-derivation and $\varphi : E \to F \otimes E$ is what we will call a δ-*connection*. That is, for any local sections $f \in \mathcal{O}_C$ and $x \in E$, we have

$$\varphi(fx) = \delta(f) \otimes x + f\varphi(x).$$

Then $\Phi(F)$ is naturally a covariant functor of F. There is a natural $\Gamma(C, \mathcal{O}_C)$-module structure on $\Phi(F)$, in which $\lambda \in \Gamma(C, \mathcal{O}_C)$ acts by

$$\lambda.(\delta, \varphi) = (\lambda\delta, \lambda\varphi).$$

There is also an evident exact sequence (10.11).

$$0 \to \mathrm{Hom}(E, F \otimes E) \to \Phi(F) \to \mathrm{Der}_{\mathcal{O}_S}(\mathcal{O}_C, F). \tag{10.11}$$

Here the map $\mathrm{Hom}(E, F \otimes E) \to \Phi(F)$ sends φ to $(0, \varphi)$ and the map $\Phi(F) \to \mathrm{Der}_{\mathcal{O}_S}(\mathcal{O}_C, F)$ sends (δ, φ) to δ.

For each open set $U \subseteq C$, setting $\underline{\Phi}(F)(U) := \Gamma(U, \underline{\Phi}(F)) = \Phi(F|_U)$, we obtain a covariant functor

$$\underline{\Phi} : (\mathcal{O}_C\text{-mod}) \to (\mathcal{O}_C\text{-mod})$$

with $\Phi(F) = \underline{\Phi}(F)(C)$, and an exact sequence

$$0 \to \underline{\mathrm{Hom}}_{\mathcal{O}_C}(E^\vee \otimes E, -) \to \underline{\Phi}(-) \to \mathrm{Der}_{\mathcal{O}_S}(\mathcal{O}_C, -) = \mathrm{Hom}_{\mathcal{O}_C}(\Omega_{C/S}, -),$$

in the sense that it is exact when applied to any $\mathcal{O}_C$-module. In fact, we claim that this is surjective on the right, and that $\underline{\Phi}$ (and hence Φ) is representable; i.e., there is a sheaf of $\mathcal{O}_C$-modules $\Upsilon_{C/S}(E)$ such that $\underline{\Phi}(-) = \underline{\mathrm{Hom}}_{\mathcal{O}_C}(\Upsilon_{C/S}(E), -)$, and $\Phi(-) = \mathrm{Hom}_{\mathcal{O}_C}(\Upsilon_{C/S}(E), -)$.

Lemma 10.35. *The functors Φ and $\underline{\Phi}$ defined above are both representable by the same sheaf of $\mathcal{O}_C$-modules $\Upsilon_{C/S}(E)$ fitting into a short exact sequence* (10.12) *inducing* (10.11)*:*

$$0 \to \Omega_{C/S} \to \Upsilon_{C/S}(E) \to \underline{\mathrm{End}}(E) \to 0. \tag{10.12}$$

Proof. For the representability we give an explicit construction, although it is possible to obtain the same result more quickly by an application of the adjoint functor theorem. As mentioned above, to prove that Φ is representable it is equivalent to prove that $\underline{\Phi}$ is representable, for which we can work locally in C. We can therefore assume $E = \mathcal{O}_C^{\oplus n}$. But then if $\delta : \mathcal{O}_C \to F$ is any derivation, it itself gives a δ-connection by $\delta^{\times n}(x_1, \ldots, x_n) = (\delta(x_1) \otimes 1, \ldots, \delta(x_n) \otimes 1)$. This gives a natural bijection between $\Phi(F)$ and $\mathrm{Der}_{\mathcal{O}_S}(\mathcal{O}_C, F) \times \mathrm{Hom}(E, F \otimes E)$ sending (δ, φ) to $(\delta, \varphi - \delta^{\times n})$. But $\mathrm{Der}_{\mathcal{O}_S}(\mathcal{O}_C, -)$ is representable by $\Omega_{C/S}$ and $\mathrm{Hom}(E, (-) \otimes E)$ is representable by $E \otimes E^\vee$. Thus $\underline{\Phi}$ is representable. To check that the sequence (10.12) in the lemma is exact, it is enough to check locally, which we have just done. □

Remark 10.36. We can view $\mathrm{Der}_{\mathcal{O}_S}(\mathcal{O}_C, \pi^*J) = \mathrm{Hom}(\Omega_{C/S}, \pi^*J)$ as the group of automorphisms of $C[\epsilon\pi^*J]$ that act as the identity on C and on π^*J (the group of 'infinitesimal automorphisms'). Similarly, we can view the group $\Phi(\pi^*J) = \mathrm{Hom}(\Upsilon_{C/S}(E), \pi^*J)$ as the group of automorphisms of the pair $(C[\epsilon\pi^*J], E + \epsilon J \otimes \pi^*E)$, the trivial square-zero extension of (C, E) by J, that act trivially on C, J, and E.

In many deformation problems, deformations and obstructions are classified by analyzing the ways to glue, and the obstruction to gluing, deformations along infinitesimal automorphisms. Thus computing the infinitesimal automorphism group of the objects under consideration goes a long way toward understand how the object can deform. Indeed, it was by calculating the infinitesimal automorphisms of a deformation of (C, E) that we arrived at the definition of Φ in the first place.

Now consider an extension (C', E') of (C, E) to $S[\epsilon J]$. Let $\pi : C \to S$ be the projection. We have a canonical projection

$$\Upsilon_{C'/S}(E') \to \Upsilon_{C/S}(E) \tag{10.13}$$

and it is easy to see that this is surjective. This induces an exact sequence

$$0 \to \pi^* J \to \mathcal{O}_C \underset{\mathcal{O}_{C'}}{\otimes} \Upsilon_{C'/S}(E') \to \Upsilon_{C/S}(E) \to 0, \tag{10.14}$$

where the morphism $\pi^* J \to \mathcal{O}_C \otimes_{\mathcal{O}_{C'}} \Upsilon_{C'/S}(E')$ is the composition

$$\pi^* J \to \mathcal{O}_C \underset{\mathcal{O}_{C'}}{\otimes} \Omega_{C'/S} \to \mathcal{O}_C \underset{\mathcal{O}_{C'}}{\otimes} \Upsilon_{C'/S}(E'). \tag{10.15}$$

As in the proof of the exactness of (10.7), the right exactness and exactness in the middle of (10.14) is formal using the universal properties. The left exactness follows easily from the exactness of (10.7) and (10.12). Indeed, $\underline{\mathrm{End}}(E')$ is a flat $\mathcal{O}_{C'}$-module, so tensoring the short exact sequence (10.12) for E' on C'/S with $\mathcal{O}_C \otimes_{\mathcal{O}'_C} (-)$, we obtain the short exact sequence

$$0 \to \mathcal{O}_C \underset{\mathcal{O}_{C'}}{\otimes} \Omega_{C'/S} \to \mathcal{O}_C \underset{\mathcal{O}_{C'}}{\otimes} \Upsilon_{C'/S}(E') \to \mathcal{O}_C \underset{\mathcal{O}_{C'}}{\otimes} \underline{\mathrm{End}}(E') \to 0.$$

Therefore using also the exactness of (10.7), the morphisms in (10.15) are both injections. The discussion shows that each extension (C', E') of (C, E) induces an extension of $\Upsilon_{C/S}(E)$ by $\pi^* J$.

Conversely, given such an extension,

$$0 \to \pi^* J \to \Upsilon' \to \Upsilon_{C/S}(E) \to 0 \tag{10.16}$$

we can recover $\mathcal{O}_{C'}$ as $\delta^{-1}\Upsilon' = \mathcal{O}_{C'} \times_{\Upsilon} \Upsilon'$, where $\delta : \mathcal{O}_C \to \Upsilon_{C/S}(E)$ is the universal derivation; i.e., derivation δ in the universal pair (δ, φ) obtained from the identity map under the identification $\mathrm{Hom}(\Upsilon_{C/S}(E), \Upsilon_{C/S}(E)) = \Phi(\Upsilon_{C/S}(E))$. Likewise, we can recover E' by first tensoring (10.16) by E and then pulling back via the universal δ-connection φ:

$$\begin{array}{ccccccccc} 0 & \longrightarrow & \pi^* J \otimes E & \dashrightarrow & E' & \dashrightarrow & E & \longrightarrow & 0 \\ & & \| & & \downarrow & & \downarrow \varphi & & \\ 0 & \longrightarrow & \pi^* J \otimes E & \longrightarrow & \Upsilon' \otimes E & \longrightarrow & \Upsilon_{C/S}(E) \otimes E & \longrightarrow & 0 \end{array}$$

Summarizing the discussion above:

Lemma 10.37. *Let $\mathcal{G}$ be the stack of pairs (C, E) where C is a nodal curve and E is a vector bundle on C. Suppose that (C, E) is an S-point of $\mathcal{F}$. Then*

$$T_{\mathcal{F}}(C, E) = \mathbf{Ext}\big(\Upsilon_{C/S}(E), \mathcal{O}_C\big).$$

In particular,

$$\begin{aligned} T^{-1}_{\mathcal{F}}(C, E) &= \mathrm{Hom}\big(\Upsilon_{C/S}(E), \mathcal{O}_C\big) \\ T^{0}_{\mathcal{F}}(C, E) &= \mathrm{Ext}^1\big(\Upsilon_{C/S}(E), \mathcal{O}_C\big). \end{aligned}$$

Note that for a nodal curve C/S (or more generally a curve with locally planar singularities), applying $\mathrm{Hom}(-, \mathcal{O}_C)$ to Lemma 10.35, and utilizing Lemmas 10.37, 10.30, and Corollary 10.23, we obtain a long exact sequence

$$0 \longrightarrow T^{-1}_{\mathsf{Fib}_{C/S}}(E) \longrightarrow T^{-1}_{\mathcal{F}}(C, E) \longrightarrow T^{-1}_{\mathcal{N}}(C/S) \longrightarrow$$

$$\longrightarrow T^{0}_{\mathsf{Fib}_{C/S}}(E) \longrightarrow T^{0}_{\mathcal{F}}(C, E) \longrightarrow T^{0}_{\mathcal{N}}(C/S) \longrightarrow$$

$$\longrightarrow \mathrm{Ext}^2(\underline{\mathrm{End}}(E), \mathcal{O}_C) \to \mathrm{Ext}^2(\Upsilon_{C/S}(E), \mathcal{O}_C) \to \mathrm{Ext}^2(\Omega_{C/S}, \mathcal{O}_C) \to \cdots$$

Note that if S is affine, then the last row above is 0.

Remark 10.38. There is an explicit treatment of infinitesimal deformations of pairs (C, E) where C is a curve and E is a *line bundle* in [80, Thms. 3.1, 4.6], with infinitesimal automorphisms, deformations, and obstructions lying respectively in $\mathrm{Ext}^i(\mathscr{P}^1_{X/S}(E), E)$, $i = 0, 1, 2$, where $\mathscr{P}^1(E)$ is the sheaf of *principal parts* of E, and fits in an exact sequence $0 \to \Omega_{C/S}(E) \to \mathscr{P}^1_{X/S}(E) \to E \to 0$. The same formulas do not give a deformation–obstruction theory for the pair (C, E) when the rank of E is larger than 1.

Remark 10.39. Illusie provides a 2-step obstruction theory for deformations of the pair (C, E). The primary obstruction is to deforming C, which lies in this case in $\mathrm{Ext}^1(\Omega_{C/S}, \pi^* J)$, and in general in $\mathrm{Ext}^2(\mathbf{L}_{C/S}, \pi^* J)$, as we have discussed (see Remark 10.34). A secondary obstruction in $\mathrm{Ext}^2(\mathscr{P}^1(E), E)$ obstructs the existence of a deformation of E that is compatible with a fixed deformation C' of C [49, Prop. IV.3.1.5]. This latter obstruction may be constructed as the cup product of the class $[C'] \in \mathrm{Ext}^1(\mathbf{L}_{C/S}, \pi^* J)$ and the *Atiyah class* [49, §§IV.2.3.6–7].

One can arrive at this 2-step obstruction theory[c] from the extension (10.12), which induces an exact sequence:

$$\mathrm{Ext}^2(\underline{\mathrm{End}}(E), \pi^* J) \to \mathrm{Ext}^2(\Upsilon_{C/S}(E), \pi^* J) \to \mathrm{Ext}^2(\Omega_{C/S}, \pi^* J).$$

The first term may of course be identified canonically with $\mathrm{Ext}^2(E, E \otimes \pi^* J)$ since E is flat.

For the sake of completeness, we also observe homogeneity:

Lemma 10.40. *The stack of pairs (C, E) where C is a curve and E is a vector bundle on E is homogeneous.*

Proof. Let $\mathcal{X}$ denote the stack of pairs (C, E) as in the statement of the lemma and let $\mathcal{N}$ be the stack of nodal curves. The forgetful map $\mathcal{X} \to \mathcal{N}$ is relatively homogeneous by Lemma 10.25 and $\mathcal{N}$ is homogeneous by Lemma 10.31 so $\mathcal{X}$ is homogeneous by Lemma 10.18. □

10.2.5. *Simultaneous deformation of curves, vector bundles, and morphisms of vector bundles*

We will consider a curve C over S, vector bundles E and F on C, and a homomorphism $\sigma : E \to F$. We ask in how many ways these data can be extended to C', E', F', and σ' over $S[\epsilon J]$ where J is a quasicoherent sheaf on S. This time, deformations will be controlled by a complex rather than by a module.

Imitating the last section we can construct a quasicoherent sheaf $\Upsilon_{C/S}(E, F)$ controlling simultaneous deformations of the two vector bundles E and F. This will be the universal example of a quasicoherent $\mathcal{O}_C$-module, equipped with a derivation

$$\delta : \mathcal{O}_C \to \Upsilon_{C/S}(E, F)$$

and δ-connections,

$$\varphi_E : E \to \Upsilon_{C/S}(E, F) \otimes E$$
$$\varphi_F : F \to \Upsilon_{C/S}(E, F) \otimes F.$$

[c]We have only verified the obstruction groups coincide, not the obstruction classes.

As it would be similar to the proof of Lemma 10.35, we will omit an explicit verification of the existence of such a universal object, as well as the construction of the natural exact sequence:

$$0 \to \Omega_{C/S} \to \Upsilon_{C/S}(E,F) \to \underline{\mathrm{End}}(E) \times \underline{\mathrm{End}}(F) \to 0.$$

Before studying the question of deforming a homomorphism $\sigma : E \to F$, we analyze how we can recover the deformation $\underline{\mathrm{Hom}}(E',F')$ of $\underline{\mathrm{Hom}}(E,F)$ from deformations of E and F, encoded as an extension of $\Upsilon_{C/S}(E,F)$ by $\pi^* J$.

If $\sigma : E \to F$ is an homomorphism of vector bundles we obtain an element

$$[\sigma,\varphi] \in \Upsilon_{C/S}(E,F) \otimes \underline{\mathrm{Hom}}(E,F)$$
$$[\sigma,\varphi] := (\mathrm{id}_{\Upsilon_{C/S}(E,F)} \otimes \sigma) \circ \varphi_E - \varphi_F \circ \sigma : E \to \Upsilon_{C/S}(E,F) \otimes F.$$

As $[\sigma,\varphi]$ depends linearly on σ, this induces a homomorphism of $\mathcal{O}_C$-modules:

$$\varphi_{\underline{\mathrm{Hom}}(E,F)} : \underline{\mathrm{Hom}}(E,F) \to \Upsilon_{C/S}(E,F) \otimes \underline{\mathrm{Hom}}(E,F).$$

Lemma 10.41. *Suppose that $\Upsilon' \in \mathbf{Ext}(\Upsilon_{C/S}(E,F), \pi^* J)$ corresponds to extensions C' of C, E' of E, and F' of F. Then there is a cartesian square of sheaves of abelian groups:*

$$\begin{array}{ccc} \underline{\mathrm{Hom}}(E',F') & \longrightarrow & \underline{\mathrm{Hom}}(E,F) \\ \downarrow{\scriptstyle \varphi_{\underline{\mathrm{Hom}}(E',F')}} & & \downarrow{\scriptstyle \varphi_{\underline{\mathrm{Hom}}(E,F)}} \\ \Upsilon' \otimes \underline{\mathrm{Hom}}(E,F) & \longrightarrow & \Upsilon \otimes \underline{\mathrm{Hom}}(E,F) \end{array}$$

Proof. For the sake of readability, we will write

$$\Upsilon = \Upsilon_{C/S}(E,F) \quad H = \underline{\mathrm{Hom}}_{\mathcal{O}_C}(E,F) \quad H' = \underline{\mathrm{Hom}}_{\mathcal{O}_{C'}}(E',F') \quad \varphi = \varphi_H$$

We also observe that we can identify $\Upsilon' = \mathcal{O}_C \otimes_{\mathcal{O}_{C'}} \Upsilon_{C'/S}(E',F')$ by Lemma 10.37 (which allows us to identify the exact sequences (10.14) and (10.16)) and then write φ' for the composition of $\varphi_{H'}$ and reduction modulo $\pi^* J$:

$$\varphi' : H' \xrightarrow{\varphi_{H'}} \Upsilon_{C'/S}(E',F') \otimes_{\mathcal{O}_{C'}} H' \to \Upsilon' \otimes_{\mathcal{O}_C} H.$$

Consider the following commutative squares:

$$\begin{array}{ccc} \mathcal{O}_{C'} & \longrightarrow & \mathcal{O}_C \\ \downarrow{\scriptstyle \rho+\epsilon\delta'} & & \downarrow{\scriptstyle \mathrm{id}+\epsilon\delta} \\ \mathcal{O}_C + \epsilon\Upsilon' & \longrightarrow & \mathcal{O}_C + \epsilon\Upsilon \end{array} \qquad \begin{array}{ccc} H' & \longrightarrow & H \\ \downarrow{\scriptstyle \rho+\epsilon\varphi} & & \downarrow{\scriptstyle \mathrm{id}+\epsilon\varphi} \\ H + \epsilon(\Upsilon' \otimes H) & \longrightarrow & H + \epsilon(\Upsilon \otimes H) \end{array}$$

We have written ρ for reduction modulo π^*J and δ' for the composition

$$\mathcal{O}_{C'} \to \Upsilon_{C'/S}(E', F') \to \mathcal{O}_C \otimes_{\mathcal{O}_{C'}} \Upsilon_{C'/S}(E', F') = \Upsilon'.$$

We have already seen that the square on the left is cartesian in the discussion of (10.16) and Lemma 10.37. The square on the right is cartesian as well, since its horizontal arrows are surjective and we can identify the kernels of both horizontal arrows with $\pi^*J \otimes H$: on the top we have applied $\otimes_{\mathcal{O}_{C'}} H'$ to the exact sequence

$$0 \to \pi^*J \to \mathcal{O}_{C'} \to \mathcal{O}_C \to 0$$

noting that H' is flat over $\mathcal{O}_{C'}$, and on the bottom we have applied $\otimes_{\mathcal{O}_C} H$ to the exact sequence (10.16), this time noting that H is flat over $\mathcal{O}_C$. Each entry in the square on the right is a module under the ring in the corresponding entry of the square on the left and the arrows in the square on the right are homomorphisms with respect to the arrows in the square on the left. It follows that the module structure on H' is induced from the fiber product. The lemma now follows, since the square

$$\begin{array}{ccc} H + \epsilon(\Upsilon' \otimes H) & \longrightarrow & H + \epsilon(\Upsilon \otimes H) \\ \downarrow & & \downarrow \\ \Upsilon' \otimes H & \longrightarrow & \Upsilon \otimes H \end{array}$$

is cartesian. □

If we fix $\sigma : E \to F$, then $\varphi_{\underline{\mathrm{Hom}}(E,F)}(\sigma) \in \Upsilon_{C/S}(E, F) \otimes \underline{\mathrm{Hom}}(E, F)$ induces a linear map:

$$\underline{\mathrm{Hom}}(E, F)^\vee \to \Upsilon_{C/S}(E, F) \qquad : \Upsilon^\bullet_{C/S}(E, F, \sigma) \tag{10.17}$$

via evaluation against $\underline{\mathrm{Hom}}(E, F)$. As indicated write $\Upsilon^\bullet_{C/S}(E, F, \sigma)$ for the complex (10.17), concentrated in degrees $[-1, 0]$.

Remark 10.42. As with $\Upsilon_{C/S}(E)$, one may arrive at the definition of $\Upsilon^\bullet_{C/S}(E, F, \sigma)$ by contemplating, with a bit of care, the automorphism group of the trivial extension of (C, E, F, σ) to $S[\epsilon J]$ (cf. Remark 10.36).

In order to state the next lemma, we recall the category $\mathbf{Ext}(K^\bullet, L)$ of extensions of a complex $K^{-1} \to K^0$ by a module L is, by definition, the category of commutative diagrams

$$\begin{array}{ccccccccc} & & & & & & K^{-1} & & \\ & & & & & \swarrow & \downarrow & & \\ 0 & \longrightarrow & L & \longrightarrow & M & \longrightarrow & K^0 & \longrightarrow & 0 \end{array}$$

in which the bottom row is exact [1, §VII.3].

Lemma 10.43. *Let $\mathcal{F}$ be the stack of quadruples (C, E, F, σ) where C is a nodal curve, E and F are vector bundles on C, and $\sigma : E \to F$ is a morphism of vector bundles. At an S-point (C, E, F, σ), we have*

$$T_{\mathcal{F}}(C, E, F, \sigma) = \mathbf{Ext}\big(\Upsilon^{\bullet}_{C/S}(E, F, \sigma), \mathcal{O}_C\big)$$

and therefore

$$\begin{aligned} T^{-1}_{\mathcal{F}}(C, E, F, \sigma) &= \mathrm{Ext}^0\big(\Upsilon^{\bullet}_{C/S}(E, F, \sigma), \mathcal{O}_C\big) \\ T^{0}_{\mathcal{F}}(C, E, F, \sigma) &= \mathrm{Ext}^1\big(\Upsilon^{\bullet}_{C/S}(E, F, \sigma), \mathcal{O}_C\big). \end{aligned}$$

Proof. As in the previous sections we will actually prove the analogous statement about extensions to $S[\epsilon J]$. Suppose that (C', E', F', σ') is an extension of (C, E, F, σ) to $S[\epsilon J]$. We have a commutative diagram:

$$\begin{array}{ccccccccc} & & & & \mathcal{O}_C \otimes_{\mathcal{O}_{C'}} \underline{\mathrm{Hom}}(E', F')^{\vee} & \xrightarrow{\sim} & \underline{\mathrm{Hom}}(E, F)^{\vee} & & \\ & & & & \downarrow & & \downarrow & & \\ 0 & \longrightarrow & \pi^* J & \longrightarrow & \mathcal{O}_C \otimes_{\mathcal{O}_{C'}} \Upsilon_{C'/S}(E', F') & \longrightarrow & \Upsilon_{C/S}(E, F) & \longrightarrow & 0 \end{array}$$

The bottom row is exact, as in the last section. As the upper horizontal arrow is an isomorphism, we obtain an element of $\mathbf{Ext}\big(\Upsilon^{\bullet}_{C/S}(E, F, \sigma), \pi^* J\big)$.

Conversely, suppose we are given an extension:

$$\begin{array}{ccccccccc} & & & & & & \underline{\mathrm{Hom}}(E, F)^{\vee} & & \\ & & & & & \beta\swarrow & \downarrow & & \\ 0 & \longrightarrow & \pi^* J & \longrightarrow & \Upsilon' & \longrightarrow & \Upsilon_{C/S}(E, F) & \longrightarrow & 0 \end{array} \tag{10.18}$$

We may construct

$$\begin{aligned} \mathcal{O}_{C'} &= \delta^{-1}\Upsilon' = \mathcal{O}_C \times_{\Upsilon} \Upsilon' \\ E' &= \varphi_E^{-1}(\Upsilon' \otimes E) = E \times_{\Upsilon \otimes E} (\Upsilon' \otimes E) \\ F' &= \varphi_F^{-1}(\Upsilon' \otimes F) = F \times_{\Upsilon \otimes F} (\Upsilon' \otimes F). \end{aligned}$$

To get $\sigma' \in \mathrm{Hom}(E', F')$, regard β as an element of $\Upsilon' \otimes \underline{\mathrm{Hom}}(E, F)$. Then the image of β in $\Upsilon_{C/S}(E, F) \otimes \underline{\mathrm{Hom}}(E, F)$ is $[\sigma, \varphi]$, by the commutativity of (10.18). Therefore $\sigma' = (\sigma, \beta)$ defines an element of

$$\underline{\mathrm{Hom}}(E', F') = \underline{\mathrm{Hom}}(E, F) \underset{\Upsilon \otimes \underline{\mathrm{Hom}}(E,F)}{\times} \Upsilon' \otimes \underline{\mathrm{Hom}}(E, F).$$

We leave the verification that these constructions are mutually inverse to the reader. □

Lemma 10.44. *Let $\mathcal{F}$ be the stack of quadruples (C, E, F, σ) where C is a nodal curve, E and F are vector bundles on C, and $\sigma : E \to F$ is a homomorphism of vector bundles. Then $\mathcal{F}$ is homogeneous.*

Proof. Let $\mathcal{N}$ be the stack of nodal curves. The forgetful map $\mathcal{F} \to \mathcal{N}$ is relatively homogeneous by Corollary 10.27 and $\mathcal{N}$ is homogeneous by Lemma 10.31. Therefore $\mathcal{F}$ is homogeneous by Lemma 10.18. □

10.2.6. *Deformations of Higgs bundles*

We may apply the method of the previous section to study deformations of a Higgs bundle $\phi : E \to E \otimes \omega_{C/S}$ on a nodal curve C over S. We find that the deformation theory of (C, E, ϕ) is controlled by the complex $\Upsilon_{C/S}(E, \sigma)$, in degrees $[-1, 0]$:

$$\underline{\mathrm{Hom}}(E, E \otimes \omega_{C/S})^\vee \to \Upsilon_{C/S}(E)$$

Lemma 10.45. *Let (C, E, ϕ) be a nodal curve of genus g over S equipped with a Higgs bundle. Then*

$$T_{\mathcal{HS}h_{\overline{M}_g}}(C, E, \phi) = \mathbf{Ext}\big(\Upsilon_{C/S}(E, \sigma), \mathcal{O}_C\big)$$

and

$$T^{-1}_{\mathcal{HS}h_{\overline{M}_g}}(C, E, \phi) = \mathrm{Ext}^0\big(\Upsilon_{C/S}(E, \sigma), \mathcal{O}_C\big)$$
$$T^{0}_{\mathcal{HS}h_{\overline{M}_g}}(C, E, \phi) = \mathrm{Ext}^1\big(\Upsilon_{C/S}(E, \sigma), \mathcal{O}_C\big).$$

Lemma 10.46. *The stack of Higgs bundles on nodal (resp. stable) curves is homogeneous.*

Proof. Let $\mathcal{F}$ be the stack of quadruples (C, E, F, σ) where C is a nodal curve over a tacit base S, E and F are vector bundles on C, and $\sigma : E \to F$ is a homomorphism. Let $\mathcal{E}$ be the stack of pairs (C, E) where C is a nodal curve and E is a vector bundle on C. Let $\mathcal{H}$ be the stack of Higgs bundles on nodal curves.

We have seen in Lemma 10.40 that $\mathcal{E}$ is homogeneous and in Lemma 10.44 that $\mathcal{F}$ is homogeneous. Then we have two projections $\mathcal{F} \to \mathcal{E}$, one sending (C, E, F, σ) to F and the latter sending (C, E, F, σ) to $\omega_{C/S} \otimes E$. We can identify $\mathcal{H}$ with the equalizer of these two maps, that is, with the fiber product $\mathcal{F} \times_{\mathcal{E} \times \mathcal{E}} \mathcal{E}$, hence is homogeneous by Lemma 10.18.

Again by Lemma 10.18, we deduce that $\mathcal{H} \to \mathcal{N}$ is relatively homogeneous, where $\mathcal{N}$ denotes the stack of nodal curves. By base change, the stack $\mathcal{HS}h_{\overline{\mathcal{M}}_g}$ of Higgs bundles on stable curves of genus g is homogeneous over $\overline{\mathcal{M}}_g$ (again by Lemma 10.18). But we know $\overline{\mathcal{M}}_g$ is homogeneous (Lemma 10.31) so we deduce that $\mathcal{HS}h_{\overline{\mathcal{M}}_g}$ is homogeneous as well. □

10.2.7. *Deformations of principal bundles*

Suppose that $X \subseteq X'$ is a square-zero extension of schemes by the ideal J and G' is a smooth algebraic group over X'. Denote by G the preimage of X in G'. Let P be an *étale* G-torsor over X, meaning that the map $G \times_X P \to P \times_X P$ is an isomorphism, and that P covers X in the étale topology (Definition A.16). We would like to classify the extensions of P to G'-torsors over X'.

Remark 10.47. An algebraic group can have étale torsors that are not torsors in the Zariski topology (see [76]), so it is important that we have specified étale torsors here. Since G is smooth, working in the fppf or fpqc topology would not yield any more torsors: flat descent implies that any G-torsor would be representable by a smooth algebraic space over X, hence would have a section étale-locally.

Remark 10.48. This discussion of G-torsors generalizes the discussion of §10.2.2: given a vector bundle E over X, we may take $G = \mathrm{GL}_n$ and let P to be the G-torsor of isomorphisms between E and $\mathcal{O}_X^n$.

Before analyzing the deformation problem, we introduce the (underlying vector bundle of the) Lie algebra of G. Let $e : X \to G$ and $e' : X' \to G'$ denote the identity sections. We set

$$\mathfrak{g} = e^* T_{G/X}$$

to be the Lie algebra of G. We will also be interested in a closely related object. To introduce it, note that the small étale sites of X and X' are equivalent (Lemma 9.10), and that if U is an étale scheme over X we denote by U' the unique (up to unique isomorphism) extension of U to an étale scheme over X'.

We write $G'_{\text{ét}}$ and $G_{\text{ét}}$ for the restrictions of G' and G to this common étale site, and we introduce $\mathfrak{g}'$ for the kernel of the projection from $G'_{\text{ét}}$ to $G_{\text{ét}}$:

$$0 \to \mathfrak{g}' \to G'_{\text{ét}} \to G_{\text{ét}} \to 0.$$

In general $\mathfrak{g}'$ does not coincide with $\mathfrak{g}$, but the following lemma relates them:

Lemma 10.49. *Let $\mathfrak{g}$ and $\mathfrak{g}'$ be as above. Then $\mathfrak{g}' \simeq \mathfrak{g} \otimes J$ as sheaves of groups on the étale site of X. In particular, $\mathfrak{g}'$ is a sheaf of commutative groups.*

Proof. We observe that $\mathfrak{g}'$ is canonically a torsor under $\mathfrak{g} \otimes J$. Indeed, for an open subset $U \subseteq X$, denoting by $G'_e(U)$ the morphisms $U' \to G'$ whose restriction to U are the identity section, then we have:

$$\begin{aligned}
\mathfrak{g}'(U)\times\mathfrak{g}'(U) &= G'_e(U') \times G'_e(U') \\
&= G'_e(U' \amalg_U U') && \text{by homogeneity of } G \\
&= G'_e(U' \amalg_U U[\epsilon J_U]) && \mathcal{O}_{U'} \times_{\mathcal{O}_U} \mathcal{O}_{U'} \simeq \mathcal{O}_{U'} \times_{\mathcal{O}_U} \mathcal{O}_U[\epsilon J], \text{ p. 312} \\
&= G_e(U') \times G_e(U[\epsilon J]) && \text{by homogeneity again} \\
&= G_e(U') \times (\mathfrak{g} \otimes J)(U)
\end{aligned}$$

since $\mathfrak{g}$ is the tangent space to G at the origin. But $\mathfrak{g}'$ also has a canonical section over X coming from e'. Therefore $\mathfrak{g}' \simeq \mathfrak{g} \otimes J$ as a $\mathfrak{g} \otimes J$-torsor. In particular, $\mathfrak{g}'$ inherits a group structure from $\mathfrak{g} \otimes J$.

There is another group structure on $\mathfrak{g}'$ by virtue of its construction as a kernel. These two group structures commute with one another, in the sense that the multiplication map of either group structure is a homomorphism with respect to the other, and the identity elements are the same. By a standard argument, this means the two group structures coincide (and that both are commutative). □

As in the earlier sections, we make some observations about the local triviality of this problem:

(i) There is an *étale* cover of X' by maps $U' \to X'$ such that, setting $U = U' \times_{X'} X$, each G_U-torsor P_U extends to a $G'_{U'}$-torsor on U'.
(ii) Any two extensions of P to X' are isomorphic on a suitable étale cover of X'.
(iii) If P' and P'' are two extensions of P and $u, v : P' \to P''$ are two isomorphisms between them, then there is a unique map $h : P' \to G'$ over X' such that $h.u = v$ (we are denoting the group action map for the torsors with a dot, and the map $h.u$ is the composition of $h \times u : P' \times P' \to G' \times P''$ with the action map $G' \times P'' \to P''$).
As u and v both reduce to the identity map on P, the map h must reduce to the constant map $P \to e$. Therefore h factors uniquely

through $\mathfrak{g}'$:

$$\begin{array}{ccc} P' & \xrightarrow{\ h\ } & G' \\ & \searrow \quad \nearrow & \\ & \mathfrak{g}' & \end{array}$$

Furthermore, the induced map $h : P' \to \mathfrak{g}'$ is equivariant with respect to the natural action of G' on P' and on $\mathfrak{g}'$ (by conjugation). Indeed, we have

$$\begin{aligned} h(g.x).u(g.x) &= v(g.x) \\ h(g.x)g.u(x) &= g.v(x) = gh(x).u(x) \end{aligned}$$

for all x in P'. It follows that $h(g.x) = gh(x)g^{-1}$ for all $g \in G'$. Finally, $h : P' \to \mathfrak{g}'$ factors through the projection $P' \to P = P'/\mathfrak{g}'$. This is because $\mathfrak{g}'$ is commutative, so $h(g.x) = gh(x)g^{-1} = h(x)$ whenever $g \in \mathfrak{g}'$. Therefore h factors through $P'/\mathfrak{g}' = P$. Thus h factors uniquely as a *G-equivariant map*

$$P \longrightarrow \mathfrak{g}' = \mathfrak{g} \otimes J.$$

Putting all of this together using the same Čech or gerbe argument from Section 10.2.2, we obtain a theory of deformations and obstructions:

Theorem 10.50. *Suppose that G is a smooth algebraic group over a scheme X, that X' is a square-zero extension of X, and that G' is an extension of G to X' by the ideal J. Let P be a G-torsor on X. There is an obstruction to extending P to a G'-torsor on X' lying in $H^2\big(X, \underline{\mathrm{Hom}}_G(P, \mathfrak{g} \otimes J)\big)$. Should this obstruction vanish, deformations are canonically a torsor on X under $H^1\big(X, \underline{\mathrm{Hom}}_G(P, \mathfrak{g} \otimes J)\big)$, and automorphisms of any given deformation are canonically isomorphic to $H^0\big(X, \underline{\mathrm{Hom}}_G(P, \mathfrak{g} \otimes J)\big)$.*

In the case of the trivial extension $X[\epsilon]$, so that $J = \mathcal{O}_X$, and $\mathfrak{g} = \mathfrak{g} \otimes J = \mathfrak{g}'$, the bundle

$$\mathfrak{p} := \underline{\mathrm{Hom}}_G(P, \mathfrak{g})$$

is known as the *adjoint bundle* of P.

Remark 10.51. The adjoint bundle can also be constructed as the quotient of $P \times \mathfrak{g}$ by the diagonal action of G. To see that these are equivalent, note that a section of $P \times_G \mathfrak{g}$ corresponds to a G-orbit in $P \times \mathfrak{g}$, which is the graph of an equivariant map $P \to \mathfrak{g}$. Thus the definition of the adjoint bundle given above is equivalent to the one given in Section 7.8.

Lemma 10.52. *The adjoint bundle $\mathfrak{p}$ of a G-torsor P over X is isomorphic to the space of G-invariant vector fields on the fibers of P over X.*

Proof. Recall that a vector field on the fibers of P over X is an X-morphism $V : P[\epsilon] \to P$ that lifts the identity on $P \subset P[\epsilon]$. To be invariant means that the diagram

$$\begin{array}{ccc} G \times P[\epsilon] & \xrightarrow{\mathrm{id}\times V} & G \times P \\ {\scriptstyle p_2}\downarrow & & \downarrow{\scriptstyle \alpha} \\ P[\epsilon] & \xrightarrow{V} & P \end{array}$$

commutes, α being the action map and p_2 the second projection. Now, the zero vector field also gives a map $P[\epsilon] \to P$, and the pair $(0, V)$ gives an X-morphism

$$P[\epsilon] \xrightarrow{(0,V)} P \times_X P \xleftarrow{\sim} G \times_X P. \tag{10.19}$$

The first map is equivariant with respect to the diagonal G-action on $P \times_X P$. Since the map $G \times_X P \to P \times_X P$ sends (g, y) to (gy, y), the second arrow is also equivariant, provided we let G act by conjugation on itself. The original condition that $P[\epsilon] \to P$ restrict to the identity on $P \subset P[\epsilon]$ means that $P \subset P[\epsilon] \to P \times_X P$ factors through the diagonal, and therefore that $P \subset P[\epsilon] \to G \times_X P \to G$ factors through the identity section of G.

Now composing (10.19) with the projection $G \times_X P \to G$, we get an equivariant map

$$P[\epsilon] = P \times X[\epsilon] \to G,$$

that factors through the identity section of G when restricted to P. This is the same as to give an equivariant map

$$P \to \underline{\mathrm{Hom}}_X(X[\epsilon], G) \times_{\underline{\mathrm{Hom}}_X(X,G)} \{e\} = T_{G/X} \times_G \{e\} = \mathfrak{g},$$

which, by definition is the same as to give a section of the adjoint bundle of P. □

Corollary 10.53. *The adjoint bundle $\mathfrak{p}$ of a G-torsor P has the structure of a sheaf of Lie algebras. If G is semisimple then the Killing form furnishes an isomorphism between $\mathfrak{p}$ and its dual.*

Proof. The point is to verify that the bracket of equivariant vector fields on P is equivariant. This can be done with a diagram chase, using the definition of the Lie bracket in Section 10.1.3, but we omit it. Then P is locally isomorphic to G so $\mathfrak{p}$ is locally isomorphic to $\mathfrak{g}$, and this isomorphism preserves the Killing form. If G is semisimple then $\mathfrak{g}$ is self dual with respect to the Killing form. □

For the next statement, let X be an S-scheme and let G be a smooth group scheme over X. Let $\mathsf{Fib}^G_{X/S}$ denote the S-stack whose sections over T are the G-torsors on X_T.

Theorem 10.54. *We have* $T_{\mathsf{Fib}^G_{X/S}}(P) = \mathrm{B}\underline{\mathrm{Hom}}_G(P, \mathfrak{g}) = \mathrm{B}\mathfrak{p}$ *where* $\mathfrak{g}$ *denotes the Lie algebra of* G. *In particular,*

$$T^{-1}_{\mathsf{Fib}^G_{X/S}}(P) = H^0\big(X, \underline{\mathrm{Hom}}_G(P, \mathfrak{g})\big) = H^0(X, \mathfrak{p})$$
$$T^{0}_{\mathsf{Fib}^G_{X/S}}(P) = H^1\big(X, \underline{\mathrm{Hom}}_G(P, \mathfrak{g})\big) = H^1(X, \mathfrak{p}).$$

Proof. Let $X' = X[\epsilon]$. Let P' be the trivial extension of P to X'. Suppose that P'' is another extension. Then $\underline{\mathrm{Isom}}^P_{G'}(P'', P')$ (isomorphisms respecting the G'-action and the maps to P) is a torsor on X' under $\underline{\mathrm{Hom}}_G(P, \mathfrak{g})$ as explained in (iii) above (X' has the same étale site as X). The same argument shows that $\underline{\mathrm{Aut}}^P_{G'}(P') = \mathfrak{p}$, so that, given any torsor Q under $\mathfrak{p} = \underline{\mathrm{Hom}}_G(P, \mathfrak{g})$, we can form the sheaf of $\mathfrak{p}$-equivariant maps from Q to P'

$$P'' = \underline{\mathrm{Hom}}_{\mathfrak{p}}(Q, P').$$

These constructions are inverse isomorphisms between $T_{\mathsf{Fib}^G_{X/S}}(P)$ and the stack of torsors under $\mathfrak{p} = \underline{\mathrm{Hom}}_G(P, \mathfrak{g})$. □

Now suppose that X is a reduced (and therefore Cohen–Macaulay) projective curve over a field k (i.e., $S = \operatorname{Spec} k$). Then by Serre duality, we have

$$T^{0}_{\mathsf{Fib}^G_{X/S}}(P) = H^1(X, \mathfrak{p}) \simeq H^0\big(X, \underline{\mathrm{Hom}}(\mathfrak{p}, \omega_X)\big)^\vee.$$

Generalizing Definition 7.26:

Definition 10.55 (G-Higgs bundle). Let X be a reduced projective curve over a field k, and let G be a smooth algebraic group over k. A *G-Higgs bundle* on X is a pair (P, Φ) where P is a G-torsor over X and $\Phi \in \mathrm{Hom}(\mathfrak{p}, \omega_X)$, where $\mathfrak{p} := \underline{\mathrm{Hom}}_G(P, \mathfrak{g})$.

Remark 10.56. If G is semisimple then $\mathfrak{p}$ is self-dual (Corollary 10.53), so a G-Higgs bundle on X can also be viewed as an element of $H^0(X, \mathfrak{p} \otimes \omega_X)$, or, in another popular notation, of $H^0(X, \operatorname{ad} P \otimes K_X)$. In other words, Definition 7.26 agrees with Definition 10.55.

In the case where P is the GL_n-torsor associated to a vector bundle E, we have $\mathfrak{p} = \underline{\mathrm{End}}(E)$, so that

$$\underline{\mathrm{Hom}}(\mathfrak{p}, \omega_X) = \underline{\mathrm{Hom}}(E, E \otimes \omega_X),$$

and $T^0_{\mathsf{Fib}^G_{X/k}}(P) = H^1(X, \mathfrak{p}) \simeq \mathrm{Hom}(\mathfrak{p}, \omega_X)^\vee = \mathrm{Hom}(E, E \otimes \omega_X)^\vee$. More generally, extending the definition of G-Higgs bundles to families of curves, and using Grothendieck duality, the discussion above proves the following corollary:

Corollary 10.57. *Let X be a family of smooth curves over S. Higgs bundles on X/S correspond to isomorphism classes of relative cotangent vectors for $\mathsf{Fib}_{X/S}$ over S. Likewise, G-Higgs bundles on X/S correspond to relative cotangent vectors over S for the stack $\mathsf{Fib}^G_{X/S}$ of principal G-bundles on X.*

10.3. *Obstruction theory*

In this section, we will consider a stack $\mathcal{F}$ (not a priori algebraic) and a lifting problem

$$\begin{array}{ccc} S & \xrightarrow{\xi} & \mathcal{F} \\ \downarrow & \nearrow & \\ S' & & \end{array} \tag{10.20}$$

in which S' is a square-zero extension of S' with ideal J. We are looking for an *obstruction theory* that can detect whether this lifting problem has a solution. The obstruction theory will be an $\mathcal{O}_S$-module $T^1_{\mathcal{F}}(\xi, J)$ that depends only on J and $\xi \in \mathcal{F}(S)$, *not on the particular extension S'*. The above lifting problem will then produce a natural obstruction $\omega \in T^1_{\mathcal{F}}(\xi, J)$ whose vanishing is equivalent to the existence of a lift.

We will see that for moduli problems that are locally unobstructed in a suitable sense (that includes all of the examples considered here) there is a natural choice for $T^1_{\mathcal{F}}(\xi, J)$ arising from the infinitesimal deformation theory. This phenomenon could be better explained in the language of torsors under abelian group stacks, but to introduce such objects would take us too far afield. We will rely instead on injective resolutions and Čech methods.

Various definitions of obstruction theories have appeared in the literature [9, §2.6], [11, Def. 4.4], [55, Def. 1.2], [77, Tag 07YG], [43, Def. 6.6], [82, Def. 3.2]. From the perspective of derived algebraic geometry, an obstruction theory arises from the promotion of a moduli problem to a derived moduli problem [57]. The definition we adopt here is closest to [43].

Definition 10.58. Let $\mathcal{F}$ be a stack on the étale site of schemes. An *obstruction theory* for $\mathcal{F}$ is a system of abelian groups $T^1(\xi, J)$ depending contravariantly on a scheme S and an element $\xi \in \mathcal{F}(S)$, and covariantly on a quasicoherent $\mathcal{O}_S$-module J, together with an obstruction map

$$\omega : \operatorname{Exal}(\mathcal{O}_S, J) \to T^1(\xi, J)$$

that is natural in S and J and has the following property: $\omega(S') = 0$ if and only if diagram (10.20) admits a lift.

Above we use the notation $\operatorname{Exal}(\mathcal{O}_S, J)$ for the $\mathcal{O}_S$-module of algebras that are extensions of $\mathcal{O}_S$ by J with $J^2 = 0$ (see e.g., [75, §1.1] for more details).

We spell out precisely the meaning of naturality in Definition 10.58. Suppose there is a commutative (not necessarily cartesian) diagram of square-zero extensions (10.21) where $R \subseteq R'$ has ideal I and $S \subseteq S'$ has ideal J.

$$\begin{array}{ccc} R & \longrightarrow & R' \\ {\scriptstyle f}\downarrow & & \downarrow \\ S & \longrightarrow & S' \end{array} \tag{10.21}$$

Then we get a homomorphism $f^*J \to I$ and therefore a morphism:

$$T^1(\xi, J) \to T^1(f^*\xi, f^*J) \to T^1(f^*\xi, I). \tag{10.22}$$

We also have obstructions $\omega(S') \in T^1(\xi, J)$ and $\omega(R') \in T^1(f^*\xi, I)$. The naturality alluded to in Definition 10.58 requires that $\omega(S')$ is carried under the morphism (10.22) to $\omega(R')$.

Our goal in this section will be to illustrate a technique for constructing an obstruction theory for a stack. For concreteness, we will consider the case where $\mathcal{F}$ is the stack of quadruples (C, E, F, σ) in which $\pi : C \to S$ is a nodal curve, E and F are vector bundles on C, and $\sigma : E \to F$ is a morphism of vector bundles. However, apart from the following observations, analogues of which frequently hold for other stacks, the particular choice of $\mathcal{F}$ will not enter into the rest of the discussion:

(1) Fix a scheme S, an S-point $\xi = (C, E, F, \sigma)$ of $\mathcal{F}$, and a square-zero extension $S \subseteq S'$ with ideal J. We obtain a square-zero extension of sheaves of commutative rings on the Zariski site, $\pi^{-1}\mathcal{O}_{S'} \to \pi^{-1}\mathcal{O}_S$,

with ideal $\pi^{-1}J$. Our problem can be phrased as the search for a square-zero extension $\mathcal{O}_{C'}$ of $\mathcal{O}_C$ by π^*J compatible with the extension $\pi^{-1}\mathcal{O}_{S'} \to \pi^{-1}\mathcal{O}_S$ and the homomorphism $\pi^{-1}J \to \pi^*J$, together with locally free $\mathcal{O}_{C'}$-modules E' and F' extending E and F and a morphism $\sigma' : E' \to F'$ extending σ.
We could replace $\pi^{-1}\mathcal{O}_{S'}$ by the extension $\mathscr{A}$ of $\pi^{-1}\mathcal{O}_S$ by π^*J obtained by pushout:

$$\begin{array}{ccccccccc} 0 & \longrightarrow & \pi^{-1}J & \longrightarrow & \pi^{-1}\mathcal{O}_{S'} & \longrightarrow & \pi^{-1}\mathcal{O}_S & \longrightarrow & 0 \\ & & \downarrow & & \downarrow & & \| & & \\ 0 & \longrightarrow & \pi^*J & \longrightarrow & \mathscr{A} & \longrightarrow & \pi^{-1}\mathcal{O}_S & \longrightarrow & 0 \end{array}$$

To find an extension of $\mathcal{O}_C$ by π^*J compatible with $\mathscr{A}$ is the same as to find a $\pi^{-1}\mathcal{O}_{S'}$-algebra extension compatible with the homomorphism $\pi^{-1}J \to \pi^*J$. Indeed, an $\mathscr{A}$-algebra extension induces a $\pi^{-1}\mathcal{O}_{S'}$-algebra extension by composition with $\pi^{-1}\mathcal{O}_{S'} \to \mathscr{A}$; conversely, if $\mathscr{B}$ is a $\pi^{-1}\mathcal{O}_{S'}$-algebra extension of $\mathcal{O}_C$ by π^*J then the map $\pi^{-1}J \to \mathscr{B}$ factors through π^*J so the map $\pi^{-1}\mathcal{O}_{S'} \to \mathscr{B}$ factors through $\mathscr{A}$.
Moreover, we may consider an arbitrary square-zero extension $\mathscr{A}$ of $\pi^{-1}\mathcal{O}_S$ by an $\mathcal{O}_C$-module $\mathscr{J}$ and define $\mathcal{F}_\xi(\mathscr{A})$ to be the category of quadruples $(\mathcal{O}_{C'}, E', F', \sigma')$ where $\mathcal{O}_{C'}$ is an extension of $\mathcal{O}_C$ by $\mathscr{J}$, compatible with $\mathscr{A}$; both E' and F' are locally free $\mathcal{O}_{C'}$-modules extending E and F, respectively; and $\sigma' : E' \to F'$ is a morphism of $\mathcal{O}_{C'}$-modules extending σ.
We observe that $\mathcal{F}_\xi$ is a homogeneous functor. This allows us to define $T^{-1}_{\mathcal{F}}(\xi, \mathscr{J})$ and $T^0_{\mathcal{F}}(\xi, \mathscr{J})$.

(2) There is a complex $\Upsilon^\bullet$ of $\mathcal{O}_C$-modules with quasicoherent cohomology such that

$$T_{\mathcal{F}}(\xi, \mathscr{J}) = \mathbf{Ext}(\Upsilon^\bullet, \mathscr{J}).$$

In other words, the infinitesimal deformations and automorphisms of ξ should be representable by a complex. In the case of interest, $\Upsilon^\bullet$ is the complex $\Upsilon^\bullet_{C/S}(E, F, \sigma)$, constructed in §10.2.5.
For the representability of the obstruction theory by a complex on S, we will also need the observation that $\Upsilon^\bullet$ is perfect as an object of the derived category. That is, locally in C it can be represented by a bounded complex of locally free modules. This is even true globally for $\Upsilon_{C/S}(E, F, \sigma)$.

(3) Whenever $\mathscr{J}$ is an *injective* $\mathcal{O}_C$-module and $\mathscr{A}$ is a square-zero extension of $\pi^{-1}\mathcal{O}_S$ by $\mathscr{J}$, there is some $\xi' \in \mathcal{F}_\xi(\mathscr{A})$. In our situation of interest, we can see that this is the case by considering the successive obstructions introduced in earlier sections. There is no local obstruction to deforming curves (Corollary 10.32). The first global obstruction (gluing isomorphism classes of local deformations) lies in

$$H^1(C, \underline{\mathrm{Ext}}^1(\Omega_{C/S}, \mathscr{J})) = 0.$$

The obstruction to finding a global deformation inside compatible local isomorphism classes lies in

$$H^2(C, \underline{\mathrm{Hom}}(\Omega_{C/S}, \mathscr{J})) = 0.$$

Therefore we can find $\mathcal{O}_{C'}$. Now the obstructions to extending E and F lie in

$$\mathrm{Ext}^2(\underline{\mathrm{End}}(E)^\vee, \mathscr{J}) = 0$$
$$\mathrm{Ext}^2(\underline{\mathrm{End}}(F)^\vee, \mathscr{J}) = 0.$$

Finally, the obstruction to extending σ, once $\mathcal{O}_{C'}$, E', and F' have been chosen, lies in

$$\mathrm{Ext}^1(\underline{\mathrm{Hom}}(E, F)^\vee, \mathscr{J}) = 0.$$

One could also verify the existence of ξ' directly by choosing local deformations arbitrarily and using the fact that $\mathscr{J}$ is a flasque sheaf to assemble them into a global deformation. Thus, if one wants to axiomatize the discussion of this section, one should require that $\mathcal{F}_\xi(\mathscr{A})$ be locally nonempty, when regarded as a CFG over the Zariski site of C.

Now we will build a global obstruction to the existence of $\xi' \in \mathcal{F}(S')$ lifting $\xi \in \mathcal{F}(S)$. The obstruction will lie in $\mathrm{Ext}^2(\Upsilon^\bullet, \pi^* J)$, which we will take as the definition of $T^1_{\mathcal{F}}(\xi, J)$.

We begin by choosing a resolution

$$0 \to \pi^* J \to \mathscr{J}^0 \to \mathscr{J}^1 \to 0$$

where $\mathscr{J}^0$ and $\mathscr{J}^1$ are sheaves of $\mathcal{O}_C$-modules and $\mathscr{J}^0$ is injective. We leave it to the reader to verify that the construction of the obstruction given here is independent of the choice of resolution.

We can push out the extension:

$$\begin{array}{ccccccccc} 0 & \longrightarrow & \pi^{-1}J & \longrightarrow & \pi^{-1}\mathcal{O}_{S'} & \longrightarrow & \pi^{-1}\mathcal{O}_S & \longrightarrow & 0 \\ & & \downarrow & & \downarrow & & \| & & \\ 0 & \longrightarrow & \mathscr{J}^0 & \longrightarrow & \mathscr{A} & \longrightarrow & \pi^{-1}\mathcal{O}_S & \longrightarrow & 0 \\ & & \downarrow & & \downarrow & & \| & & \\ 0 & \longrightarrow & \mathscr{J}^1 & \longrightarrow & \mathscr{B} & \longrightarrow & \pi^{-1}\mathcal{O}_S & \longrightarrow & 0 \end{array}$$

Note $\mathscr{B}$ is canonically isomorphic to $\pi^{-1}\mathcal{O}_S + \epsilon\mathscr{J}^1$ because the map $\pi^{-1}J \to \mathscr{J}^1$ is zero.

The commutative diagram above induces a map:

$$\mathcal{F}_\xi(\mathscr{A}) \to \mathcal{F}_\xi(\mathscr{B}) = T_{\mathcal{F}}(\xi, \mathscr{J}^1). \tag{10.23}$$

The identification with $T_{\mathcal{F}}(\xi, \mathscr{J}^1)$ comes by virtue of the canonical isomorphism $\mathscr{B} \simeq \pi^{-1}\mathcal{O}_S + \epsilon\mathscr{J}^1$.

Note that $\mathcal{F}_\xi(\mathscr{B})$ contains a canonical zero element (corresponding to the zero element of $T_{\mathcal{F}}(\xi, \mathscr{J}^1)$), namely the image of ξ under the map $\mathcal{F}_\xi(\pi^{-1}\mathcal{O}_S) \to \mathcal{F}_\xi(\mathscr{B})$ induced from the homomorphism $\pi^{-1}\mathcal{O}_S \to \pi^{-1}\mathcal{O}_S + \epsilon\mathscr{J}^1 = \mathscr{B}$, so we can speak of the kernel of (10.23). By definition, this is the fiber product of groupoids $0 \times_{\mathcal{F}_\xi(\mathscr{B})} \mathcal{F}_\xi(\mathscr{A})$.

Lemma 10.59. *There is a canonical identification between* $\mathcal{F}_\xi(\pi^{-1}\mathcal{O}_{S'})$ *and the kernel of* (10.23).

Proof. The diagram

$$\begin{array}{ccc} \pi^{-1}\mathcal{O}_{S'} & \longrightarrow & \mathscr{A} \\ \downarrow & & \downarrow \\ \pi^{-1}\mathcal{O}_S & \longrightarrow & \mathscr{B} \end{array}$$

is cartesain (the bottom arrow is the canonical splitting of the projection $\mathscr{B} \to \pi^{-1}\mathcal{O}_S$). Therefore the homogeneity of $\mathcal{F}_\xi$ implies that

$$\begin{array}{ccc} \mathcal{F}_\xi(\pi^{-1}\mathcal{O}_{S'}) & \longrightarrow & \mathcal{F}_\xi(\mathscr{A}) \\ \downarrow & & \downarrow \\ \mathcal{F}_\xi(\pi^{-1}\mathcal{O}_S) & \longrightarrow & \mathcal{F}_\xi(\mathscr{B}) \end{array}$$

is also cartesian. But $\mathcal{F}_\xi(\pi^{-1}\mathcal{O}_S) = 0$, by definition, so this identifies $\mathcal{F}_\xi(\pi^{-1}\mathcal{O}_{S'})$ with the kernel of (10.23). $\square$

As $\mathscr{J}^0$ is injective, $\mathcal{F}_\xi(\mathscr{A}) \neq \varnothing$ (observation 3 on p. 330), so that $\mathrm{Ext}^1(\Upsilon^\bullet, \mathscr{J}^0)$ acts simply transitively on the set $|\mathcal{F}_\xi(\mathscr{A})|$ of isomorphism classes in $\mathcal{F}_\xi(\mathscr{A})$ (by observation (2) on p. 330). As $\mathscr{B}$ is a (canonically) split extension of $\pi^{-1}\mathcal{O}_S$, we have a (canonical) identification $\mathrm{Ext}^1(\Upsilon^\bullet, \mathscr{J}^1) = |\mathcal{F}_\xi(\mathscr{B})|$. Now we have an exact sequence in the top row below, with compatible actions illustrated in the bottom row:

$$\begin{array}{ccccccc}
\mathrm{Ext}^1(\Upsilon^\bullet, \pi^{-1}J) & \longrightarrow & \mathrm{Ext}^1(\Upsilon^\bullet, \mathscr{J}^0) & \longrightarrow & \mathrm{Ext}^1(\Upsilon^\bullet, \mathscr{J}^1) & \longrightarrow & \mathrm{Ext}^2(\Upsilon^\bullet, \pi^{-1}J) \\
\circlearrowright & & \circlearrowright & & \| & & \\
|\mathcal{F}_\xi(\pi^{-1}\mathcal{O}_{S'})| & \longrightarrow & |\mathcal{F}_\xi(\mathscr{A})| & \longrightarrow & |\mathcal{F}_\xi(\mathscr{B})| & &
\end{array}$$

Note that $\mathcal{F}_\xi(\pi^{-1}\mathcal{O}_{S'})$ may be empty. As the action of $\mathrm{Ext}^1(\Upsilon^\bullet, \mathscr{J}^0)$ on $|\mathcal{F}_\xi(\mathscr{A})|$ is faithful and transitive, we have that the image of $|\mathcal{F}_\xi(\mathscr{A})|$ is an $\mathrm{Ext}^1(\Upsilon^\bullet, \mathscr{J}^0)$ coset in $\mathrm{Ext}^1(\Upsilon^\bullet, \mathscr{J}^1)$. In other words, it gives a well-defined element of $\mathrm{Ext}^2(\Upsilon^\bullet, \pi^{-1}J)$ obstructing the existence of an element of $\mathcal{F}_\xi(\pi^{-1}\mathcal{O}_{S'})$. We therefore define

$$T^1_{\mathcal{F}}(S, J) := \mathrm{Ext}^2(\Upsilon^\bullet, \pi^{-1}J).$$

By construction, $T^1_{\mathcal{F}}(S, J)$ is functorial with respect to S (contravariant) and J (covariant) and the obstruction class is natural.

Definition 10.60 (Representable deformation-obstruction theory). Let $\mathcal{F}$ be a stack in the étale topology on schemes. We will say that a deformation-obstruction theory $T^i_{\mathcal{F}}$, $i = -1, 0, 1$, is *representable* at an S-point ξ if there is a complex of locally free sheaves $\mathbf{E}^\bullet$ on S, such that for any $f : T \to S$ and any quasicoherent sheaf J on T, we have a natural (in T and in J) identification

$$T^i_{\mathcal{F}}(f^*\xi, J) = \mathrm{Ext}^i(f^*\mathbf{E}^\bullet, J)$$

for $i = -1, 0, 1$. We will say it is *finitely presentable* if the vector bundles in the complex $\mathbf{E}^\bullet$ may be chosen to have finite rank.

We say that the deformation theory is *locally representable* if it is representable at all points valued in affine schemes. We say it is *locally finitely presentable* if it is finitely presentable at all points valued in affine schemes.

Lemma 10.61. *Let $\mathcal{F}$ be the stack of quadruples (C, E, F, σ) where C is a nodal curve, E and F are vector bundles over C, and $\sigma : E \to F$ is a morphism of vector bundles. Then the obstruction theory for $\mathcal{F}$ introduced above is locally finitely presentable.*

Proof. Let A be a commutative ring and let $\xi = (C, E, F, \sigma)$ be an A-point of $\mathcal{F}$. We want to show that there is a complex of locally free A-modules $\mathbf{E}^\bullet$ representing $T^i_{\mathcal{F}}(\xi, J)$ for $i = -1, 0, 1$ and all A-modules J. We assume first that A is noetherian.

Let $\Xi^\bullet$ be a bounded above complex of locally free sheaves on C that is dual to $\Upsilon^\bullet$. Then

$$T^i_{\mathcal{F}}(\xi, J) = \operatorname{Ext}^i(\Upsilon^\bullet, \pi^* J) = H^i(C, \Xi^\bullet \otimes \pi^* J).$$

Therefore, by [47, III.12.2], there is a complex $L^\bullet$ of finite rank vector bundles on S such that[d]

$$T^i_{\mathcal{F}}(\xi, J) = h^i(L^\bullet \otimes_A J).$$

But now we may take $\mathbf{E}^\bullet$ to be a complex dual to $L^\bullet$ and obtain

$$\operatorname{Ext}^i(\mathbf{E}^\bullet, J) = h^i(L^\bullet \otimes_A J)$$

as required.

Now we pass to the general case. We can always write A as a filtered colimit of commutative rings of finite type $A = \varinjlim A_i$. As $\mathcal{F}$ is locally of finite presentation (Corollary 11.15) there is some index j and some $\xi_j = (C_j, E_j, F_j, \sigma_j) \in \mathcal{F}(A_j)$ inducing (C, E, F, σ). Now, $T^i_{\mathcal{F}}(\xi, J) = T^i_{\mathcal{F}}(\xi_j, J)$, with J regarded as an A_j-module via the map $A_j \to A$. Therefore

$$T^i_{\mathcal{F}}(\xi, J) = T^i_{\mathcal{F}}(\xi_j, J) = \operatorname{Ext}^i_{A_j}(\mathbf{E}^\bullet, J) = \operatorname{Ext}^i_A(A \underset{A_j}{\otimes} \mathbf{E}^\bullet, J)$$

so $T_{\mathcal{F}}$ is locally representable at ξ. □

Lemma 10.62. *Suppose that $\mathcal{X}$ is a stack with a locally finitely presentable obstruction theory. Then, for any field-valued point ξ of $\mathcal{X}$, the vector spaces $T^{-1}_{\mathcal{X}}(\xi)$, $T^0_{\mathcal{X}}(\xi)$, and $T^1_{\mathcal{X}}(\xi)$ are finite dimensional.*

Proof. This is immediate, because the vector spaces in question may be identified with cohomology groups of the dual complex of a complex of finite rank vector bundles. □

[d]Loc. cit. requires that $\Xi^\bullet$ be a quasicoherent sheaf, but the proof works for a complex as long as one takes the total complex of the Čech double complex at the bottom of p. 282.

11. Artin's criterion for algebraicity

Intuition from analytic moduli spaces suggests that moduli spaces should locally be embedded as closed subspaces of their tangent spaces. For this to apply in an algebraic context, locally must be interpreted to mean *étale-locally* for schemes and algebraic spaces, and *smooth-locally* for algebraic stacks.

Artin gives criteria under which a stack is locally cut out by polynomial equations inside its tangent space, thereby ensuring the stack is algebraic. Since Artin's original formulation, there have been a number of improvements [26, 43, 44, 58, 68]. The statement we give here is close to the form given by Hall [43], but with some hypotheses strengthened for the sake of transparency:

Theorem 11.1 (Artin's criterion [43, Thm. A]). *Let S be an excellent scheme and let $\mathcal{X}$ be a CFG over the category S/S of S-schemes. Then $\mathcal{X}$ is an algebraic stack over $(\mathsf{S}/S)_{\mathrm{et}}$ that is locally of finite presentation over S if and only if it has the following properties:*

(1) $\mathcal{X}$ is a stack in the étale topology (Definition 3.31).
(2) $\mathcal{X}$ is homogeneous (Definition 10.12).
(3) $\mathcal{X}$ has finite dimensional tangent and automorphism spaces (Section 10.1).
(4) $\mathcal{X}$ is integrable (Definition 11.17).
(5) $\mathcal{X}$ is locally of finite presentation (Sections 9.6 and 11.2).
(6) $\mathcal{X}$ has a locally finitely presentable obstruction theory (Definition 10.60).

Remark 11.2. The assumption (3) actually follows from (5) (see Lemma 10.62); however, we include the hypothesis (3) since it is useful from an expository perspective. In particular, a theorem of Schlessinger–Rim (see Theorem 11.10) uses the hypotheses (2) and (3) on a CFG.

Remark 11.3. Hall only requires the existence of a multistep obstruction theory, which is an *a priori* weaker hypothesis than (6). *A posteriori*, every algebraic stack has a cotangent complex, whence a single step obstruction theory. In our case, we actually constructed the obstruction theory for Higgs bundles in pieces in Section 10.2, but we were able to assemble it into a single-step obstruction theory in Section 10.3.

Remark 11.4. Property (5) is sometimes phrased, '$\mathcal{X}$ is limit preserving'.

Remark 11.5. For Property (4), some would say '$\mathcal{X}$ is effective' or 'formal objects of $\mathcal{X}$ may be effectivized' or 'formal objects of $\mathcal{X}$ can be algebraized' (in the literature, the term 'algebraized' is sometimes reserved for algebraization over a scheme of finite type; see Definition 11.22). We picked up the term 'integrable' from [14], the intuition being that infinitesimal arcs can be integrated into formal arcs, in the manner that tangent vectors are integrated to curves on smooth manifolds.

The rest of this section will be devoted to explaining these properties and verifying them for the stack of Higgs bundles and related moduli problems; at the end, we give a brief explanation of how these properties combine to imply a stack is algebraic.

11.1. *The Schlessinger–Rim criterion*

We have seen that homogeneity is a necessary condition for representability by an algebraic stack. The Schlessinger–Rim criterion implies that it is sufficient for *prorepresentability.*

By abstract nonsense, a covariant functor is prorepresentable if and only if it preserves finite limits. Generally, it is not very practical to check that functors on Artin rings preserve arbitrary finite limits, ultimately due to the restrictions on descent for flat modules along non-flat morphisms in Theorem 10.14. Fortunately, Schlessinger was able to prove that homogeneity, that is, respect for a restricted class of limits, along with a finite dimensional tangent space, are sufficient to imply a functor is prorepresentable [74]. In this section, we will discuss Rim's generalization of Schlessinger's result to groupoids [1, Exp. VI].

Suppose that $\mathcal{X}$ is a CFG on Λ-schemes, where Λ is a complete noetherian local ring with residue field k, and $\xi : \operatorname{Spec} k \to \mathcal{X}$ is a k-point. Let $\mathscr{C}_\Lambda$ be the category of local artinian Λ-algebras with residue field k. One may interpret $\mathscr{C}_\Lambda^{\mathsf{op}}$ as the category of infinitesimal extensions of $\operatorname{Spec} k$. Let $\mathcal{X}_\xi$ be the fibered category on $\mathscr{C}_\Lambda^{\mathsf{op}}$ whose fiber over a ring A in $\mathscr{C}_\Lambda^{\mathsf{op}}$ consists of all $\eta \in \mathcal{X}(A)$ whose image via the projection $A \to k$ is ξ. The CFG $\mathcal{X}_\xi$ gives a formal picture of $\mathcal{X}$ near the point ξ.

Remark 11.6. Note that the restriction of the étale topology to $\mathscr{C}_\Lambda^{\mathsf{op}}$ is trivial, as every morphism in $\mathscr{C}_\Lambda^{\mathsf{op}}$ has a section over the residue field, and sections of étale maps extend infinitesimally. Therefore all covers in $\mathscr{C}_\Lambda^{\mathsf{op}}$ have sections, every presheaf is a sheaf, and every CFG is a stack. Therefore we can use the terms 'stack' and 'CFG' interchangeably over $\mathscr{C}_\Lambda^{\mathsf{op}}$.

Remark 11.7. Typically, we will start with a CFG $\mathcal{X}$ over S. This induces a CFG $\mathcal{X}_{\mathscr{C}_\Lambda^{\mathsf{op}}}$ over $\mathscr{C}_\Lambda^{\mathsf{op}}$ by restricting to the full subcategory obtained by taking objects over the spectrum of such a ring. Given an object $\xi \in \mathcal{X}(\operatorname{Spec} k)$, the category $\mathcal{X}_\xi$ over $\mathscr{C}_\Lambda^{\mathsf{op}}$ has objects the pairs (η, ϕ) where $\phi : \xi \to \eta$ is a morphism of $\mathcal{X}$ (necessarily cartesian) lying above an infinitesimal extension $\operatorname{Spec} k \to \operatorname{Spec} A$, with A in $\mathscr{C}_\Lambda^{\mathsf{op}}$:

$$\begin{array}{ccc} \xi & \longrightarrow & \eta \\ \downarrow & & \downarrow \\ \operatorname{Spec} k & \longrightarrow & \operatorname{Spec} A \end{array}$$

Morphisms in $\mathcal{X}_\xi$ are defined in the obvious way.

We write $\widehat{\mathscr{C}}_\Lambda$ for the category of complete local Λ-algebras with residue field k that are formally of finite type over Λ; i.e., completions of rings of finite type over Λ.

Recall the notion of a groupoid object of $\widehat{\mathscr{C}}_\Lambda^{\mathsf{op}}$ from Definition C.2 and its associated CFG C.6. Note that this coincides with the associated stack from Definition C.12, since the topology of $\widehat{\mathscr{C}}_\Lambda^{\mathsf{op}}$ is trivial.

Definition 11.8 ([1, Def. VI.2.11]). We say a category fibered in groupoids over $\mathscr{C}_\Lambda^{\mathsf{op}}$ is *prorepresentable* if it is representable by (i.e., equivalent to) the CFG associated to a groupoid object $U_1 \overset{s}{\underset{t}{\rightrightarrows}} U_0$ in $\widehat{\mathscr{C}}_\Lambda^{\mathsf{op}}$ (Example C.16). The groupoid is said to be *smooth* if the morphisms s and t satisfy the formal criterion for smoothness (Definition 9.12).

Remark 11.9. In the above discussion, it is important to note that we are considering stacks over $\mathscr{C}_\Lambda^{\mathsf{op}}$, not $\widehat{\mathscr{C}}_\Lambda^{\mathsf{op}}$. In particular, if $\mathcal{X}$ is representable by a groupoid object $U_1 \overset{s}{\underset{t}{\rightrightarrows}} U_0$ in $\widehat{\mathscr{C}}_\Lambda^{\mathsf{op}}$, the morphism $U_0 \to \mathcal{X}$ (Example C.16) is not in general defined by an object over U_0 (as U_0 is not in general in $\mathscr{C}_\Lambda^{\mathsf{op}}$). However, if $\mathfrak{m}$ is the maximal ideal of U_0, and V_k is the vanishing locus of $\mathfrak{m}^j$ in U_0, we can write $U_0 = \varinjlim_k V_k$ with each V_k in $\mathscr{C}_\Lambda^{\mathsf{op}}$ and there is a family of compatible morphisms in $V_k \to \mathcal{X}$ defined by objects of $\mathcal{X}$ over V_k—in other words, an element of $\varprojlim_k \mathcal{X}(U_0/\mathfrak{m}^j)$. Such an element is called a formal element of $\mathcal{X}$ over U_0 (see Definition 11.17). We will return to this topic again in §11.3. Note finally that if $U_1 \overset{s}{\underset{t}{\rightrightarrows}} U_0$ is a smooth groupoid object, then the morphism $U_0 \to \mathcal{X}$ is formally smooth (Proposition C.18).

The following theorem says, essentially, that a stack looks like an algebraic stack in a formal neighborhood of a point if and only if it is homogeneous:

Theorem 11.10 (Schlessinger [74, Thm. 2.11], Rim [1, Thm. VI.2.17]). *Let $\mathcal{X}$ be a fibered category over $\mathscr{C}_\Lambda^{\mathsf{op}}$ such that $\mathcal{X}(k)$ is a single point. Then $\mathcal{X}$ is prorepresentable by a smooth groupoid object of $\widehat{\mathscr{C}}_\Lambda^{\mathsf{op}}$ if and only if $\mathcal{X}$ is homogeneous and $T_{\mathcal{X}}^{-1}$ and $T_{\mathcal{X}}^{0}$ are finite dimensional k-vector spaces.*

One direction of the implications in the theorem is clear: it is quite easy to see that the stack associated to a smooth groupoid object $U_1 \underset{t}{\overset{s}{\rightrightarrows}} U_0$ of $\widehat{\mathscr{C}}_\Lambda^{\mathsf{op}}$ is homogeneous, since the stacks associated to the U_i are homogeneous. The finite dimensionality of $T_{\mathcal{X}}^{-1}(\xi)$ and $T_{\mathcal{X}}^{0}(\xi)$ follows from the finite dimensionality of the tangent spaces of stacks represented by objects of $\widehat{\mathscr{C}}_\Lambda^{\mathsf{op}}$, which consists of objects formally of finite type.

11.2. *Local finite presentation*

The definition of morphisms locally of finite presentation was given in §9.6. The proof of the following lemma is formal:

Lemma 11.11. *(i) Suppose that $g : \mathcal{Y} \to \mathcal{Z}$ is locally of finite presentation. Then $f : \mathcal{X} \to \mathcal{Y}$ is locally of finite presentation if and only if $gf : \mathcal{X} \to \mathcal{Z}$ is locally of finite presentation.*
(ii) The base change of a morphism that is locally of finite presentation is also locally of finite presentation.

There is a repertoire of techniques for proving moduli problems are locally of finite presentation to be found in [36]. We combine these with Lemma 11.11 to prove that the stack of Higgs bundles is locally of finite presentation.

Lemma 11.12. *The stack of proper nodal curves is locally of finite presentation.*

Proof. Let $\mathcal{N}$ denote the stack of proper nodal curves. Suppose that a commutative ring A is the filtered colimit of commutative rings A_i. Put $S = \operatorname{Spec} A$ and $S_i = \operatorname{Spec} A_i$. Let C be an element of $\mathcal{N}(A)$. That is, C is a flat family of nodal curves over A. We want to show that C is induced by base change from a nodal curve C_i over A_i for some i, and that (up to increasing the index i) this curve is unique up unique isomorphism.

First of all, C is of finite presentation over S. By [36, Thm. (8.8.2) (ii)], there is an index i and a scheme C_i of finite presentation over A_i such that $C = C_i \times_{S_i} S$. It follows from [36, Thm. (8.8.2) (i)] that C is unique up to unique isomorphism and increase of the index i. By [36, Thm. (8.10.5) (xii)], we can arrange for C_i to be proper over S_i by replacing i with a larger index.

Now C has a cover by open subsets U, each of which admits an étale map $U \to V$, where $V = \operatorname{Spec} A[x,y]/(xy - t_U)$ for some $t_U \in A$. Refining the cover, we can assume that the open subsets U are affine, and hence of finite presentation over A. As C is quasicompact, this cover can be assumed finite, so by increasing i, we can assume that t_U appears in A_i for all i. By increasing i still further, we can assume each U is the preimage in C of an open subset $U_i \subseteq C_i$ [36, Prop. (8.6.3)] and that these open subsets cover C_i [36, Thm. (8.10.5) (vi)]. Now $V = V_i \times_{S_i} S$ so by [36, Thm. (8.8.2) (i)], the map $U \to V$ is induced from a map $U_i \to V_i$ over S_i, at least after increasing i still further. By [37, Prop. (17.7.8) (ii)], we can ensure that the map $U_i \to V_i$ is étale, at least after increasing i. Then C_i is a family of nodal curves over S_i, and the proof is complete. □

Lemma 11.13. *Let $\mathcal{E}$ be the stack of pairs (C, E) where C is a proper nodal curve and E is a vector bundle on C and let $\mathcal{N}$ be the stack of proper nodal curves. The projection $\mathcal{E} \to \mathcal{N}$ is locally of finite presentation.*

Proof. As before, A is the filtered colimit of commutative rings A_i. We suppose that (C, E) is an A-point of $\mathcal{E}$ and that C is induced from a nodal curve C_i over A_i. We want to show that, up to increasing i, the vector bundle E is induced from a unique (up to unique isomorphism) vector bundle E_i over C_i. As E is of finite presentation, we may increase i to obtain a quasicoherent sheaf of finite presentation E_i over C_i inducing E by pullback [36, Thm. (8.5.2) (ii)]. The uniqueness of E_i follows from [36, Thm. (8.5.2) (i)]. Increasing i still further, we can ensure that E_i is a vector bundle [36, Prop. (8.5.5)]. □

Lemma 11.14. *Let $\mathcal{F}$ be the stack of tuples (C, E, F, σ) where C is a nodal curve, E and F are vector bundles on C, and $\sigma : E \to F$ is a morphism of vector bundles. Let $\mathcal{G}$ be the stack of triples (C, E, F) as above, and $\mathcal{F} \to \mathcal{G}$ the projection forgetting σ. Then $\mathcal{F}$ is locally of finite presentation over $\mathcal{G}$.*

Proof. We assume $A = \varinjlim A_i$ is a filtered colimit of commutative rings and that $(C, E, F, \sigma) \in \mathcal{F}(A)$ and $(C_i, E_i, F_i) \in \mathcal{G}(A_i)$ induces $(C, E, F) \in \mathcal{G}(A)$. By an immediate application of [36, Thm. (8.5.2) (i)], we discover that, after increasing i, we can find $\sigma_i : E_i \to F_i$ inducing σ and that σ_i is unique up to further increasing i. □

Combining Lemmas 11.11, 11.12, 11.13, and 11.14, we obtain

Corollary 11.15. *The stack of tuples (C, E, F, σ) where C is a nodal curve, E and F are vector bundles on C, and $\sigma : E \to F$ is a morphism of vector bundles is locally of finite presentation.*

Corollary 11.16. *The stack of Higgs bundles is locally of finite presentation.*

Proof. This is deduced from the previous corollary by the same argument as in Lemma 10.46. □

11.3. *Integration of formal objects*

Let A be a complete noetherian local ring with maximal ideal $\mathfrak{m}$. A formal A-point of X is an object of the inverse limit $\varprojlim_k X(A/\mathfrak{m}^j)$. It can be checked easily that for any scheme X, the function

$$X(A) \to \varprojlim X(A/\mathfrak{m}^j) \tag{11.1}$$

is a bijection. It is only slightly more difficult to verify this gives an equivalence when X is an algebraic stack, provided that one interprets the limit of groupoids correctly. One efficient way of describing the limit is as the category of sections of $\mathcal{X}$ over the subcategory

$$\operatorname{Spec}(A/\mathfrak{m}) \to \operatorname{Spec}(A/\mathfrak{m}^2) \to \operatorname{Spec}(A/\mathfrak{m}^3) \to \cdots$$

of $\mathscr{C}_\Lambda^{\mathsf{op}}$.

Lifting a formal A-point of X to an A-point of X may be seen as an analogue of integrating a tangent vector to a curve. We must require a formal point, as opposed to merely a tangent vector, because not every tangent vector can be integrated on a singular space.[e]

Definition 11.17 (Integrating formal points). Let $\mathcal{X}$ be a fibered category over the category of schemes and let A be a complete noetherian local

[e]Many formulations of Artin's criterion only require (11.1), when applied to a stack $\mathcal{X}$, to have dense image. This strengthens the analogy to integrating tangent vectors, since there is not a unique curve with a given tangent vector. However, it is generally no more difficult to prove (11.1) is an equivalence than it is to prove it has dense image.

ring with maximal ideal $\mathfrak{m}$. By a *formal A-point* of $\mathcal{X}$ we mean an object of the category

$$\varprojlim_j \mathcal{X}(A/\mathfrak{m}^j).$$

We say that a formal A-point of $\mathcal{X}$ can be *algebraized*, or that it can be *effectivized*, or that it is *integrable* if it lies in the essential image of the functor

$$\mathcal{X}(A) \to \varprojlim_j \mathcal{X}(A/\mathfrak{m}^j).$$

If every formal A-point of $\mathcal{X}$ can be algebraized, for every complete noetherian local ring A, then we say *formal objects of* $\mathcal{X}$ *can be algebraized*, or *can be effectivized*, or *are integrable.*

For a long time, the main algebraization theorem was Grothendieck's existence theorem, which asserts that formal objects of the stack of coherent sheaves on a proper scheme can be algebraized:

Theorem 11.18 (Groth. existence [24, Thm. 8.4.2], [33, Thm. (5.1.4)]). *Let X be a proper scheme over $S = \operatorname{Spec} A$ with A a complete noetherian local ring with maximal ideal $\mathfrak{m}$. For each j, let $S_j = \operatorname{Spec} A/\mathfrak{m}^j$ and let $X_j = X \times_S S_j$. Then*

$$\operatorname{Coh}(X) \to \varprojlim_j \operatorname{Coh}(X_j)$$

is an equivalence of categories.

Very recently, Bhatt [13] and Hall and Rydh [45] have proved strong new integration theorems extending Grothendieck's. Since Grothendieck's existence theorem will suffice for the stack of Higgs bundles on curves, we will not need to state these new results.

Lemma 11.19. *Formal families of proper nodal curves can be algebraized.*

Proof. Let $\mathcal{N}$ denote the stack of proper nodal curves. Let A be a complete noetherian local ring with maximal ideal $\mathfrak{m}$. Set $A_j = A/\mathfrak{m}^{j+1}$ and $S_j = \operatorname{Spec} A_j$ and suppose that $C_j \in \mathcal{N}(S_j)$ are the components of a formal A-point of $\mathcal{N}$. Then C_0 is a curve over the field A_0. Pick a very ample line bundle L_0 on C_0 with $H^1(C_0, L_0) = 0$. Extend L_0 inductively to a compatible system of line bundles L_j on each C_j: By Lemma 10.22, the obstruction to extending L_j to L_{k+1} lies in

$$H^2(C, \pi^*(\mathfrak{m}^j/\mathfrak{m}^{j+1})) = \mathfrak{m}^j/\mathfrak{m}^{j+1} \otimes H^2(C_0, \mathcal{O}_{C_0}) = 0.$$

Then $V_j = \pi_* L_j$ is a (trivial) vector bundle on S_j. Moreover, by Lemma 10.19, the obstruction to extending a section of L_j to a section of L_{k+1} lies in

$$H^1(C, \pi^*(\mathfrak{m}^j/\mathfrak{m}^{j+1}) \otimes L_0) = \mathfrak{m}^j/\mathfrak{m}^{j+1} \otimes H^1(C_0, L_0) = 0$$

so $V_j\big|_{S_\ell} = V_\ell$ for $\ell \leq k$. There is therefore a (trivial) vector bundle V on S whose restriction to S_j is V_j for all k. The complete linear series of the L_j give a system of closed embeddings $C_j \to \mathbf{P}(V_j)$. We may regard the structure sheaves $\mathcal{O}_{C_j}$ as a compatible system of quotients of the structure sheaf of $\mathbf{P}_{A_j}(V_j)$, so by Grothendieck's existence theorem they can be algebraized to a quotient $\mathcal{O}_C$ of the structure sheaf of $\mathbf{P}_A(V)$. Let C be the corresponding closed subscheme of $\mathbf{P}_A(V)$.

By construction C is proper over $\operatorname{Spec} A$. It is also flat by the infinitesimal criterion for flatness [22, Ex. 6.5]. Therefore it is a family of nodal curves (see Rem. 10.29). □

Remark 11.20. The above argument can be used more generally to show that formal families of proper schemes can integrated if the central fiber X has $H^2(X, \mathcal{O}_X) = 0$. See [75, Thm. 2.5.13] or [41, Thm.4] for different ways of organizing the ideas.

Lemma 11.21. *Let $\mathcal{F}$ be the stack of quadruples (C, E, F, σ) where C is a nodal curve, E and F are vector bundles on C, and $\sigma : E \to F$ is a morphism of vector bundles. Then formal objects of $\mathcal{F}$ can be integrated.*

Proof. Suppose A is a complete noetherian local ring with maximal ideal $\mathfrak{m}$, set $A_j = A/\mathfrak{m}^{j+1}$. Given a formal family $(C_j, E_j, F_j, \sigma_j) \in \mathcal{F}(A_j)$, we seek an element $(C, E, F, \sigma) \in \mathcal{F}(A)$ inducing it. We may find C by the Lemma 11.19. Note that C is projective over A and the E_j and F_j are each formal families of coherent sheaves over C, so by Grothendieck's existence theorem (Theorem 11.18) they can be algebraized to coherent sheaves E and F over C. Moreover, both E and F are flat, by the infinitesimal criterion for flatness, so they are vector bundles. One more application of Grothendieck's existence theorem algebraizes the family of homomorphisms of coherent sheaves $\sigma_j : E_j \to F_j$ to a homomorphism $\sigma : E \to F$ and the lemma is complete. □

11.4. *Artin's theorems on algebraization and approximation*

The question of integrability concerns objects of a stack $\mathcal{X}$ lying over the spectrum of a complete local algebra over a field. With the Schlessinger–Rim theorem and integration, we can factor any morphism $\operatorname{Spec} k \to \mathcal{X}$ through a map $\operatorname{Spec} A \to \mathcal{X}$ such that A is a complete noetherian local ring and is *formally smooth* over $\mathcal{X}$. This is tantalizingly close to showing $\mathcal{X}$ is an algebraic stack: we need too find a factorization that is genuinely smooth over $\mathcal{X}$.

The distinction between smoothness and formal smoothness is finiteness of presentation, so we need to find a finite type ring B *that is still formally smooth* over $\mathcal{X}$ at the k-point, and induces A by completion at a point (this will give smoothness of the map $B \to \mathcal{X}$ at the given k-point of B; to extend to smoothness on an open neighborhood of the k-point, see §11.6).

This may be the subtlest part of the proof of Artin's criterion. It is resolved by Artin's *approximation theorem*, proved originally by Artin [8, Thm. 1.12] with some technical hypotheses, and in its current form by B. Conrad and J. de Jong [19, Thm. 1.5] using a spectacular theorem of Popescu [67, Thm. 1.3].

The algebraization theorem (Theorem 11.23) asserts that under our assumption that $\mathcal{X}$ is locally of finite presentation (Theorem 11.1(5)), given $\xi \in \mathcal{X}(A)$ in an appropriate complete local ring with residue field k, one can find a finite type k-algebra B with a marked point (maximal ideal $\mathfrak{n}$), whose completion at the marked point is A, and an element $\eta \in \mathcal{X}(B)$ that agrees to a specifiable finite order with ξ. That is one may select j beforehand and then find η such that the restriction of η to $B/\mathfrak{n}^j \simeq A/\mathfrak{m}^j$ agrees with the restriction ξ_j of ξ. Even though $\eta \in \mathcal{X}(B)$ does not necessarily restrict to $\xi \in \mathcal{X}(A)$, it will differ from ξ only up to an automorphism of A that is the identity modulo $\mathfrak{m}^j$. In particular, it will still be formally smooth at the closed point (see the proof of Theorem 11.23).

We begin by introducing notation. Let k be a field and let Λ be a complete noetherian local ring with residue field k. Recall, we denote by $\mathscr{C}_\Lambda$ the category of local artinian Λ-algebras with residue field k, and by $\widehat{\mathscr{C}}_\Lambda$ the category of complete local Λ-algebras with residue field k that are formally of finite type over Λ.

Definition 11.22 (Algebraization over a scheme of finite type). Let $\mathcal{X}$ be a CFG over S and let $(A, \mathfrak{m}) \in \widehat{\mathscr{C}}_\Lambda$. We say a formal element $\hat{\eta}$ of $\mathcal{X}$ over $\operatorname{Spec} A$, i.e., an object of $\varprojlim_j \mathcal{X}(A/\mathfrak{m}^j)$, can be *algebraized over a scheme of finite type* if there exist:

- a finitely generated Λ-algebra B,
- a k-point $s : \operatorname{Spec} k \to \operatorname{Spec} B$ corresponding to a maximal ideal

$\mathfrak{n} \subseteq B$, and,
- an object $\theta \in \mathcal{X}(B)$,

such that

(1) $A = \widehat{B}_{\mathfrak{n}}$, the completion of B at $\mathfrak{n}$, and,
(2) the formal element $\hat{\eta}$ is isomorphic to the image of θ under the map

$$\mathcal{X}(B) \to \varprojlim_j \mathcal{X}(A/\mathfrak{m}^j)$$

induced by the morphism $B \to B_{\mathfrak{n}} \to B_{\mathfrak{n}}/\mathfrak{n}^j = A/\mathfrak{m}^j$.

Artin's algebraization theorem asserts the following:

Theorem 11.23 (Artin's algebraization theorem). *Let $\mathcal{X}$ be a CFG that is locally of finite presentation over an excellent DVR or field Λ. Let $(A, \mathfrak{m})$ be a complete local Λ-algebra with residue field k that is formally of finite type over Λ. If $\bar{\eta} \in \mathcal{X}(A)$ then there is a finite type Λ-algebra B with a maximal ideal $\mathfrak{n}$ at whose completion B is isomorphic to A, along with $\eta \in \mathcal{X}(B)$ inducing $\bar{\eta} \in \mathcal{X}(A)$.*

The proof is almost an immediate consequence of Artin's approximation theorem:

Theorem 11.24 (Artin's approx. [8, Thm. 1.12], [19, Thm. 1.5]). *Let $\mathcal{X}$ be a CFG that is locally of finite presentation over an excellent DVR or field Λ. Let $(A, \mathfrak{m})$ be a complete local Λ-algebra with residue field k that is formally of finite type over Λ. Given $\bar{\eta} \in \mathcal{X}(A)$, there is a finitely generated Λ-algebra B with k-point $s : \operatorname{Spec} k \to \operatorname{Spec} B$ corresponding to a maximal ideal $\mathfrak{n}$ and $\widehat{B}_{\mathfrak{n}} = A$, such that for any positive integer j there is an element $\eta \in \mathcal{X}(B)$ (depending on j) such that the images of η and $\bar{\eta}$ are isomorphic in $\mathcal{X}(A/\mathfrak{m}^j)$.*

Proof of Theorem 11.23 from Theorem 11.24. Take B and η as in the Approximation theorem, with $j = 2$. In other words, η_2 and $\hat{\eta}_2$ are isomorphic in $\mathcal{X}(A/\mathfrak{m}^2)$. Now, inductively, using formal smoothness, we can use diagram (11.2)

$$\begin{array}{ccc} \operatorname{Spec} A/\mathfrak{m}^j & \xrightarrow{\psi_{j-1}} & \operatorname{Spec} A \\ \downarrow & \overset{\psi_j}{\nearrow} & \\ \operatorname{Spec} A/\mathfrak{m}^{j+1} & & \Big\downarrow \bar{\eta} \\ \downarrow & & \\ \operatorname{Spec} B & \xrightarrow{\quad\eta\quad} & \mathcal{X} \end{array} \tag{11.2}$$

to construct for all j morphisms $\psi_j : \operatorname{Spec} A/\mathfrak{m}^{j+1} \to \operatorname{Spec} A$ so that $\bar{\eta}\psi_j \cong \eta_j$. This induces a morphism $\psi : \operatorname{Spec} A \to \operatorname{Spec} A$ so that $\bar{\eta}\psi \cong \eta$. But ψ is an isomorphism modulo $\mathfrak{m}^2$, so it must be surjective, and a surjective endomorphism of a noetherian ring is an isomorphism (e.g., [75, Lem. C.5]). □

11.5. *Algebraicity of the stack of Higgs bundles*

In Lemma 10.44, Lemma 10.43, Lemma 11.21, Corollary 11.15, and Lemma 10.61, we have verified the conditions of Theorem 11.1 for the stack $\mathcal{F}$ that parameterizes proper nodal curves equipped with a homomorphism of vector bundles:

Theorem 11.25. *Let $\mathcal{F}$ be the stack of quadruples (C, E, F, σ) where C is a nodal curve, E and F are vector bundles, and $\sigma : E \to F$ is a morphism of vector bundles. Then $\mathcal{F}$ is an algebraic stack.*

Analogous arguments show that the stack of Higgs bundles satisfies the axioms, hence is algebraic. Alternately, one may consider the stacks $\mathcal{E}_1$ of pairs (C, E) where C is a nodal curve and E is a vector bundle on C and $\mathcal{E}_2$ of triples (C, E, F) where C is a nodal curve and E and F are vector bundles on C. (Note that $\mathcal{E}_2 = \mathcal{E}_1 \times_{\mathcal{M}} \mathcal{E}_1$.) Then we can construct the stack of Higgs bundles as the fiber product $\mathcal{E}_1 \times_{\mathcal{E}_2} \mathcal{F}$ where the map $\mathcal{E}_1 \to \mathcal{E}_2$ sends (C, E) to $(C, E, E \otimes \omega_C)$. We therefore have

Corollary 11.26. *The stack of Higgs bundles on proper nodal curves is an algebraic stack.*

11.6. *Outline of the proof of Artin's criterion*

We will briefly summarize the proof of Theorem 11.1. Since assumption Theorem 11.1(1) is that $\mathcal{X}$ is a stack, the key point is to establish the existence of a smooth representable covering of $\mathcal{X}$ by a scheme. The basic idea of the proof is to begin with an arbitrary point $\xi_0 \in \mathcal{X}(k)$, valued in a field k, and find a smooth neighborhood $U \to \mathcal{X}$ of this point by enlarging $\operatorname{Spec} k$ until it is smooth over $\mathcal{X}$. In more concrete terms, U will be a versal deformation of ξ_0. Repeating this for every point of $\mathcal{X}$ and taking a disjoint union of the different U gives a smooth cover of $\mathcal{X}$ by a scheme.

We now explain in more detail how the arguments in the previous sections imply the existence of the schemes U.

Versality at a point

By the Schlessinger–Rim theorem (see Theorem 11.10), the homogeneity of $\mathcal{X}$ (Theorem 11.1(2)) and the finite dimensionality of $T^{-1}_{\mathcal{X}}(\xi_0)$ and $T^0_{\mathcal{X}}(\xi_0)$ (Theorem 11.1(3)) guarantee that $\mathcal{X}$ is prorepresentable at ξ_0. That is, there is a formal groupoid $\widehat{V}_1 \rightrightarrows \widehat{V}_0$, with $\widehat{V}_i = \operatorname{Spec} \widehat{R}_i$ for complete noetherian local rings $\widehat{R}_i$, whose associated CFG (Definition C.13) agrees with $\mathcal{X}$ on infinitesimal extensions of ξ_0.

This gives a *formal* morphism $\widehat{V}_0 \to \mathcal{X}$ that is formally smooth (Example C.16). In other words, we have compatible elements $\xi_j \in \mathcal{X}(\operatorname{Spec} \widehat{R}_0/\mathfrak{m}^{j+1})$, where $\mathfrak{m}$ is the maximal ideal of $\widehat{R}_0$. The assumption that formal objects of $\mathcal{X}$ integrate uniquely (Theorem 11.1(4)) guarantees that this formal morphism comes from a genuine morphism $\widehat{V}_0 \to \mathcal{X}$, i.e., from an element of $\xi \in \mathcal{X}(\operatorname{Spec} \widehat{R}_0)$.

Now the map $\widehat{V}_0 \to \mathcal{X}$ is formally smooth at ξ_0, but it is not of finite type. To remedy this we can use the local finite presentation of $\mathcal{X}$ (Theorem 11.1(5)), which ensures that $\widehat{V}_0 \to \mathcal{X}$ must factor through some scheme V of finite type. Unfortunately, it is not clear we can exert any control over V, even formally, to guarantee it is formally smooth over $\mathcal{X}$. Fortunately, we may rely on Artin's algebraization theorem (Theorem 11.23) to ensure that V is still formally smooth over $\mathcal{X}$ at the central point.

This does not yet guarantee that the map $V \to \mathcal{X}$ is smooth: we only have formal smoothness at one point. The next step will be to show formal smoothness at a point implies formal smoothness nearby.

Versality in a neighborhood

Write $V = \operatorname{Spec} R$. We know that R is of finite type and we now have a map $\eta : V \to \mathcal{X}$ that we know to be formally smooth at a point lifting ξ. All that is left is to find an open neighborhood of this point at which the map is actually smooth. For this we use the obstruction theory $T^1_{\mathcal{X}}$ (Theorem 11.1(6)).

Hall shows that the existence of a locally presentable obstruction theory representing the automorphisms, deformations, and obstructions of $\mathcal{X}$ implies that there is a *relative* obstruction theory $T^1_{V/\mathcal{X}}$ for V over $\mathcal{X}$ and that $T^1_{V/\mathcal{X}}$ is a *coherent functor.*

As V is formally smooth over $\mathcal{X}$ at ξ_0, we know that $T^1_{V/\mathcal{X}}(\eta, J) = 0$ for any quasicoherent sheaf J on V supported at ξ_0. By a theorem of Ogus and Bergman [65, Thm. 2.1], coherent functors over noetherian rings satisfy an analogue of Nakayama's lemma, which guarantees that $T^1_{W/\mathcal{X}}(\eta\big|_W, J) = 0$ for all quasicoherent sheaves on W, where W is the localization of V at the point ξ_0. Making use of the local finite presentation of $\mathcal{X}$ (Theorem 11.1(5)), Hall shows that this implies $T^1_{U/\mathcal{X}}(\eta\big|_U, J) = 0$ for all quasicoherent J on an open subset $U \subseteq V$ containing W. Then $U \to \mathcal{X}$ is locally of finite presentation and satisfies the formal criterion for smoothness.

Bootstrapping to representability

To conclude that $U \to \mathcal{X}$ is smooth, we only need to show that it is representable by algebraic spaces. This is proved by observing that the hypotheses of Theorem 11.1 on $\mathcal{X}$ imply relative versions of themselves for the map $U \to \mathcal{X}$ (cf. Lemma 10.18, Lemma 11.11, and the relative obstruction theory $T^1_{U/\mathcal{X}}$ mentioned above). The same hypotheses then hold for any base change $U_Z \to Z$ via any map $Z \to \mathcal{X}$. Taking Z to be a scheme and viewing U_Z as a sheaf in the étale topology on schemes over Z, this reduces the problem to showing that U_Z is an algebraic space. Now we can try to prove the theorem for $\mathcal{X} = U_Z$: In effect, Theorem 11.1 is reduced to the case where $\mathcal{X}$ is a *sheaf* as opposed to a stack.

Now we have a scheme U and a map $U \to \mathcal{X}$ that is formally smooth and locally of finite presentation, and we are faced with the same problem: to show $U \to \mathcal{X}$ is representable by algebraic spaces. But this time, the relative diagonal of U over $\mathcal{X}$ is injective, so that if we iterate the process one more time, we discover once again that our task is to prove U_Z is an algebraic space. But $U_Z = U \times_{\mathcal{X}} Z$ is a subsheaf of $U \times Z$, since the diagonal of $\mathcal{X}$ is injective. Therefore taking $U \times Z$ to be the new base scheme and replacing $\mathcal{X}$ with $U \times Z$, we discover we may assume further that $\mathcal{X}$ is even a *subsheaf* of the base scheme S.

But now, if $W \to \mathcal{X}$ is any map, we have

$$W \times_{\mathcal{X}} U = W \times_S U$$

since $\mathcal{X}$ is a subsheaf of S. Since $U \to S$ is schematic—it is a morphism of schemes, after all—so must be $U \to \mathcal{X}$.

A. Sheaves, topologies, and descent

We collect a few more technical topics surrounding the subject of descent. In particular, we show that the presence of nontrivial automorphisms *always* prevents an algebraic stack from having a representable sheaf of isomorphism classes (Corollary A.7), we give a few more technically efficient ways of thinking about descent (§A.2), we discuss a bit more about saturations of pretopologies (§A.3) and we recollect and example of Raynaud showing that genus 1 curves do not form a stack unless one admits algebraic spaces into the moduli problem (§A.4).

A.1. *Torsors and twists*

The point of this section is to explain how nontrivial automorphisms give rise to nontrivial families, and prevent a stack from having a representable sheaf of isomorphism classes. In other words, moduli functors parameterizing isomorphism classes of objects with nontrivial automorphism groups are never representable.

A.1.1. *Locally trivial families and cohomology*

In general, one can build a locally trivial family with fiber X over a base S from cohomology classes in $H^1\big(S, \mathrm{Aut}(X)\big)$ ([77, Tag 02FQ]). If one represents the cohomology class with a Čech cocycle then it is a recipe for assembling the family from trivialized families on open subsets. The cocycle condition is precisely the one necessary to ensure that the family can be glued together, while the coboundaries act via isomorphisms. The cocyle condition appears in the definition of the gluing condition for a stack in §3.3 for precisely this reason.

A.1.2. *Locally trivial families from torsors over the base*

Example 1.6 describes a special case of a standard method for constructing locally trivial families is via automorphisms, and covers of the base. Namely, given schemes X_1 and S, and an étale principal Γ-bundle $\widetilde{S} \to S$ for some discrete group Γ, then for each

$$\phi : \Gamma \to \mathrm{Aut}(X_1)$$

one can construct a (étale) locally trivial family X/S with fibers isomorphic to X_1 by setting $X = (X_1 \times \widetilde{S})/\Gamma$, where the quotient is via the (free) diagonal action induced by ϕ. Alternately, this can be constructed as the space of equivariant morphisms from $\widetilde{S}$ to X_1. Using the previous remark, this defines a map

$$\mathrm{Hom}(\Gamma, \mathrm{Aut}(X_1)) \to H^1(S, \mathrm{Aut}(X_1))$$

(of course, the cohomology should be taken in a topology in which $\widetilde{S}$ is a torsor). When $\widetilde{S} \to S$ is a universal cover, we obtain a map $\mathrm{Hom}(\pi_1(S), \mathrm{Aut}(X_1)) \to H^1(S, \mathrm{Aut}(X_1))$.

Remark A.1. In the discussion above (§A.1.2), it is possible for the isotrivial family X to be trivial, even if the homomorphism ϕ is nontrivial. We will highlight one situation in which the family X can be assured to be nontrivial. Let X_1 and $\widetilde{S} \to S$ be as in the discussion above (§A.1.2). Let $G = \mathrm{Aut}(X_1)$, and let $G' \leq G$ be a discrete subgroup. From §A.1.1 and A.1.2 above, we obtain a commutative diagram

$$\begin{array}{ccccc} \mathrm{Hom}(\pi_1(S), G') & \hookrightarrow & \mathrm{Hom}(\pi_1(S), G) & \longrightarrow & \mathrm{Hom}(\pi_1(S), \pi_0(G)) \\ \| & & \downarrow & & \| \\ H^1(S, G') & \longrightarrow & H^1(S, G) & \longrightarrow & H^1(S, \pi_0(G)) \end{array} \tag{A.1}$$

where the horizontal maps are the natural maps, and the vertical equalities on the left and right come from the fact that G' and $\pi_0(G)$ are totally disconnected. In summary, we can conclude that a locally trivial family X/S obtained from a homomorphism $\phi : \pi_1(S) \to G'$ will be a nontrivial family so long as the image of ϕ in $\mathrm{Hom}(\pi_1(S), \pi_0(G))$ is nontrivial. Note that the kernel of the natural map $H^1(S, G') \to H^1(S, G)$ corresponds to principal G' bundles that can be equivariantly embedded in $G \times S$; i.e., the kernel is given by sections over S of the coset space $(G/G') \times S$.

Example A.2 (Disconnected automorphism groups). If $\mathrm{Aut}(X_1)$ is disconnected, then there exist nontrivial locally trivial families X/S with fibers isomorphic to X_1. This follows immediately from the previous remark, so long as one can find a space S with $\pi_1(S) = \mathbb{Z}$, or at least that has a nontrivial principal $\mathbb{Z}$-bundle. Indeed, for any $\alpha \in \mathrm{Aut}(X_1)$ not in the connected component of the identity, one would take $G' = \langle \alpha \rangle$ and set X/S to be the locally trivial family corresponding to the map $\mathbb{Z} \to \langle \alpha \rangle$ by $1 \mapsto \alpha$. In the complex analytic setting, we can simply take $S = \mathbb{C}^*$. However, we can arrange for this even in the category of schemes by joining a pair of rational curves at two points; this has fundamental group $\mathbb{Z}$ in the sense that it has a simply connected covering space (even in the Zariski topology!) with a simply transitive action action of $\mathbb{Z}$.

Example A.3 (Finite automorphism groups). If $\mathrm{Aut}(X_1)$ is finite and nontrivial (or more generally, has a finite subgroup not contained in the connected component of the identity), one can easily construct similar examples using finite covers. Indeed, then $G = \mathrm{Aut}(X_1)$ contains a nontrivial finite cyclic subgroup $G' = \mu_n$ (not contained in the connected component of the identity). Let $S = \mathbb{C}^*$ and $\widetilde{S} \to S$ be the cyclic cover $z \mapsto z^n$, which is an étale principal μ_n-bundle. Then, as in (A.1) of Remark A.1 above, we obtain a diagram

$$\begin{array}{ccccccc}
\mathrm{Hom}(\mu_n, G') = & \mathrm{Hom}(\pi_1(S), G') & \hookrightarrow & \mathrm{Hom}(\pi_1(S), G) & \to & \mathrm{Hom}(\pi_1(S), \pi_0(G)) \\
\searrow & \| & & \downarrow & & \| \\
& H^1(S, G') & \longrightarrow & H^1(S, G) & \longrightarrow & H^1(S, \pi_0(G)).
\end{array}$$

Therefore, a generator of $\mu_n = \mathrm{Hom}(\mu_n, G')$ determines a nontrivial locally trivial family given explicitly by $X = (X_1 \times \mathbb{C}^*)/\mu_n \to S = \mathbb{C}^*/\mu_n = \mathbb{C}^*$, where the quotient is by the diagonal action under the identifications $\mu_n \leq \mathrm{Aut}(X_1)$, and $\mu_n \leq \mathrm{Aut}(\mathbb{C}^*)$ acting by a primitive n-th root of unity.

Example A.4 (Isotrivial families of curves). For every g there exists a relative curve $\pi : X \to S$ of genus g that is isotrivial, but not isomorphic to a trivial family. For $g = 0$, any nontrivial ruled surface $X \to S$ provides an example. The previous Example A.3, and Example 1.6, provide nontrivial isotrivial families for $g \geq 1$. For instance, for $g \geq 2$, one could begin with a hyperelliptic curve $\{y^2 = f(x)\}$ carrying the nontrivial action of μ_2 sending y to $-y$. The construction in Example A.3 yields the family $\{ty^2 = f(x)\}$, where t is the coordinate on $S = \mathbb{C}^*$.

A.1.3. *Twisting by a torsor*

All of the previous examples are special cases of the general process of twisting by a torsor.

Example A.5 (Twist by a torsor). Suppose that Z is an S-point of a stack $\mathcal{X}$ and that the automorphisms group of Z is a smooth group scheme G over S (in fact, a flat group scheme is enough if $\mathcal{X}$ is an algebraic stack). Let P be a G-torsor.

As G is smooth, P is smooth over S, and hence has sections over some étale cover of S. Therefore P is covering in the étale topology. (Note that $P \to S$ it is not necessarily an étale morphism, but this is not an obstacle to using it for descent, since we may use smooth descent; alternatively, see Appendix A.3.)

By descent, $\mathcal{X}(S)$ may be identified with the G-equivariant objects of $\mathcal{X}(P)$. In particular, let Z_P in $\mathcal{X}(P)$ be the pullback of Z to P. Then $Z \in \mathcal{X}(S)$ corresponds to Z_P with the trivial action of G. However, we can also ask G to act on $Z_P = P \times_S Z$ by the given action of G on P and by automorphisms on Z, giving an object $Y := P \times_G Z := (P \times_S Z)/G$ in $\mathcal{X}(S)$ by descent. We call Y the twist of Z by the torsor P.

We note that the torsor P can be recovered from a twist Y of Z as the sheaf $\mathscr{I}som_{\mathcal{X}}(Z, Y)$. Thus the twist is nontrivial if and only if the torsor P was. Moreover, one has an equivalence of categories between the full subcategory of $\mathcal{X}(S)$ consisting of twists of Z and the category $\mathrm{B}G(S)$ consisting of G-torsors on S.

The following theorem was communicated to us by Jason Starr [78]. It implies that if $\mathcal{X}$ is a stack in the fppf topology in which there are objects with nontrivial automorphisms, and X is its associated sheaf of isomorphism classes, then X cannot be representable by a scheme (Corollary A.7).

Theorem A.6. *Let G be an algebraic group of finite type over an algebraically closed field k. Then there is a scheme S of finite type over k and a nontrivial G-torsor over S.*

Proof. We suppose that G is an algebraic group of finite type and that $H^1(S, G) = 0$ for every scheme S. We wish to show that $G = 0$. We will break the proof into several steps.

Step 1: *G is connected.* We will show that G is connected by showing that any homomorphism $\mathbb{Z} \to G$ has image in the connected component of the identity. To this end, as in Example A.2, let S be a k-scheme with a nontrivial $\mathbb{Z}$-torsor P. For concreteness, let us take S to be an irreducible rational curve with a single node (otherwise smooth) and take P to be an infinite chain of copies of the normalization of S, attached at nodes. Then using the right hand side of the commutative diagram (A.1), we may conclude every homomorphism $\mathbb{Z} \to G$ composes to a morphism $\mathbb{Z} \to G \to \pi_0(G)$, with trivial image, and we are done.

We now assume always that G is connected, and proceed to consider $G_{\mathrm{red}} \subseteq G$, the maximal reduced closed subscheme (necessarily a subgroup since we work over an algebraically closed field, e.g., [77, Tag 047R]). By Chevalley's theorem (see for instance [18, Thm. 1.1] for a modern treatment), there is an exact sequence

$$1 \to G_{\mathrm{aff}} \to G_{\mathrm{red}} \to A \to 0$$

where G_{aff} is smooth connected and affine, and A is an abelian variety. Our next goal will be to show that $G_{\mathrm{red}} = 0$; we will do this in several steps.

Step 2: $H^1(S, G_{\mathrm{red}}) = 0$ *for all* S. For any scheme S, we have an exact sequence:

$$\mathrm{Hom}(S, G) \to \mathrm{Hom}(S, G/G_{\mathrm{red}}) \to H^1(S, G_{\mathrm{red}}) \to H^1(S, G).$$

We have assumed that $H^1(S, G) = 0$. Furthermore, $G \to G/G_{\mathrm{red}}$ is smooth (since G_{red} is) and G/G_{red} is artinian, so G admits a (not necessarily homomorphic) section over G/G_{red}. This implies $\mathrm{Hom}(S, G) \to \mathrm{Hom}(S, G/G_{\mathrm{red}})$ is surjective, so $H^1(S, G_{\mathrm{red}}) = 0$.

Step 3: G_{aff} *is unipotent.* Consider the exact sequence

$$\mathrm{Hom}(\mathbb{P}^1, G_{\mathrm{red}}) \to \mathrm{Hom}(\mathbb{P}^1, A) \to H^1(\mathbb{P}^1, G_{\mathrm{aff}}) \to \mathrm{Hom}(\mathbb{P}^1, G_{\mathrm{red}}) = 0.$$

As A is an abelian variety, every map from $\mathbb{P}^1$ to A is constant and therefore lifts to G_{red}. It follows that $H^1(\mathbb{P}^1, G_{\mathrm{aff}}) = 0$. Let $U \subseteq G_{\mathrm{aff}}$ be the unipotent radical [15, 11.21, p.157]. Consider the exact sequence:

$$H^1(\mathbb{P}^1, G_{\mathrm{aff}}) \to H^1(\mathbb{P}^1, G_{\mathrm{aff}}/U) \to H^2(\mathbb{P}^1, U).$$

Since U is an iterated extension of $\mathbb{G}_a$s [15, Chap. V, Cor. 15.5 (ii), p. 205], we have that $H^2(\mathbb{P}^1, U) = 0$, whence $H^1(\mathbb{P}^1, G_{\mathrm{aff}}/U) = 0$.

Let $T \subseteq G_{\mathrm{aff}}/U$ be a maximal torus, and let W be its Weyl group. As G_{aff}/U is reductive by definition [15, Chap. IV, 11.21, p.158], there is an injection (in fact a bijection!)

$$H^1(\mathbb{P}^1, T)/W \to H^1(\mathbb{P}^1, G_{\mathrm{aff}}/U) = 0$$

by a theorem of Grothendieck [31, Thm. 1.1] (note that Grothendieck works analytically, but his proof of the injectivity is valid algebraically [60, Thm. 0.3]). Thus $H^1(\mathbb{P}^1, T)$ is finite; although we do not need it, note that since the Weyl group coinvariants of $H^1(\mathbb{P}^1, T)$ are trivial if and only if $H^1(\mathbb{P}^1, T)$ is, we can actually conclude immediately that $H^1(\mathbb{P}^1, T) = 0$. But $T \simeq \mathbb{G}_m^r$ for some r, and $H^1(\mathbb{P}^1, \mathbb{G}_m) = \mathbb{Z}$. Therefore $r = 0$. That is, the maximal torus T of G_{aff}/U is trivial, so that G_{aff} is unipotent [15, Chap. IV, Cor. 11.5, p. 148].

Step 4: G_{red} *is affine.* Now that we know G_{aff} is unipotent, we argue that $A = 0$. Let $S = \mathbb{A}^1 \smallsetminus \{0\}$. Consider the exact sequence

$$H^1(S, G_{\mathrm{red}}) \to H^1(S, A) \to H^2(S, G_{\mathrm{aff}}).$$

Since S is affine, and G_{aff} is unipotent and therefore an iterated extension of quasicoherent sheaves (associated to $\mathbb{G}_a$s), we have $H^2(S, G_{\mathrm{aff}}) = 0$. This implies $H^1(S, A) = 0$. Choose an integer n relatively prime to the characteristic of k. Consider the exact sequence

$$\mathrm{Hom}(S, A) \xrightarrow{[n]} \mathrm{Hom}(S, A) \to H^1(S, A[n]) \to H^1(S, A)$$

where $[n]$ denotes multiplication by n and $A[n]$ is the n-torsion subgroup. Since S is rational, every map $S \to A$ is constant. In particular, $\mathrm{Hom}(S, A) \xrightarrow{[n]} \mathrm{Hom}(S, A)$ is surjective. Therefore $H^1(S, A[n])$ injects into $H^1(S, A) = 0$, and so is also 0. Now, $A[n] \simeq (\mathbb{Z}/n\mathbb{Z})^{2g}$ where g is the dimension of A. We know that $H^1(S, \mathbb{Z}/n\mathbb{Z}) = \mathbb{Z}/n\mathbb{Z}$, so we deduce that $g = 0$, and therefore $A = 0$.

Step 4: $G_{\mathrm{red}} = 0$. Now we know that $G_{\mathrm{red}} = G_{\mathrm{aff}}$ is affine and unipotent. We can choose an injective homomorphism $G_{\mathrm{red}} \subseteq G'$ where G' is smooth affine and reductive (i.e., embed it in an appropriate GL_n). Consider the G_{red}-torsor G' over G'/G_{red}. This must be trivial, since $H^1(G'/G_{\mathrm{red}}, G_{\mathrm{red}}) = 0$ from Step 2, so $G' \simeq G'/G_{\mathrm{red}} \times G_{\mathrm{red}}$ as a scheme. But G' is affine, so this implies G'/G_{red} is affine. Therefore G_{red} is reductive, by Matsushima's criterion [72, Thm. A]. As it is also unipotent, this means $G_{\mathrm{red}} = 0$.

Step 5: $G = 0$. Since $G_{\mathrm{red}} = 0$, this means that $G = G/G_{\mathrm{red}}$ is the spectrum of an artinian local ring. But we can once again choose a closed embedding $G \subseteq G'$ where G' is smooth and affine (for example, let G act on its ring of regular functions). The quotient G'/G is also reduced and affine. But $H^1(G'/G, G) = 0$ by assumption, so that $G' \simeq G'/G \times G$ as a scheme. As G' is reduced, this means G is reduced, and therefore $G = 0$. □

Corollary A.7. *Suppose that X is the presheaf of isomorphism classes in an algebraic stack $\mathcal{X}$ such that all objects over algebraically closed fields have automorphism groups that are algebraic groups of finite type over the field. If X is a sheaf in the fppf topology then $X \simeq \mathcal{X}$. In particular, if $\mathcal{X}$ admits an object over an algebraically closed field k with a nontrivial algebraic automorphism group of finite type over k, then X is not representable by a scheme.*

Proof. Let x be a k-point of $\mathcal{X}$, where k is algebraically closed. Let G be the stabilizer group of x. We obtain a monomorphism $\mathrm{B}G \to \mathcal{X}_k$. If P is any G-torsor over a scheme S then we obtain a twist $S \to \mathrm{B}G \to \mathcal{X}$ (Example A.5). By definition, the twists agree locally, so that the induced maps to

X coincide locally. But we have assumed X is a sheaf, so that the maps agree globally as well. But the torsor can be recovered, up to isomorphism, from the twist, so all G-torsors over all k-schemes are trivial. Therefore by Theorem A.6, G is trivial, so no point of $\mathcal{X}$ has a nontrivial stabilizer group and $\mathcal{X} \simeq X$. □

A.2. *More on descent*

We highlight a few alternate formulations of the descent properties (Section 3.3.2). The definitions we present here are more efficient than those given Section 3.3.2, and often lead to more streamlined proofs, but they come at a cost of abstraction.

A.2.1. *Descent using gluing data*

In Definition 3.25 we needed a cleavage in order to be able to write things like $X_i\big|_{S_{ij}}$. To do this properly requires keeping track of a number of canonical isomorphisms, which were intentionally elided in Definition 3.25. Reliance on a cleavage can be avoided by explicitly choosing a restriction at each step, instead of insisting on a canonical choice from the beginning. Although this definition avoids the technical deficiencies of Definition 3.25, it only exacerbates the proliferation of indices. Nevertheless, we will see in §A.2.2 that it points the way towards a definition that is both technically correct and pleasantly efficient.

Definition A.8 (Descent datum via gluing data). Let $(\mathsf{S}, P\mathscr{T})$ be a presite and let $\{S_i \to S\}$ be a cover of S in S. A *descent datum* with respect to this cover consists of the following data:

(i) objects $X_i \in \mathcal{M}(S_i)$, $X_{ij} \in \mathcal{M}(S_{ij})$, and $X_{ijk} \in \mathcal{M}(S_{ijk})$ for all indices i, j, k of the cover;
(ii) morphisms

$$X_{ij} \to X_i \quad X_{ij} \to X_j \quad X_{ijk} \to X_{ij} \quad X_{ijk} \to X_{ik} \quad X_{ijk} \to X_{jk}$$

in $\mathcal{M}$ respectively covering the canonical projections

$$S_{ij} \to S_i \quad S_{ij} \to S_j \quad S_{ijk} \to S_{ij} \quad S_{ijk} \to S_{ik} \quad S_{ijk} \to S_{jk}.$$

The category of descent data with respect to $\{S_i \to S\}$, denoted $\mathcal{M}(S_\bullet)$, has as objects the descent data as defined above. A morphism $(X_\bullet) \to (Y_\bullet)$ consists of morphisms

$$X_i \to Y_i \qquad X_{ij} \to Y_{ij} \qquad X_{ijk} \to X_{ijk}$$

for all indices i, j, k, commuting with the sructural morphisms (a more precise formulation of this condition will appear below in Remark A.11).

Remark A.9. An observant reader may be wondering where the cocycle condition (3.4) is hiding in Definition A.8. It is built into the commutativity diagram (A.2),

$$\begin{array}{ccccc} & & X_i & & \\ & \nearrow & \uparrow & \nwarrow & \\ X_{ij} & \leftarrow & X_{ijk} & \rightarrow & X_{ik} \\ \downarrow & \swarrow & \downarrow & \searrow & \downarrow \\ X_j & \longleftarrow & X_{jk} & \longrightarrow & X_k \end{array} \tag{A.2}$$

which itself is forced by the commutativity of the corresponding diagram with each X replaced by an S, and the fact that every arrow in $\mathcal{M}$ is cartesian. The morphism α_{ij} of Definition 3.27 is the composition

$$X_i\big|_{S_{ij}} \xleftarrow{\sim} X_{ij} \xrightarrow{\sim} X_j\big|_{S_{ij}}$$

with α_{ik} and α_{jk} defined similarly. The cocycle condition is the identity of the compositions

$$X_i\big|_{S_{ijk}} \xleftarrow{\sim} X_{ij}\big|_{S_{ijk}} \xrightarrow{\sim} X_j\big|_{S_{ijk}} \xleftarrow{\sim} X_{jk}\big|_{S_{ijk}} \xrightarrow{\sim} X_k\big|_{S_{ijk}}$$
$$X_i\big|_{S_{ijk}} \xrightarrow{\sim} X_{jk}\big|_{S_{ijk}} \xrightarrow{\sim} X_k\big|_{S_{ijk}}$$

by virtue of the restriction of Diagram (A.2) to S_{ijk}.

If X is an object of $\mathcal{M}(S)$ then using Axiom (i) of a category fibered in groupoids (Definition 2.8) we can find induced objects $X_i \in \mathcal{M}(S_i)$, $X_{ij} \in \mathcal{M}(S_{ij})$, and $X_{ijk} \in \mathcal{M}(S_{ijk})$ for all indices i, j, k. We also obtain the required morphisms among these by many applications of Axiom (ii) of a category fibered in groupoids. By an application of the axiom of choice, we can do this for all objects of $\mathcal{M}(S_\bullet)$ and obtain a functor $\mathcal{M}(S) \to \mathcal{M}(S_\bullet)$.

Definition A.10 (Effective descent datum via gluing data). A descent datum $X \in \mathcal{M}(S_\bullet)$ for $\mathcal{M}$ with respect to a cover $\{S_i \to S\}$ is said to be effective if it lies in the essential image of $\mathcal{M}(S) \to \mathcal{M}(S_\bullet)$.

Remark A.11. We give a technical reformulation of Definition A.8, motivated by the idea that the data we are keeping track of can be organized concisely by the nerve of subcovers consisting of three open sets; i.e., 2-simplices. Suppose that the cover $\{S_i \to S\}$ is indexed by a set I. Let $\mathbf{\Delta}$

be the category of nonempty, totally ordered, finite sets of cardinality ≤ 3, equipped with a morphism to I. That is, an object of $\mathbf{\Delta}$ is a pair (T, f) where T is a totally ordered finite set with $1 \leq |T| \leq 3$ (i.e., T is either $\{0\}$, $\{0 < 1\}$, or $\{0 < 1 < 2\}$ up to isomorphism) and $f : T \to I$ is a function. A morphism $(T, f) \to (T', f')$ is an order preserving function $g : T \to T'$ such that $f' \circ g = f$.

We introduce abbreviations for certain objects of $\mathbf{\Delta}$. Every object of $\mathbf{\Delta}$ is isomorphic to one of the following:

(i) for $i \in I$, we write i for the object $(\{0\}, (0 \mapsto i))$ of $\mathbf{\Delta}$;
(ii) for $i, j \in I$, we write ij for the object $(\{0 < 1\}, (0 \mapsto i, 1 \mapsto j))$;
(iii) for $i, j, k \in I$, we write ijk for the object $(\{0 < 1 < 2\}, (0 \mapsto i, 1 \mapsto j, 2 \mapsto k))$.

The cover $\{S_i \to S\}$ induces a functor $S_\bullet : \mathbf{\Delta}^{\mathsf{op}} \to \mathsf{S}$ sending x to S_x. In other words, it sends i to S_i and ij to S_{ij} and ijk to S_{ijk}. The morphisms are all the canonical projections among fiber products.

Now a descent datum, in the sense of Definition A.8 is a functor $X_\bullet : \mathbf{\Delta}^{\mathsf{op}} \to \mathcal{M}$ such that $\pi X_\bullet = S_\bullet$:

$$\begin{array}{ccc} & & \mathcal{M} \\ & \overset{X_\bullet}{\nearrow} & \downarrow \pi \\ \mathbf{\Delta} & \xrightarrow[S_\bullet]{} & \mathsf{S} \end{array}$$

A morphism of descent data $X_\bullet \to Y_\bullet$ is a natural transformation of functors that projects to the identity natural transformation of $S_\bullet$.

A.2.2. *Descent using sieves*

A.2.2.1. Sieves on topological spaces In fact, there is an even more efficient formulation of the definition of a sheaf. Recall that the if V is an object of O_X then h_V is the functor represented by V. It is convenient to work with the related functor, which we again denote by h_V, defined as $h_V(W) = \{\iota_{W,V}\}$ if $W \subseteq V$ and $h_V(W) = \varnothing$ otherwise.

Suppose that $U \in \mathsf{O}_X$. For any open cover $\mathcal{U}$ of U, define a presheaf $h_{\mathcal{U}}$:

$$h_{\mathcal{U}}(W) = \bigcup_{V \in \mathcal{U}} h_V(W) = \begin{cases} \{\iota_{W,X}\} & \exists\ V \in \mathcal{U} \text{ such that } W \subseteq V \\ \varnothing & \text{else.} \end{cases}$$

If $\mathscr{F} : \mathsf{O}_X^{\mathsf{op}} \to (\mathsf{Set})$ is a presheaf then, as $h_{\mathcal{U}} \subseteq h_X$ (e.g., if $W \subseteq U$ for

$U \in \mathcal{U}$, then we use $W \subseteq U \subseteq X$), we get a morphism

$$\mathscr{F}(X) \xleftarrow{\sim} \mathrm{Hom}(h_X, \mathscr{F}) \to \mathrm{Hom}(h_{\mathcal{U}}, \mathscr{F}). \tag{A.3}$$

The bijectivity of the arrow on the left is Yoneda's lemma.

The interested reader may prove the following proposition:

Proposition A.12. *Let $\mathscr{F}$ be a presheaf.*

(i) $\mathscr{F}$ is separated if and only if (A.3) *is an injection for all open covers $\mathcal{U}$ of all $U \in \mathsf{O}_X$.*
(ii) $\mathscr{F}$ is a sheaf if and only if (A.3) *is a bijection for all open covers $\mathcal{U}$ of all $U \in \mathsf{O}_X$.*

A.2.2.2. Sieves in general

Definition A.13. Let S be a category and S an object of S. A sieve of S is a subcategory of S/S that is fibered in groupoids over S/S.

In other words, a sieve of S is a subcategory $\mathcal{R} \subseteq \mathsf{S}/S$ such that whenever $S'' \to S'$ is a morphism in S/S and S' is in $\mathcal{R}$ then S'' is also in $\mathcal{R}$.

In Example 2.14 we saw how to associate a sieve $\mathcal{R}$ to any family of maps $\{S_i \to S\}$. By a *covering sieve* we mean a sieve associated to a covering family in the pretopology. In fact, all of the axioms of a Grothendieck topology can be formulated purely in terms of sieves [2, Exp. II, Déf. 1.1], but that will not concern us here.

Definition A.14 (Descent datum via sieves). Let S be a presite, let $\pi : \mathcal{M} \to \mathsf{S}$ be a category fibered in groupoids over S, and let $\mathcal{R}$ be a covering sieve of $S \in \mathsf{S}$. A *descent datum* for $\mathcal{M}$ with respect to $\mathcal{R}$ is a functor $X : \mathcal{R} \to \mathcal{M}$ lifting the canonical projection $\mathcal{R} \subseteq \mathsf{S}/S \to \mathsf{S}$ (sending an object $S' \to S$ of S/S to S'):

$$\begin{array}{ccccc} & & & & \mathcal{M} \\ & \nearrow^{X} & & & \downarrow \\ \mathcal{R} & \hookrightarrow & \mathsf{S}/S & \longrightarrow & \mathsf{S}. \end{array}$$

Descent data over the sieve $\mathcal{R}$ are the objects of a category $\mathcal{M}(\mathcal{R})$ where a morphism $X \to Y$ is a natural transformation projecting to the identity natural transformation of the projection $\mathcal{R} \to \mathsf{S}$.

To construct the functor $\mathcal{M}(S) \to \mathcal{M}(\mathcal{R})$ for a sieve $\mathcal{R}$ of S, observe that $\mathcal{M}(S) \simeq \mathcal{M}(\mathsf{S}/S)$ by the 2-Yoneda lemma (on the left we mean the fiber of $\mathcal{M}$ over S and on the right we mean the category of morphisms from S/S to $\mathcal{M}$). Composing this equivalence with the restriction $\mathcal{M}(\mathsf{S}/S) \to \mathcal{M}(\mathcal{R})$ induced from the inclusion $\mathcal{R} \subseteq \mathsf{S}/S$ induces

$$\mathcal{M}(S) \simeq \mathcal{M}(\mathsf{S}/S) \to \mathcal{M}(\mathcal{R}).$$

Definition A.15 (Effective descent datum via sieves). A descent datum $X \in \mathcal{M}(\mathcal{R})$ for $\mathcal{M}$ with respect to a sieve $\mathcal{R}$ is said to be *effective* if it lies in the essential image of $\mathcal{M}(S) \to \mathcal{M}(\mathcal{R})$.

A.3. *Grothendieck topologies*

Grothendieck pretopologies seem quite natural from the definition of a sheaf, but have the deficiency that many different pretopologies can give rise to the same category of sheaves. This is not unlike the way different bases of a topological space should be considered equivalent. The topology associated to a pretopology is the finest pretopology that gives the same category of sheaves (see [24, Rem. 2.25, Def. 2.47, Prop. 2.49]). In this section, we give an idea of how this works, providing a review of [24, §2.3.5], although here for brevity we define the topology directly, without a discussion of refinements of pretopologies.

Note that typically a Grothendieck topology is defined in terms of sieves, not coverings; the point is that the sieves associated to a pretopology are the same as the sieves associated to the topology obtained from the pretopology [24, Prop. 2.48], and therefore both induce the same Grothendieck topology in the sense of sieves.

For clarity of the discussion in this section, we will refer to coverings $\{S_\alpha \to S\}$ in a pretopology $\mathscr{T}$ on a category S as *basic coverings*. We start with a definition (cf. [24, Def. 2.45]):

Definition A.16 (Coverings with respect to $\mathscr{T}$). Let $\mathscr{T}$ be a pretopology on a category S. We call a family of morphisms $\{S_\alpha \to S\}$ *covering* (with respect to $\mathscr{T}$) if there is a basic covering family $\{T_\beta \to S\}$ such that, for each β, there is an α such that the morphism $T_\beta \to S$ factors through S_α. The covering family $\{T_\beta \to S\}$ is called a basic covering *refinement* of $\{S_\alpha \to S\}$.

Remark A.17. Note that the covering $\{S_\alpha \to S\}$ in Definition A.16 is a covering in the sense of Definition 6.2.

Lemma A.18. *Assume that* S *has all fiber products. Let* $\mathscr{T}$ *be a pretopology on* S *and let* $\mathscr{T}'$ *be the collection of all covering families with respect to* $\mathscr{T}$*. Then* $\mathscr{T}'$ *is a pretopology on* S.

Proof. Property (PT0) is automatic, since S has all fiber products. Likewise (PT3) is immediate.

If $\{S_\alpha \to S\}$ has a basic covering refinement $\{T_\beta \to S\}$ and $S' \to S$ is any morphism then $\{S_\alpha \times_S S' \to S'\}$ has the basic covering refinement $\{T_\beta \times_S S' \to S'\}$, hence is in $\mathscr{T}'$. This proves (PT1).

Suppose that $\{S_\alpha \to S\}$ has a basic covering refinement $\{T_\beta \to S\}$, and each S_α has a family $\{S_{\alpha\gamma} \to S_\alpha\}$ with a basic covering refinement $\{T_{\alpha\delta} \to S_\alpha\}$. For each β, choose an $\alpha(\beta)$ and a factorization of $T_\beta \to S$ through $S_{\alpha(\beta)}$. Then the maps $T_{\alpha(\beta)\delta} \times_{S_{\alpha(\beta)}} T_\beta \to T_\beta$ are covering. As the T_β cover S, the property (PT2) implies that the $T_{\alpha(\beta)\delta}$ cover S. Therefore the $T_{\alpha(\beta)\delta}$ give a basic covering refinement of the family $S_{\alpha\gamma} \to S$. □

Definition A.19 (Grothendieck topology). A pretopology $\mathscr{T}$ is called *a (Grothendieck) topology* if $\mathscr{T}' = \mathscr{T}$, in the notation of Lemma A.18. If $\mathscr{T}$ is a pretopology then $\mathscr{T}'$ is called the *associated topology* to $\mathscr{T}$.

Remark A.20. In [24, Def. 2.52] what we call a topology is called a saturated pretopology.

Example A.21. Let $\mathscr{T}$ be the pretopology on topological spaces where the basic covering families are open covers. Then every surjective local isomorphism has a section over a suitable open cover, so surjective local isomorphisms are covering in the associated saturated topology.

Example A.22. Consider the étale pretopology on schemes, defined in Example 3.7. Every smooth surjection admits a section over some étale cover, so every smooth surjection is covering in the associated topology to the étale pretopology.

The following lemma shows that a pretopology and its associated topology have the same sheaves. Passage to the associated topology may therefore significantly expand the class of morphisms with respect to which one can use descent.

Lemma A.23 ([24, Prop. 2.49, Prop. 2.53(iii)]). *If* $\mathscr{T}$ *is a pretopology on* S *then a presheaf on* S *is a sheaf with respect to* $\mathscr{T}$ *if and only if it is a sheaf with respect to* $\mathscr{T}'$.

Proof. Since $\mathscr{T} \subseteq \mathscr{T}'$, it is immediate that sheaves in $\mathscr{T}'$ are sheaves in $\mathscr{T}$. For the converse, suppose that $\mathscr{F}$ is a sheaf with respect to $\mathscr{T}$ and let $\{S_\alpha \to S\}$ be a covering family (with respect to $\mathscr{T}'$).

Construct a presheaf $\mathscr{F}'$ over S by the following formula:

$$\mathscr{F}'(R) = \mathrm{eq}\Big(\prod_\alpha \mathscr{F}(U_\alpha \times_S R) \rightrightarrows \prod_{\alpha,\beta} \mathscr{F}(U_\alpha \times_S U_\beta \times_S R)\Big).$$

There is a natural map $\varphi : \mathscr{F} \to \mathscr{F}'$, which we would like to show is an isomorphism. Both $\mathscr{F}$ and $\mathscr{F}'$ are sheaves in the pretopology $\mathscr{T}$, so this is a local problem in $\mathscr{T}$. We can therefore replace S by a basic cover from $\mathscr{T}$. Since $\{U_\alpha \to U\}$ has a refinement by a basic cover, we can assume that there is a section $\sigma : S \to U_\alpha$ for some α.

We can use σ to construct an inverse ψ to the map $\varphi : \mathscr{F} \to \mathscr{F}'$. Indeed, if $\xi \in \mathscr{F}'(S)$ then let ξ_α be its projection on the α component. Then $\sigma^*(\xi_\alpha) \in \mathscr{F}(S)$ and we set $\psi(\xi) = \sigma^*(\xi_\alpha)$. It is immediate that $\psi\varphi(\eta) = \eta$ for all $\eta \in \mathscr{F}(S)$.

We check that $\varphi\psi(\xi) = \xi$. What we need to check is that, for all indices β,

$$\xi_\beta = \sigma^*(\xi_\alpha)\big|_{U_\beta}.$$

By assumption, ξ is equalized by the maps to $\prod_{\alpha,\beta} \mathscr{F}(U_{\alpha\beta})$, where $U_{\alpha\beta} = U_\alpha \times_S U_\beta$. Therefore $\xi_\alpha\big|_{U_{\alpha\beta}} = \xi_\beta\big|_{U_{\alpha\beta}}$. But $(\sigma, \mathrm{id}_{U_\beta})$ determines a section of $U_{\alpha\beta}$ over U_β, so we determine that

$$\xi_\beta = \sigma^*(\xi_\beta\big|_{U_{\alpha\beta}}) = \sigma^*(\xi_\alpha\big|_{U_{\alpha\beta}}) = \sigma^*(\xi_\alpha)\big|_{U_\beta},$$

as required. □

A.4. *An example of ineffective descent*

The category fibered in groupoids $\mathcal{M}_1$ is not an étale stack! As we will see, it is possible to create a descent datum for genus 1 curves that is not effective. This is really a deficiency of our definition of $\mathcal{M}_1$ in §2.2.2, by which $\mathcal{M}_1$ parameterizes *schematic* families of smooth, proper curves of genus 1. The proper thing to do would be to include in our moduli problem smooth proper families of *algebraic spaces* (Definition B.6) whose fibers are curves of genus 1.

To construct an ineffective descent datum for $\mathcal{M}_1$, it will help to notice that every descent datum for S can at least be descended to a *sheaf* on the big étale site of S that is locally on S representable by a family of genus 1 curves. This sheaf is precisely the algebraic space we should have admitted into the moduli problem for $\mathcal{M}_1$. Our task in this section is to construct such a sheaf X that is locally in S representable by genus 1 curves, but is not globally representable by a scheme.

To begin, note that if $X \to S$ is any family of smooth curves of genus 1 over a base S, then the relative Jacobian $J \to S$ of $X \to S$ is a family of abelian schemes of dimension 1 over S. This construction is local on S, so that even if X is merely a sheaf over S that is *locally* representable by a family of smooth curves of genus 1, one obtains a descent datum for a family of elliptic curves over S. But the descent datum comes with a compatible family of ample line bundles (coming from the origin of the group structure) so by Theorem 3.38, it can be descended to a family of elliptic curves over S.

Furthermore, J acts on X making X into a J-torsor. Therefore X is classified up to isomorphism by an element $[X] \in H^1(S, J)$ (this can be étale or flat cohomology). Raynaud shows that, provided S is quasicompact, this element $[X]$ is torsion if and only if X is projective over S [70, Cor. XIII 2.4 ii)], and that, provided S is normal, X is projective over S if and only if it is representable by a scheme [70, Prop. XIII 2.6]. In fact, Raynaud proves these statements more generally about torsors under abelian varieties:

Theorem A.24 ([70, Cor. XIII 2.4 ii)]). *Assume that S is a quasicompact scheme and let J be a projective abelian variety over S. Then a J-torsor X is projective if and only if its class in $H^1(S, J)$ is torsion.*

Theorem A.25 ([70, Prop. XIII 2.6 i)]). *Let S be a quasicompact, normal scheme and J an abelian variety. Then a J-torsor X is representable by a scheme if and only if it is projective.*

Sketch of a proof of Theorem A.24. First, suppose the class represented by X is torsion, say $n[X] = 0$. As $n[X]$ is represented by $\left[X/J[n]\right]$, where $J[n]$ is the n-torsion of J, we have a finite map

$$X \to X/J[n] \simeq J.$$

This implies X is projective over S, as J is.

Conversely, if X is projective over S, then a relatively ample line bundle L on X over S induces a relatively ample line bundle L' on J [70, Lem. XI 1.6]. To see this, note that the relative Néron–Severi group $\mathrm{NS}_{X/S}$ of X over S is isomorphic to $\mathrm{NS}_{J/S}$, so that L determines a class in $\mathrm{NS}_{J/S}$. Any line bundle M on J determines a line bundle

$$\mu^* M \otimes p_1^* M^\vee \otimes p_2^* M^\vee \otimes e^* M \tag{A.4}$$

(where $\mu : J \times J \to J$ is the addition map, p_i are the projections, and e is the composition of the projection $J \to S$ and the zero section) on $J \times_S J$ that depends only on the Néron–Severi class of M. Restricting this to the diagonal of $J \times J$ recovers a line bundle on J whose image in the Néron–Severi group is twice that of M. Altogether, this gives a map:

$$\mathrm{NS}_{X/S} \simeq \mathrm{NS}_{J/S} \to \mathrm{Pic}_{J/S}.$$

Applying this to L yields a line bundle L' on J. Moreover, under a local isomorphism between J and X, the Néron–Severi class of L' is double that of L. Thus L' is relatively ample on J over S.

In general, a relatively ample line bundle M on J induces an isogeny $J \to \hat{J}$ of abelian schemes over S (where $\hat{J}$ is the dual abelian variety) by way of (A.4), and hence a morphism $H^1(S, J) \to H^1(S, \hat{J})$. The image of $[X]$ is the class $[\hat{X}]$, where $\hat{X}$ is the $\hat{J}$-torsor consisting of line bundles on X that lie in the same Néron–Severi class as L' [70, Cor. XIII 1.2 ii)]. By assumption this torsor is trivial (the line bundle $L^{\otimes 2}$ provides a section), so $[\hat{X}] = 0$. On the other hand, we have an exact sequence

$$H^1(S, K) \to H^1(S, J) \to H^1(S, \hat{J}),$$

where K is the kernel of $J \to \hat{J}$, so $[X]$ lies in the image of $H^1(S, K)$. Since L is ample, K is finite, so $H^1(S, K)$ is torsion, and therefore so is $[X]$. □

Sketch of a proof of a special case of Theorem A.25. We will prove a special case of Theorem A.25, following the proof of [70, Thm. V 3.10], that will suffice to construct our example. We assume that S is the spectrum of a noetherian local ring with closed point s and generic point η. Let X be a J-torsor over S. Choose an effective Cartier divisor $D \subseteq X$ whose complement $U \subseteq X$ is quasi-affine and meets the closed fiber of X. We will argue that the line bundle $L = \mathcal{O}_X(D)$ must be ample.

We make a few observations:

(a) *The J-orbit of U is all of X.* This is because U meets X_s and J_s acts transitively on X_s, so JU contains X_s. But every point of X specializes to a point of X_s, since X is proper over S, and JU is open, hence contains all of X.

(b) *Let M be the line bundle*

$$M = (p_1 + p_2 + p_3)^* L \otimes (p_1 + p_3)^* L^\vee \otimes (p_2 + p_3)^* L^\vee \otimes p_3^* L \quad \text{(A.5)}$$

on $J \times_S J \times_S X$. There is some positive integer n such that $M^{\otimes n}$ is the pullback of a line bundle on $J \times_S J$. As $J \times_S J$ is normal, it is sufficient to verify this over the generic point of S [37, Cor. (21.4.13), Erratum 4.53, p.361]. We may therefore assume that $S = \eta$ is the spectrum of a field. If $X(\eta) \neq \varnothing$ then $X \simeq J$ and this is a version of the theorem of the cube (see [63, §5, Cor. 6 and §6, Thm., Cor. 2 p.58]); in this case $n = 1$. If X does not have a section over η then it will certainly have one over some finite extension $p : \eta' \to \eta$, so p^*M is the pullback of a line bundle M' on $J_{\eta'} \times_{\eta'} J_{\eta'}$. If η' has degree n over η then the norm of p^*M is $M^{\otimes n}$ and this is the pullback of the line bundle $\mathrm{Norm}_{\eta'/\eta}(M')$ on $J_\eta \times_\eta J_\eta$.

We replace D with nD so that (A.5) is the pullback of a line bundle on $J \times_S J$ without passing to a tensor power.

(c) Restricting (A.5) to a point $(g, -g)$ of J yields an isomorphism of line bundles on X:

$$T_g^* L \otimes T_{-g}^* L \simeq L^{\otimes 2}$$

where T_g denotes translation by g.

To prove the ampleness of L, we must show that, for any point x of X, and any open neighborhood V of x in X, there is some index n and some $f \in \Gamma(X, L^{\otimes n})$ such that the open set $X_f \subseteq X$ defined by the nonvanishing of f is contained in V [32, Thm. (4.5.2)]. Suppose first that $x \in U$. Let z be the tautological section of $\mathcal{O}_X(D)$ that vanishes exactly along D. As U is quasi-affine, there is some affine open neighborhood W of x in $V \cap U$, defined by the nonvanishing of a function h on U. Then there is some $n \geq 1$ such that $z^n h$ extends to all of X. That is, we may regard $z^n h$ as a section without poles of $\mathcal{O}_X(nD)$ and $W_h = W$.

If x is not in U, we can at least find a $g \in J$ such that $T_g(x) \in U$ (by observation (a), above). Applying the argument above to $T_g(x)$ and $T_g(V)$, we can find a section $f \in \Gamma(X, L^{\otimes n})$ such that $x \in W_f \subseteq T_g(V)$ by the argument above. We have some freedom in the choice of g, and by avoiding a closed set of possibilities, we can ensure that $T_{-g}(x) \in U$ as well. Then we can consider $f' = T_g^* f \otimes T_{-g}^* z^n$ as a section of $T_g^* L^{\otimes n} \otimes T_{-g}^* L^{\otimes n} \simeq L^{\otimes 2n}$ (by (c), above). Now, $W_{f'} = T_g^{-1}(W_f) \cap T_{-g}^{-1}(W_z) = T_g^{-1}(W_f) \cap T_{-g}^{-1}(U)$. By our choice of g, we have $x \in W_{f'}$ and $W_{f'} \subseteq T_g^{-1} T_g(V) = V$, as required. ☐

Finally, [70, XIII 3.2] proves that there is a genus 1 curve over a normal noetherian local ring of dimension 2 whose image in $H^1(S, J)$ is non-torsion, establishing the existence of the desired family $X \to S$. Another version of the construction (and a slightly stronger conclusion) can be found in [85]. We will summarize Raynaud's construction.

We begin with a discrete valuation ring R with algebraically closed residue field. Let $T = \operatorname{Spec} R$ and assume that T has a connected, étale double cover $T' \to T$. Let π be a uniformizer for T. For any object Y over T, we will write Y' for its base change to T'. We write τ for the generic point of T and t for the special point; τ' is the generic point of T' and t_1 and t_2 are the two points in the fiber of T' over t.

Let E be an elliptic curve over T and let $V \subseteq E$ be the complement of the zero section in E. Let W be the quotient of V by the inversion in the group law of E, so $W \simeq \mathbb{A}^1_T$. Let $\gamma : W \to W$ be multiplication by π and let Z be the normalization in V of the composition

$$V \to W \xrightarrow{\gamma} W.$$

All we will use about Z is the following lemma, whose proof is straightforward:

Lemma A.26. *The map $f : V \to Z$ is an isomorphism over $\tau \in T$ and constant over $t \in T$. Furthermore, Z is normal.*

Let s be the unique closed point of Z and let $S = \operatorname{Spec} \mathcal{O}_{Z,s}$. Let η' be the generic point of S' and let s_1 and s_2 the two closed points lying above s. Let U_i be the complement in S' of s_i, and let U_{12} be their intersection. Then $U_1 \cup U_2 = S'$, so we use the Mayer–Vietoris sequence to find a class in $H^1(S', E')$:

$$H^0(U_1, E') \times H^0(U_2, E') \to H^0(U_{12}, E') \to H^1(S', E').$$

Now, $U_{12} = \{\eta'\}$ so $H^0(U_{12}, E') = E'(\eta')$. Recall that we have a map $E' \to Z'$ that is an isomorphism over the generic points. Therefore we have a canonical element ξ of $E'(\eta')$ corresponding to the inclusion of the generic point.

An element of $H^0(U_i, E')$ can be seen as a rational map from E' to itself over T' that restricts to a constant map over t_i. Any such map must in fact factor through a section of E' over T'. Therefore we have

$$H^0(U_i, E') = E'(T').$$

Consider the image of ξ in $H^1(S', E')$. This cannot possibly be torsion, for if it were then it would have a multiple in the image of $H^0(U_1, E') \times H^0(U_2, E')$. That is impossible, because no multiple of the identity map on an elliptic curve is a difference of constant maps. Therefore we have found a non-torsion element in $H^1(S', E')$. Since S' is étale over the normal scheme S, it is normal, and therefore by Theorem A.24, the corresponding sheaf is a descent datum for an elliptic curve over S' that is not projective, hence not effective by Theorem A.25.

Raynaud goes a bit further and shows that the base for the descent may be chosen to be local. Indeed, letting q denote the projection from S' to S, we have

$$H^1(S', E') = H^1(S, q_* E').$$

Now, $q_* E' = q_* q^* E'$ is the Weil restriction of scalars of E' via the finite, étale map q, hence is an abelian scheme of dimension 2 over S. It comes with a canonical inclusion $E \subseteq q_* E'$ whose quotient is another elliptic curve F (in fact a quadratic twist of E). Then we have an exact sequence

$$H^1(S, E) \to H^1(S, q_* E') \to H^1(S, F)$$

so the non-torsion class $\xi \in H^1(S', E') = H^1(S, q_* E)$ determines a non-torsion class either in $H^1(S, F)$ or in $H^1(S, E)$. Either way, we obtain a non-effective descent datum for genus 1 curves over S.

B. The many meanings of algebraicity

We work over the presite S of étale covers of schemes (over some fixed base scheme). Many authors have given different definitions of algebraicity. Deligne and Mumford required a schematic diagonal and an étale cover by a scheme [20, Def. (4.5)], but insisted their definition was not the right one except for quasiseparated stacks. Knutson required all algebraic spaces to be quasiseparated [53, Ch. 2, Def. 1.1]. Artin gave his definition only for stacks that are locally of finite presentation and required a diagonal representable by algebraic spaces (in the sense of Knutson) and a smooth cover by a scheme [9, Def. (5.1)]. Laumon and Moret-Bailly defined an 'algebraic stack (understood quasiseparated)' by adding quasiseparation to Artin's conditions [54, Def. (4.1)]. The Stacks Project uses Artin's conditions, but without requiring the algebraic spaces to be quasiseparated [77, Tag 026N].

Recall that in Definition 6.27 we defined algebraic stacks as the smallest class of stacks on the category of schemes that includes all schemes and includes all stacks that admit smooth covers by stacks in the class. Our definition is equivalent to the Stacks Project's, but usage in the literature varies widely.

B.1. *Stacks with flat covers by schemes and stacks over the fppf site*

The étale topology is not the only natural topology on schemes. From the perspective of descent, at least for quasicoherent sheaves, the flat topologies (fppf and fpqc) might be even more natural. The fpqc topology presents certain technical issues, owing to the absence of a sheafification functor, so we will not discuss it.

There are two ways one might try to replace the étale topology with the fppf topology in our discussion of algebraic stacks. We might try to limit the class of stacks by insisting they be stacks in the fppf, as opposed to just the étale, topology. Or, we might enlarge the class of stacks under consideration by permitting them to have fppf, as opposed to necessarily smooth, covers by schemes. It turns out that either modification yields the same class of algebraic stacks. We will sketch the main ideas behind this result.

We start by stating the following lemma whose proof we omit as it is well-known and not difficult.

Lemma B.1. *The class of flat morphisms of schemes is stable under composition and base change and is local to the source and target in the fppf topology.*

The lemma implies that one could develop a theory of flat-adapted algebraic stacks in the étale topology. The following theorem of Artin explains that to do so would yield nothing new:

Theorem B.2 ([9, Thm. (6.1)], [54, Thm. 10.1], [77, Tag 06DB]). *Let $\mathcal{X}$ be a stack in the fppf topology on the category of schemes. If there exists morphism*

$$U \xrightarrow{P} \mathcal{X}$$

from an algebraic space U that is representable by algebraic spaces, faithfully flat, and of finite presentation, then there is such a morphism P that is smooth and surjective. In particular, $\mathcal{X}$ is an SP-algebraic stack in the sense of Definition 6.27.

Proof. We give a rough sketch of the proof, following Artin. By an 'induction on stackiness', it is sufficient to assume that the diagonal of $\mathcal{X}$ is representable by algebraic spaces.

We consider the following moduli problem $\mathcal{V}$. Choose a cover of U by a disjoint union U_0 of affine schemes (which is certainly possible, since U is a scheme). Since $U_0 \times_{\mathcal{X}} U_0$ is an algebraic space, we can choose a disjoint union of affine schemes U_1 and a smooth cover $U_1 \to U_0 \times_{\mathcal{X}} U_0$. Let s and t denote the two projections from U_1 to U_0. Let U_2 be a smooth cover of the space of triples $(\alpha, \beta, \gamma) \in U_1 \times U_1 \times U_1$ such that $s(\alpha) = s(\gamma)$, $t(\alpha) = s(\beta)$, and $t(\beta) = t(\gamma)$ and $\beta \circ \alpha = \gamma$ as isomorphisms between objects of $\mathcal{X}$.

For any scheme S, we define an S-point of $\mathcal{W}$ to be

(i) the choice of a finite union of components $V_i \subseteq U_i$ for each i such that $V_\bullet$ forms a subgroupoid of $U_\bullet$,

(ii) a finite, locally free, surjective S-scheme Z with a distinguished basis $\mathcal{O}_Z \simeq \mathcal{O}_S^d$, and

(iii) a morphism of groupoids $Z_\bullet \to V_\bullet$:

$$\begin{array}{ccccccc} Z_2 & \substack{\longrightarrow\\ \longrightarrow\\ \longrightarrow} & Z_1 & \substack{\longrightarrow\\ \longrightarrow} & Z_0 & \longrightarrow & S \\ \downarrow & & \downarrow & & \downarrow & & \\ V_2 & \substack{\longrightarrow\\ \longrightarrow\\ \longrightarrow} & V_1 & \substack{\longrightarrow\\ \longrightarrow} & V_0 & \longrightarrow & \mathcal{X} \end{array}$$

(iv) where we have set $Z_0 = Z$ and $Z_i = Z_{i-1} \times_S Z$ for $i \geq 1$.

We argue that $\mathcal{V}$ is representable by a disjoint union of affine schemes, indexed by the choice of $V_\bullet$. The algebra structure on $\mathcal{O}_Z$ is determined by its structure constants and various identities among them, hence is parameterized by an affine scheme. The maps $Z_i \to V_i$ are determined by various elements of $\mathcal{O}_Z$ and relations among them (since the V_i are affine schemes). For each commutativity condition we have a pair of maps $Z_i \to V_{i-1}$ that we wish to coincide. That is a closed condition (since affine schemes are separated).

Any S-point of $\mathcal{V}$ determines a descent datum for a morphism $S \to \mathcal{X}$ in the fppf topology. Since $\mathcal{X}$ is a stack in the fppf topology, this descends to a morphism $S \to \mathcal{X}$ and we obtain a morphism of groupoids $\mathcal{V} \to \mathcal{X}$. We have just seen that $\mathcal{V}$ is representable by a disjoint union of affine schemes, so it remains to verify this map is smooth.

Now let $\mathcal{W} \subseteq \mathcal{V}$ be the open substack where $Z_0 \to V_0 \times_{\mathcal{X}} S$ is a local complete intersection morphism. To see that $\mathcal{W}$ is indeed open in $\mathcal{V}$, note that there are open subsets $W_i \subseteq Z_i$ where the maps $W_i \to V_i$ are local complete intersection morphisms. Since the Z_i are proper over S, the image in S of the complement of W_i is closed, so the condition that the fiber of $Z_i \to V_i$ be a closed immersion and a local complete intersection morphisms is open on S.

Now we verify that $\mathcal{W}$ is locally of finite presentation, formally smooth, and surjective over $\mathcal{X}$.

To see that $\mathcal{W} \to \mathcal{X}$ is locally of finite presentation, one observes that once a morphism $S \to \mathcal{X}$ is specified, a lift to $\mathcal{W}$ involves only a finite amount of additional data.

Next we verify the smoothness, for which we can use the infinitesimal criterion. Given a lifting problem

$$\begin{array}{ccc} S & \longrightarrow & \mathcal{W} \\ \downarrow & \nearrow & \downarrow \\ S' & \longrightarrow & \mathcal{X} \end{array}$$

in which S is affine and S' is an infinitesimal extension of S, we have, by definition a morphism of groupoids $Z_\bullet \to V_\bullet$, with the Z_i finite and locally free over S, that we would like to extend to $Z'_\bullet \to V_\bullet$ with Z'_i finite and locally free over S'.

In this case, the map $Z_0 \to U_0$ is a local complete intersection morphism, and Z_0 is affine, so there is no obstruction to extending it to a morphism $Z'_0 \to U_0$ with Z'_0 finite and locally free over S'. This induces a pair of morphisms $Z'_1 \rightrightarrows V_0$, hence a map $Z'_1 \to V_0 \times_{\mathcal{X}} V_0$. Now, $V_1 \to V_0 \times_{\mathcal{X}} V_0$ is smooth, so the map $Z_1 \to V_0 \times_{\mathcal{X}} V_0$ lifts to V_1. Now let $R \subseteq V_1 \times V_1 \times V_1$ be the set of triples (α, β, γ) such that the equation $\beta \circ \alpha = \gamma$ makes sense and holds in $\mathcal{X}$. Then we obtain $Z'_2 \to R$ and V_2 is smooth over R, so $Z'_2 \to R$ lifts to V_2. An infinitesimal deformation of a local complete intersection morphism is still a local complete intersection morphism, and an infinitesimal deformation of a closed embedding is still a closed embedding, so we have produced the required extension.

Finally, we have to check $\mathcal{W} \to \mathcal{X}$ is surjective. Let k be the spectrum of an algebraically closed field and let $S = \operatorname{Spec} k$. Let ξ be a k-point of $\mathcal{X}$. The fiber of U_0 over S is a nonempty algebraic space T. Choose a smooth cover P of T by a scheme. This scheme is certainly flat over k, so it has a dense open subset where it is Cohen–Macaulay [77, Tag 045U]. Pick a point p of P where P is Cohen–Macaulay, and let Z_0 be the vanishing locus of a regular sequence at p. Now let V_0 be a component of U_0 that contains the image of p under the composition $Z_0 \to P \to T \to U_0$. Then $Z_0 \to V_0 \times_{\mathcal{X}} S$ is a local complete intersection morphism by construction. Furthermore, we obtain a map $Z_1 \to V_0 \times_{\mathcal{X}} V_0 \subseteq U_1$. But recall that Z_1 has just one point, by construction, so we choose a component V_1 of U_1 whose image in $U_0 \times_{\mathcal{X}} U_0$ contains the image of Z_0. Since Z_1 is artinian, there is a lift of

$Z_1 \to V_0 \times_{\mathcal{X}} V_0$ to V_1. Then we repeat the same process to get $Z_2 \to V_2$ and we conclude. □

Remark B.3. Note that the condition that $\mathcal{X}$ have separated and quasi-compact diagonal does not depend on the presentation.

Corollary B.4. *Suppose that G is a* flat *group scheme over S, acting on an S-scheme X. Then the stack $[X/G]$ (see §C.1) is algebraic. If G is quasiseparated over S (e.g., quasiprojective) then $[X/G]$ is a quasiseparated algebraic stack.*

Proof. The cover $X \to [X/G]$ is a G-torsor, hence is an fppf cover. Therefore from the theorem, $[X/G]$ is an algebraic stack. We may identify $X \times_{[X/G]} X$ with $X \times_S G$, under which identification the diagonal map becomes the inclusion $(\mathrm{id}_X, e) : X \to X \times_S G$, with e denoting the identity section of G over S. This morphism is certainly representable and separated (it is an injective morphism of schemes). It is quasicompact if G is quasiseparated over S, since a section of a quasiseparated morphism is quasicompact [77, Tag 03KP]. □

Next we consider the question of stacks in the fppf topology. Note first that an algebraic stack in the fppf topology is clearly an algebraic stack in the étale topology, by restriction. Now we show the converse:

Corollary B.5. *Algebraic stacks are stacks in the fppf toplogy.*

Proof. Suppose that $\mathcal{X}$ is an algebraic stack. Let $\mathcal{X}'$ be the fppf stackification. By induction, we can assume that we have already shown the diagonal of $\mathcal{X}$ is a relative fppf sheaf, which means that $\mathcal{X} \to \mathcal{X}'$ is injective.

Now let $U \to \mathcal{X}$ be a smooth cover. We argue that $U \to \mathcal{X}'$ is representable by algebraic spaces. Indeed, if $S \to \mathcal{X}'$ is any morphism, we can find an fppf cover $T \to S$ such that $T \to \mathcal{X}'$ lifts to $\mathcal{X}$. Then $T \times_{\mathcal{X}'} U = T \times_{\mathcal{X}} U$ since $\mathcal{X} \subseteq \mathcal{X}'$, and $T \times_{\mathcal{X}} U$ is an algebraic space. But the map

$$T \times_{\mathcal{X}'} U \to S \times_{\mathcal{X}} U$$

is the base change of the fppf cover $T \to U$, so $S \times_{\mathcal{X}} U$ has an fppf cover by an algebraic space. It is therefore an algebraic space, as required.

Furthermore, $S \times_{\mathcal{X}'} U \to S$ is a smooth cover, since smoothness can be verified locally in the fppf topology and $T \times_{\mathcal{X}'} U = T \times_{\mathcal{X}} U \to T$ is a smooth cover. Therefore $U \to \mathcal{X}'$ is a smooth cover, and as this factors through $\mathcal{X}$, the map $\mathcal{X} \to \mathcal{X}'$ is a smooth cover. Now both $\mathcal{X}$ and $\mathcal{X}'$ are stacks in the étale topology, and $\mathcal{X} \to \mathcal{X}'$ was already seen to be injective, so $\mathcal{X} \to \mathcal{X}'$ is an isomorphism. □

B.2. *Other definitions of algebraicity*

The following definition collects some of the most common meanings attributed to algebraicity of a stack on the étale site of schemes, in roughly chronological order. After giving the definition, we analyze the relationships among them, as well as to our Definition 6.27.

Definition B.6 (Algebraic stack). Let $\mathcal{X}$ be a category fibered in groupoids over $\mathsf{S} = \mathsf{S}_{\mathrm{et}}$.[f] We define various notions of algebraic stack using the table below. Namely, we call $\mathcal{X}$ a *Deligne–Mumford algebraic stack (resp. Knutson algebraic space, resp. Artin algebraic stack, etc.)* if the diagonal $\Delta : \mathcal{X} \longrightarrow \mathcal{X} \times \mathcal{X}$ satisfies the condition specified in the second column of the table on page 371, and there is a scheme U and a surjection $p : U \longrightarrow \mathcal{X}$ that satisfies the conditions in the third column. The morphism $p : U \to \mathcal{X}$ is called a *presentation* of $\mathcal{X}$. A *morphism* between any such stacks is a morphism of the underlying CFGs.

[f]Much of the literature works over the étale site of affine schemes. Since every scheme has an étale (even Zariski) cover by affine schemes, the notions of stacks agree.

[g]Note that our definition of representability is different, but equivalent to some of the notions used in the literature (see Lemma B.11). The notions of quasicompact and separated morphisms of algebraic spaces defined in §9 carry over directly to all of the notions of algebraic spaces discussed here. Also, in the sources, the various notions of algebraic spaces are defined for sheaves, rather than stacks. Here, for uniformity, we have simply added the injectivity hypothesis on the diagonal (see Lemma 4.8).

[h]The morphism $p : U \to \mathcal{X}$ is either assumed to be schematic, or is representable by the same class of algebraic spaces for which the diagonal is representable by virtue of Lemma B.12. In §9, we defined surjective, étale, and smooth morphisms of algebraic spaces; these definitions carry over directly to all of the notions of algebraic spaces discussed here.

[i]Deligne and Mumford caution that their definition is correct only for quasiseparated stacks [20, Footnote (1), p. 98].

[j]Quasicompactness of the diagonal immediately implies quasicompactness of the map in [53, Def. 1.1(c)]. We leave the converse to the reader. See also [53, Tech. Detail 1.p], cf. Stacks Project Algebraic Spaces.

[k]Artin gives his definition only under an additional assumption of local finite presentation.

[l]Laumon and Moret-Bailly ask only for a Laumon–Moret-Bailly algebraic space U and a smooth surjection onto $\mathcal{X}$ [54, Def. (4.1)], but then U has an étale cover by a scheme, so the definition is equivalent.

[m]The Stacks Project requires its algebraic spaces and algebraic stacks to be sheaves in the fppf topology [77, Tag 025Y, 026O]. This yields an equivalent definition by [77, Tag 076M] in the case of algebraic spaces and by [77, Tag 076U] in the case of algebraic stacks (see §B.1).

We call $\mathcal{X}$...	if the diagonal $\Delta : \mathcal{X} \to \mathcal{X} \times \mathcal{X}$ is...[g]	and there is a scheme U and a surjective morphism $p : U \longrightarrow \mathcal{X}$ that is ...[h]
a *Deligne–Mumford algebraic (DM algebraic) stack [20, Def. (4.6)]*[i]	schematic	étale.
a *Knutson algebraic space [53, Def. II.1.1]*	injective and quasicompact[j]	schematic and étale.
an *Artin algebraic stack [9, Def. (5.1)]*[k]	representable by Knutson algebraic spaces	smooth.
a *Laumon–Moret-Bailly (LMB) algebraic space [54, Def. (1.1)]*	injective, schematic, and quasicompact	étale.
a *Laumon–Moret-Bailly Deligne–Mumford (LMB DM) stack [54, Def. (4.1)]*	representable by Laumon–Moret-Bailly algebraic spaces, separated, and quasicompact	étale.[l]
a *Laumon–Moret-Bailly (LMB) algebraic stack [54, Def. (4.1)]*	representable by Laumon–Moret-Bailly algebraic spaces, separated, and quasicompact	smooth.[l]
a *Fantechi Deligne–Mumford (F DM) stack [23, Def. 5.2]*		schematic and étale.
a *Fantechi (F) algebraic stack [23, Def. 5.2]*		schematic and smooth.
a *stacks project (SP) algebraic space [77, Tag 025Y]*[m]	injective and schematic	étale.
a *stacks project Deligne–Mumford (SP DM) stack [77, Tag 03YO]*[m]	representable by SP algebraic spaces	étale.
a *stacks project (SP) algebraic stack [77, Tag 026O]*[m]	representable by SP algebraic spaces	smooth.

Warning B.7. There is an unfortunate, confusing point in the nomenclature introduced in Definition B.6. Deligne and Mumford defined an algebraic stack to be what we have, for the sake of historical verisimilitude, called a 'Deligne–Mumford algebraic stack' above. Artin defined algebraic stacks more inclusively, and the modern terminology is more inclusive still. Meanwhile, the term Deligne–Mumford stack has come to refer to algebraic stacks with unramified diagonal. As the term 'algebraic' has become ever more inclusive, so has 'Deligne–Mumford', so that now the class of 'Deligne–Mumford stacks', while contained in the class of algebraic stacks, unfortunately includes some stacks that are not 'Deligne–Mumford algebraic stacks' in the sense we defined them here. The relationship among the definitions is clarified in Figure 1.

Remark B.8. A smooth (resp. étale) morphism of algebraic stacks $\pi : \mathcal{Y} \to \mathcal{X}$ that is representable by algebraic spaces is surjective if and only if it is covering in the étale topology. Indeed, π is surjective if and only if its base change to any scheme is surjective, if and only if its base change to any scheme is covering, if and only if it is covering. In short, for Definition B.6, in the table on page 371 we could replace the heading '*... and there is a scheme U and a surjective morphism $p : U \to \mathcal{X}$ ...*' with the heading '*... and there is a scheme U and a cover $p : U \to \mathcal{X}$ ...*'. Indeed, in the definition, we have either stipulated that p is at least representable by algebraic spaces, or we obtain this from the condition on the diagonal (see Lemma B.12).

B.3. *Remarks on representability*

Note that in order to be able to speak about morphisms representable by the classes of algebraic stacks in Definition B.6, we need to know that these categories admit fiber products.

Lemma B.9 (cf. [77, Tags 02X2 and 04T2]). *All of the classes of stacks in Definition B.6 admit fiber products and these coincide with fiber products taken on the underlying CFGs.*

Proof. To deal with the entries in Definition B.6 that involve a schematic cover (DM stack, K algebraic space, LMB algebraic space, FDM stack, F algebraic stack, SP algebraic space), suppose that $\mathcal{X} \to \mathcal{Z}$ and $\mathcal{Y} \to \mathcal{Z}$ are morphisms of stacks of the appropriate type. Choose smooth schematic covers (or étale schematic covers, as the case warrants) $X \to \mathcal{X}$, $Y \to \mathcal{Y}$, and $Z \to \mathcal{Z}$ by schemes X, Y, and Z. Set $X_Z = X \times_{\mathcal{Z}} Z$ and $Y_Z = Y \times_{\mathcal{Z}} Z$, as in the diagram below:

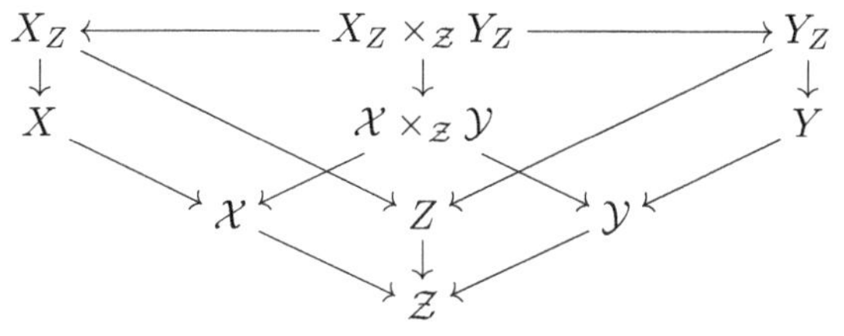

The projections $X_Z \to \mathcal{X}$ and $Y_Z \to \mathcal{Y}$ are both smooth (or étale, according to the case) and covering (by base change and composition; Lemma 6.5). Now,

$$X_Z \underset{\mathcal{Z}}{\times} Y_Z = (X_Z \times Y_Z) \underset{Z \times Z}{\times} (Z \underset{\mathcal{Z}}{\times} Z)$$

so $X_Z \times_{\mathcal{Z}} Y_Z$ is a fiber product of schemes, hence is a scheme. Moreover, the map $X_Z \times_{\mathcal{Z}} Y_Z \to \mathcal{X} \times_{\mathcal{Z}} \mathcal{Y}$ is a fibered product of smooth (or étale, as the case warrants) coverings hence is a smooth (or étale) covering. Finally, via composition and base change, Lemma 6.8 implies the maps $X_Z \to \mathcal{X}$ and $Y_Z \to \mathcal{Y}$ are schematic, so the same applies to $X_Z \times_{\mathcal{Z}} Y_Z \to \mathcal{X} \times_{\mathcal{Z}} \mathcal{Y}$ from Lemma 6.8(3).

Now this implies that fiber products of SP algebraic spaces are SP algebraic spaces so that the same argument can be repeated, with 'representable by algebraic spaces' substituted for 'schematic' and 'algebraic space' substituted for 'scheme'. This proves the lemma for SP algebraic stacks and SP DM stacks.

All of the remaining classes in Definition B.6 can be characterized as algebraic stacks with additional conditions on the diagonal. The conditions imposed on the diagonal are all properties that are preserved by fibered products of morphisms (note that properties preserved by composition and fibered product are preserved by fibered products of morphisms; see the proof of Lemma 6.8(3)), and the diagonal morphism of a fibered product is the fibered product of the diagonal morphisms. □

Remark B.10. Using Definition 6.6, we now may speak of morphisms representable by any of the classes of stacks in Definition B.6.

Although we will not need to do so, it is sometimes convenient to test representability by various classes of stacks using smaller classes. The following lemma shows that the class of affine schemes suffices to address most representability questions.

Lemma B.11. *A morphism of stacks $f : \mathcal{X} \to \mathcal{Y}$ is representable by a class of algebraic spaces in Definition B.6 (resp. schematic) if and only if for every affine scheme $S = \operatorname{Spec} A$, the fibered product $\mathcal{X} \times_{\mathcal{Y}} S$ is representably by an algebraic space in that class (resp. a scheme).*

Proof. All of the properties involved in the various definitions of algebraic space satisfy étale descent, so they can be verified étale locally. Since algebraic spaces have étale covers by affine schemes (Theorem B.13) we can check if a morphism is representable by algebraic spaces in any sense by testing with affine schemes. The same argument with Zariski descent and schemes takes care of the respected case. □

If $\mathcal{X}$ is a CFG then conditions on the diagonal correspond by base change to conditions on the fiber product $S \times_{\mathcal{X} \times \mathcal{X}} S$ for all schemes S and all pairs of morphisms $x, y : S \to \mathcal{X}$. Recall from Example 4.6 that this fiber product may be identified with $\mathscr{I}som_{\mathcal{X}}(x, y)$ and from Lemma 4.8 that a CFG has injective diagonal if and only if it is equivalent to a sheaf.

Lemma B.12 ([54, Cor. 3.13], [12, Prop. 5.12], [77, Tag 045G]). *Let $\mathcal{M}$ be a CFG over over S. The following conditions are equivalent:*

(1) The diagonal morphism $\mathcal{M} \xrightarrow{\Delta} \mathcal{M} \times \mathcal{M}$ is representable by SP algebraic spaces (resp. representable by K algebraic spaces, resp. schematic);
(2) For all S in S, and all x, y in $\mathcal{M}(S)$, the presheaf $\mathscr{I}som(x, y)$ on S is representable by an SP algebraic S-space (resp. K algebraic S-space, resp. S-scheme);
(3) For all S in S and all x in $\mathcal{M}(S)$, the morphism $x : S \to \mathcal{M}$ (guaranteed by the Yoneda lemma) is representable by SP algebraic spaces (resp. representable by K algebraic spaces, resp. schematic);
(4) For every SP algebraic space (resp. K algebraic space, resp. scheme) S, we have that every morphism $S \to \mathcal{M}$ is SP-representable (resp. K representable, resp. schematic).

B.4. *Overview of the relationships among the definitions of algebraicity*

Our definition of an algebraic stack can also be characterized in a similar manner to Definition B.6.

Theorem B.13. *Each type of stack $\mathcal{X}$ in the first column below is characterized by the condition on the diagonal in the second column and the existence of a surjection $p : U \to \mathcal{X}$ satisfying the condition in the third column.*

A stack $\mathcal{X}$ is...	*if the diagonal $\Delta : \mathcal{X} \to \mathcal{X} \times \mathcal{X}$ is...*	*and there is a scheme U and a surjective morphism $p : U \longrightarrow \mathcal{X}$ that is ...*
an algebraic space (Definition 6.28)	*injective*	*schematic and étale.*
a Deligne–Mumford stack (Definition 6.28)	*unramified*	*representable by algebraic spaces and smooth.*
a Deligne–Mumford stack (Definition 6.28)		*representable by algebraic spaces and étale.*
an algebraic stack (Definition 6.28)		*representable by algebraic spaces and smooth.*

Sketch. Suppose $\mathcal{X}$ is an algebraic stack. Considering the iterative nature of Definition 6.27, it is clear that $\mathcal{X}$ has a smooth cover $P : U \to \mathcal{X}$ by a scheme U. We need to show that P is representable by algebraic spaces. This morphism must have injective relative diagonal (since $U \times_{\mathcal{X}} U$ is a sheaf of sets and $U \to U \times_{\mathcal{X}} U \to U \times U$ is injective), so it is representable by algebraic spaces, as required.

This argument applies also to show that Deligne–Mumford stacks have étale covers by algebraic spaces.

Suppose now that $\mathcal{X}$ is an algebraic space. Then the diagonal of $\mathcal{X}$ is injective and representable by algebraic spaces. In particular, it is locally quasifinite and separated, so by separated, locally quasifinite descent [77, Tag 02W8], it is schematic.

To complete the proof, we need to show that algebraic spaces and algebraic stacks with unramified diagonals have *étale* covers by schemes. We will show these statements simultaneously (using an 'induction on stackiness'), mostly following [77, Tag 06N3].

We argue that every finite-type point of $\mathcal{X}$ has an étale neighborhood that is a scheme. This will suffice, since if x is a geometric point of $\mathcal{X}$, we can find a smooth map $U \to \mathcal{X}$ where U is an affine scheme and x lifts to U. Then this lift has a specialization to a closed point of U, which induces a point y of $\mathcal{X}$ of finite type. Any étale neighborhood of y will also contain x.

Let $x = \operatorname{Spec} k$ be a point of $\mathcal{X}$ of finite type. We argue first that there is a factorization $x \to y \to \mathcal{X}$ where y is *unramified* over $\mathcal{X}$. Let $R = x \times_{\mathcal{X}} x$. Then R is flat over x (since x is the spectrum of a field) and of finite type (since the composition $u \times_{\mathcal{X}} u \to u \times_{\mathcal{X}} \mathcal{X}$ is the base change of a morphism $u \to \mathcal{X}$ of finite type).

Since R is of finite type over x via the first projection, its geometric fiber has finitely many components. Replacing k with a finite extension, we can therefore assume that the connected components of R are geometrically connected. Let $R_0 \subset R$ be the connected component of the diagonal section. Note that $R_0 \to x \times x$ is unramified and its fiber over the diagonal $x \to x \times x$ is injective on geometric points. But every geometric fiber is either empty or is a torsor under the fiber over the diagonal, so that every fiber is injective on geometric points. It follows that $R_0 \to x \times x$ is a injection, so that it defines an equivalence relation on x.

Let $y = x/R_0$ be the quotient of x by this equivalence relation. Since $R_0 \subset R$ is open, R_0 is flat over x and therefore this is an algebraic space (Theorem B.2) equipped with a map $y \to \mathcal{Z}$. In fact, y must be the spectrum of a field: choose a smooth cover $\operatorname{Spec} A \to R_0$ by a scheme, and let ℓ be the equalizer of the two maps $k \to A$. Then ℓ is a field (since an element of k is equalized by two homomorphisms if and only if its inverse is). Moreover, we obtain a map $y \to \operatorname{Spec} \ell$ that is covering and injective in the fppf topology. Therefore it is an isomorphism (see [77, Tag 0B8A]).

Now we argue that $y \to \mathcal{X}$ is unramified. Indeed, the diagonal map $y \to y \times_{\mathcal{X}} y$ pulls back via the fppf cover $R_0 = x \times_{\mathcal{X}} x \to y \times_{\mathcal{X}} y$ to $x \times_y x = R_0$, which is open in R. Therefore $y \to y \times_{\mathcal{X}} y$ is an open embedding, so y is unramified over $\mathcal{X}$.

Now choose a smooth morphism $U \to \mathcal{X}$ containing y in its image, with $U = \operatorname{Spec} A$ affine. Let $V = U \times_{\mathcal{X}} y$. Then V is a smooth algebraic space over y and $V \to U$ is unramified. If $\mathcal{X}$ is an algebraic space then we have seen that $U \to \mathcal{X}$ is schematic, so that V is a scheme; in that case we write $W = V$. In general, then we know the diagonal of $\mathcal{X}$ is at least representable by algebraic spaces, so V is an algebraic space. By 'induction on stackiness' we may assume that there is a scheme W and an étale map $W \to V$ whose image in $\mathcal{X}$ contains the image of y.

Up to an étale extension of ℓ, we can assume that y is the image of an ℓ-point of W, which we denote w. We take u to be its image in U. The map $W \to U$ is unramified, so the maximal ideal $\mathfrak{m}$ of w in $\mathcal{O}_W$ is generated by the image of the maximal ideal $\mathfrak{n}$ of u in U. We can therefore choose functions $f_1, \ldots, f_d \in \mathcal{O}_{U,u}$ whose images in $\mathcal{O}_{W,w}$ form a basis for $\mathfrak{m}/\mathfrak{m}^2$. Replacing W by an open neighborhood of w, the vanishing locus of $f_1, \ldots, f_d$ in W will be $\{w\}$. In particular, the vanishing locus is unramified over y. But the locus where $U \to \mathcal{X}$ is unramified is open, so that there is an open neighborhood U' of $u \in U$ where $U' \to \mathcal{X}$ is unramified. Since it is also smooth, it is étale, as required. □

The implications among all of the definitions are described in Figure 1. They can all essentially be explained by putting various conditions on the diagonal of an algebraic stack. In this sense, from an expository perspective, the definition of an algebraic stack is the basic definition, and the rest can easily be obtained from this. In practice this is somewhat misleading, however, since one must first define an algebraic space to define an algebraic stack.

B.5. *Relationships among definitions of algebraic spaces*

The following implications hold for algebraic spaces:

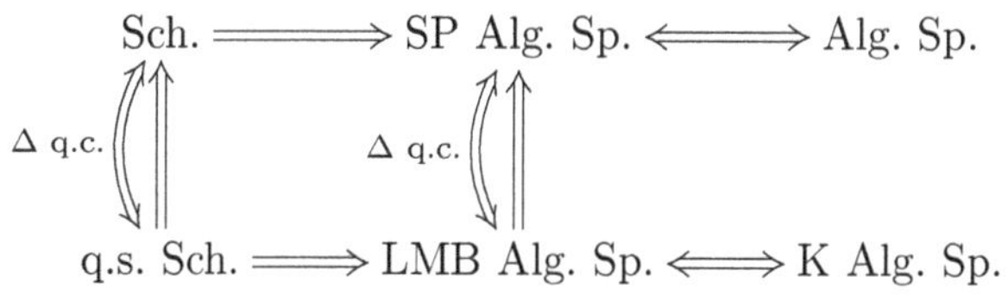

An arrow with a label indicates that the implication holds under the additional assumption indicated on the diagonal. A two headed arrow implies that the definitions are equivalent under the given assumption on the diagonal.

The only arrows that require justification are the equivalence between LMB algebraic spaces and K algebraic spaces, and the equivalence between algebraic spaces and SP algebraic spaces.

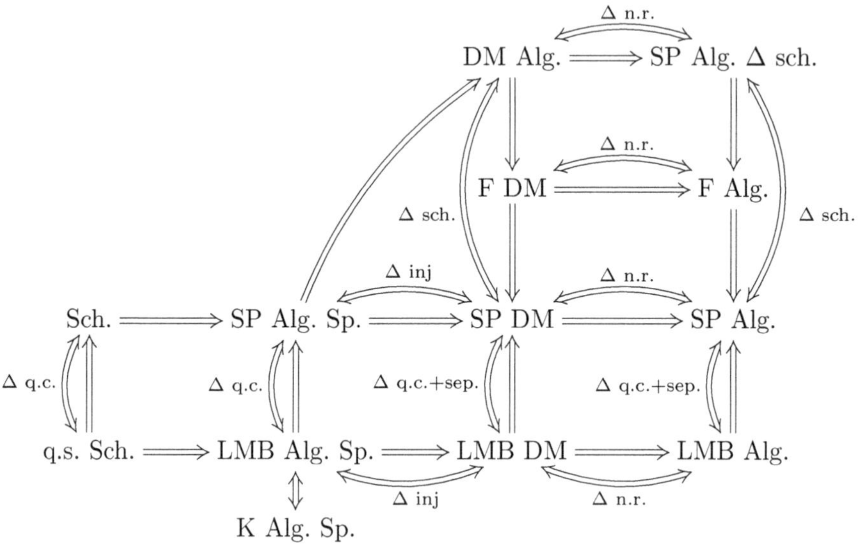

Fig. 1. An arrow from one entry to another signifies that the class of objects at the arrow's tail are also of the type at its head. An arrow with a label means that an object of the type at the tail satisfying the additional condition named in the label is also of the type at the head. A double-headed arrow should be interpreted as a pair of arrows pointing in both directions with the same label; in other words, the condition in the label makes the conditions at its ends equivalent.

Lemma B.14. *LMB algebraic spaces are the same as K algebraic spaces and SP algebraic spaces are the same as algebraic spaces.*

Proof. It is clear that LMB algebraic spaces are K algebraic spaces and that SP algebraic spaces are algebraic spaces. For the converse, we only need to show that the diagonal is schematic. This is [77, Tag 046K]. In fact, the diagonal of an algebraic space is injective, hence separated and locally quasifinite. Therefore the diagonal is schematic, by separated, locally quasifinite descent [77, Tag 02W8]. □

B.6. *Relationships among the definitions of algebraic stacks*

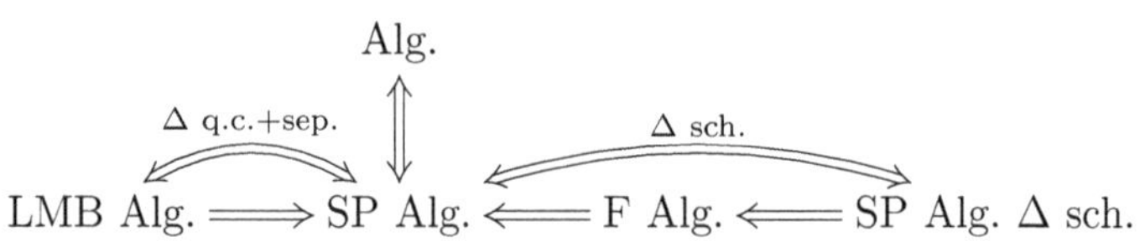

Most of the implications are immediate, so we only make a few comments.

Lemma B.15. *The diagonal morphism of an algebraic stack is representable by algebraic spaces.*

Proof. One can check this using Theorem B.13 and a slight modification of the proof of [77, Tag 04XS]. □

Lemma B.16. *Algebraic stacks are the same as SP algebraic stacks.*

Proof. From Theorem B.13, we only need to show that the diagonal morphism is representable by SP algebraic spaces. Since we have shown already that algebraic spaces are SP algebraic spaces, we conclude using the previous lemma. □

Lemma B.17. *An SP algebraic stack with quasicompact and separated diagonal is an LMB algebraic stack.*

Proof. Suppose that $\mathcal{X}$ is SP algebraic, with quasicompact and separated diagonal. By Lemma B.15, the diagonal morphism for $\mathcal{X}$ is representable by algebraic spaces. An algebraic space is an LMB algebraic space if and only if its diagonal is quasicompact (see Section B.5). Since the diagonal of $\mathcal{X}$ is separated, the double diagonal is a closed embedding and, a fortiori, quasicompact. Hence the diagonal of $\mathcal{X}$ is representable by LMB algebraic spaces, as required. □

Remark B.18. We expect there are F algebraic stacks that do not have schematic diagonal, but we do not know of any example. We also expect that there are SP algebraic stacks that do not admit a smooth schematic morphism from a scheme (and therefore are not F algebraic stacks), but we do not know of any example. See Example B.21 for an SP algebraic stack that does not have schematic diagonal.

B.7. *Relationships among the definitions of Deligne–Mumford stacks*

For Deligne–Mumford stacks, we have implications:

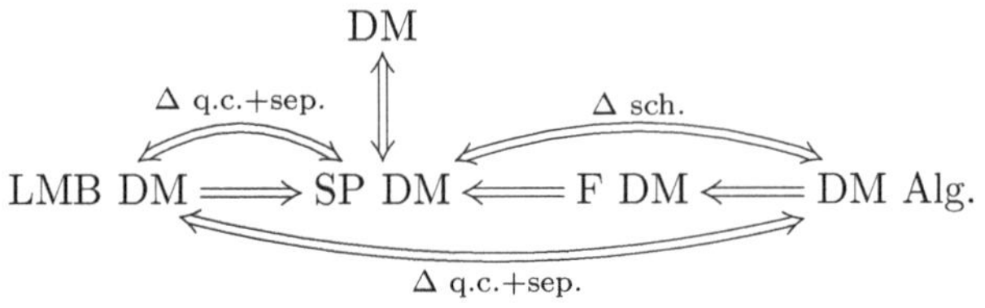

Most of the implications are immediate, but we focus on the main points:

Lemma B.19. *DM stacks are the same as SP DM stacks.*

Proof. From Theorem B.13 we only need to show that a DM stack has diagonal representable by algebraic spaces (as these are the same as SP algebraic spaces). However, by Theorem B.13 it is immediate that DM stacks are algebraic stacks, and we have seen in Lemma B.15 that the diagonal of an algebraic stack is representable by algebraic spaces. □

Lemma B.20 ([54, Lem. 4.2]). *All LMB DM stacks have schematic diagonal.*

Proof. Any locally quasifinite, separated morphism that is representable by algebraic spaces is schematic [77, Tag 03XX]. □

The rest of the implications are obvious from Theorem B.13 and the two lemmas above.

Example B.21. We will construct an SP Deligne–Mumford stack without schematic diagonal (i.e., an SP DM stack that is not a DM algebraic stack). Let G be $\mathbb{A}^1$ with a doubled origin. This can be regarded as a group scheme over $\mathbb{A}^1$ by distinguishing one of the two origins as the identity element. If $X \to \mathbb{A}^1$ is a morphism of schemes, with X_0 the fiber over the origin, then a G-torsor on X is a $\mathbb{Z}/2\mathbb{Z}$-torsor on X_0.

We will show that the map $p : \mathbb{A}^1 \to \mathrm{B}G$ is not schematic. Indeed, if $Z \to \mathrm{B}G$ is any morphism then the base change of p is the total space of the corresponding torsor. Therefore we have to find a scheme Z over $\mathbb{A}^1$ and a G-torsor over Z that is not representable by a scheme.

To find such a torsor, choose a scheme W and an étale double cover $W' \to W$ that does not have a section Zariski-locally. Let $Z = \mathbb{A}^1_W$ (so that $Z_0 = W$) and let P be the G-torsor over Z corresponding to $W' \to Z_0$.

Remark B.22. In regards to the example above, we expect that there are also F DM stacks that do not have schematic diagonal (and are therefore not DM algebraic stacks), but we do not know of any example. We also expect that there are SP DM stacks that do not admit a smooth schematic morphism from a scheme (and therefore are not F DM stacks), but we do not know of any example. Example B.21 shows there are SP algebraic stacks that do not have schematic diagonal, and are therefore not DM algebraic stacks.

B.8. *Stacks with unramified diagonal*

The implications in Figure 1 regarding unramified diagonal all follow immediately from the following lemma:

Lemma B.23. *An algebraic stack with unramified diagonal is a DM stack.*

Proof. Now that we have the identification between algebraic stacks and SP algebraic stacks, and DM stacks and SP DM stacks, this is [77, Tag 06N3]. □

B.9. *The adapted perspective*

We show how all of the stacks that have appeared in this chapter can be described as stacks adapted to a given presite.

B.9.1. *Stacks adapted to the étale presite of schemes*

We have already seen the following. A stack adapted to the étale presite with injective diagonal is the same thing as an algebraic space. A stack adapted to the étale presite with injective and quasicompact diagonal is the same thing as an LMB algebraic space. A stack adapted to the étale presite is the same thing as an F DM stack. A stack adapted to the étale presite with quasicompact and separated diagonal is the same thing as an LMB DM stack (Lemma B.20).

B.9.2. *Stacks adapted to the étale presite of algebraic spaces*

Stacks adapted to the étale presite of algebraic spaces induce Deligne–Mumford stacks on the étale presite of schemes. Indeed, given a stack in the étale topology on algebraic spaces one obtains a stack on the étale presite of schemes by restriction. Conversely, a stack on the étale presite of schemes extends uniquely to the étale presite of algebraic spaces since every algebraic space has an étale cover by schemes. Using again that every algebraic space has an étale cover by schemes, the definition of stack adapted to the étale presite of algebraic spaces agrees with the characterization of Deligne–Mumford stacks in Theorem B.13.

B.9.3. *Stacks adapted to the smooth presite of algebraic spaces*

Stacks adapted to the smooth presite of algebraic spaces induce algebraic stacks by restricting from the category of algebraic spaces to schemes.

B.9.4. *Stacks adapted to the fppf presite of algebraic spaces*

A stack adapted to the fppf presite of algebraic spaces induces an algebraic stack on the étale presite of schemes, by restricting from the category of algebraic spaces to the category of schemes (see Theorem B.2).

B.9.5. *An example of a stack that is not adapted to a presite*

Example B.24. Consider the logarithmic abelian varieties of Kajiwara, Kato, and Nakayama [50–52], which are sheaves in the étale topology and possess logarithmically étale covers by logarithmic schemes but have no such étale covers (logarithmically étale maps are a more general class of morphisms including all étale maps but also some blowups and other non-flat morphisms). However, the logarithmic étale topology is not subcanonical and there is no subcanonical topology in which logarithmic abelian varieties are covered by logarithmic schemes.

B.10. *Conditions on the relative diagonal of a stack and bootstrapping*

As we have seen, most of the variations on the definition of an algebraic stacks outlined in Definition B.6 can be obtained from the definition of an algebraic stack by imposing conditions on the diagonal. Here we aim to extend the bootstrapping result of Proposition 6.31 to these other cases. In other words, our goal here is to show that a stack that is *relatively algebraic* over an algebraic stack (according to any of the definitions in Definition B.6) is itself algebraic.

For the next lemma, recall that for a morphism $f : \mathcal{X} \to \mathcal{Y}$ of algebraic stacks, the diagonal Δ_f is representable by algebraic spaces ([77, Tag 04XS]).

Lemma B.25. *Let* **P** *be a property of morphisms of algebraic spaces that is stable under composition and base change. Let* $\mathcal{X}, \mathcal{Y}, \mathcal{Z}$ *be algebraic stacks* *over* S.

(1) *If* $\mathcal{X} \xrightarrow{f} \mathcal{Y}$ *has property* $\mathbf{P}$ *for* Δ_f *and* $\mathcal{Y}' \to \mathcal{Y}$ *is any morphism, then the morphism* $\mathcal{X} \times_{\mathcal{Y}} \mathcal{Y}' \xrightarrow{f'} \mathcal{Y}'$ *obtained from the fibered product has property* $\mathbf{P}$ *for* $\Delta_{f'}$.

(2) *If* $\mathcal{X} \xrightarrow{f} \mathcal{Y}$ *and* $\mathcal{Y} \xrightarrow{g} \mathcal{Z}$ *are morphisms such that* Δ_f *and* Δ_g *have property* $\mathbf{P}$, *then for the composition* $\mathcal{X} \xrightarrow{g \circ f} \mathcal{Y}$, *the diagonal morphism* $\Delta_{g \circ f}$ *has property* $\mathbf{P}$.

(3) *If* $\mathcal{X} \xrightarrow{f} \mathcal{X}'$ *and* $\mathcal{Y} \xrightarrow{g} \mathcal{Y}'$ *are morphisms over a stack* $\mathcal{Z}$ *such that* Δ_f *and* Δ_g *have property* $\mathbf{P}$, *then for the fibered product morphism* $\mathcal{X} \times_{\mathcal{Z}} \mathcal{Y} \xrightarrow{f \times_{\mathrm{id}_{\mathcal{Z}}} g} \mathcal{X}' \times_{\mathcal{Z}} \mathcal{Y}'$ *the diagonal morphism* $\Delta_{f \times_{\mathrm{id}_{\mathcal{Z}}} g}$ *has property* $\mathbf{P}$.

(4) *A morphism* $f : \mathcal{X} \to \mathcal{Y}$ *has property* $\mathbf{P}$ *for* Δ_f *if and only if for every scheme* S *and every morphism* $S \to \mathcal{Y}$, *the base change* $f' : \mathcal{X} \times_{\mathcal{Y}} S \to S$ *has property* $\mathbf{P}$ *for* $\Delta_{f'}$.

(5) *Suppose that property* $\mathbf{P}$ *satisfies the following condition: If* $X \xrightarrow{f} Y$ *and* $Y \xrightarrow{g} Z$ *are morphisms of algebraic spaces such that* g *and* $f \circ g$ *satisfy property* $\mathbf{P}$, *then* f *satisfies property* $\mathbf{P}$. *Then if* $\mathcal{X} \xrightarrow{f} \mathcal{Y}$ *and* $\mathcal{Y} \xrightarrow{g} \mathcal{Z}$ *are morphisms such that* Δ_g *and* $\Delta_{g \circ f}$ *have property* $\mathbf{P}$, *then* Δ_f *has property* $\mathbf{P}$.

Proof. This is essentially contained in [77, Tag 04YV]. (1) follows from the 2-cartesian diagrams:

$$\begin{array}{ccc} \mathcal{X}' & \xrightarrow{f'} & \mathcal{Y}' \\ \downarrow & & \downarrow \\ \mathcal{X} & \xrightarrow{f} & \mathcal{Y} \end{array} \qquad\qquad \begin{array}{ccc} \mathcal{X}' & \xrightarrow{\Delta_{f'}} & \mathcal{X}' \times_{\mathcal{Y}'} \mathcal{X}' \\ \downarrow & & \downarrow \\ \mathcal{X} & \xrightarrow{\Delta_f} & \mathcal{X} \times_{\mathcal{Y}} \mathcal{X} \end{array} \tag{B.1}$$

The diagram on the left induces the diagram on the right, and it then follows from base change that $\Delta_{f'}$ has property $\mathbf{P}$.

(2) follows from (1) using the 2-commutative diagram below for the

morphisms $f : \mathcal{X} \to \mathcal{Y}$ and $g : \mathcal{Y} \to \mathcal{Z}$. The square is a 2-fibered product.

$$\begin{array}{ccccc} & & \overset{\Delta_{g\circ f}}{\frown} & & \\ \mathcal{X} & \xrightarrow{\Delta_f} & \mathcal{X} \times_{\mathcal{Y}} \mathcal{X} & \longrightarrow & \mathcal{X} \times_{\mathcal{Z}} \mathcal{X} \\ & \searrow & \downarrow & & \downarrow \\ & & \mathcal{Y} & \xrightarrow{\Delta_g} & \mathcal{Y} \times_{\mathcal{Z}} \mathcal{Y}. \end{array} \tag{B.2}$$

(3) is essentially [39, Rem. (1.3.9) p.33], which observes that the conclusion follows from (1) and (2), together with the fact that given morphisms $f : \mathcal{X} \to \mathcal{X}'$ and $g : \mathcal{Y} \to \mathcal{Y}'$ over a stack $\mathcal{Z}$, the product $\mathcal{X} \times_{\mathcal{Z}} \mathcal{Y} \overset{f \times_{\mathrm{id}_{\mathcal{Z}}} g}{\longrightarrow} \mathcal{X}' \times_{\mathcal{Z}} \mathcal{Y}'$ is given by the composition of morphisms obtained from fibered product diagrams:

$$\mathcal{X} \times_{\mathcal{Z}} \mathcal{Y} \xrightarrow{f \times_{\mathrm{id}_{\mathcal{Z}}} \mathrm{id}_{\mathcal{Y}}} \mathcal{X}' \times_{\mathcal{Z}} \mathcal{Y} \xrightarrow{\mathrm{id}_{\mathcal{X}'} \times_{\mathrm{id}_{\mathcal{Z}}} g} \mathcal{X}' \times_{\mathcal{Z}} \mathcal{Y}'.$$

(4) If Δ_f has property **P**, this follows from (1). Conversely, assume that for every scheme S and every morphism $S \to \mathcal{Y}$, the base change $\mathcal{X} \times_{\mathcal{Y}} S \to S$ has property **P** for its diagonal. By definition, for $\Delta_f : \mathcal{X} \to \mathcal{X} \times_{\mathcal{Y}} \mathcal{X}$ to have property **P** means that the base change $S \times_{\mathcal{X} \times_{\mathcal{Y}} \mathcal{X}} \mathcal{X} \to S$ has property **P**. But

$$S \underset{\mathcal{X} \times_{\mathcal{Y}} \mathcal{X}}{\times} \mathcal{X} = S \underset{\mathcal{X}_S \times_S \mathcal{X}_S}{\times} \mathcal{X}_S$$

where $\mathcal{X}_S = \mathcal{X} \times_{\mathcal{Y}} S$ and by assumption, the diagonal of $\mathcal{X}_S \to S$ has property **P**.

(5) This follows from diagram (B.2), and the previous parts of the lemma. □

Corollary B.26. *For any of the classes* C *of objects introduced in Definition B.6, if* $\mathcal{Y}$ *is of class* C *and* $f : \mathcal{X} \to \mathcal{Y}$ *is representable by objects of* C *then* $\mathcal{X}$ *is of class* C.

Proof. Lemma 6.32 covers the case of algebraic stacks. For F algebraic stacks one can easily adapt the proof of Lemma 6.32. The remaining classes C of stacks introduced in Definition B.6 can can be obtained by imposing various conditions on the diagonal of an algebraic stack, all of which are stable under composition and base change. Fix a class of such stacks, and call the necessary conditions on the diagonal condition **P**. In particular, $\mathcal{Y}$ is of class C means that the diagonal Δ_π of the structure map $\pi : \mathcal{Y} \to \mathsf{S}$ has property **P**. If $f : \mathcal{X} \to \mathcal{Y}$ is representable by objects of C, then from

Lemma B.25(4), we have that Δ_f has property **P**. Then by Lemma B.25(2), we have that $\Delta_{\pi\circ f}$ has property **P**. In other words, $\mathcal{X}$ is of class C. $\square$

Remark B.27. The arguments above show the following, as well ([77, Tag 04YV]). Let $f : \mathcal{X} \to \mathcal{Y}$ be a morphism of algebraic stacks. The morphism f is representable by LMB algebraic stacks (resp. LMB DM stacks, resp. LMB algebraic spaces) if and only if Δ_f is quasicompact and separated (resp. quasicompact, separated, and unramified, resp. quasicompact, separated, and injective).

C. Groupoids and stacks

Stacks are often studied via groupoid objects. In this section we discuss torsors, groupoid objects, and stacks arising as quotients of groupoid objects. In the end we show that groupoid objects adapted to a presite, i.e., those where the source and target maps are coverings in the presite, are essentially the same thing as stacks adapted to the presite.

C.1. *Torsors and group quotients*

Let X be in S/S, and let G be a sheaf of groups over S acting on the right on X:

$$X \times_S G \xrightarrow{\ \sigma\ } X$$

We define a CFG over S/S, $[X/G]$ in the following way. The objects over an S-scheme $f : S' \to S$ are diagrams

$$\begin{array}{ccc} P' & \longrightarrow & X_{S'} \\ \downarrow & & \\ S' & & \end{array}$$

where P' is a $G_{S'}$-torsor (principal bundle) over S', and $P' \to X_{S'}$ is a $G_{S'}$-equivariant morphism. Morphisms are defined by pullback. There is a morphism $[X/G] \to S$ given by forgetting everything except the S-scheme $f : S' \to S$, and there is an S-morphism $q : X \to [X/G]$ given by the trivial G_X-bundle

$$\begin{array}{ccc} X \times_S G & \xrightarrow{\mathrm{id} \times \sigma} & X \times_S X \\ {\scriptstyle pr_1}\downarrow & & \\ X. & & \end{array}$$

This induces a 2-cartesian diagram

$$\begin{array}{ccc} X \times_S G & \xrightarrow{\sigma} & X \\ {\scriptstyle pr_1}\downarrow & & \downarrow{\scriptstyle q} \\ X & \xrightarrow{q} & [X/G] \end{array} \tag{C.1}$$

that is a co-equalizer for $X \times_S G \overset{\sigma}{\underset{pr_1}{\rightrightarrows}} X$ (i.e., initial in the category of stacks for the diagram (C.1)); in other words, $[X/G]$ is a quotient in the category of stacks for the action of G on X, in the sense that any G-equivariant map out of X factors through it. Note that if there exists a scheme X/G that is a quotient in the category of schemes for the action of G on X (i.e., a co-equalizer in the category of *schemes*), then there is a morphism

$$[X/G] \to X/G.$$

Of particular importance is the trivial action of G on $X = S$. The quotient $[S/G]$ is denoted $\mathrm{B}G$. As a CFG, $\mathrm{B}G$ consists of the pairs (S', P') where $S' \in \mathsf{S}/S$ and P' is a G-torsor over S.

Remark C.1. There is a similar construction for left group actions; in the notation above, we would have the stack $[G\backslash X]$.

C.2. *Groupoid objects and groupoid quotients*

C.2.1. *Groupoid objects*

Let Gpd denote the *category* (not 2-category) of small groupoids. There are functors Obj and Mor from (Gpd) to (Set) sending a groupoid X, respectively, to its set X_0 of objects and its set X_1 of morphisms. There are two canonical morphisms $s, t : X_1 \to X_0$ sending a morphism to its source and target. More data are required to specify a groupoid, but these are often left tacit and the groupoid is usually denoted by a pair of morphisms of sets $X_1 \overset{t}{\underset{s}{\rightrightarrows}} X_0$.

A groupoid object in a category S is defined by taking X_1, X_0, and all of the morphisms involved in specifying the groupoid to lie in S (as opposed to Set). If this is the case, then we obtain functors $\mathrm{Hom}_{\mathsf{S}}(-, X_i)$ from S^{op} to (Set). This can all be said more concisely as follows:

Definition C.2 (Groupoid object). A *groupoid object* of a category S is a functor $\mathscr{X} : \mathsf{S}^{\mathrm{op}} \to (\mathsf{Gpd})$, along with objects X_0 and X_1 in S, respectively representing the composition of functors $\mathsf{Obj} \circ \mathscr{X}$ and $\mathsf{Mor} \circ \mathscr{X}$ from S^{op} to (Set). The groupoid object $\mathscr{X}$ is often denoted $X_1 \underset{s}{\overset{t}{\rightrightarrows}} X_0$. A *morphism of groupoids objects* is a morphism (natural transformation) of functors.

Remark C.3. There are morphisms $s, t : X_1 \to X_0$ associated to a the groupoid object $\mathscr{X}$ in S obtained from the source and target maps associated to a groupoid. Moreover, associated to a morphism $\mathscr{X} \to \mathscr{Y}$ of groupoid objects in S are morphisms $X_0 \to Y_0$ and $X_1 \to Y_1$.

Example C.4 (Constant groupoid object). Let X be any object of S. Define $X_0 = X_1 = X$. Then $X_1 \underset{\mathrm{Id}}{\overset{\mathrm{Id}}{\rightrightarrows}} X_0$ is a groupoid object of S. We call such a groupoid *constant* and denote it, abusively, by the same letter X.

Example C.5 (Action groupoid). Let G be a group object of S acting on the left on an object X_0. Define $\mathsf{Obj}\mathscr{X}(U) = \mathrm{Hom}(U, X_0)$ and let $\mathsf{Mor}\mathscr{X}(U)$ be the set of all triples (g, x, y) where $x, y \in \mathsf{Obj}X(U)$ and $g \in \mathrm{Hom}(U, G)$ is a U-point of G such that $gx = y$. The composition of (g, x, y) and (g', y, z) is the triple $(g'g, x, z)$. This is known as the *action groupoid.* It is also typically denoted by $G \times X_0 \underset{\sigma}{\overset{pr_2}{\rightrightarrows}} X_0$.

Example C.6. One can immediately associate a category fibered in groupoids to any groupoid object. Indeed, suppose that $\mathscr{X}$ is a groupoid object of S. Construct a category $\mathcal{X}$ whose objects are pairs (U, ξ) where U is an object of S and $\xi \in \mathsf{Obj}\mathscr{X}(U)$. A morphism $(U, \xi) \to (V, \eta)$ consists of a morphism $f : U \to V$ of S and a morphism $\phi : \xi \to f^*\eta$ of $\mathscr{X}(U)$. The composition of $(f, \phi) : (U, \xi) \to (V, \eta)$ and $(g, \psi) : (V, \eta) \to (W, \zeta)$ is $(gf, \phi \circ f^*\psi)$. It is easy to verify that this category is fibered in groupoids over S with the projection sending (U, ξ) to U.

When S has a topology, this groupoid is rarely a stack, although it is a prestack if the topology is subcanonical. If S is subcanonical and $X_1 \underset{s}{\overset{t}{\rightrightarrows}} X_0$ is a groupoid object, then it is common to denote the associated prestack by $[\ X_1 \underset{s}{\overset{t}{\rightrightarrows}} X_0\ ]^{\mathrm{pre}}$. There is an abstract process of *stackification* of prestacks, analogous to sheafification of presheaves, by which a CFG $\mathcal{X}$ is replaced by the initial stack receiving a map from $\mathcal{X}$. In the situation of groupoid objects, this stack is typically denoted by $[\ X_1 \underset{s}{\overset{t}{\rightrightarrows}} X_0\ ]$ (see e.g., [12, Def. 3.11, Def. 4.10] for more details on this approach). Moreover, there is a morphism $X_0 \to [\, X_1 \overset{t}{\underset{s}{\rightrightarrows}} X_0 \,]$ that makes the following diagram

$$\begin{array}{ccc} X_1 & \xrightarrow{t} & X_0 \\ {\scriptstyle s}\downarrow & & \downarrow \\ X_0 & \longrightarrow & [X_1 \overset{t}{\underset{s}{\rightrightarrows}} X_0] \end{array} \tag{C.2}$$

2-cartesian ([77, Tag 04M8]) and essentially a 2-coequalizer for $X_1 \overset{t}{\underset{s}{\rightrightarrows}} X_0$ (see [77, Tag 04MA] for more details on the precise meaning of this). In other words, the stack provides a "quotient" for the groupoid.

Rather than undertake an explanation of this construction and the attendant 2-universal property, we will give a direct construction of the stack associated to a groupoid object.

Remark C.7. In the case of left group action, the stack $[\, G \times X_0 \overset{pr_2}{\underset{\sigma}{\rightrightarrows}} X_0 \,]$ is equivalent to the stack $[G \backslash X_0]$.

C.2.2. *Augmented groupoids*

Definition C.8 (Augmented groupoid object). A groupoid object $\mathscr{X}$ of a category S is said to be *augmented* toward an object X of S when it is equipped with a morphism $\mathscr{X} \to X$. A groupoid $\mathscr{X}$ augmented toward X is often denoted $X_1 \rightrightarrows X_0 \to X$.

If $\mathscr{X} \to X$ and $\mathscr{Y} \to Y$ are augmented groupoids, a morphism of augmented groupoids from $\mathscr{X} \to X$ to $\mathscr{Y} \to Y$ is a commutative diagram of groupoid objects as in (C.3):

$$\begin{array}{ccc} \mathscr{X} & \longrightarrow & \mathscr{Y} \\ \downarrow & & \downarrow \\ X & \xrightarrow{f} & Y. \end{array} \tag{C.3}$$

Remark C.9. To augment a groupoid $\mathscr{X}$ torwards X, it is equivalent to give a morphism $f : X_0 \to X$ such that $fs = ft$.

Definition C.10 (Cartesian morphism of augmented groupoid objects). If S admits fibered products, a morphism of augmented groupoid objects as in (C.3) is called *cartesian* if $X_0 \to X \times_Y Y_0$ and $X_1 \to X \times_Y Y_1$ are isomorphisms.

Example C.11. Suppose that $q : X_0 \to X$ is a morphism in a category admitting fiber products. We define a groupoid as follows: $\mathsf{Obj}\mathscr{X}(U) = \mathrm{Hom}(U, X_0)$ and $\mathsf{Mor}\mathscr{X}(U)$ is the set of pairs $(f, g) \in \mathrm{Hom}(U, X_0)$ such that $qf = qg$. In other words, $\mathsf{Mor}\mathscr{X}$ is represented by $X_1 = X_0 \times_X X_0$. The composition of the pair (f, g) and (g, h) is, by definition, the pair (f, h), and the identity of $f \in \mathsf{Obj}\mathscr{X}(U)$ is the pair (f, f). The groupoid $X_1 \rightrightarrows X_0$ is augmented toward X by construction.

Let $\mathsf{Gpd}_\mathsf{S}^+$ denote the category of augmented groupoid objects of S, with cartesian morphisms. The projection sending an augmented groupoid object $(\mathscr{X} \to X)$ to X makes $\mathsf{Gpd}_\mathsf{S}^+$ into a CFG over S.

C.2.3. *Stacks associated to groupoid objects*

Now suppose that S is equipped with a pretopology.

Definition C.12 (Presentation of an augmented groupoid obejct). Let S be a presite admitting fibered products. Let $X_1 \rightrightarrows X_0 \to X$ be a groupoid of S augmented toward X. We call it a *presentation* of X if $X_0 \to X$ is covering (Definition 6.2) and the canonical map $X_1 \to X_0 \times_X X_0$ is an isomorphism.

Definition C.13 (Category associated to a groupoid object). Let S be a presite admitting fibered products. Let $\mathscr{X} = (X_1 \rightrightarrows X_0)$ be a groupoid object of S. We construct a CFG, $\mathcal{X}$, called the *CFG associated to a groupoid object.* The objects of $\mathcal{X}$ are triples $(U, \mathscr{U}, \xi)$ where U is an object of S, where $\mathscr{U} \to U$ is a presentation of U, and where $\xi : \mathscr{U} \to \mathscr{X}$ is a morphism of groupoid objects.

A morphism in $\mathcal{X}$ from $(U, \mathscr{U}, \xi)$ to $(V, \mathscr{V}, \eta)$ is a cartesian morphism (f, φ) of augmented groupoids from $(\mathscr{U} \to U)$ to $(\mathscr{V} \to V)$ such that $\eta \circ \varphi = \xi$ as morphisms of groupoids $\mathscr{U} \to \mathscr{X}$.

The morphism $\mathcal{X} \to \mathsf{S}$ is given by sending $(U, \mathscr{U}, \xi)$ to U.

The following lemma asserts that the category $\mathcal{X}$ is a CFG; the proof is straightforward, so it is omitted.

Lemma C.14. *If S admits fiber products, then the category $\mathcal{X}$ over S constructed above is a CFG.*

For the following lemma, let Cov denote the category whose objects are covering morphisms $X \to U$ in S and whose morphisms are cartesian squares. If S has fiber products then the projection $\mathsf{Cov} \to \mathsf{S}$ sending $(X \to U)$ to U makes Cov into a CFG over S (Example 3.21).

Lemma C.15. *Assume that S is a subcanonical presite with fiber products. Let $\mathcal{X}$ be the CFG constructed as above in Definition C.13. If Cov is a stack over S then so is $\mathcal{X}$.*

Proof. We give just a sketch. The idea is to use Lemma 4.9. We start with an object S of S and the canonical morphism $\mathsf{S}/S \to \mathsf{S}$. Given a cover $\mathcal{R} = \{U_i \to S\}$ of S and a morphism $\mathcal{R} \to \mathcal{X}$ we need to show how to obtain the lift $\mathsf{S}/S \to \mathcal{X}$. So, given groupoid objects $\mathscr{U}_i$ over each U_i, along with compatible data over the double and triple fiber products U_{ij} and U_{ijk}, these descend to a groupoid object $\mathscr{S}$ over S by descending the objects $\mathsf{Obj}\mathscr{U}_i$ and $\mathsf{Mor}\mathscr{U}_i$ of Cov and the morphisms between them (using that Cov forms a stack and that morphisms between representable objects form sheaves). Then the maps $\mathscr{U}_i \to \mathscr{X}$ descend to $\mathscr{S} \to \mathscr{X}$ by descending the maps on objects and morphisms, again using the subcanonicity of the presite. This gives the desired morphism $\mathsf{S}/S \to \mathcal{X}$. □

Example C.16. If $\mathscr{X}$ is the groupoid object $U_1 \overset{s}{\underset{t}{\rightrightarrows}} U_0$ and $\mathcal{X}$ is its associated stack, there is a canonical map $U_0 \to \mathcal{X}$, which is covering. We construct the triple $(U_0, \mathscr{U}_0, \mathscr{U}_0 \to \mathscr{X})$ giving this morphism as follows. The presentation $\mathscr{U}_0 \to U_0$ is given by

$$U_1 \times_{U_0} U_1 \overset{pr_1}{\underset{comp}{\rightrightarrows}} U_1 \xrightarrow{\ s\ } U_0;$$

here the maps for the fibered product are the source and target maps repsectively, and the bottom arrow *comp* is the composition morphism taking a pair (α, β) in $U_1 \times_{U_0} U_1$ (over some S) to the composition $\beta \circ \alpha$ in U_1 (over S). This is a presentation of U_0 since s is covering (the projections $s, t : U_1 \to U_0$ are always covering in a groupoid object because of the 'identity map' section $U_0 \to U_1$).

To describe $\mathscr{U}_0 \to \mathscr{X}$, it is convenient to describe $\mathscr{U}_0$ as follows. For any scheme S, $\mathscr{U}_0(S)$ is the groupoid in which the objects are the morphisms of $\mathscr{X}(S)$. The morphisms from $(\xi \to \zeta)$ to $(\eta \to \omega)$ in $\mathscr{U}_0(S)$ are the commutative squares

$$\begin{array}{ccc} \xi & \longrightarrow & \zeta \\ \downarrow & & \| \\ \eta & \longrightarrow & \omega. \end{array}$$

That is, there are no morphisms unless $\zeta = \omega$. Clearly $\mathsf{Obj}\mathscr{U}_0(S) = \mathrm{Hom}(S, U_1)$ and $\mathsf{Mor}\mathscr{U}_0(S) = \mathrm{Hom}(S, U_1 \times_{U_0} U_1)$. There is a canonical map $\mathscr{U}_0 \to \mathscr{X}$ sending an object $(\xi \to \zeta)$ to ξ.

Finally, we show that $U_0 \to \mathcal{X}$ is covering. For a scheme S, a morphism $S \to \mathcal{X}$ is a triple $(S, \mathscr{S}, \xi : \mathscr{S} \to \mathscr{X})$, where $\mathscr{S} = S_1 \overset{s'}{\underset{t'}{\rightrightarrows}} S_0 \to S$ is a presentation. The morphism $\mathscr{S} \to \mathscr{X}$ induces a morphism $S_0 \to U_0$. The fact that $S_0 \to S$ is covering means that there is a cover $\{T_\alpha \to S\}$ of S that lifts to S_0, and therefore composing gives morphisms $T_\alpha \to U_0$. This gives the cover $\{T_\alpha \to S\}$ of S whose compositions $T_\alpha \to S \to \mathcal{X}$ lift to U_0.

Remark C.17. In fact, there is a cocartesian diagram in the 2-category of stacks:

$$\begin{array}{ccc} U_1 & \xrightarrow{s} & U_0 \\ {\scriptstyle t}\downarrow & & \downarrow \\ U_0 & \longrightarrow & \mathcal{X} \end{array}$$

We will not use this property, so we do not give a proof (see e.g., [12, Def. 3.11, Def. 4.10]).

C.3. *Adapted groupoid objects and adapted stacks*

Here we show that a stack associated to a groupoid object having source and target that are coverings in the presite is the same as a stack adapted to the presite; i.e., it is an algebraic stack. Moreover, the groupoid object induces a presentation of the stack.

Proposition C.18 ([12, Prop. 5.21]). *Let* S *be a subcanonical presite admitting fibered products. Let* $U_1 \overset{t}{\underset{s}{\rightrightarrows}} U_0$ *be a groupoid object, and set* $\mathcal{X} = [\, U_1 \overset{t}{\underset{s}{\rightrightarrows}} U_0 \,]$. *If* $\mathbf{P}$ *is any property of morphisms in* S *that is stable under base change and local (on the target), then:*

(1) Assuming the diagonal $\mathcal{X} \to \mathcal{X} \times \mathcal{X}$ *is* S*-representable, it has property* $\mathbf{P}$ *if and only if* $(s,t) : U_1 \to U_0 \times U_0$ *has property* $\mathbf{P}$.

(2) Assuming the morphism $U_0 \to \mathcal{X}$ *(corresponding to the identity of* U_0*) is* S*-representable, it has property* $\mathbf{P}$ *if and only if* s *(or* t*) has property* $\mathbf{P}$.

Proof. For the first claim, $U_1 \to U_0 \times U_0$ is the base change of the diagonal via $U_0 \times U_0 \to \mathcal{X} \times \mathcal{X}$, so the former inherits property **P** from the latter. Conversely, if $S \to \mathcal{X} \times \mathcal{X}$ is any morphism, let

$$\begin{array}{ccc} V & \longrightarrow & W \\ \downarrow & & \downarrow \\ T & \longrightarrow & S \end{array}$$

be the base change of the cartesian diagram

$$\begin{array}{ccc} U_1 & \longrightarrow & U_0 \times U_0 \\ \downarrow & & \downarrow \\ \mathcal{X} & \xrightarrow{\Delta} & \mathcal{X} \times \mathcal{X}. \end{array}$$

Then $V \to W$ has property **P** by base change. But $W \to S$ is a cover, since it is the base change of $U_0 \times U_0 \to \mathcal{X} \times \mathcal{X}$, so property **P** descends to $T \to S$. This applies to any morphism $S \to \mathcal{X} \times \mathcal{X}$, so Δ has property **P**.

For the second claim, there is a cartesian diagram

$$\begin{array}{ccc} U_1 & \xrightarrow{t} & U_0 \\ {\scriptstyle s}\downarrow & & \downarrow \\ U_0 & \longrightarrow & \mathcal{X} \end{array} \tag{C.4}$$

so s and t inherit property **P** from $U_0 \to \mathcal{X}$. Conversely, suppose $S \to \mathcal{X}$ is any morphism. Let

$$\begin{array}{ccc} V & \longrightarrow & W \\ \downarrow & & \downarrow \\ T & \longrightarrow & S \end{array}$$

be the base change of (C.4). Then $V \to W$ is **P** by base change. But $W \to S$ is covering (by base change; see Example C.16), so **P** descends to $T \to S$. This applies to any $S \to \mathcal{X}$, so $U_0 \to \mathcal{X}$ is **P**, as required. □

Proposition C.19 ([12, Prop. 4.19, 5.19]). *Let* S *be a subcanonical presite admitting fibered products. Let* $\mathcal{X}$ *be a stack, let* $P : U \to \mathcal{X}$ *be an* S*-representable morphism from an object* U *of* S*. Then there is an associated groupoid object* $U \times_{\mathcal{X}} U \underset{pr_1}{\overset{pr_2}{\rightrightarrows}} U$ *in* S *(see e.g., [12, Prop. 3.5]). If moreover the morphism* $P : U \to \mathcal{X}$ *is a cover in the sense of Definition 5.2,* *then $\mathcal{X}$ is equivalent to $[\, U \times_{\mathcal{X}} U \overset{pr_2}{\underset{pr_1}{\rightrightarrows}} U \,]$, and the projections pr_1, pr_2 are covers in the presite.*

Proof. Set $U_0 = U$ and $U_1 = U \times_{\mathcal{X}} U$. Let $\mathcal{U}$ be the stack associated to the groupoid object $\mathscr{U} = U_1 \rightrightarrows U_0$. We construct a map $\mathcal{U} \to \mathcal{X}$ by descent. Let $Z \to \mathcal{U}$ be any morphism, where Z is a scheme. By definition, this corresponds to a groupoid presentation of $\mathscr{Z}$ of Z and a cartesian morphism $\mathscr{Z} \to \mathscr{U}$. By composition, this gives a map to the constant groupoid object $\mathcal{X}$, and this descends uniquely to a map $Z \to \mathcal{X}$. This is easily shown to be functorial in Z, hence gives a morphism $\mathcal{U} \to \mathcal{X}$.

Now we argue that $\mathcal{U} \to \mathcal{X}$ is an isomorphism if U_0 covers $\mathcal{X}$. Indeed, the map $U \to \mathcal{X}$ factors through $\mathcal{U}$, so $\mathcal{U} \to \mathcal{X}$ is surjective. On the other hand, $U_0 \times_{\mathcal{U}} U_0 \to U_0 \times_{\mathcal{X}} U_0$ is an isomorphism. This is the pullback under the cover $U_0 \times_{\mathcal{X}} U_0 \to \mathcal{U} \times_{\mathcal{X}} \mathcal{U}$ of the diagonal $\mathcal{U} \to \mathcal{U} \times_{\mathcal{X}} \mathcal{U}$. Therefore the relative diagonal of $\mathcal{U} \to \mathcal{X}$ is an isomorphism, which is to say that $\mathcal{U} \to \mathcal{X}$ is injective. Combined with the surjectivity, this means $\mathcal{U} \to \mathcal{X}$ is an isomorphism.

The statement that pr_1 and pr_2 are covers can be obtained from the 2-cartesian diagram (C.2). □

Definition C.20 (Groupoid object adapted to a presite). Let S be a subcanonical presite admitting fibered products. We say a groupoid object $U_1 \overset{t}{\underset{s}{\rightrightarrows}} U_0$ is *adapted to the presite* if s and t are covers in the presite and the natural morphism $U_0 \to \mathcal{X} = [\, U_1 \overset{t}{\underset{s}{\rightrightarrows}} U_0 \,]$ is S-representable.

Corollary C.21. *The stack associated to a groupoid object adapted to a presite is a stack adapted to the presite; in particular, it is algebraic. Conversely, a stack adapted to a presite is the stack associated to a groupoid object adapted to the presite.*

Acknowledgments

These notes provide an elaboration on lectures the first author gave at the summer school *The Geometry, Topology and Physics of Moduli Spaces of Higgs Bundles* at the Institute for Mathematical Sciences at the National University of Singapore in July of 2014. He would like to thank the organizers for their invitation, and the IMS at NUS for their hospitality. He would also like to thank Tony Pantev for discussions on the moduli stack of Higgs bundles. Both authors would like to thank the referees and Sebastian Bozlee for their comments on earlier versions of the chapter. The first author was partially supported by NSF grant DMS-1101333, a Simons Foundation Collaboration Grant for Mathematicians (317572), and NSA Grant H98230-16-1-005. The second author was partially supported by an NSA Young Investigator's Grant H98230-14-1-0107.

References

1. *Groupes de monodromie en géométrie algébrique. I.* Lecture Notes in Mathematics, Vol. 288. Springer-Verlag, Berlin-New York, 1972. Séminaire de Géométrie Algébrique du Bois-Marie 1967–1969 (SGA 7 I), Dirigé par A. Grothendieck. Avec la collaboration de M. Raynaud et D. S. Rim.
2. *Théorie des topos et cohomologie étale des schémas.* Lecture Notes in Mathematics, Vol. 269. Springer-Verlag, Berlin-New York, 1972. Séminaire de Géométrie Algébrique du Bois-Marie 1963–1964 (SGA 4 I), Dirigé par A. Grothendieck. Avec la collaboration de M. Artin et J.-L. Verdier.
3. *Théorie des topos et cohomologie étale des schémas. Tome 2.* Lecture Notes in Mathematics, Vol. 270. Springer-Verlag, Berlin-New York, 1972. Séminaire de Géométrie Algébrique du Bois-Marie 1963–1964 (SGA 4), Dirigé par M. Artin, A. Grothendieck et J. L. Verdier. Avec la collaboration de N. Bourbaki, P. Deligne et B. Saint-Donat.
4. *Théorie des topos et cohomologie étale des schémas. Tome 3.* Lecture Notes in Mathematics, Vol. 305. Springer-Verlag, Berlin-New York, 1973. Séminaire de Géométrie Algébrique du Bois-Marie 1963–1964 (SGA 4), Dirigé par M. Artin, A. Grothendieck et J. L. Verdier. Avec la collaboration de P. Deligne et B. Saint-Donat.
5. *Revêtements étales et groupe fondamental (SGA 1).* Documents Mathématiques (Paris) [Mathematical Documents (Paris)], 3. Société Mathématique de France, Paris, 2003. Séminaire de géométrie algébrique du Bois Marie 1960–61. [Algebraic Geometry Seminar of Bois Marie 1960-61], Directed by A. Grothendieck, With two papers by M. Raynaud, Updated and annotated reprint of the 1971 original [Lecture Notes in Math., 224, Springer, Berlin; MR0354651 (50 #7129)].
6. Jarod Alper. Good moduli spaces for Artin stacks. *Ann. Inst. Fourier (Grenoble)*, 63(6):2349–2402, 2013.
7. Allen B. Altman and Steven L. Kleiman. Compactifying the Picard scheme. *Adv. in Math.*, 35(1):50–112, 1980.
8. M. Artin. Algebraic approximation of structures over complete local rings. *Inst. Hautes Études Sci. Publ. Math.*, (36):23–58, 1969.
9. M. Artin. Versal deformations and algebraic stacks. *Invent. Math.*, 27:165–189, 1974.
10. Arnaud Beauville, M. S. Narasimhan, and S. Ramanan. Spectral curves and the generalised theta divisor. *J. Reine Angew. Math.*, 398:169–179, 1989.

11. K. Behrend and B. Fantechi. The intrinsic normal cone. *Invent. Math.*, 128(1):45–88, 1997.
12. Kai Behrend, Brian Conrad, Dan Edidin, William Fulton, Barbara Fantechi, Lothar Göttsche, and Andrew Kresch. *Algebraic stacks.* 2014. available online from Kresch's webpage.
13. Bhargav Bhatt. Algebraization and Tannaka duality, 2014. Available online: `arXiv: 1404.7483`.
14. Bhargav Bhatt and Daniel Halpern-Leinster. Tannaka duality revisited. `arXiv: 1507.01925`.
15. Armand Borel. *Linear algebraic groups*, volume 126 of *Graduate Texts in Mathematics.* Springer-Verlag, New York, second edition, 1991.
16. Siegfried Bosch, Werner Lütkebohmert, and Michel Raynaud. *Néron models*, volume 21 of *Ergebnisse der Mathematik und ihrer Grenzgebiete (3) [Results in Mathematics and Related Areas (3)].* Springer-Verlag, Berlin, 1990.
17. Sebastian Casalaina-Martin. A tour of stable reduction with applications. In *A celebration of algebraic geometry*, volume 18 of *Clay Math. Proc.*, pages 65–117. Amer. Math. Soc., Providence, RI, 2013.
18. Brian Conrad. A modern proof of Chevalley's theorem on algebraic groups. *J. Ramanujan Math. Soc.*, 17(1):1–18, 2002.
19. Brian Conrad and A. J. de Jong. Approximation of versal deformations. *J. Algebra*, 255(2):489–515, 2002.
20. P. Deligne and D. Mumford. The irreducibility of the space of curves of given genus. *Inst. Hautes Études Sci. Publ. Math.*, (36):75–109, 1969.
21. R. Donagi and T. Pantev. Langlands duality for Hitchin systems. *Invent. Math.*, 189(3):653–735, 2012.
22. David Eisenbud. *Commutative algebra*, volume 150 of *Graduate Texts in Mathematics.* Springer-Verlag, New York, 1995. With a view toward algebraic geometry.
23. Barbara Fantechi. Stacks for everybody. In *European Congress of Mathematics, Vol. I (Barcelona, 2000)*, volume 201 of *Progr. Math.*, pages 349–359. Birkhäuser, Basel, 2001.
24. Barbara Fantechi, Lothar Göttsche, Luc Illusie, Steven L. Kleiman, Nitin Nitsure, and Angelo Vistoli. *Fundamental algebraic geometry*, volume 123 of *Mathematical Surveys and Monographs.* American Mathematical Society, Providence, RI, 2005. Grothendieck's FGA explained.
25. Daniel Ferrand. Conducteur, descente et pincement. *Bull. Soc. Math. France*, 131(4):553–585, 2003.
26. Hubert Flenner. Ein Kriterium für die Offenheit der Versalität. *Math. Z.*, 178(4):449–473, 1981.
27. Eberhard Freitag and Reinhardt Kiehl. *Étale cohomology and the Weil conjecture*, volume 13 of *Ergebnisse der Mathematik und ihrer Grenzgebiete (3) [Results in Mathematics and Related Areas (3)].* Springer-Verlag, Berlin, 1988. Translated from the German by Betty S. Waterhouse and William C. Waterhouse, With an historical introduction by J. A. Dieudonné.
28. D. Gaitsgory. Grothendieck topologies and deformation theory. II. *Compositio Math.*, 106(3):321–348, 1997.

29. Jean Giraud. *Cohomologie non abélienne*. Springer-Verlag, Berlin-New York, 1971. Die Grundlehren der mathematischen Wissenschaften, Band 179.
30. Phillip Griffiths and Joseph Harris. *Principles of algebraic geometry*. Wiley Classics Library. John Wiley & Sons, Inc., New York, 1994. Reprint of the 1978 original.
31. A. Grothendieck. Sur la classification des fibrs holomorphes sur la sphre de Riemann. *Amer. J. Math.*, 79:121–138, 1957.
32. A. Grothendieck. Éléments de géométrie algébrique. II. Étude globale élémentaire de quelques classes de morphismes. *Inst. Hautes Études Sci. Publ. Math.*, (8):222, 1961.
33. A. Grothendieck. Éléments de géométrie algébrique. III. Étude cohomologique des faisceaux cohérents. I. *Inst. Hautes Études Sci. Publ. Math.*, (11):167, 1961.
34. A. Grothendieck. Éléments de géométrie algébrique. III. Étude cohomologique des faisceaux cohérents. II. *Inst. Hautes Études Sci. Publ. Math.*, (17):91, 1963.
35. A. Grothendieck. Éléments de géométrie algébrique. IV. Étude locale des schémas et des morphismes de schémas. II. *Inst. Hautes Études Sci. Publ. Math.*, (24):231, 1965.
36. A. Grothendieck. Éléments de géométrie algébrique. IV. Étude locale des schémas et des morphismes de schémas. III. *Inst. Hautes Études Sci. Publ. Math.*, (28):255, 1966.
37. A. Grothendieck. Éléments de géométrie algébrique. IV. Étude locale des schémas et des morphismes de schémas IV. *Inst. Hautes Études Sci. Publ. Math.*, (32):361, 1967.
38. A. Grothendieck and J. A. Dieudonné. *Eléments de géométrie algébrique. I*, volume 166 of *Grundlehren der Mathematischen Wissenschaften [Fundamental Principles of Mathematical Sciences]*. Springer-Verlag, Berlin, 1971.
39. A. Grothendieck and J. A. Dieudonné. *Eléments de géométrie algébrique. I*, volume 166 of *Grundlehren der Mathematischen Wissenschaften [Fundamental Principles of Mathematical Sciences]*. Springer-Verlag, Berlin, 1971.
40. Alexander Grothendieck. *Fondements de la géométrie algébrique. [Extraits du Sém-inaire Bourbaki, 1957–1962.]*. Secrétariat mathématique, Paris, 1962.
41. Alexander Grothendieck. Géométrie formelle et géométrie algébrique. In *Séminaire Bourbaki, Vol. 5*, pages Exp. No. 182, 193–220, errata p. 390. Soc. Math. France, Paris, 1995.
42. Alexander Grothendieck. Technique de descente et théorèmes d'existence en géometrie algébrique. I. Généralités. Descente par morphismes fidèlement plats. In *Séminaire Bourbaki, Vol. 5*, pages Exp. No. 190, 299–327. Soc. Math. France, Paris, 1995.
43. Jack Hall. Openness of versality via coherent functors, 2012. Available online: `arXiv:1206.4179`.
44. Jack Hall and David Rydh. Artin's criteria for algebraicity revisited, 2013. Available online: `arXiv:1306.4599`.
45. Jack Hall and David Rydh. Coherent Tannaka duality and algebraicity of hom-stacks, 2014. Available online: `arXiv:1405.7680`.

46. Joe Harris and Ian Morrison. *Moduli of curves*, volume 187 of *Graduate Texts in Mathematics*. Springer-Verlag, New York, 1998.
47. Robin Hartshorne. *Algebraic geometry*. Springer-Verlag, New York-Heidelberg, 1977. Graduate Texts in Mathematics, No. 52.
48. Daniel Huybrechts and Manfred Lehn. *The geometry of moduli spaces of sheaves*. Cambridge Mathematical Library. Cambridge University Press, Cambridge, second edition, 2010.
49. Luc Illusie. *Complexe cotangent et déformations. I*. Lecture Notes in Mathematics, Vol. 239. Springer-Verlag, Berlin-New York, 1971.
50. Takeshi Kajiwara, Kazuya Kato, and Chikara Nakayama. Logarithmic abelian varieties. *Nagoya Math. J.*, 189:63–138, 2008.
51. Takeshi Kajiwara, Kazuya Kato, and Chikara Nakayama. Logarithmic abelian varieties. I. Complex analytic theory. *J. Math. Sci. Univ. Tokyo*, 15(1):69–193, 2008.
52. Takeshi Kajiwara, Kazuya Kato, and Chikara Nakayama. Logarithmic abelian varieties, III: logarithmic elliptic curves and modular curves. *Nagoya Math. J.*, 210:59–81, 2013.
53. Donald Knutson. *Algebraic spaces*. Lecture Notes in Mathematics, Vol. 203. Springer-Verlag, Berlin-New York, 1971.
54. Gérard Laumon and Laurent Moret-Bailly. *Champs algébriques*, volume 39 of *Ergebnisse der Mathematik und ihrer Grenzgebiete. 3. Folge. A Series of Modern Surveys in Mathematics [Results in Mathematics and Related Areas. 3rd Series. A Series of Modern Surveys in Mathematics]*. Springer-Verlag, Berlin, 2000.
55. Jun Li and Gang Tian. Virtual moduli cycles and Gromov-Witten invariants of algebraic varieties. *J. Amer. Math. Soc.*, 11(1):119–174, 1998.
56. Max Lieblich. Remarks on the stack of coherent algebras. *Int. Math. Res. Not.*, pages Art. ID 75273, 12, 2006.
57. Jacob Lurie. *Derived algebraic geometry*. 2004. Thesis (Ph.D.) –Massachusetts Institute of Technology. Available online: `math.harvard.edu/~lurie/papers/DAG.pdf`.
58. Jacob Lurie. Representability theorems, 2012. Available online: `math.harvard.edu/~ lurie/papers/DAG-XIV.pdf`.
59. Saunders Mac Lane and Ieke Moerdijk. *Sheaves in geometry and logic*. Universitext. Springer-Verlag, New York, 1994. A first introduction to topos theory, Corrected reprint of the 1992 edition.
60. Johan Martens and Michael Thaddeus. Compactifications of reductive groups as moduli stacks of bundles. *Compos. Math.*, 152(1):62–98, 2016.
61. John Milnor. *Introduction to algebraic K-theory*. Princeton University Press, Princeton, N.J.; University of Tokyo Press, Tokyo, 1971. Annals of Mathematics Studies, No. 72.
62. David Mumford. Picard groups of moduli problems. In *Arithmetical Algebraic Geometry (Proc. Conf. Purdue Univ., 1963)*, pages 33–81. Harper & Row, New York, 1965.
63. David Mumford. *Abelian varieties*, volume 5 of *Tata Institute of Fundamental Research Studies in Mathematics*. Published for the Tata Institute of

Fundamental Research, Bombay; by Hindustan Book Agency, New Delhi, 2008. With appendices by C. P. Ramanujam and Yuri Manin, Corrected reprint of the second (1974) edition.
64. Nitin Nitsure. Moduli space of semistable pairs on a curve. *Proc. London Math. Soc. (3)*, 62(2):275–300, 1991.
65. Arthur Ogus and George Bergman. Nakayama's lemma for half-exact functors. *Proc. Amer. Math. Soc.*, 31:67–74, 1972.
66. Martin C. Olsson. Hom-stacks and restriction of scalars. *Duke Math. J.*, 134(1):139–164, 2006.
67. Dorin Popescu. General Néron desingularization and approximation. *Nagoya Math. J.*, 104:85–115, 1986.
68. J. P. Pridham. Representability of derived stacks. *J. K-Theory*, 10(2):413–453, 2012.
69. Daniel G. Quillen. *Homotopical algebra*. Lecture Notes in Mathematics, No. 43. Springer-Verlag, Berlin-New York, 1967.
70. Michel Raynaud. Faisceaux amples sur les schémas en groupes et les espaces homogènes. *C. R. Acad. Sci. Paris Sér. A-B*, 262:A1313–A1315, 1966.
71. Michel Raynaud. Spécialisation du foncteur de Picard. *Inst. Hautes Études Sci. Publ. Math.*, (38):27–76, 1970.
72. R. W. Richardson. Affine coset spaces of reductive algebraic groups. *Bull. London Math. Soc.*, 9(1):38–41, 1977.
73. Daniel Schaub. Courbes spectrales et compactifications de Jacobiennes. *Math. Z.*, 227(2):295–312, 1998.
74. Michael Schlessinger. Functors of Artin rings. *Trans. Amer. Math. Soc.*, 130:208–222, 1968.
75. Edoardo Sernesi. *Deformations of algebraic schemes*, volume 334 of *Grundlehren der Mathematischen Wissenschaften [Fundamental Principles of Mathematical Sciences]*. Springer-Verlag, Berlin, 2006.
76. Jean-Pierre Serre. Espaces fibrés algébriques (d'après André Weil). In *Séminaire Bourbaki, Vol. 2*, pages Exp. No. 82, 305–311. Soc. Math. France, Paris, 1995.
77. The Stacks Project Authors. Stacks Project. `stacks.math.columbia.edu`, 2015.
78. Jason Starr. Algebraic groups without torsors. MathOverflow. URL: http://mathoverflow.net/q/253519 (version: 2016-11-01).
79. Bertrand Toën and Gabriele Vezzosi. Homotopical algebraic geometry. II. Geometric stacks and applications. *Mem. Amer. Math. Soc.*, 193(902):x+224, 2008.
80. Jie Wang. Deformations of pairs (X, L) when X is singular. *Proc. Amer. Math. Soc.*, 140(9):2953–2966, 2012.
81. André Weil. Théorie des points proches sur les variétés différentiables. In *Géométrie différentielle. Colloques Internationaux du Centre National de la Recherche Scientifique, Strasbourg, 1953*, pages 111–117. Centre National de la Recherche Scientifique, Paris, 1953.
82. Jonathan Wise. Obstruction theories and virtual fundamental classes, 2011.

83. Jonathan Wise. The deformation theory of sheaves of commutative rings. *J. Algebra*, 352:180–191, 2012.
84. Jonathan Wise. The deformation theory of sheaves of commutative rings II. *Ann. Sc. Norm. Super. Pisa Cl. Sci. (5)*, 14(2):339–359, 2015.
85. Wouter Zomervrucht. Ineffective descent of genus one curves, 2015. Available online: `arXiv:1501.04304`.

www.ingramcontent.com/pod-product-compliance
Ingram Content Group UK Ltd.
Pitfield, Milton Keynes, MK11 3LW, UK
UKHW020352150926
8624IPUK00004B/123

9 789813 229082